LARGE SAMPLE METHODS IN STATISTICS

AN INTRODUCTION

WITH APPLICATIONS

LARGE
SAMPLE
METHODS
IN
STATISTICS
AN INTRODUCTION
WITH APPLICATIONS

Pranab K. Sen
Julio M. Singer

CHAPMAN & HALL
New York • London

First published in 1993 by
Chapman & Hall
29 West 35th Street
New York, NY 10001-2299

Published in Great Britain by
Chapman & Hall
2-6 Boundary Row
London SE1 8HN

Library of Congress Cataloging-in-Publication Data

Sen, Pranab Kumar, 1937–
   Large sample methods in statistics : an introduction with applications / by
Pranab K. Sen, Julio M. Singer.
     p. cm.
   Includes bibliographical references and index.
   ISBN 0-412-04221-5
    1. Asymptotic distribution (Probability theory) 2. Stochastic processes.
I. Singer, Julio da Motta, 1950– . II. Title.
QA273.6.S46 1993
519.5'2--dc20   92-46163
                      CIP

British Library Cataloguing in Publication Data also available.

To our inspiring mothers
Kalyani Sen and Edith Singer

Contents

Preface

Students and investigators working in Statistics, Biostatistics or Applied Statistics in general are constantly exposed to problems which involve large quantities of data. Since in such a context, exact statistical inference may be computationally out of reach and in many cases not even mathematically tractable, they have to rely on approximate results. Traditionally, the justification for these approximations was based on the convergence of the first four moments of the distributions of the statistics under investigation to those of some normal distribution. Today we know that such an approach is not always theoretically adequate and that a somewhat more sophisticated set of techniques based on the convergence of characteristic functions may provide the appropriate justification. This need for more profound mathematical theory in statistical large sample theory is even more evident if we move to areas involving dependent sequences of observations, like Survival Analysis or Life Tables; there, some use of martingale structures has distinct advantages. Unfortunately, most of the technical background for the understanding of such methods is dealt with in specific articles or textbooks written for an audience with such a high level of mathematical knowledge, that they exclude a great portion of the potential users.

This book is intended to cover this gap by providing a solid justification for such asymptotic methods, although at an intermediate level. It focuses primarily on the basic tools of conventional large sample theory for independent observations, but also provides some insight to the rationale underlying the extensions of these methods to more complex situations involving dependent measurements. The main thrust is on the basic concepts of convergence and asymptotic distribution theory for a large class of statistics commonly employed in diverse practical problems. Chapter 1 describes the type of problems considered in the text along with a brief summary of some basic mathematical and statistical concepts required for a good understanding of the remaining chapters. Chapters 2 and 3 contain the essential tools needed to prove asymptotic results for independent sequences of random variables as well as an outline of the possible extensions to cover the dependent sequence case. Chapter 4 explores the relationship between order statistics and empirical distribution functions with respect to their asymptotic properties and illustrates their use in some applications. Chapter 5 discusses some general results on the asymptotics of estimators and test statistics; their actual application to Categorical Data and Regression Analysis is illustrated in Chapters 6 and 7, respectively. Finally, Chapter 8 deals with an introductory exposition of the technical background required to deal with the asymptotic theory for statistical functionals. The objective here is to provide some motivation and the general flavor of the problems in this area, since a rigorous treatment would require a much higher level of mathematical background, than the one we contemplate. The eight chapters were initially conceived for a one-semester course for second year students in Biostatistics or Applied Statistics doctoral programs as well as for last year undergraduate or first year graduate programs in Statistics. A more realistic view, however, would restrict the material for such purposes to the first five chapters along with a glimpse into Chapter 8. Chapters 6 and 7 could be included as supplementary material in Categorical Data and Linear Models courses, respectively. Since the text includes a number of practical examples, it may be useful as a reference text for investigators in many areas requiring the use of Statistics.

The authors would like to thank the numerous students who took Large Sample Theory courses at the Department of Biostatistics, University of North Carolina at Chapel Hill and Department of Statistics, University of São Paulo, providing important contributions to the design of this text. We would also like to thank Ms. Denise Morris, Ms. Mónica Casajús and Mr. Walter Vicente Fernandes for their patience in the typing of the manuscript. The editorial assistance provided by Antonio Carlos Lima with respect to handling TEX and LATEX was crucial to the completion of this project. We are also grateful to Dr. José Galvão Leite and Dr. Bahjat Qaqish for their enlightening comments and careful revision of portions of the manuscript. Finally we must acknowledge the Cary C. Boshamer Foundation, University of North Carolina at Chapel Hill as well as Conselho Nacional de Desenvolvimento Científico e Tecnológico, Brazil and Fundação de Amparo à Pesquisa do Estado de São Paulo, Brazil for providing financial support during the years of preparation of the text.

CHAPTER 1

Objectives and Scope: General Introduction

1.1 Introduction

Large sample methods in Statistics constitute the general methodology underlying fruitful simpler statistical analyses of data sets involving a large number of observations. Drawing statistical conclusions from a given data set involves the choice of suitable statistical models relating to the observations which incorporate some random (stochastic) or chance factors whereby convenient probability laws can be adopted in an appropriate manner. It is with respect to such postulated probability laws that the behavior of some sample statistics (typically, an estimator in an estimation problem or a test statistic in a hypothesis testing problem) needs to be studied carefully so that the conclusions can be drawn with an adequate degree of precision. If the number of observations is small and/or the underlying probability model is well specified, such stochastic behavior can be evaluated in an exact manner. However, with the exception of some simple underlying probability laws (such as the normal or Poisson distributions), the exact sampling distribution of a statistic may become very complicated as the number of observations in the sample becomes large. Moreover, if the data set actually involves a large number of observations, there may be a lesser need to restrict oneself to a particular probability law, and general statistical conclusions as well can be derived by allowing such a law to be a member of a broader class. In other words, one may achieve more robustness with respect to the underlying probability models when the number of observations is large. On the other hand, there are some natural (and minimal) requirements for a statistical method to qualify as a valid large sample method. For example, in the case of an estimator of a parameter, it is quite natural to expect that as the sample size increases, the estimator should be closer to the parameter in some meaningful sense; in the literature, this property is known as the **consistency** of estimators. Similarly, in testing a null hypothesis, a test should be able to detect the falsehood of the null hypothesis (when it is not true) with more and more confidence when the sample size becomes large; this relates to the consistency of statistical tests. In either case, there is a need to study general regularity conditions under which such stochastic convergence of sample statistics holds. A second natural requirement for a large sample procedure is to ensure that the corresponding exact sampling distribution can be adequately approximated by a simpler one (such as the normal, Poisson, or chi-squared distributions), for which extensive tables and charts are available to facilitate the actual applications. In the literature, this is known as **convergence in distribution** or **central limit theory**. This alone is a very important topic of study and is saturated with applications in diverse problems of statistical inference. Third, in a given setup, there are usually more than one procedure satisfying the requirements of consistency and convergence in distribution. In choosing an appropriate one within such a class, a natural criterion is optimality in some well-defined sense. In the estimation problem, this optimality criterion may relate to **minimum variance** or **minimum risk** (with respect to a suitable loss function), and there are vital issues in choosing such an optimality criterion and assessing its adaptability to the large sample case. In the testing problem, a test should be **most powerful**, but, often, such an optimal test may not exist (especially in the multiparameter testing problem), and hence, alternative optimality (or desirability) criteria are to be examined. This branch of statistical inference dealing with **asymptotically optimal procedures** has been a very active area of productive research during the past fifty years, and yet there is room for further related developments! Far more important is the enormous scope of applications of these asymptotically optimal procedures in various problems, the study of which constitutes a major objective of the current book. In a data set, observations generally refer to some measurable characteristics which conform to either a continuous/discrete scale or even to a categorical setup where the categories may or may not be ordered in some well-defined manner. Statistical analysis may naturally depend on the basic nature of such observations. In particular, the analysis of **categorical data models** and their ramifications may require some special attention, and in the literature, analysis of qualitative data (or **discrete multivariate analysis**) and **generalized linear models** have been successfully linked to significant applications in a variety of situations. Our general objectives include the study of large sample methods pertinent to such models as well.

Pedagogically, large sample theory has its natural roots in probability theory, and the past two decades have witnessed a phenomenal interaction between the theory of random (stochastic) processes and large sample statistical theory. In particular, weak convergence (or invariance principles) of stochastic processes and related results have paved the way for a vast

simplification of the asymptotic distribution theory of various statistical estimators and test statistics; a complete coverage of this recent development
is outside the scope of the current book. There are, however, other monographs [viz., Serfling (1980), LeCam (1986) and Pfanzagl (1982)] which
deal with this aspect in much more detail and at considerably higher level
of sophistication. Our goal is entirely different. We intend to present the
basic large sample theory with a minimum coating of abstraction and at a
level commensurate with the usual graduate programs in Applied Statistics
and Biostatistics; nevertheless, our book can also be used for Mathematical Statistics programs (with a small amount of supplementary material
if planned at a more advanced stage). As such, a measure theoretic orien-
tation is minimal, and the basic theory is always illustrated with suitable
examples, often picked up from important practical problems in some areas
of Applied Statistics (especially, Biostatistics). In other words, our main objective is to present the essentials of large sample theory of Statistics with
a view toward its application to a variety of problems that generally crop
up in other areas.

To stress our basic motivation, we start with an overview of the applications of large sample theory in various statistical models. This is done in
Section 1.2. Section 1.3 deals with a brief description of the basic coverage
of the book, and Section 1.4 with a review of some background mathematical tools.

1.2 Large sample methods: an overview of applications

In a very broad sense, the objective of **statistical inference** is to draw
conclusions about some characteristics of a certain population of interest
based on the information obtained from a representative sample thereof.
In general, the corresponding strategy involves the selection of an appropriate family of (stochastic) models to describe the characteristics under
investigation, an evaluation of the compatibility of such models with the
available data (goodness of fit) and the subsequent estimation of or tests
of hypotheses about the parameters associated with the chosen family of
models. Such models may have different degrees of complexity and depend
on assumptions with different degrees of restrictiveness. Let us examine
some examples.

Example 1.2.1: Consider the problem of estimating the average height μ
of a population based on a random sample of n individuals. Let Y denote
the height of a randomly selected individual and F the underlying distribution function. In such a context, three alternative (stochastic) models
include the following (among others):

a) F is assumed symmetric and continuous with mean μ and finite variance σ^2 and the observations are the heights $Y_1, \ldots, Y_n$ of the n individuals in the sample;

b) F is assumed normal with mean μ and known variance σ^2 and the observations are the heights $Y_1, \ldots, Y_n$ of the n individuals in the sample;

c) The assumptions on F are as in either (a) or (b) but the observations correspond to the numbers of individuals falling within each of m height intervals (grouped data). This is, perhaps, a more realistic model, since, in practice, we are only capable of coding the height measurements to a certain degree of accuracy (i.e., to the nearest millimeter). ∎

Example 1.2.2: Consider the OAB blood classification system (where the O allele is recessive and the A and B alleles are codominant). Let p_O, p_A, p_B and p_{AB} respectively denote the probabilities of occurrence of the phenotypes OO, (AA,AO), (BB, BO) and AB in a given population; also, let q_O, q_A and q_B respectively denote the probabilities of occurrence of the O, A and B alleles in that population. This genetic system is said to be in Hardy-Weinberg equilibrium if the following relations hold: $p_O = q_O^2$, $p_A = q_A^2 + 2q_O q_A$, $p_B = q_B^2 + 2q_O q_B$ and $p_{AB} = 2q_A q_B$. A problem of general concern to geneticists is to test whether a given population satisfies the Hardy-Weinberg conditions based on the evidence provided by a sample of n observational units for which the observed phenotype frequencies are n_O, n_A, n_B and n_{AB}. Under the assumption of random sampling, an appropriate (stochastic) model for such a setup corresponds to the **multinomial model** specified by the following probability function:

$$p(n_O, n_A, n_B, n_{AB}) = \frac{n!}{n_O! n_A! n_B! n_{AB}!} p_O^{n_O} p_A^{n_A} p_B^{n_B} p_{AB}^{n_{AB}},$$

with $p_O + p_A + p_B + p_{AB} = 1$. ∎

Example 1.2.3 (Linear model): A manufacturing company produces electric lamps wherein the cross section of the coil (say, X) may be one of $s \, (\geq 2)$ possible choices, designated as $x_1, \ldots, x_s$, respectively. It is conjectured that the level x_i may have some influence on the mean life of a lamp. For each level x_i, consider a set of n lamps taken at random from a production lot, and let Y_{ij} denote the life length (in hours, say) corresponding to the jth lamp in the ith lot ($i = 1, \ldots, s; j = 1, \ldots, n$). It may be assumed that the Y_{ij} are independent for different i ($= 1, \ldots, s$) and j ($= 1, \ldots, s$). Further, assume that Y_{ij} has a distribution function F_i, defined on $\mathbb{R}^+ = [0, \infty)$, $i = 1, \ldots, s$. In this setup, it is quite conceivable that F_i depends in some way on the level x_i, $i = 1, \ldots, s$. Thus, as in Example

1.2.1, we may consider a variety of models relating to the F_i, among which we pose the following:

a) F_i is normal with mean μ_i and variance σ_i^2, $i = 1, \ldots, s$, where the σ_i^2 may or may not be the same.

b) F_i is continuous (and symmetric) with median μ_i, $i = 1, \ldots, s$ but its form is not specified.

c) Although F_i may satisfy (a) or (b), since the Y_{ij} are recorded in class intervals (of width one hour, say), we have to consider effectively appropriate (ordered) categorical data models.

In model (a), if we assume further that $\sigma_1^2 = \cdots = \sigma_s^2 = \sigma^2$, we have the **classical (normal theory) multisample location model**; in (b), if we let $F_i(y) = F(y - \mu_i)$, $i = 1, \ldots, s$, we have the so called **nonparametric multisample location model**. Either (a) or (b) may be made more complex when we drop the assumption of homogeneity of the variances σ_i^2 or allow for possible scale perturbations in the shift model, i.e., if we let $F_i(y) = F\{(y - \mu_i)/\sigma_i\}$, $i = 1, \ldots, s$, where the σ_i^2 are not necessarily the same.

For the multisample location model (normal or not), we may incorporate the dependence on the levels of X by writing $\mu_i = \mu_i(x_i)$, $i = 1, \ldots, s$; furthermore, if we assume that the $\mu_i(x)$ are linear functions of known coefficients z_{ik}, $k = 1, \ldots, q - 1 \le s$, which may depend on x_i, we may write $\mu_i = \sum_{k=0}^{q-1} z_{ik}\beta_k$, $i = 1, \ldots, s$, where β_k, $k = 0, \ldots, q - 1$, are the (unknown) **regression parameters**. This leds us to the **linear model**:

$$Y_{ij} = \sum_{k=0}^{q-1} z_{ik}\beta_k + \varepsilon_{ij}, \quad i = 1, \ldots, s, \quad j = 1, \ldots, n. \tag{1.2.1}$$

For example, we may take $z_{ik} = x_i^k$, $k = 0, \ldots, q - 1$, so that (1.2.1) reduces to the **polynomial regression model**. More generally, $z_{ik} = g_k(x_i)$, where $g_k(x)$, $k = 0, \ldots, q - 1$, are functions of known form; typically the $g_k(x)$ are orthogonal polynomials. If, in addition, we assume that (a) holds with $\sigma_1^2 = \cdots = \sigma_s^2 = \sigma$, we have the **classical (normal theory) homoscedastic linear model**; the corresponding **heteroscedastic** model is obtained when the σ_i^2, $i = 1, \ldots, s$, are not all equal. When, as in (b), the form of the F_i's is not specified, (1.2.1) is termed a **nonparametric linear model**. Note that in either case, the specification coefficients z_{ik}, $i = 1, \ldots, s$, $k = 0, \ldots, q - 1$, may be made a bit more complex by letting $x_1, \ldots, x_s$ represent s combinations of alloys and cross sections of the coil; here we may introduce the alloy and cross-section main effects as well as their interactions into the model.   ∎

Example 1.2.4 (Dilution bioassay): Consider a bioassay problem involving a standard preparation (S) and a new preparation (T). Let Y_S denote the **dose** of S needed to produce the response; also, let Y_T denote the dose of T required to produce the same response. Assume that Y_S has the (tolerance) distribution F_S defined on $\mathbb{R}^+$ and that Y_T follows the distribution F_T, also defined on $\mathbb{R}^+$. Thus, $F_S(0) = 0 = F_T(0)$ and $F_S(\infty) = 1 = F_T(\infty)$. In many assays, it is conceived that the test preparation T behaves as if it were a **dilution** (or concentration) of the standard preparation. In such a case, for $x \in \mathbb{R}^+$ we may set $F_T(x) = F_S(\rho x)$, where ρ (> 0) is termed the **relative potency** of the test preparation with respect to the standard one. Note that for $\rho = 1$, the two distribution functions are the same, so that the two preparations are equipotent; for $\rho > 1$, the test preparation produces the same response with a smaller dose than the standard one, and hence is more potent, whereas for $\rho < 1$, the opposite conclusion holds.

If we assume that F_S corresponds to a normal distribution with mean μ_S and variance σ_S^2, then F_T is normal with mean $\mu_T = \rho^{-1}\mu_S$ and variance $\sigma_T^2 = \rho^{-2}\sigma^2$. Thus, $\mu_S/\mu_T = \rho$ and $\sigma_S/\sigma_T = \rho$. Moreover, because of the positivity of the dose, a normal tolerance distribution may not be very appropriate, unless μ_S/σ_S and μ_T/σ_T are large. As such, often, it is advocated that instead of the dose, one should work with dose transformations called **dose metameters** or **dosages**. For example, if we take $Y_T^* = \log Y_T$ and $Y_S^* = \log Y_S$, the response distributions, denoted by F_T^* and F_S^*, respectively satisfy

$$F_S^*(y) = F_T^*(y - \log \rho), \quad -\infty < y < \infty, \qquad (1.2.2)$$

which brings in directly the relevance of the linear models. ∎

Example 1.2.5 (Quantal bioassay): Suppose that a drug (or a toxic substance) is administered at s different dose levels, say $d_1, \ldots, d_s$, $s \geq 2$, $0 \leq d_1 < \cdots < d_s$; at each level, n subjects are tried and let $n_1, \ldots, n_s$ denote the number of subjects having positive response [here the response Y is all $(Y = 1)$ or nothing $(Y = 0)$]. At the dose level d_j, we denote by $\pi_j = \pi(d_j)$ the probability of a positive response. Then $\pi_j = P\{Y = 1 \mid d_j\} = 1 - P\{Y = 0 \mid d_j\}$, $1 \leq j \leq s$, and the joint distribution of $n_1, \ldots, n_s$ is given by

$$\prod_{j=1}^{s} \binom{n}{n_j} \pi_j^{n_j} (1 - \pi_j)^{n-n_j}, \quad 0 \leq n_j \leq n, \quad 1 \leq j \leq s.$$

It is quite conceivable that there is an underlying tolerance distribution, say F, defined on $[0, \infty)$ and a threshold value, say T_0, such that whenever the actual response level exceeds T_0, one has $Y = 1$ (i.e., a positive response) and $Y = 0$, otherwise. Moreover, we may also quantify the effect of the dose level d_j by means of a suitable function, say $\beta(d_j)$, $1 \leq j \leq s$, so that we may write

$$\pi_j = \pi(d_j) = 1 - F\{T_0 - \beta(d_j)\}, \quad 1 \leq j \leq s.$$

With this formulation, we are now in a more flexible situation wherein we may assume suitable regularity conditions on F and $\beta(d_j)$, leading to appropriate statistical models which can be more convenient for statistical analysis. For example, taking $x_j = \log d_j$, we may put $\beta(d_j) = \beta^*(x_j) = \beta_0^* + \beta_1^* x_j$, $1 \leq j \leq s$, where β_0^* and β_1^* are unknown and, as such,

$$1 - \pi_j = F(T_0 - \beta_0^* - \beta_1^* x_j), \quad 1 \leq j \leq s.$$

It is common to assume that the tolerance follows a logistic or a normal distribution. In the first case, we have $F(y) = \{1 + e^{-y/\sigma}\}^{-1}$, $-\infty < y < \infty$, with σ (> 0) denoting a scale factor. Then, $F(y)/\{1 - F(y)\} = \exp(y/\sigma)$ or $y = \sigma \log[F(y)/\{1 - F(y)\}]$, implying that for $1 \leq j \leq s$,

$$\log \frac{\pi_j}{1 - \pi_j} = \log \frac{1 - F\{T_0 - \beta_0^* - \beta_1^* x_j\}}{F\{T_0 - \beta_0^* - \beta_1^* x_j\}} = \alpha + \beta x_j, \quad (1.2.3)$$

where $\alpha = (\beta_0^* - T_0)/\sigma$ and $\beta = \beta_1^*/\sigma$. The quantity $\log\{\pi_j/(1 - \pi_j)\}$ is termed a **logit** at dose level d_j. Similarly, when F is normal, we have

$$\pi_j = 1 - F(T_0 - \beta_0^* - \beta_1^* x_j) = \Phi(\alpha + \beta x_j), \quad 1 \leq j \leq s,$$

where $\Phi(y) = (\sqrt{2\pi})^{-1} \int_{-\infty}^{y} \exp\{-x^2/2\} \, dx$. Therefore

$$\Phi^{-1}(\pi_j) = \alpha + \beta x_j, \quad 1 \leq j \leq s, \quad (1.2.4)$$

and the quantity $\Phi^{-1}(\pi_j)$ is termed a **probit** or **normit** at the dose level d_j.

Both the logit and probit models may be employed in more general situations where we intend to study the relationship between a dichotomous response and a set of explanatory variables along the lines of the linear model (1.2.1). For the logit case, such extensions are known as **logistic regression models**. In a broader sense, the models discussed above may be classified as **generalized linear models**. ∎

One of the important issues in statistical inference relates to the assessment of the validity of the model assumptions. In general, such assessments are rather delicate since the techniques available for such purposes (i.e., the **Kolmogorov-Smirnov** goodness-of-fit test for models like (b) in Example 1.2.1 or (a) in Example 1.2.3 have little power to detect departures from the assumed model. Also, such validity assessments may only be verified by approximate methods for large samples (i.e., the chi-squared goodness-of-fit test for situations like the one in Example 1.2.2). In the light of these facts, less-restrictive models like those similar to model (a) in Example 1.2.1 or

(b) in Example 1.2.3 are usually more appealing, since they cover a broader range of situations. However, as we shall discuss in the sequel [viz. Chap. 7], they are linked to some problems relative to the derivation of exact statistical properties of estimators and tests of hypotheses about the associated parameters.

In general, given a suitable family of (stochastic) models, there are several alternative methods for estimating and/or testing hypotheses about the parameters of interest. They are usually generated by some kind of heuristic rationale which include **least-squares (LS)**, **maximum likelihood (ML)** or, more generally, **M-methods** (since they are directed at the minimization of appropriate discrepancy functions of the observed data) for obtaining estimators, and the **likelihood ratio (LR)** principle or some nonparametric methods for testing hypotheses. For example, the LS method applied to model (a) in Example 1.2.1 or the ML method applied to model (b) yield the sample mean $\overline{Y}_n$ as the estimator for μ; other methods lead to the sample median or to sample trimmed means. Also, to test hypotheses of the form $H : \mu = \mu_0$, where μ_0 is some specified constant, the LR principle generates the sample average $\overline{Y}_n$ as the test statistic; other alternatives include Wilcoxon's signed rank method which leads to test statistics based on the signed ranks of the observed data. An application of the ML method to the problem described in Example 1.2.2 produces estimators of the gene frequencies defined as solutions $\hat{q}_O$, $\hat{q}_A$ and $\hat{q}_B$ to the equations

$$
2\hat{q}_O(\hat{q}_A + 2\hat{q}_A\hat{q}_O)^{-1}n_A
$$
$$
-2(\hat{q}_B + 2\hat{q}_O)^{-1}n_B + \hat{q}_A^{-1}n_{AB} - 2\hat{q}_O^{-1}n_O = 0,
$$
$$
2\hat{q}_O(\hat{q}_B + 2\hat{q}_B\hat{q}_O)^{-1}n_B
$$
$$
-2(\hat{q}_A + 2\hat{q}_O)^{-1}n_A + \hat{q}_B^{-1}n_{AB} - 2\hat{q}_O^{-1}n_O = 0,
$$
$$
\hat{q}_O + \hat{q}_A + \hat{q}_B = 1. \tag{1.2.5}
$$

Alternative methods of producing estimators for q_O, q_A and q_B are the gene counting method and Bernstein's method described in Elandt-Johnson (1971, p.395) for example.

Given the availability of competing methods of estimation and hypothesis testing in most practical situations, an important aspect of statistical inference relates to the choice of the best alternative among them. In this direction, the **efficiency** of such estimators and test statistics must be evaluated with respect to some desired optimality criteria such as the **minimum mean squared error**, **minimum variance unbiased estimation** or **minimax principles** for estimation and **uniformly most powerful**, or **locally most powerful principles** for hypothesis testing.

In most cases of practical interest, this evaluation depends on the statistical properties of the competing estimators or test statistics (i.e., some measure of variability or the probability distribution itself). Thus, the investigation of such statistical properties plays an important role in the process of statistical inference.

In some cases, the specification of the parent distribution will lead directly to the required statistical properties. If, for example, the assumptions underlying model (b) in Example 1.2.1 hold, then $\overline{Y}_n$ follows a $N(\mu, \sigma^2/n)$ distribution and it can be shown that it is optimal in the sense of having smaller variance in the class of unbiased estimators of μ. Also, the test statistic $z = \sqrt{n}(\overline{Y}_n - \mu_0)/\sigma$ follows a $N(0,1)$ distribution and is optimal in the sense that it is uniformly most powerful among the tests of size α for testing $H : \mu = \mu_0$ against one-sided alternatives $A : \mu > \mu_0$ or $A : \mu < \mu_0$. In the situation depicted in Example 1.2.2, the ML estimators of the gene frequencies are only obtainable via iterative procedures [solutions to (1.2.5)] and the derivation of their exact distribution is mathematically intractable. In the setup of Example 1.2.4, a similar problem holds when we are interested in estimating the relative potency parameter ρ from the responses of two groups of animals, the first submitted to the standard preparation and the second to the new one. Here we need to take into account the information contained in both the sample averages and sample variances, and as noted by Finney (1978, chap.2), the standard results are not applicable. In the same context, one may even question the proposed normality for F_S^* or F_T^* and, in such a case, the dosage-invariant estimating and testing procedures suggested by Sen (1963) come into the picture; such procedures do not require normality, but their exact statistical properties are difficult to obtain. There are other cases, like model (a) in Example 1.2.1 or model (b) in Example 1.2.3, in which the assumptions on the underlying distribution are too vague to permit the derivation of the exact distribution of many of the relevant estimators and test statistics. This is also the case even for some highly structured models like (c) in Example 1.2.1 or Example 1.2.3. Finally, there are situations, as under model (a) in Example 1.2.3, where transformed variables like $Y_{ij}^* = \log Y_{ij}$ are considered in order to stabilize the variance and allow for the use of standard analysis of variance procedures. Since the transformation may also affect the form of the underlying distributions, the exact statistical properties of the analytical techniques may not be directly assessed.

In summary, with the exception of some very specific simple cases, the exact statistical properties of many useful estimators and test statistics are not directly obtainable for many models of practical interest. Therefore, methods for approximating the probability distributions (or some appropriate summary measures) of such procedures are of special concern in the field of statistical inference. The main tools for such approximations are based on the statistical properties of large samples and are investigated under the general denomination of **Large Sample Methods** (or **Asymptotic Statistical Theory**). Although such large sample procedures are only directed at the development of approximate results, they are associated with the following advantages:

i) They allow for some flexibility in the selection of models (large sample properties of estimators and test statistics are usually less dependent on the particular functional form of the underlying distribution, and thus, are relatively robust against departures from some assumed model).

ii) In many cases, they allow for some relaxation of the assumptions of independence and identical distribution of the sample elements, usually required in small sample procedures.

iii) They generally produce simple and well-studied limiting distributions such as the normal, chi-squared, Weibull, etc.

iv) They provide ample flexibility in the selection of estimators and test statistics and thus allow for the choice of those with more appealing interpretation.

Another important topic covered by large sample methods relates to the evaluation of the rates of convergence of the distribution of interest to some limiting distribution. Although such an issue is usually associated with some rather complicated technical problems, it should not be under-emphasized, since the applied statistician must deal with finite sample sizes in practice.

The recent advances in the usage of computers for data management purposes have triggered a general interest in the development of statistical methods designed to deal with large data sets. As a consequence, many new techniques have been proposed which are rather more complicated (and realistic) than those that can be handled by exact methods. Therefore, large sample methods in Statistics have become increasingly necessary tools for both theoretically or practically oriented statisticians. With this basic consideration, we proceed to present an outline of the material covered in this text (i.e., the general methodology), emphasizing potential applications and the interrelationships among several probabilistic concepts required for such purposes.

1.3 The organization of this book

This book is primarily meant for graduate students in (applied or theoretical) Statistics as well as statisticians who intend to make good use of large sample theory in their professional activities. As prerequisites, certain basic results in real analysis, matrix algebra and other areas are presented (without proofs) in the last section of this chapter. A reader having adequate mathematical background may skip that section.

Chapter 2 deals with the two basic forms of **stochastic approximation** useful for large samples; the first concerns the approximation of a sequence of random variables by another random variable (the properties of which are presumably known), and the second is related to the approximation of a sequence of distribution functions by another distribution function (also with presumably known properties). Within the first case, we distinguish three different modes: **convergence in probability, convergence in rth mean** and **almost sure convergence**, which essentially address different dimensions of stochastic approximation. The second form of approximation refers to the so-called convergence in distribution (or weak convergence). It is implied by the other three modes of convergence above and, therefore, is the weakest among them. It is the most important for statistical applications, however, since the related limiting distribution function generally may be employed in the construction of confidence intervals and significance tests about the parameters of interest.

The first three modes of convergence mentioned above are usually employed to show that certain sequences of statistics (such as ML or LS estimators, for example) converge to specified constants (which may be viewed as degenerate random variables). Many statistics of practical interest may be expressed as sums of the underlying random variables (as the sample mean $\overline{Y}_n$ in the context of the estimation of μ in Example 1.2.1) and, in such cases, the most important results on their stochastic convergence to certain constants are known under the general denomination of **Laws of Large Numbers**. The proofs of such results generally rely on some probability inequalities (like **Markov, Chebyshev**, etc.) or on the **Borel-Cantelli Lemma** on the convergence of series involving the probabilities of certain sequences of events. Although most of these results are usually directed at independently distributed underlying random variables, extensions are available to the dependent case, covering a broad class of **martingales, submartingales** and **reverse (sub)martingales**.

The concept of **stochastic order** of random variables [the $O_P(\cdot)$ and $o_P(\cdot)$ representations] is a related issue of great importance in the evaluation of the magnitude of stochastic approximations. Also, the idea of convergence of series of random variables deserves some attention given its application in the study of **random walks** and **Sequential Analysis**.

Chapter 3 is devoted to the study of weak convergence of sequences of statistics which may be expressed as (normalized) sums of the underlying random variables. Although many of the related issues constitute special cases of the problems discussed in Chapter 2, they deserve special attention in view of the broad range of applications; as mentioned above, many statistics of practical importance may be expressed in this form. In this context, a topic of interest concerns the different techniques generally employed to prove **weak convergence**; among these, the most useful involves convergence of the associated **characteristic functions** or **moment generating functions**. The main results on this topic are known as **Central Limit Theorems** and essentially demonstrate that such sequences are asymptotically normal, i.e., that they converge weakly to the normal distribution, which occupies a central position in Statistics. These theorems may be proved under different assumptions on the moments and on the dependence structure of the underlying random variables and usually rely on the characteristic function technique.

Also, in many situations where the statistics of interest may be expressed as functions of other statistics for which weak convergence has been established, the results by **Sverdrup** and **Slutsky** are useful tools to show that weak convergence holds for themselves. Finally it is convenient to address the extensions of all the above concepts to the multivariate case; in particular, consideration must be given to the **Cramér-Wold device**, which is useful in the process of reduction of multivariate problems to their univariate counterparts.

In connection with the above discussion, a question of interest refers to the convergence of the moments of the random variables under consideration given that weak convergence (or some other mode of convergence) has been established; here an important theorem, due to Cramér, plays a major role.

Two special cases merit a detailed treatment; the first is related to **quadratic forms** of **asymptotically normal** statistics and the second to some **transformations** directed at obtaining asymptotical distributions which do not depend on unknown parameters, the so-called **variance stabilizing transformations**. An important associated topic is the well-known **Delta-method**, which is an extremely useful technique for obtaining the asymptotic distributions of many (smooth) functions of asymptotically normal statistics.

For most practical purposes, the above considerations must be complemented with some discussion on the corresponding rates of convergence. In this direction, a bound on the error of approximation of the Central Limit Theorem is provided by the **Berry-Esséen Theorem**; alternatively, such an evaluation may be carried out via **Gram-Charlier** or **Edgeworth expansions**. This second approach, however, requires more restrictive assumptions on the moments of the underlying random variables than those needed for the Central Limit Theorem to hold.

The topic considered in Chapter 4 refers to the application of the tools examined in the previous chapters to order statistics (including sample quantiles and extreme values) and **empirical distribution functions**. This investigation is relevant not only because of the intrinsic importance of such concepts in statistical inference (they form the basis of some useful estimation procedures) but also because many of these specific large sample methods may be successfully employed for other statistical procedures. The interchangeability of the two concepts plays an important role in the study of their asymptotic properties since it allows the use of different technical approaches. The asymptotic properties of the sample quantiles follow from direct applications of the methods of Chapters 2 and 3. The **extreme values** (order statistics) merit special attention; an important result known as the **Gnedenko Theorem** indicates that the asymptotic distribution of the sample minimum (or maximum) is restricted to one of three possible distributions, depending on the behavior of the tails of the distribution of the underlying random variables. Another related result refers to the asymptotic independence of the sample minimum and maximum.

The most useful applications of large sample methods discussed above in the evaluation of the statistical behavior of the empirical distribution function are associated with the **Kolmogorov-Smirnov statistic**. They are mainly directed at some useful inequalities for providing bounds on the probabilities of large deviations or at almost sure convergence (via the well-known **Glivenko-Cantelli Theorem**). A detailed treatment of the weak convergence of the Kolmogorov-Smirnov statistic depends on the stochastic processes approach considered in Chapter 8.

Chapter 5 lays out specific details for the application of large sample methods to some general estimation and hypothesis testing criteria. These include the methods of Moments, Least Squares, Maximum Likelihood, as well as M- and R- methods, for estimation and Wald, Score and likelihood ratio methods for hypothesis testing. The main interest is in examining the asymptotic behavior of the statistics generated by each procedure in the light of the concepts discussed previously and then contrasting them via some appropriate optimality criteria (in order to choose the best, in some sense).

In the case of estimation procedures, such criteria essentially reduce to the concepts of **asymptotic unbiasedness** and **asymptotic efficiency**; with respect to the latter, the most appealing procedure relates to distributional optimality, although other alternatives are available. Here, asymptotic normality plays a major role and estimators are essentially compared with respect to the variance (or covariance matrices) of the corresponding limiting (normal) distributions. In connection with such ideas, the concept of (absolutely) efficient estimators may be developed with a similar spirit to that of the Cramér-Rao Inequality for small samples; such estimators are termed **Best Asymptotically Normal (BAN)**.

The situation is slightly more complicated in the case of testing procedures, since not only **Type 1** and **Type 2 error probabilities** but also specific type of alternative hypothesis must be taken into consideration. Although many different approaches to the evaluation of the **asymptotic efficiency** of test statistics are available, one of the most useful is due to **Pitman**; under this approach, the asymptotic power functions are evaluated with respect to alternatives which are, in some sense, close to the null hypothesis and permit an efficiency comparison in terms of the corresponding noncentrality parameters.

Chapter 6 and 7 are directed at the specific application of asymptotic methods in some of the most commonly used families of (statistical) models: categorical data models and linear models. The corresponding large sample results are related to those discussed in Chapter 5 under a more general setting; however, some particularities associated with such models justify a somewhat more specific analysis. The inclusion of a detailed treatment of large sample methods for categorical data models in this text may be justified not only on the grounds of their importance in statistical methodology but also because they provide good examples for the application of almost all asymptotic concepts discussed in earlier chapters. In particular, they are useful in emphasizing the asymptotic equivalence of statistical procedures such as maximum likelihood estimation for **product multinomial distributions** and the corresponding **weighted least-squares methods**.

The linear regression models pose special problems related to the fact that the observed random variables are not identically distributed. Also, in many cases the exact functional form of the underlying distribution is not completely specified and, in such situations, the asymptotic results are of major interest. Least-squares methods are attractive under these conditions, since they may be employed in a rather general setup. However, other alternatives, including generalized and weighted least-squares procedures as well as robust M-estimation procedures have also been given due attention.

In this context, generalized linear models (in the spirit referred to in Example 1.2.5) deserve special mention. The study of their asymptotic properties involves methodology having roots in the general theory of ML estimation discussed in Chapter 5 but also bear similarity with that of generalized least squares for linear models.

The last topic covered in Chapter 7 deals with nonparametric regression; we essentially describe the difficulties in establishing asymptotic results for some of the usual techniques employed in this field.

Chapter 8 is devoted to some basic, but more complicated technical problems which have been deferred from earlier chapters. The concept of stochastic convergence and convergence in law, as developed in Chapters 2 and 3 for the so-called finite dimensional cases, have gone through some extensive generalizations in the recent past. This involves weak convergence in metric spaces which is definitely beyond the intended level of presentation. Nevertheless, a unified introduction to such weak invariance principles with due emphasis on the partial sum and empirical distributional processes is presented along with some applications. Weak convergence of **statistical functionals** (which are differentiable in a sense) and the **Bahadur-Kiefer representation** for sample quantiles have been considered in the same vein. Finally, some insight into the role of weak convergence in nonparametrics is given. Although less technical than required, this chapter may provide good motivation for the use of these novel techniques in various applied problems.

In the rest of the section we present a summary of the principal abbreviations and notations followed in the text.

Abbreviation	Meaning
a.e.	almost everywhere
c.f.	characteristic function
d.f.	distribution function
iff	if and only if
i.i.d.	independent and identically distributed
lhs	left hand-side
p.d.f.	probability density function
p.s.d.	positive semi-definite
rhs	right hand-side
s.t.	such that
r.v.	random variables (vectors)

Let r and s be real numbers. Then $r \wedge s$ and $r \vee s$ denote respectively $\min(r, s)$ and $\max(r, s)$; also $[r]$ denotes the largest integer contained in r. We shall use the notation

$$\widehat{\beta} = \arg\min f(\mathbf{x}, \beta)$$

to indicate that $\widehat{\beta}$ is the value of the "parameter" vector β that minimizes $f(\mathbf{x}, \beta)$, considered as a function of β. Similarly, we may define

$$\widehat{\beta} = \arg\max f(\mathbf{x}, \beta).$$

In general, boldface lowercase (uppercase) characters denote vectors (matrices) and $\mathbf{x}^t$ denotes the transpose of $\mathbf{x}$. Also, $\text{tr}(\mathbf{A})$ and $|\mathbf{A}|$ denote the **trace** and the **determinant** of a matrix $\mathbf{A}$; $\text{ch}_i(\mathbf{A})$ denotes the ith characteristic root of $\mathbf{A}$ (in descending order). The symbols Φ and φ are usually reserved for the d.f. and the p.d.f. of the normal distribution; we also let $X \sim N(\mu, \sigma^2)$, $X \sim \text{Bin}(n, \pi)$, $X \sim \text{Unif}(\theta_1, \theta_2)$ and $X \sim \text{Exp}(\lambda)$ respectively indicate that a random variable X follows the normal, binomial, uniform and exponential distributions with the corresponding parameters in parentheses. The notation $\chi_p^2(\delta)$ indicates the noncentral chi-squared distribution with p degrees of freedom and noncentrality parameter $\delta \neq 0$; the central ($\delta = 0$) chi-squared distribution is denoted by χ_p^2. The characteristic function associated with a d.f. F is usually denoted ϕ_F.

Given a sample $X_1, \ldots, X_n$, $\overline{X}_n = n^{-1} \sum_{i=1}^{n} X_i$ denotes the sample average and $S_n^2 = (n-1)^{-1} \sum_{i=1}^{n} (X_i - \overline{X}_n)^2$ denotes the sample variance.

Theorems are sequentially numbered within sections and those that are referred to by proper names will be accompanied by the reference number in parentheses [e.g., Courant Theorem (1.4.2)].

Other notations will be introduced as they appear for the first time.

1.4 Basic tools and concepts

The notation $y = f(x)$ is used to mean that y is a function of x for a certain set of values of x, that is, for each x of the **domain** A, $f(x)$ associates a corresponding value of y of the **range** B; this may be also represented by $f: A \rightarrow B$. If both A and B are collections of real numbers, f is called a **real-valued function** of a **real variable**. On the other hand, f is said to be a **vector-valued function** of dimension m of p **real variables** if A and B are collections of p and m real vectors, respectively; the value of the function f of dimension m at the point $\mathbf{x} = (x_1, \ldots, x_p)^t \in A \subset \mathbb{R}^p$ is denoted by $\mathbf{f}(\mathbf{x}) = [f_1(\mathbf{x}), \ldots, f_m(\mathbf{x})]^t$.

Let $f(x)$, $x \in [a, b)$, be a real-valued function. If for every $\varepsilon > 0$, there exists an $\eta > 0$, such that for some c^-

$$|f(x) - c^-| < \varepsilon \quad \text{whenever} \quad b - \eta < x < b, \qquad (1.4.1)$$

then c^- is termed the **left-hand limit** of f at b. The **right-hand limit** c^+ is defined similarly. If both c^- and c^+ exist and $c^- = c^+ = c$, then $f(x)$ is said to be **continuous** at b. Also, $f(x)$ is said to be continuous over the interval (a, b) if it is continuous at each $x \in (a, b)$; if the interval is $[a, b]$, we also need f to be continuous to the left (right) at b (a). Some care is needed to have this continuity over $[a, b]$ when a is $-\infty$ or b is $+\infty$. In other words, we say that f is continuous at x_0 if, given $\varepsilon > 0$, there exists $\delta = \delta(x_0, \varepsilon)$ such that $|x - x_0| < \delta$ implies $|f(x) - f(x_0)| < \varepsilon$. If, for a given interval I (open or closed), δ does not depend on x_0, we say that f is **uniformly continuous** over I. Note that if f is continuous over a closed interval, it is uniformly continuous over that interval.

Let the function $f(x)$ be defined on (a, b). If for every $x_0 \in (a, b)$ the limit

$$g(x_0) = \lim_{\Delta x \to 0} \left\{ \frac{1}{\Delta x} [f(x_0 + \Delta x) - f(x_0)] \right\} \qquad (1.4.2)$$

exists, then $g(x)$ is called the **derivative** of $f(x)$ and is denoted by $f'(x)$. A function $f(x)$ is said to be **differentiable** if $f'(x)$ exists. Note that continuity of f may not ensure its differentiability. We may rewrite (1.4.2) as

$$f(x) - f(x_0) = (x - x_0)f'(x_0) + o(x - x_0) \quad \text{as} \quad x \to x_0, \qquad (1.4.3)$$

where $o(x - x_0)$ denotes a quantity that is negligible when compared to $(x - x_0)$ as $x \to x_0$ [the notation $o(\cdot)$ and $O(\cdot)$ will be elaborated on in Section 2.2]. This, in turn, provides two important results:

i) The **first mean value theorem** in Differential Calculus: If $f(x)$ is continuous over $[a, b]$ and differentiable for $a < x < b$, then

$$f(b) - f(a) = (b - a)f'(x^*) \quad \text{for some} \quad x^* \in (a, b). \qquad (1.4.4)$$

ii) The **fundamental theorem** of Integral Calculus: For $b > a$

$$f(b) - f(a) = \int_a^b g(x)dx, \quad \text{where} \quad g(x) = f'(x). \qquad (1.4.5)$$

The derivative of $f'(x)$, denoted by $f^{(2)}(x)$, may be defined as in (1.4.2), wherein we replace f by f' and the chain rule applies to define the nth derivative $f^{(n)}(x)$, for every $n \geq 1$ (whenever it exits). In some cases, we also use the common notation f'' and f''' for $f^{(2)}$ and $f^{(3)}$, respectively. Let $f(x)$, $a \leq x \leq b$, be a continuous function and have continuous derivatives up to the $(k + 1)$th order for some $k \geq 0$. Then, for every $x \in [a, b]$, $x_0 \in (a, b)$,

$$f(x) = f(x_0) + \sum_{j=1}^{k} \frac{(x - x_0)^j}{j!} f^{(j)}(x_0) + R_k(x, x_0), \qquad (1.4.6)$$

where

$$R_k(x, x_0) = \frac{(x - x_0)^{k+1}}{(k + 1)!} f^{(k+1)}(hx_0 + (1 - h)x) \qquad (1.4.7)$$

for some $0 < h < 1$. This is known as the **Taylor expansion** (up to the kth order) with a remainder term. There is no need to take $x = x_0$ in (1.4.6), while the case $k = 0$ is treated in (1.4.4); also, $k = +\infty$ leads to the **Taylor series expansion**.

Now let f be defined on (a, b) and suppose that there exists an interior point c, such that $f(c) = \sup\{f(x) : a \leq x \leq b\}$ or $f(c) = \inf\{f(x): a \leq x \leq b\}$, i.e., f has a **relative (local) maximum** or **minimum** at c. If the derivative $f'(c)$ exists, then $f'(c) = 0$. Moreover, if $f^{(2)}(c)$ also exists and $f^{(2)}(c) \neq 0$, then:

i) $f^{(2)}(c) < 0$ implies that $f(c)$ is a relative maximum;

ii) $f^{(2)}(c) > 0$ implies that $f(c)$ is a relative minimum.

Consider now the computation of the limit (1.4.2) of $f(x) = f_1(x)/f_2(x)$. If, at any point x_0, $f_1(x)$ and $f_2(x)$ both have limits, say c_1 and c_2, where $c_2 \neq 0$, then the limit of $f(x)$ at x_0 is c_1/c_2. If $c_2 = 0$, but $c_1 \neq 0$, by working with $1/f(x) = f_2(x)/f_1(x)$, we may claim that $f(x) \to \text{sign}(c_1)\infty$ as $x \to x_0$. However, there is an indeterminate form when both c_1 and c_2 are zero and in such a case we may apply **L'Hôpital's rule**: assume that f_1 and f_2 have derivatives f_1' and f_2', respectively [at each point of an open interval (a, b)], and that as $x \to a^+$, $f_1(x) \to 0$, $f_2 \to 0$, but $f_2'(x) \neq 0$ for each $x \in (a, b)$ and $f_1'(x)/f_2'(x)$ has a limit as $x \to a^+$. Then

$$\lim_{x \to a^+} \{f_1(x)/f_2(x)\} = \lim_{x \to a^+} \{f_1'(x)/f_2'(x)\}. \tag{1.4.8}$$

Consider now a real-valued function $f(\mathbf{x})$ of p real variables. The (first order) **partial derivatives** of f with respect to x_i, $i = 1, \ldots, p$ at the point $\mathbf{x} = (x_1, \ldots, x_p)^t$ is defined as

$$f_i'(\mathbf{x}) = \frac{\partial}{\partial x_i} f(\mathbf{x})$$

$$= \lim_{\Delta x_i \to 0} \frac{f(x_1, \ldots, x_{i-1}, x_i + \Delta x_i, x_{i+1}, \ldots, x_p) - f(x_1, \ldots, x_p)}{\Delta x_i}$$

$$\tag{1.4.9}$$

when the limit exists. The **gradient** of f at the point $\mathbf{x}$ is the p-vector of (first order) partial derivatives:

$$\dot{\mathbf{f}}(\mathbf{x}) = \frac{\partial}{\partial \mathbf{x}} f(\mathbf{x}) = \left[\frac{\partial}{\partial x_1} f(\mathbf{x}), \frac{\partial}{\partial x_2} f(\mathbf{x}), \ldots, \frac{\partial}{\partial x_p} f(\mathbf{x}) \right]^t. \tag{1.4.10}$$

Along the same lines, we may define the second order partial derivatives of f with respect to x_i, x_j, $i, j = 1, \ldots, p$ at the point $\mathbf{x}$ [denoted by $(\partial^2/\partial x_i \partial x_j)f(\mathbf{x})$] by replacing $f'(\mathbf{x})$ for $f_i(\mathbf{x})$ in (1.4.9). The extension to higher order partial derivatives is straightforward. Within this context, the **Hessian** of f at the point $\mathbf{x}$ is defined as the $(p \times p)$ matrix of second order partial derivatives:

$$\mathbf{H}(\mathbf{x}) = \frac{\partial^2}{\partial \mathbf{x} \partial \mathbf{x}^t} f(\mathbf{x}) = \frac{\partial}{\partial \mathbf{x}} \left\{ \left[\frac{\partial}{\partial \mathbf{x}} f(\mathbf{x}) \right]^t \right\}$$

$$= \begin{pmatrix} \frac{\partial^2}{\partial x_1^2} f(\mathbf{x}) & \frac{\partial^2}{\partial x_1 \partial x_2} f(\mathbf{x}) & \cdots & \frac{\partial^2}{\partial x_1 \partial x_p} f(\mathbf{x}) \\ \frac{\partial^2}{\partial x_2 \partial x_1} f(\mathbf{x}) & \frac{\partial^2}{\partial x_2^2} f(\mathbf{x}) & \cdots & \frac{\partial^2}{\partial x_2 \partial x_p} f(\mathbf{x}) \\ \vdots & \vdots & \cdots & \vdots \\ \frac{\partial^2}{\partial x_p \partial x_1} f(\mathbf{x}) & \frac{\partial^2}{\partial x_p \partial x_2} f(\mathbf{x}) & \cdots & \frac{\partial^2}{\partial x_p^2} f(\mathbf{x}) \end{pmatrix} . \quad (1.4.11)$$

Now let $f(\mathbf{x})$, $\mathbf{x} \in A \subseteq \mathbb{R}^p$, be a continuous function and have continuous partial derivatives up to $(k+1)$th order for some $k \geq 1$. Then, for every $\mathbf{x} \in A$, $\mathbf{x}_0 \in A$, we have the following (multivariate) Taylor expansion:

$$f(\mathbf{x}) = f(\mathbf{x_0}) + \sum_{j=1}^{k} \frac{1}{j!} \sum_{i_1=1}^{p} \cdots \sum_{i_j=1}^{p} \frac{\partial^j}{\partial x_{i_1} \cdots \partial x_{i_j}} f(\mathbf{x}) \prod_{l=1}^{j} (x_{i_l} - x_{0_{i_l}})$$
$$+ R_k(\mathbf{x}, \mathbf{x}_0), \quad (1.4.12)$$

where $R_k(\mathbf{x}, \mathbf{x}_0)$ is given by

$$\frac{1}{(k+1)!} \sum_{i_1=1}^{p} \cdots \sum_{i_{k+1}=1}^{p} \frac{\partial^{k+1}}{\partial x_{i_1} \cdots \partial x_{i_{k+1}}} f[h\mathbf{x}_0 + (1-h)\mathbf{x}] \prod_{l=1}^{k+1} (x_{i_l} - x_{0_{i_l}})$$
$$(1.4.13)$$

for some $0 < h < 1$. In particular, the second order Taylor expansion is extremely useful in many applications; in matrix notation, it may be expressed as

$$f(\mathbf{x}) = f(\mathbf{x}_0) + (\mathbf{x} - \mathbf{x}_0)^t \frac{\partial}{\partial \mathbf{x}} f(\mathbf{x}) \Big|_{\mathbf{x}=\mathbf{x}_0}$$
$$+ (\mathbf{x} - \mathbf{x}_0)^t \frac{\partial^2}{\partial \mathbf{x} \partial \mathbf{x}^t} f(\mathbf{x}) \Big|_{\mathbf{x}=\mathbf{x}_0} (\mathbf{x} - \mathbf{x}_0) + o(||\mathbf{x} - \mathbf{x}_0||^2).$$
$$(1.4.14)$$

Here the notation $f(\mathbf{x})|_{\mathbf{x}=\mathbf{x}_0}$ is used to emphasize the fact that the function $f(\mathbf{x})$ is evaluated at $\mathbf{x} = \mathbf{x}_0$.

In many applications, we have an integral of the form

$$F(t) = \int_a^\infty f(x, t) dx; \quad (1.4.15)$$

in such cases t is usually called a "parameter." Suppose that $f(x, t)$ is continuous in x and t and has a (partial) derivative $\partial f / \partial t$ which is also continuous in x and t (for $x \in [a, \infty)$ and $t \in [c, d]$), such that

$$\int_a^\infty f(x, t) dx \quad \text{and} \quad \int_a^\infty \frac{\partial}{\partial t} f(x, t) dx \quad (1.4.16)$$

converge (the second one, uniformly in t). Then $F(t)$ has a continuous derivative for $t \in [c, d]$, and

$$F'(t) = \int_a^\infty \frac{\partial}{\partial t} f(x, t) dx. \qquad (1.4.17)$$

In regard to the uniformity condition in (1.4.16), we may introduce the notion of **uniform integrability**; we say that $h(t) = \{h(x, t), \ x \in \mathbb{R}\}$ is uniformly (in t) integrable, if

$$\int_{|x|>c} |h(x, t)| dx \to 0 \quad \text{as} \quad c \to \infty, \quad \text{uniformly in } t. \qquad (1.4.18)$$

In some cases, the limits of integration in (1.4.15) are themselves functions of the "parameter t"; suppose, for example, that $a(t)$ and $b(t)$ have continuous derivatives for $t \in [c, d]$ so that $F(t) = \int_{a(t)}^{b(t)} f(x, t) dx$. Then, under the same conditions above, we have

$$F'(t) = f[b(t), t]b'(t) - f[a(t), t]a'(t) + \int_{a(t)}^{b(t)} \frac{\partial f}{\partial t}(x, t) dx. \qquad (1.4.19)$$

There are some other properties of functions which will be introduced in later chapters in the appropriate context. We may, however, mention here a few functions of special importance to our subsequent analysis:

i) **Exponential function:**

$$f(x) = \exp(x) = 1 + x + \frac{x^2}{2!} + \frac{x^3}{3!} + \cdots; \qquad (1.4.20)$$

ii) **Logarithmic function:**

$$f(x) = \log x, \quad \text{i.e.,} \quad x = \exp\{f(x)\} \qquad (1.4.21)$$

and also

$$f(x) = \log(1 + x) = x - \frac{1}{2}x^2 + \frac{1}{3}x^3 - \frac{1}{4}x^4 + \cdots; \qquad (1.4.22)$$

iii) **Gamma function:**

$$f(x) = \Gamma(x) = \int_0^\infty e^{-y} y^{x-1} dy, \quad x \geq 0. \qquad (1.4.23)$$

Note that if $x = n+1$ is a positive integer, then $\Gamma(x) = \Gamma(n+1) = n!$ ($= 1 \times 2 \times \cdots \times n$). Also, for $n \geq 5$, we have the **Stirling approximation** to the factorials:

$$n! \approx (2\pi)^{1/2} n^{n+1/2} \exp(-n - \frac{1}{12n} - \cdots). \qquad (1.4.24)$$

Let us now consider certain inequalities which are very useful in statistical applications.

Cauchy-Schwarz Inequality: Let $\mathbf{a} = (a_1, \ldots, a_n)^t$ and $\mathbf{b} = (b_1, \ldots, b_n)^t$ be two n-vectors. Then

$$(\mathbf{a}^t \mathbf{b})^2 = \left(\sum_{i=1}^{n} a_i b_i \right)^2 \leq \left(\sum_{i=1}^{n} a_i^2 \right) \left(\sum_{i=1}^{n} b_i^2 \right) = (\mathbf{a}^t \mathbf{a})(\mathbf{b}^t \mathbf{b}), \qquad (1.4.25)$$

where the equality holds when $\mathbf{a} \propto \mathbf{b}$. The integral version of (1.4.25) is given by

$$
\begin{aligned}
< f_1, f_2 >^2 &= \left(\int_A f_1(x) f_2(x) d\mu(x) \right)^2 \\
&\leq \left(\int_A f_1^2(x) d\mu(x) \right) \left(\int_A f_2^2(x) d\mu(x) \right) \\
&= < f_1, f_1 >< f_2, f_2 >;
\end{aligned}
\qquad (1.4.26)
$$

where f_1 and f_2 are real functions defined on a set A and are square integrable with respect to a measure μ (on A).

Hölder Inequality: Let $\mathbf{a} = (a_1, \ldots, a_n)^t$ and $\mathbf{b} = (b_1, \ldots, b_n)^t$ be n-vectors and let r and s be positive numbers such that $r^{-1} + s^{-1} = 1$. Then

$$|\mathbf{a}^t \mathbf{b}| = \left| \sum_{i=1}^{n} a_i b_i \right| \leq \left(\sum_{i=1}^{n} |a_i|^r \right)^{1/r} \left(\sum_{i=1}^{n} |b_i|^s \right)^{1/s}, \qquad (1.4.27)$$

where the equality holds only if a_i, b_i are of the same sign and $|b_i| \propto |a_i|^{r-1}$. As in (1.4.26), an integral version of (1.4.27) is easy to write. Note that if $r = s = 2$, (1.4.27) reduces to (1.4.25). A direct corollary to (1.4.27) is the following.

Minkowski Inequality: For $k \geq 1$ and $\mathbf{a}, \mathbf{b}$ as in (1.4.27),

$$\left[\sum_{i=1}^{n} |a_i + b_i|^k \right]^{1/k} \leq \left(\sum_{i=1}^{n} |a_i|^k \right)^{1/k} + \left(\sum_{i=1}^{n} |b_i|^k \right)^{1/k}. \qquad (1.4.28)$$

A closely related inequality is the following.

C_r Inequality: For every $r > 0$, a and b real,

$$|a + b|^r \leq C_r \{ |a|^r + |b|^r \}, \qquad (1.4.29)$$

where

$$C_r = \begin{cases} 1, & 0 \leq r \leq 1 \\ 2^{r-1}, & r > 1. \end{cases} \qquad (1.4.30)$$

The inequality extends to any m (≥ 1) numbers $a_1, \ldots, a_m$, where C_r is then equal to m^{r-1}, $r > 1$, and 1 for $0 \leq r \leq 1$.

Arithmetic mean/Geometric mean/Harmonic mean (AM/GM/ HM) Inequality: Let $a_1, a_2, \ldots, a_n$ be non-negative numbers. Then, for every $n \ (\geq 1)$,

$$AM = \frac{1}{n}\sum_{i=1}^{n} a_i \geq GM = \left(\prod_{i=1}^{n} a_i\right)^{1/n} \geq HM = \left\{\frac{1}{n}\sum_{i=1}^{n}\frac{1}{a_i}\right\}^{-1}, \quad (1.4.31)$$

where the equality sign holds only when $a_1 = a_2 = \cdots = a_n$. An integral version of (1.4.31) can be worked on as in (1.4.26), provided the arithmetic mean exists.

Entropy Inequality: Let $\sum_{i=1}^{n} a_i$ and $\sum_{i=1}^{n} b_i$ be convergent sequences of positive numbers, such that $\sum_{i=1}^{n} a_i \geq \sum_{i=1}^{n} b_i$. Then

$$\sum_{i\geq 1} a_i \log(b_i/a_i) \leq 0, \quad (1.4.32)$$

where the equality holds when and only when $a_i = b_i$, for all $i \geq 1$.

A real-valued random variable X is characterized by its **distribution function** (d.f.)

$$F_X(x) = P\{X \leq x\}, \quad x \in \mathbb{R}. \quad (1.4.33)$$

The d.f. F_X may be continuous everywhere or may have jump discontinuities. Two important classes of d.f.'s are the so-called **absolutely continuous type** and the **discrete type**. In the former case, $F_X(x)$ has a derivative $f_X(x) = (d/dx)F_X(x)$ almost everywhere (a.e.) which is also termed the **probability density function** (p.d.f.) of X. In the latter case, $F_X(x)$ is a step function and its jump discontinuities represent the probability masses $f_X(x)$ associated with the discrete points $\{x\}$; in this setup, $f_X(x)$ is often called the **probability function**. We consider here some of the d.f.'s mostly adopted in statistical inference.

i) **Binomial distribution**. It is defined in terms of a positive integer n and a parameter $\pi \ (0 \leq \pi \leq 1)$; the r.v. X can only assume the integer values $0, 1, \ldots, n$, with probability

$$P\{X = x\} = f_X(x) = \binom{n}{x}\pi^x(1-\pi)^{n-x}, \quad x = 0, 1, \ldots, n. \quad (1.4.34)$$

Therefore, we have

$$F_X(y) = \sum_{x=0}^{\min([y],n)} \pi^x(1-\pi)^{n-x}, \quad y \geq 0.$$

For this law, one has

$$EX = \sum_{x=0}^{n} xf_X(x) = n\pi \quad \text{and} \quad \mathrm{Var}X = n\pi(1-\pi). \quad (1.4.35)$$

ii) **Negative binomial distribution.** It is defined in terms of a positive integer r and a parameter π $(0 < \pi \leq 1)$; the r.v. X can assume the values $0, 1, \ldots$ with probability

$$P\{X = x\} = f_X(x) = \binom{r + x - 1}{x} \pi^{r-1}(1 - \pi)^x, \quad x = 0, 1, \ldots.$$

(1.4.36)

Here we have

$$EX = \frac{r(1 - \pi)}{\pi} \quad \text{and} \quad \operatorname{Var} X = \frac{r(1 - \pi)}{\pi^2}.$$

(1.4.37)

iii) **Poisson distribution.** The r.v. X takes on the value x $(= 0, 1, 2, \ldots)$ with probability

$$P\{X = x\} = f_X(x) = e^{-\lambda}\lambda^x/x!, \quad x \geq 0,$$

(1.4.38)

where λ (> 0) is a parameter. For this discrete distribution, one obtains that

$$EX = \operatorname{Var} X = \lambda.$$

(1.4.39)

iv) **Uniform $[0, 1]$ distribution.** The p.d.f. $f_X(x)$ is defined as

$$f_X(x) = \begin{cases} 1, & x \in [0, 1] \\ 0, & \text{otherwise} ; \end{cases}$$

(1.4.40)

so that $F_X(x) = x$, $0 \leq x \leq 1$, $F_X(x) = 0$, $x \leq 0$ and $F_X(x) = 1$, $x \geq 1$.

$$EX = \frac{1}{2} \quad \text{and} \quad \operatorname{Var}(X) = \frac{1}{12}.$$

(1.4.41)

v) **Gamma distribution.** Its p.d.f. is specified by

$$f_X(x) = c^p\{\Gamma(p)\}^{-1}e^{-cx}x^{p-1}, \quad x \geq 0,$$

(1.4.42)

where $c > 0$ and $p > 0$ are associated parameters. The d.f. F_X is the specified by the **incomplete gamma function**

$$F_X(x) \quad = \int_0^x f_X(y)dy = \int_0^{cx} e^{-y}y^{p-1}dy \Big/ \int_0^\infty e^{-y}y^{p-1}dy$$

$$= I_p(cx), \quad x \geq 0.$$

(1.4.43)

It is easy to verify that

$$EX = pc^{-1} \quad \text{and} \quad \operatorname{Var} X = pc^{-2}.$$

(1.4.44)

The gamma distribution with parameters $p = k/2$, $k \geq 1$, integer and $c = 1/2$ is known as the (central) **chi-squared** distribution with k degrees of freedom.

vi) **(Negative) exponential distribution.** It is a special case of the gamma distribution with $p = 1$, i.e.,

$$f_X(x) = ce^{-cx}, \quad x \geq 0, \quad c > 0. \tag{1.4.45}$$

vii) **Beta distribution.** The p.d.f. is defined by

$$f_X(x) = \frac{\Gamma(p+q)}{\Gamma(p)\Gamma(q)} x^{p-1}(1-x)^{q-1}, \quad 0 \leq x \leq 1, \tag{1.4.46}$$

where $p\ (> 0)$ and $q\ (> 0)$ are associated parameters. The d.f. $F_X(x)$ is, therefore, the incomplete beta function. It is easy to show that

$$EX = \frac{p}{p+q} \quad \text{and} \quad \text{Var}X = \frac{pq}{(p+q)^2(p+q+1)}. \tag{1.4.47}$$

viii) **Double exponential distribution.** The p.d.f. is given by

$$f_X(x) = \frac{\lambda}{2} e^{-\lambda|x-\theta|}, \quad -\infty < x < \infty, \tag{1.4.48}$$

where θ (real) and $\lambda\ (> 0)$ are the associated parameters. It is easy to verify that

$$EX = \theta \quad \text{and} \quad \text{Var}X = 2\lambda^{-2}. \tag{1.4.49}$$

ix) **Cauchy distribution.** The p.d.f. is given by

$$f_X(x) = (\lambda/\pi)\{\lambda^2 + (x-\theta)^2\}^{-1}, \quad -\infty < x < \infty, \tag{1.4.50}$$

where θ (real) and $\lambda\ (> 0)$ are location and scale parameters, respectively. It is easy to verify that EX and $\text{Var}X$ do not exist.

x) **Normal distribution.** The p.d.f. is given by

$$\varphi(x) = \frac{1}{\sqrt{2\pi}\sigma} e^{-(x-\mu)^2/2\sigma^2}, \quad -\infty < x < \infty, \tag{1.4.51}$$

where μ(real) and $\sigma > 0$ are location and scale parameters, respectively. It is easily seen that

$$EX = \mu \quad \text{and} \quad \text{Var}X = \sigma^2. \tag{1.4.52}$$

The normal distribution occupies a central position in the theory of large sample theory. In particular, we note that if the independent random variables $X_1, \ldots, X_n$ follow normal distributions with $EX_i = 0$ and $\text{Var}X_i = \sigma^2\ i = 1, \ldots, n$ then

$$Q = \sum_{i=1}^{n} X_i^2/\sigma^2$$

follows a chi-squared distribution with n degrees of freedom; if $EX_i = \mu_i$, $i = 1, \ldots, n$, then Q follows the so-called **noncentral chi-squared**

distribution with n degrees of freedom and noncentrality parameter $\sum_{i=1}^{n} \mu_i^2/\sigma^2$. Consider the case where $\mathrm{E}X_i = 0$, $i = 1, \ldots, n$, and let X_0 be a random variable following a normal distribution with $\mathrm{E}X_0 = \delta$ and $\mathrm{Var}X_0 = 1$; then,

$$t = \sqrt{n}X_o/\sqrt{Q}$$

follows a central (noncentral) **Student t distribution** with n degrees of freedom whenever $\delta = 0$ ($\delta \neq 0$); δ is termed the noncentrality parameter. Finally, if we let $X_1, \ldots, X_{n_1}$ and $Y_1, \ldots, Y_{n_2}$ denote two sets of independent normal random variables with $\mathrm{E}X_i = \mu$, $\mathrm{E}Y_j = 0$, $\mathrm{Var}X_i = \mathrm{Var}Y_j = \sigma^2$, $i = 1, \ldots, n_1$, $j = 1, \ldots, n_2$ and let $Q_1 = \sum_{i=1}^{n_1} X_i^2/\sigma^2$ and $Q_2 = \sum_{j=1}^{n_2} Y_j^2/\sigma^2$, then

$$F = n_2 Q_1/n_1 Q_2$$

follows a central (noncentral) **F distribution** with n_1 degrees of freedom in the numerator and n_2 degrees of freedom in the denominator whenever $\mu = 0$ ($\mu \neq 0$); μ^2/σ^2 is termed the noncentrality parameter.

xi) **Logistic distribution.** The p.d.f. is given by

$$f_X(x) = e^{-(x-\alpha)/\beta}/\{1 + e^{-(x-\alpha)/\beta}\}^2, \quad -\infty < x < \infty, \quad (1.4.53)$$

where $\alpha \in \mathbb{R}$ and $\beta > 0$ are location and scale parameters, respectively. It is easy to see that

$$\mathrm{E}X = \alpha \quad \text{and} \quad \mathrm{Var}\,X = \beta^2 \pi^2/3. \quad (1.4.54)$$

There are several functions which are associated with a d.f. F. Among these, the **moment generating function** (m.g.f.) and **characteristic function** (c.f.) deserve special mention. The m.g.f. $M_F(t)$ of a d.f. F is defined by

$$M_F(t) = \mathrm{E}_F(e^{tX}) = \int e^{tx} dF_X(x), \quad t \in \mathbb{R} \quad (1.4.55)$$

whenever the integral on the right-hand side exists. $M_F(t)$ provides the moments of F by successive differentiation, that is, for every positive integer k,

$$\mathrm{E}_F(X^k) = \mu_k' = (d^k/dt^k)M_F(t)\big|_{t=0}$$

and, hence, it bears its name. Unfortunately, $M_F(t)$ may not always exist [viz. the Cauchy distribution in (1.4.50)]. For this reason, often, it is more convenient to work with the c.f., which is defined as follows. For t real, the c.f. $\phi_F(t)$ corresponding to the d.f. F is given by

$$\begin{aligned} \phi_F(t) &= \int e^{it} dF_X(x) = \int \cos tx \, dF_X(x) + i \int \sin tx \, dF_X(x) \\ &= \phi_F^{(1)}(t) + i\phi_F^{(2)}(t), \end{aligned}$$

where $i = \sqrt{-1}$. Since $|e^{itx}| = 1$, for all $x \in \mathbb{R}$, $\phi_F(t)$ exists for all F. Note that $\phi_F(0) = 1$, $|\phi_F(t)| \leq 1$ and $\phi_F(t)$ is uniformly continuous on the entire real line $\mathbb{R}$. Further, note that $\phi_F(-t)$ is the c.f. of the r.v. $(-1)X$ [whose d.f. is $1 - F(-x)$], so that if X has a d.f. F_X symmetric about 0 [i.e., $F_X(-x) + F_X(x) = 1$, for all x]; then $\phi_F(-t) = \phi_F(t)$, so that $\phi_F(t)$ is real. In general, by (1.4.55), $\phi_F(t)$ is a complex-valued function. Exercises (1.4.1–1.4.9) are set to compute the c.f.'s of all the d.f. presented above. As we see, they all have different c.f.'s. In fact, there is a one-to-one correspondence between d.f.'s and their characteristic functions. Two different d.f.'s F and G cannot have the same c.f., and conversely two different c.f.'s cannot relate to a common d.f.. Characteristic functions have nice convergence properties: we shall present such results in Chapter 2. If moments of F up to a certain order exist, then the c.f. $\phi_F(t)$ can be expanded in a form, which provides a very handy tool for analysis. We present this basic result in the following:

Theorem 1.4.1: *Let Y be a random variable with distribution function F such that $\nu_{k+\delta} = \mathrm{E}|Y|^{k+\delta} < \infty$ for some integer $k \geq 1$ and $0 < \delta < 1$. Also let $\mu_r = \mathrm{E}Y^r$, $r = 1, \ldots, k$. Then*

$$\phi_F(t) = \mathrm{E}(e^{itY}) = 1 + it\mu_1 + \cdots + (it)^k \mu_k / k! + R_k(t),$$

where $|R_k(t)| \leq c|t|^{k+\delta} \nu_{k+\delta}$ with $c < \infty$ independent of t.

Proof: First consider the expansion

$$e^{iu} = 1 + iu + \cdots + (iu)^k / k! + (iu)^k (e^{iu\xi} - 1)/k! \qquad (1.4.56)$$

where $0 < \xi < 1$. For $|u| < 2$, let $k = 1$ and note that

$$|e^{iu} - 1| = |u(i - ue^{i\xi u}/2)| \leq 2|u| = 4|u/2| \leq 4|u/2|^\delta. \qquad (1.4.57)$$

For $|u| \geq 2$, observe that

$$|e^{iu} - 1| \leq 2 \leq 2|u/2|^\delta \leq 4|u/2|^\delta. \qquad (1.4.58)$$

From (1.4.57) and (1.4.58) we may conclude that

$$|e^{iu} - 1| \leq 4|u/2|^\delta = 2^{2-\delta}|u|^\delta. \qquad (1.4.59)$$

Using the same expansion as above, we may write

$$\phi_F(t) = \mathrm{E}(e^{itY}) = 1 + it\mathrm{E}Y + \cdots + (it)^k \mathrm{E}Y^k / k! + R_k(t),$$

where

$$R_k(t) = \frac{(it)^k}{k!} \int_{-\infty}^{+\infty} y^k (e^{it\xi y} - 1) dF(y)$$

and, using (1.4.59), we have

$$
\begin{aligned}
|R_k(t)| &\leq \frac{|t|^k}{k!} \int_{-\infty}^{+\infty} |y|^k |e^{it\xi y} - 1| dF(y) \\
&\leq \frac{|t|^k}{k!} \int_{-\infty}^{+\infty} |y|^k 2^{2-\delta} |ty|^\delta dF(y) \\
&= \frac{2^{2-\delta}|t|^{k+\delta}}{k!} \int_{-\infty}^{+\infty} |y|^{k+\delta} + dF(y) \\
&= C|t|^{k+\delta} \nu_{k+\delta}.
\end{aligned}
\tag{1.4.60}
$$

∎

Let us also present some parallel results for the multivariate case which will be considered in subsequent chapters. The d.f. F of a random vector $\mathbf{X} = (X_1, \ldots, X_p)^t$, $p \geq 1$, is defined as

$$
F(\mathbf{x}) = P\{\mathbf{X} \leq \mathbf{x}\}, \quad \mathbf{x} \in I\!\!R^p.
\tag{1.4.61}
$$

The c.f. $\phi_F(\mathbf{t})$ of $\mathbf{X}$ (or F), a function of $\mathbf{t} = (t_1, \cdots, t_p)^t$, is given by:

$$
\phi_F(\mathbf{t}) = \int \exp(i\mathbf{t}^t\mathbf{x}) dF(\mathbf{x}).
\tag{1.4.62}
$$

The one-to-one correspondence between F and ϕ_F remains intact. Of particular importance is the class of **multivariate normal** d.f.'s which can be uniquely represented by the c.f.'s:

$$
\phi_\Phi(\mathbf{t}) = \exp\left\{ i\mathbf{t}^t\boldsymbol{\mu} - \frac{1}{2}\mathbf{t}^t\boldsymbol{\Sigma}\mathbf{t} \right\}, \quad \mathbf{t} \in I\!\!R^p,
\tag{1.4.63}
$$

where $\boldsymbol{\mu} = (\mu_1, \ldots, \mu_p)^t$ is the mean vector and $\boldsymbol{\Sigma} = ((\sigma_{ij}))$ is the dispersion or variance-covariance matrix corresponding to the d.f. Φ. In general, $\boldsymbol{\Sigma}$ is positive semi definite (p.s.d.); if $\boldsymbol{\Sigma}$ is positive definite (p.d.), then the d.f. Φ is of full rank and has a p.d.f. given by

$$
\varphi(\mathbf{x}) = (2\pi)^{-p/2}|\boldsymbol{\Sigma}|^{-1/2} \exp\left\{ -\frac{1}{2}(\mathbf{x} - \boldsymbol{\mu})^t \boldsymbol{\Sigma}^{-1}(\mathbf{x} - \boldsymbol{\mu}) \right\}, \quad \mathbf{x} \in I\!\!R^p.
\tag{1.4.64}
$$

Note that the p.d.f. does not exist if $|\boldsymbol{\Sigma}| = 0$, but through the c.f. in (1.4.62) we may still characterize the multinormal distribution.

In the above context, as well as in other situations, for a matrix $\mathbf{A}$ we may define a **generalized inverse** as any matrix $\mathbf{A}^-$ such that

$$
\mathbf{A}\mathbf{A}^-\mathbf{A} = \mathbf{A}.
\tag{1.4.65}
$$

Note that if we let $\mathbf{H} = \mathbf{A}^-\mathbf{A}$, then $\mathbf{H}^2 = \mathbf{H}\mathbf{H} = \mathbf{A}^-\mathbf{A}\mathbf{A}^-\mathbf{A} = \mathbf{A}^-\mathbf{A} = \mathbf{H}$, so that $\mathbf{H}$ is **idempotent**, and hence

$$\text{rank}(\mathbf{A}) = \text{rank}(\mathbf{H}) = \text{trace}(\mathbf{H}). \tag{1.4.66}$$

In various statistical applications, we may need to partition a symmetric matrix $\mathbf{A}$ as $\begin{pmatrix} \mathbf{A}_{11} & \mathbf{A}_{12} \\ \mathbf{A}_{21} & \mathbf{A}_{22} \end{pmatrix}$, where $\mathbf{A}_{11}$ and $\mathbf{A}_{22}$ are square matrices. If $\mathbf{A}$ and $\mathbf{A}_{11}$ are p.d., then we may verify that

$$\mathbf{A}^{-1} = \begin{pmatrix} \mathbf{A}_{11}^{-1} + \mathbf{A}_{11}^{-1}\mathbf{A}_{12}\mathbf{A}_{22.1}^{-1}\mathbf{A}_{21}\mathbf{A}_{11}^{-1} & -\mathbf{A}_{11}\mathbf{A}_{12}\mathbf{A}_{22.1}^{-1} \\ -\mathbf{A}_{22.1}^{-1}\mathbf{A}_{21}\mathbf{A}_{11}^{-1} & \mathbf{A}_{22.1}^{-1} \end{pmatrix}, \tag{1.4.67}$$

where

$$\mathbf{A}_{22.1} = \mathbf{A}_{22} - \mathbf{A}_{21}\mathbf{A}_{11}^{-1}\mathbf{A}_{12}. \tag{1.4.68}$$

Whenever $\mathbf{A}$ is $m \times m$, the determinantal equation $|\mathbf{A} - \lambda\mathbf{I}| = 0$ has m roots, say, $\lambda_1, \ldots, \lambda_m$, which are termed the **characteristic roots** or **eigenvalues** of $\mathbf{A}$. If $\mathbf{A}$ is p.s.d., then all the λ_i's are non-negative, and we take them in the order $\lambda_1 \geq \lambda_2 \geq \cdots \geq \lambda_p \geq 0$. Then, corresponding to each λ_i, there is a vector $\mathbf{x}_i$ (called the **characteristic vector** or **eigenvector**), such that

$$\mathbf{A}\mathbf{x}_i = \lambda_i\mathbf{x}_i, \quad i = 1, \ldots, p. \tag{1.4.69}$$

Hence, there exists an orthogonal matrix $\mathbf{B}$, such that

$$\mathbf{B}^t\mathbf{A}\mathbf{B} = \Lambda = \text{diag}(\lambda_1, \ldots, \lambda_p), \tag{1.4.70}$$

where $\text{diag}(\mathbf{x})$ denotes a diagonal matrix with the elements of $\mathbf{x}$ along the main diagonal. This decomposition is useful for studying quadratic forms

$$Q = \mathbf{x}^t\mathbf{A}\mathbf{x}, \quad \mathbf{A} \quad \text{p.s.d. symmetric.} \tag{1.4.71}$$

Writing $\mathbf{x} = \mathbf{B}\mathbf{y}$, we obtain from (1.4.70) and (1.4.71) that

$$Q = \mathbf{y}^t\mathbf{B}^t\mathbf{A}\mathbf{B}\mathbf{y} = \mathbf{y}^t\Lambda\mathbf{y} = \sum_{i=1}^{p} \lambda_i y_i^2. \tag{1.4.72}$$

From (1.4.72) and using the ordering of the λ_i's we obtain that

$$\lambda_p \mathbf{y}^t\mathbf{y} \leq Q \leq \lambda_1 \mathbf{y}^t\mathbf{y}. \tag{1.4.73}$$

In fact, (1.4.73) extends to the following more general result:

Theorem 1.4.2 (Courant): *Let $\mathbf{A}$ and $\mathbf{B}$ be two p.s.d. matrices with $\mathbf{B}$ nonsingular. Then if λ_i denotes the ith characteristic root of $\mathbf{A}\mathbf{B}^{-1}$ for $i = 1, \ldots, p$, we have*

$$\text{ch}_p(\mathbf{A}\mathbf{B}^{-1}) = \lambda_p = \inf_{\mathbf{x}} \frac{\mathbf{x}^t\mathbf{A}\mathbf{x}}{\mathbf{x}^t\mathbf{B}\mathbf{x}} \leq \sup_{\mathbf{x}} \frac{\mathbf{x}^t\mathbf{A}\mathbf{x}}{\mathbf{x}^t\mathbf{B}\mathbf{x}} = \lambda_1 = \text{ch}_1(\mathbf{A}\mathbf{B}^{-1}).$$
$$\tag{1.4.74}$$

1.5 Concluding notes

We have stressed the basic motivation for the use of large sample theory through five practical examples. We believe that the reader should look into a variety of other examples appearing in their own professional work and justify the need for such procedures. We also believe that some general acquaintance with the basic theory of Statistics [at the level of Hogg and Craig (1970), for example] will enable the reader to scan through Section 1.3. The material in Section 1.4 was mainly included for the sake of self-containment, but we stress that a revisit to this mathematical background area may really be very beneficial for a smooth transit through the rest of the book. Some further useful results are presented as exercises.

1.6 Exercises

Exercise 1.4.1: For the binomial law (1.4.34), show that the characteristic function is $\phi(t) = (\pi e^{it} + 1 - \pi)^n$, $t \in \mathbb{R}$. Use this to verify (1.4.35).

Exercise 1.4.2: Obtain the characteristic function for the Poisson law (1.4.38) and use it to verify (1.4.39).

Exercise 1.4.3: For the Unif[0, 1] distribution, obtain the characteristic function and verify (1.4.41).

Exercise 1.4.4: For the gamma distribution (1.4.42), derive the characteristic function. Hence or otherwise, derive the characteristic function for the chi-squared distribution with k degrees of freedom.

Exercise 1.4.5: Derive the characteristic function for the beta distribution (1.4.46) and hence, or otherwise, verify (1.4.47).

Exercise 1.4.6: For the double-exponential distribution (1.4.48) derive the characteristic function and verify (1.4.49).

Exercise 1.4.7: Show that the characteristic function for the Cauchy distribution does not satisfy the condition for the existence of its first two moments.

Exercise 1.4.8: For the normal distribution (1.4.51), show that the characteristic function is $\phi(t) = \exp(it\mu - t^2\sigma^2/2)$, $t \in \mathbb{R}$.

Exercise 1.4.9: Derive the characteristic function for the $F(p, n-p)$ distribution from that in Exercise 1.4.5.

CHAPTER 2

Stochastic Convergence

2.1 Introduction

In problems of statistical inference (on the basis of a representative sample), statistical decision rules are usually based on suitable estimators (of parameters of interest), test statistics (for appropriate statistical hypotheses) and other plausible functions of sample observations. These **statistics** may be real- or vector-valued random functions or may even be more general random functions, such as **stochastic processes**. Being stochastic in nature, these statistics are characterized by chance fluctuations (governed by the underlying model and other factors) which generally exhibit a diminishing trend with the increase in the size of the sample on which they are based. A minimal requirement for a good statistical decision rule is its increasing reliability with increasing sample sizes (termed **consistency**). Thus, for an estimator, consistency relates to an increasing closeness to its population counterpart with increasing sample sizes. However, in view of the stochastic nature of the estimator, this closeness needs to be defined and interpreted in a meaningful and precise manner incorporating the stochastic nature of the fluctuation of the estimator around the parameter it estimates. Generally, a **distance function** (or **norm**) of this stochastic fluctuation is incorporated in the formulation of this closeness, and consistency refers to the convergence of this norm to 0 in a well-defined manner. Similarly, dealing with a test of significance (for a statistical hypothesis), consistency refers to the increasing degree of confidence in the ability of the test to reject the null hypothesis when it is not true as the sample size increases. In the framework of a more general statistical decision rule, consistency refers to decreasing **risks** of making incorrect statistical decisions with increasing sample sizes. Basically, the consistency of a statistical decision rule rests on some **convergence properties** of an associated sequence of statistics, and this concept is intimately related to the dual one of **stochastic convergence** in the classical theory of probability.

For a sequence $\{f_n\}$ of real numbers, we say that f_n converges to a limit f as n goes to ∞, if for every positive ε, there exists a positive integer n_0

$[= n_0(\varepsilon)]$, such that

$$|f_n - f| < \varepsilon, \quad n \geq n_0. \tag{2.1.1}$$

Equivalently, we may state this in terms of the **Cauchy property** that

$$|f_{n+m} - f_n| < \varepsilon, \quad n \geq n_0, \, m \geq 1. \tag{2.1.2}$$

In a more general case, the f_n may be q-vectors, for some $q \geq 1$, denoted here by $\mathbf{f}_n$. In this case, if $\mathbf{x} = (x_1, \ldots, x_q)^t \in \mathbb{R}^q$ we may introduce the **Euclidean norm**

$$\|\mathbf{x}\| = (\mathbf{x}^t \mathbf{x})^{1/2} = \left(\sum_{i=1}^{q} x_i^2 \right)^{1/2} \tag{2.1.3}$$

and (2.1.1) and (2.1.2) remain intact if we replace $|f_n - f|$ and $|f_{n+m} - f_n|$ by $\|\mathbf{f}_n - \mathbf{f}\|$ and $\|\mathbf{f}_{n+m} - \mathbf{f}_n\|$, respectively. Instead of the simple Euclidean norm, we may even consider other norms, such as

$$\textbf{max-norm}: \|\mathbf{x}\|_\infty = \max\{|x_j| : 1 \leq j \leq q\}, \tag{2.1.4}$$

$$\textbf{quadratic norm}: \|\mathbf{x}\|_{\mathbf{W}} = (\mathbf{x}^t \mathbf{W} \mathbf{x})^{1/2}, \tag{2.1.5}$$

where $\mathbf{W}$ is a given positive definite (p.d.) matrix; the particular choice of $\mathbf{W} = \mathbf{I}$ leads to the Euclidean norm. We may even proceed one step further and consider a more general situation where $f_n = \{f_n(t), t \in T\}$ is a function defined on an interval T (viz., $T = [0, 1]$ or $[0, \infty)$). In such a case, for a function $x = \{x(t), t \in T\}$, an appropriate norm may be

$$\|x\|_{\sup} = \sup\{|x(t)| : t \in T\} \tag{2.1.6}$$

or some ramifications of it. [When $x(t)$ is a q-vector, we may as well replace $|x(t)|$ in (2.1.6) by one of the norms in (2.1.4) or (2.1.5)]. Again, the definition of the convergence of $\{f_n\}$ to a limit f ($= \{f(t), t \in T\}$), given in (2.1.1) or (2.1.2), stands valid when we replace the simple norm in (2.1.1) by (2.1.6). The notion of convergence, as has been outlined above, can also be formulated in terms of a general **metric space**. However, we shall refrain ourselves from such a more abstract treatment, and confine ourselves to the particular cases treated above.

In dealing with a sequence $\{f_n\}$ of random elements (statistics), we may, however, note that no matter how large an n is chosen, the inequality $|f_n - f| < \varepsilon$ or $\|\mathbf{f}_n - \mathbf{f}\| < \varepsilon$ may not hold with probability equal to one, but may do so with probability arbitrarily close to one when n is indefinitely large. Moreover, in the case of non-stochastic elements, (2.1.1) may also be equivalently written as

$$\sup_{N \geq n} |f_N - f| < \varepsilon, \quad n \geq n_0, \tag{2.1.7}$$

where again $|f_N - f|$ may be replaced by $\|\mathbf{f}_N - \mathbf{f}\|$, or other norms too. On the other hand, for the case of stochastic elements, convergence can be defined in terms of the marginal event $\|\mathbf{f}_n - \mathbf{f}\|$ or in terms of the simultaneous events in (2.1.7), and this subtle difference in the mode of convergence leads us to formulate two allied concepts: **weak (or stochastic) convergence (or convergence in probability)** and **strong (or almost sure or almost certain) convergence**; these will be elaborated in the subsequent sections of this chapter. Dealing with non stochastic elements, both these modes of convergence are equivalent to the original mode of convergence laid down earlier. However, in the case of stochastic elements, weak convergence may not necessarily ensure strong convergence and, generally, more stringent regularity conditions may be necessary for the latter to hold. There are some other modes of convergence (notably, **complete convergence** and **convergence in the rth mean**, for $r > 0$) which will also be considered. The interrelationships of these modes of convergence will be studied too.

Typically, in a statistical estimation problem, f_n may refer to an estimator (possibly vector valued), so that f_n is a random element, whereas f, the corresponding parameter, is non stochastic. But, in a more general setup, we may encounter a situation where $\{f_n\}$ is a stochastic sequence converging to a (possibly) stochastic f (in some suitable sense). Hence, to preserve generality, we will consider the various modes of stochastic convergence treating both $\{f_n\}$ and f as possibly random elements. These concepts are systematically presented in Section 2.2. Mutual implications of these modes of convergence are also discussed in the same section. Probability inequalities and the classical **Laws of Large Numbers (LLN)** play a vital role in this context, and these are treated in full generality in Section 2.3. These LLN and probability inequalities are usually posed for sums or averages of independent (and, often, identically distributed) random variables (or vectors). In many statistical problems, however, we may have non standard situations where the statistics may not be expressible as linear functions of independent random variables (i.e., they may be non linear and may have non independent components). To cope with such situations, some extensions of the LLN and probability inequalities to possibly dependent random elements are also considered in Section 2.4. In particular, much emphasis has been laid down on the so-called **martingale, reverse martingale** and related sequences. A handful of examples has also been set to illustrate their potential use in problems of statistical inference. The last section deals with some rather isolated topics having some use in statistical inference, and a more detailed description of them is given there.

2.2 Modes of stochastic convergence

The concepts of stochastic convergence as outlined in the previous section may be formalized in the following definitions. In this context, we consider a set A of elements $\{x_i\}$ along with a **distance function** $d(x_i, x_j)$ with the properties that

  i) $d(x_i, x_i) = 0$,

 ii) $d(x_i, x_j) > 0$ if $x_i \neq x_j$,

iii) $d(x_i, x_j) = d(x_j, x_i)$,

 iv) $d(x_i, x_j) + d(x_j, x_k) \geq d(x_i, x_k)$,

for all i, j, k. Note that the norms introduced in (2.1.1), (2.1.2), (2.1.3), (2.1.4) and (2.1.5) satisfy these conditions.

Definition 2.2.1 (Convergence in probability): *A sequence $\{T_n\}$ of random elements is said to converge in probability to a (possibly degenerate) random element T, if for every (given) positive numbers ε and η, there exists a positive integer $n_0 = n_0(\varepsilon, \eta)$, such that*

$$P\{d(T_n, T) > \varepsilon\} < \eta, \quad n \geq n_0. \tag{2.2.1}$$

This mode of convergence is usually denoted by

$$T_n - T \xrightarrow{\text{P}} 0 \tag{2.2.2}$$

($T_n - T$ converges in probability to 0). In the case where T is non stochastic, we may write

$$T_n \xrightarrow{\text{P}} T. \tag{2.2.3}$$

Definition 2.2.2 [Almost sure (or strong) convergence]: *A sequence $\{T_n\}$ of random elements is said to converge almost surely (a.s.) or strongly to a (possibly degenerate) random element T, if for every (given) positive ε and η, there exists a positive integer $n_0 = n_0(\varepsilon, \eta)$, such that*

$$P\{d(T_N, T) > \varepsilon \text{ for some } N \geq n\} < \eta, \quad n \geq n_0. \tag{2.2.4}$$

In symbols, we write this as

$$T_n - T \xrightarrow{\text{a.s.}} 0, \tag{2.2.5}$$

($T_n - T$ converges almost surely to 0) and, if T is non stochastic, (2.2.5) may also be written as $T_n \xrightarrow{\text{a.s.}} T$.

   To clarify the difference between (2.2.1) and (2.2.4), let $A_n = [d(T_n, T) > \varepsilon]$, for $n \geq n_0$; then the stochastic convergence in (2.2.1) entails that

$$P(A_n) \to 0 \quad \text{as} \quad n \to \infty. \tag{2.2.6}$$

On the other hand, the a.s. convergence in (2.2.4) entails that

$$P\left\{\bigcup_{N \geq n} A_N\right\} \to 0 \quad \text{as} \quad n \to \infty \qquad (2.2.7)$$

and, hence, (2.2.4) is a stronger mode of convergence than (2.2.1). Since $P\left\{\bigcup_{N \geq n} A_N\right\} \geq P(A_n)$, it is quite clear that (2.2.5) ensures (2.2.2); however, the converse may not be true. We shall make more comments on it later.

To illustrate these modes of convergence, let us consider some simple examples. Let $X_1, \ldots, X_n, \ldots$ be a sequence of independent and identically distributed (i.i.d.) random variables (r.v.) having a distribution function (d.f.) F, defined on the real line $\mathbb{R}$, such that the mean μ and variance σ^2, defined below, are finite.

$$\mu = \int_{\mathbb{R}} x \, dF(x) \quad \text{and} \quad \sigma^2 = \int_{\mathbb{R}} x^2 \, dF(x) - \mu^2. \qquad (2.2.8)$$

Based on a sample $(X_1, \ldots, X_n)$ of size n, we consider the estimators

$$T_{n1} = \overline{X}_n = n^{-1} \sum_{i=1}^{n} X_i, \quad T_{n2} = n^{-1} \sum_{i=1}^{n} (X_i - \overline{X}_n)^2. \qquad (2.2.9)$$

Also, let

$$F_n(x) = n^{-1} \sum_{i=1}^{n} I_{\{X_i \leq x\}}, \quad x \in \mathbb{R}, \qquad (2.2.10)$$

where I_A denotes the indicator function of the set A, be the **empirical** (or **sample) distribution function (EDF)**. Consider the following:

i) $T_n = T_{n1}$, $T = \mu$ and $d(x, y) = |x - y|$;

ii) $\mathbf{T}_n = (T_{n1}, T_{n2})^t$, $\mathbf{T} = (\mu, \sigma^2)^t$, and $d(\mathbf{x}, \mathbf{y}) = \{(\mathbf{x} - \mathbf{y})^t (\mathbf{x} - \mathbf{y})\}^{1/2}$;

iii) $T_n = \{F_n(x), x \in \mathbb{R}\}$, $T = \{F(x), x \in \mathbb{R}\}$ and $d(x, y) = \|x - y\|_{\sup}$ in (2.1.6).

In the first case, $\{T_n\}$ corresponds to a sequence of real-valued r.v.'s, in the second case to a sequence of vector-valued r.v.'s and, in the last, to a sequence of random functions or stochastic processes. In the first two cases, we have the usual Euclidean norm for $d(\cdot)$, whereas in the last case, in view of the infinite dimensionality, the Euclidean distance does not work out well and we have adapted a more meaningful metric introduced in (2.1.6). As we shall see in subsequent sections, in all three cases, $T_n - T \to 0$ in probability/a.s., as $n \to \infty$. In passing, we may remark that unless the functional form of F is assumed to be known (and it involves T_n as a sufficient statistic), the EDF F_n contains more information than the statistic

T_n. Thus, one may be interested in the EDF itself for a better feeling of the underlying d.f., and hence, may like to deal with such a stochastic process rather than finite dimensional random vectors for T_n. This subtle point will be made clearer in a later chapter where the significance of nonparametric methods relative to their parametric counterparts will be discussed.

We take this opportunity to introduce some notations which will be consistently used in this as well as in later chapters. Consider two sequences $\{a_n\}$ and $\{b_n\}$ of real numbers. Then, we introduce the following:

Definition 2.2.3: *We say that $a_n = O(b_n)$ if there exists a finite positive number K and a positive integer $n(K)$, such that*

$$|a_n/b_n| \leq K, \quad n \geq n(K). \tag{2.2.11}$$

In particular, $a_n = O(1)$ means that $|a_n| \leq K$, for some finite, positive K for n sufficiently large, i.e., $\{a_n\}$ is eventually bounded.

Definition 2.2.4: *We say that $a_n = o(b_n)$ if for any $\varepsilon > 0$, there exists a positive integer $n(\varepsilon)$, such that*

$$|a_n/b_n| < \varepsilon, \quad n \geq n(\varepsilon). \tag{2.2.12}$$

In particular $a_n = o(1)$ means that $a_n \to 0$ as $n \to \infty$.

Definition 2.2.5: *For a sequence $\{X_n\}$ of random variables, if for every $\eta > 0$ there exist a positive constant $K(\eta)$ $(< \infty)$ and a positive integer $n(\eta)$, such that*

$$P\{|X_n| \leq K(\eta)\} \geq 1 - \eta, \quad n \geq n(\eta), \tag{2.2.13}$$

then we say that $X_n = O_p(1)$. We also say that if (2.2.13) holds, then $\{X_n\}$ is bounded in probability.

The definition in (2.2.13) readily extends to the case of random vectors $\{\mathbf{X}_n\}$; all we need is to replace $|X_n|$ by the Euclidean norm $\|\mathbf{X}_n\|$. In this case, we shall write $\mathbf{X}_n = O_p(\mathbf{1})$ to designate the vector case.

Definition 2.2.6: *If for a sequence $\{X_n\}$ of random variables and another sequence $\{b_n\}$ (of possibly random variables), for every $\eta > 0$, $\varepsilon > 0$, there exists a positive integer $n(\varepsilon, \eta)$, such that*

$$P\{|X_n/b_n| \leq K(\eta)\} \geq 1 - \eta, \quad n \geq n(\eta), \tag{2.2.14}$$

then we say that $X_n = O_p(b_n)$. The definition immediately extends to the vector case by replacing the norm in (2.2.14) by the usual Euclidean norm.

Note that (2.2.13) is a special case of (2.2.14) when we allow b_n to be bounded in probability. Further, with the metric in (2.1.6), all these notations can also be extended to random functions.

Definition 2.2.7: *For a sequence $\{X_n\}$ of random variables, if for every $\eta > 0$, $\varepsilon > 0$, there exists a positive integer $n(\varepsilon, \eta)$, such that*

$$P\{|X_n| > \eta\} < \varepsilon, \quad n \geq n(\varepsilon, \eta), \tag{2.2.15}$$

then we say that $X_n = o_p(1)$. In other words, $X_n = o_p(1)$ is equivalent to saying that $X_n \xrightarrow{\text{p}} 0$. The definition extends directly to the vector case by adapting the Euclidean norm. Here we write $\mathbf{X}_n = o_p(\mathbf{1})$. Also, the metric in (2.1.6) may be used to extend it to the case of random functions.

Definition 2.2.8: *If for a sequence $\{X_n\}$ of random variables and another sequence $\{b_n\}$ (of possibly random variables), for every $\eta > 0$, $\varepsilon > 0$, there exists a positive integer $n(\varepsilon, \eta)$, such that*

$$P\{|X_n/b_n| > \eta\} < \varepsilon, \quad n \geq n(\varepsilon, \eta), \tag{2.2.16}$$

then we say that $X_n = o_p(b_n)$; the definition also extends to the vector case under the Euclidean norm, and to the functional case, under (2.1.6).

Further, note that

$$X_n = o_p(b_n) \quad \text{and} \quad b_n = O_p(c_n) \Rightarrow X_n = o_p(c_n). \tag{2.2.17}$$

Thus, (2.2.15) can be obtained from (2.2.16) and (2.2.13) on the $\{b_n\}$.

Definition 2.2.9: *If, for any sequence of real numbers $\{x_n\}$, such that $x_n \to L$ as $n \to \infty$, we have $a(x_n) = O[b(x_n)]$, in the sense of (2.2.14), then we say that $a(x) = O[b(x)]$ as $x \to L$. This definition can be fitted to (2.2.14) by letting $a_n = a(x_n)$ and $b_n = b(x_n)$, and allowing $x_n \to L$.*

Definition 2.2.10: *If, for any sequence of real numbers $\{x_n\}$, such that $x_n \to L$ as $n \to \infty$, we have $a(x_n) = o[b(x_n)]$, in the sense of (2.2.16), then we say that $a(x) = o[b(x)]$, as $x \to L$.*

Definition 2.2.11: *If, in the setup of (2.2.13), we have*

$$P\{|X_N| > K(\eta) \text{ for some } N \geq n\} < \eta, \quad n \geq n(\eta), \tag{2.2.18}$$

then we say that $X_n = O(1)$ a.s. Similarly, if we replace (2.2.14) by

$$P\{|X_N/b_N| > K(\eta) \text{ for some } N \geq n\} < \eta, \quad n \geq n(\eta), \tag{2.2.19}$$

then we say that $X_n = 0(b_n)$ a.s.

Definition 2.2.12: *If (2.2.15) is replaced by*

$$P\{|X_N| > \eta \text{ for some } N \geq n\} < \varepsilon, \quad n \geq n(\varepsilon, \eta), \qquad (2.2.20)$$

then we say that $X_n = o(1)$ a.s. Similarly, replacing (2.2.16) by

$$P\{|X_N/b_N| > \eta \text{ for some } N \geq n\} < \varepsilon, \quad n \geq n(\varepsilon, \eta), \qquad (2.2.21)$$

we say that $X_n = o(b_n)$ a.s.

Finally, we may note that (2.2.18) through (2.2.21) all extend to the vector case under the adaptation of the Euclidean norm, and also to the case of random functions under the metric in (2.1.6).

By virtue of the notations introduced above, we have the following:

$$T_n - T \xrightarrow{\text{P}} 0 \Leftrightarrow d(T_n, T) = o_p(1), \qquad (2.2.22)$$

$$T_n - T \xrightarrow{\text{a.s.}} 0 \Leftrightarrow d(T_n, T) = o(1) \text{ a.s.} \qquad (2.2.23)$$

Thus, both the convergence in probability and almost sure convergence results can be stated in the framework of **stochastic orders** of the distance $d(T_n, T)$ and this holds in the vector case as well as for random functions too. Consequently, one may reduce the problem to that involving the sequence $\{d(T_n, T)\}$ of real valued r.v.'s and to verify (2.2.15) or (2.2.20) for this sequence.

There is a third mode of convergence which has interest of its own and which also provides a simpler way of establishing (2.2.2). It will be advantageous for us to introduce this mode of stochastic convergence for the real- or vector-valued statistics first, and then to extend the definition to the more general case.

Definition 2.2.13 [Convergence in the rth mean $(r > 0)$]: *A sequence $\{T_n\}$ of random variables (or vectors) is said to converge in the rth mean (for some $r > 0$) to a (possibly degenerate) random variable (or vector) T, if for any given positive ε, there exists a positive integer $n_0 = n_0(\varepsilon)$, such that $E\|T_n - T\|^r < \varepsilon$, for every $n \geq n_0$, or in other words, if*

$$E\|T_n - T\|^r \to 0 \quad \text{as} \quad n \to \infty, \quad \text{i.e.,} \quad E\|T_n - T\|^r = o(1). \qquad (2.2.24)$$

Here, we adopt the Euclidean distance in (2.1.3). In symbols, we put

$$T_n - T \xrightarrow{r\text{th}} 0. \qquad (2.2.25)$$

To cover the general case of random functions, we conceive of a metric $d(\cdot, \cdot)$, such as the one in (2.1.6), and we say that $T_n - T \xrightarrow{r\text{th}} 0$ (in the topology d) if for every (given) $\varepsilon > 0$, there exists a positive integer $n_0 = n_0(\varepsilon)$, such that

$$E(\{d(T_n, T)\}^r) < \varepsilon, \quad n \geq n_0 \quad (r > 0). \qquad (2.2.26)$$

Here also, if T is non stochastic, we may write $T_n \xrightarrow{\text{rth}} T$.

Before we proceed to study the mutual implications of these modes of convergence, we present another concept of convergence which plays a fundamental role in large sample theory.

Definition 2.2.14 [Convergence in distribution (or weak convergence)]: *A sequence $\{T_n\}$ of random variables with distribution functions $\{F_n\}$ is said to converge in distribution (or in law) to a (possibly degenerate) random variable T with a distribution function F, if for every (given) $\varepsilon > 0$, there exists an integer $n_0 = n_0(\varepsilon)$, such that at every point of continuity (x) of F,*

$$|F_n(x) - F(x)| < \varepsilon, \quad n \geq n_0. \tag{2.2.27}$$

This mode of convergence is denoted by either

$$F_n \xrightarrow{w} F, \quad T_n \xrightarrow{D} T \quad \text{or} \quad \mathcal{L}(T_n) \to \mathcal{L}(T). \tag{2.2.28}$$

If the $\mathbf{T}_n$ are vectors, we may, similarly, work with the multivariate distribution functions (i.e., $F_n(\mathbf{x}) = P\{\mathbf{T}_n \leq \mathbf{x}\}$ and $F(\mathbf{x}) = P\{\mathbf{T} \leq \mathbf{x}\}$) and (2.2.27) still works out neatly. However, when the T_n are random functions, this definition may run into some technical difficulties, and additional conditions may be needed to ensure (2.2.28). We shall discuss that in Chapter 8.

Recall that, in Definition 2.2.14, T_n and T need not be defined on the same probability space. Further, $d(T_n, T)$ may not make any sense since, in fact, T_n may not converge to T in any mode; rather, the distribution of T_n converges to that of T, and in (2.2.28), $\mathcal{L}(T_n) \to \mathcal{L}(T)$ actually signifies that. It may be noted further that if T is a degenerate r.v. (at the point c), then

$$F(t) = \begin{cases} 0 & \text{if } t < c \\ 1 & \text{if } t \geq c. \end{cases} \tag{2.2.29}$$

Therefore, if (2.2.27) holds and if F is a degenerate d.f. at a point c, then we obtain that F_n converges to the degenerate d.f. F, and hence, as $n \to \infty$,

$$F_n(t) \to 0 \text{ or } 1 \text{ accordingly as } t \text{ is } < \text{ or } > c, \tag{2.2.30}$$

so that $T_n \xrightarrow{P} c$. This leads us to the following:

Proposition 2.2.1: *For a sequence $\{T_n\}$ of random variables or vectors, convergence in law (or distribution) to a random variable or vector T, degenerate at a point c, ensures that $T_n \to c$, in probability, as $n \to \infty$.*

We shall find this proposition very useful in what follows. We proceed now to study the mutual implications of the various modes of convergence;

in practice, this gives us a variety of tools to establish a particular mode of convergence. We first consider the following (probability) inequality which has also innumerous other applications.

Chebyshev Inequality: *Let U be a non-negative r.v. with a finite mean $\mu = EU$. Then, for every $t > 0$,*

$$P\{U > t\mu\} \leq t^{-1}. \qquad (2.2.31)$$

Proof: Note that

$$
\begin{aligned}
\mu = EU &= \int_0^{+\infty} u\, dP(U \leq u) \\
&= \int_0^{t\mu} u\, dP(U \leq u) + \int_{t\mu}^{+\infty} u\, dP(U \leq u) \\
&\geq \int_{t\mu}^{+\infty} u\, dP(U \leq u) \\
&\geq t\mu \int_{t\mu}^{+\infty} dP(U \leq u) = t\mu P\{U > t\mu\}, \qquad (2.2.32)
\end{aligned}
$$

and hence, dividing both sides by $t\mu$, we obtain (2.2.31) from (2.2.32). ∎

Proposition 2.2.2: *Almost sure convergence implies convergence in probability.*

Proof: The proof of this proposition is given in (2.2.6) and (2.2.7). ∎

Proposition 2.2.3: *Convergence in the rth mean for some $r > 0$ implies convergence in probability.*

Proof: Let $U_n = |T_n - T|^r$, where $r > 0$. Then by (2.2.31), we obtain that for every $t > 0$, $P\{U_n > tEU_n\} \leq t^{-1}$, i.e.,

$$P\left\{|T_n - T| > t^{1/r}[E|T_n - T|^r]^{1/r}\right\} \leq t^{-1}. \qquad (2.2.33)$$

Thus, for given $\varepsilon > 0$ and $\eta > 0$, choose n_0 and t, such that

$$t^{-1} < \eta, \quad \text{and} \quad t^{1/r}[E|T_n - T|^r]^{1/r} < \varepsilon, \quad n \geq n_0, \qquad (2.2.34)$$

which is always possible since $E|T_n - T|^r \to 0$ as $n \to \infty$. Thus, from (2.2.33), we obtain that $P\{|T_n - T| > \varepsilon\} \leq \eta$, for every $n \geq n_0$, and hence, $T_n - T$ converges in probability to 0, as $n \to \infty$. ∎

Note that for Proposition 2.2.3, r is arbitrary (but positive). Further, the converse implication may not be always true. To illustrate this point, let $X_n \sim \text{Bin}(n, \pi)$ and $T_n = n^{-1}X_n$. Note that for all $\pi \in (0, 1)$,

$$\text{E}T_n = \pi \quad \text{and} \quad \text{E}(T_n - \pi)^2 = n^{-1}\pi(1 - \pi) \leq (4n)^{-1}. \qquad (2.2.35)$$

Thus, using Proposition 2.2.3, we immediately obtain that T_n converges to π in the second mean, and hence, $T_n \xrightarrow{\text{P}} \pi$. Consider next the parameter $\theta = \pi^{-1}$, and its natural estimator $Z_n = T_n^{-1}$. Since $T_n \xrightarrow{\text{P}} \pi$ and θ is a continuous function of π (except at $\pi = 0$), it immediately follows that $Z_n \xrightarrow{\text{P}} \theta$. However, $P\{X_n = 0\} = (1 - \pi)^n > 0$ for every $n \geq 1$ (although it converges to 0 as $n \to \infty$) and, hence, for any $r > 0$, $\text{E}Z_n^r$ fails to be finite. Therefore, Z_n does not converge to θ, in the rth mean, for any $r > 0$. This simple example also shows that the tails of a distribution may be negligible, but their contributions to moments of the distribution may not be so. We shall discuss more about this in connection with **uniform integrability** in a later section. In some simple situations, we have a sufficient condition under which convergence in probability implies convergence in rth mean, and we present this below.

Proposition 2.2.4: *Convergence in probability for almost surely bounded random variables implies convergence in the rth mean, for every $r > 0$.*

Proof: Note that a r.v. X is said to be a.s. bounded, if there exists a positive constant $c\ (< \infty)$, such that $P\{|X| \leq c\} = 1$. Suppose now that $T_n - T \xrightarrow{\text{P}} 0$ and that $P\{|T_n - T| \leq c\} = 1$, for some $c : 0 < c < \infty$. Then observe that for any $\varepsilon > 0$, we may write

$$\text{E}|T_n - T|^r = \int_{|t| \leq c} |t|^r dP\{|T_n - T| \leq t\}$$

$$= \int_{|t| \leq \varepsilon} |t|^r dP\{|T_n - T| \leq t\} + \int_{c \geq |t| > \varepsilon} |t|^r dP\{|T_n - T| \leq t\}$$

$$\leq \varepsilon^r P\{|T_n - T| \leq \varepsilon\} + c^r P\{|T_n - T| > \varepsilon\}. \qquad (2.2.36)$$

Since $T_n - T \xrightarrow{\text{P}} 0$ and ε is arbitrary, the rhs of (2.2.36) can be made arbitrarily small by choosing ε small and n adequately large. This ensures that $|T_n - T|$ converges to 0 in the rth mean. Since $r\ (> 0)$ is arbitrary, the proof is complete.   ∎

As an illustration, we go back to the binomial example cited before. Note that $P\{0 \leq T_n \leq 1\} = 1$, for every $n \geq 1$, whereas $\pi \in [0, 1]$. Hence, it is easy to verify that $P\{|T_n - \pi| \leq 1\} = 1$, for every $n \geq 1$. Also, we have seen earlier that $T_n \xrightarrow{\text{P}} \pi$. Thus, by Proposition 2.2.4, we conclude that $T_n - \pi \xrightarrow{r\text{th}} 0$, for every $r > 0$. Incidentally, in proving convergence in probability, here, we started with convergence in the second mean, but because of the boundedness of $T_n - \pi$, we ended up with convergence in

rth mean for every $r > 0$.

Next, we explore the situation where the basic r.v.'s may not be a.s. bounded. Toward this, we may note that a transformation $|T_n - T|$ to $W_n = g(|T_n - T|)$, for some monotone and bounded $g(\cdot)$, may lead to an affirmative answer. In this context, we present the following:

Theorem 2.2.1: *Let $w = g(t)$, $0 \leq t < \infty$, be a monotone nondecreasing, continuous and bounded function, $g(0) = 0$. Let $W_n = g(|T_n - T|)$, $n \geq 1$. Then, $T_n - T \xrightarrow{\text{P}} 0$ iff $W_n \xrightarrow{r\text{th}} 0$, for some $r > 0$.*

Proof: Note that under the assumed regularity conditions on w, for every $\varepsilon > 0$ there exists an $\eta > 0$ such that $\eta \to 0$ as $\varepsilon \to 0$ and $0 \leq t \leq \varepsilon$ is equivalent to $w(t) < \eta$. Now, suppose that $T_n - T \xrightarrow{\text{P}} 0$. Then, $W_n = g(|T_n - T|)$ also converges in probability to 0. Further, by construction, W_n is a.s. bounded and, hence, by Proposition 2.2.4, $W_n \xrightarrow{\text{P}} 0$ which implies that $W_n \xrightarrow{r\text{th}} 0$, for any $r > 0$. Alternatively, if $W_n \xrightarrow{r\text{th}} 0$, for some $r > 0$, then by Proposition 2.2.3, $W_n \xrightarrow{\text{P}} 0$ and this, in turn, implies that $|T_n - T| \xrightarrow{\text{P}} 0$, completing the proof. ∎

In practice, we may choose W_n by letting

$$W_n^r = |T_n - T|^r / \{1 + |T_n - T|^r\}$$

for an appropriate r (> 0); a typical choice of r is 2. Note that W_n is then a.s. bounded by 1, and, for $r = 2$, the evaluation of the moments of W_n may also be quite convenient. We illustrate this point with the estimation of $\theta = \pi^{-1}$ related to the binomial example cited before. As in the discussion after (2.2.35), we define $Z_n = T_n^{-1} = n/X_n$. Then

$$
\begin{aligned}
W_n^2 &= \frac{(Z_n - \theta)^2}{1 + (Z_n - \theta)^2} = \frac{(n - \pi^{-1}X_n)^2}{X_n^2 + (n - \pi^{-1}X_n)^2} \\
&= \frac{(n\pi - X_n)^2}{\pi^2 X_n^2 + (n\pi - X_n)^2} = \frac{(T_n - \pi)^2}{(T_n - \pi)^2 + \pi^2 T_n^2} \leq 1.
\end{aligned}
$$

$$(2.2.37)$$

Thus, $(T_n - \pi) \xrightarrow{2\text{nd}} 0$ implies $W_n \xrightarrow{2\text{nd}} 0$, and a direct application of Theorem 2.2.1 yields that $Z_n - \theta \xrightarrow{\text{P}} 0$.

We should not, however, overemphasize the role of Theorem 2.2.1. In general, it may not be very easy to show that for an arbitrary r (> 0), $W_n \xrightarrow{r\text{th}} 0$, and simpler proofs of the stochastic convergence may be available. We shall continue to explore these tools in what follows.

Next, we study the mutual implications of convergence in rth mean for various ordered r. Before we present the proof of the next proposition, we consider another basic inequality which has an abundance of uses in statistical problems.

Jensen Inequality: *Let X be a r.v. and $g(x)$, $x \in \mathbb{R}$, be a convex function such that $Eg(X)$ exists. Then,*

$$g(EX) \leq Eg(X), \qquad (2.2.38)$$

with the equality sign holding only when g is linear almost everywhere.

Proof. Let $a < b < c$ be three arbitrary points on $\mathbb{R}$, and define

$$\alpha = \frac{c-b}{c-a} \quad \text{and} \quad \beta = 1 - \alpha = \frac{b-a}{c-a}. \qquad (2.2.39)$$

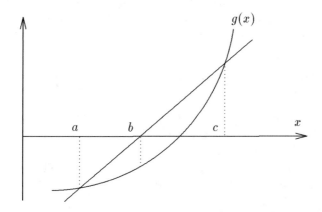

Figure 2.2.1. *Representation of the convex function $g(x)$*

Note that $b = \alpha a + \beta c$. Also, by the definition of a convex function, we have

$$
\begin{aligned}
g(b) &= g(\alpha a + \beta b) \leq \alpha g(a) + \beta g(c) \\
&= (c-a)^{-1}[(c-b)g(a) + (b-a)g(c)], \qquad (2.2.40)
\end{aligned}
$$

and after slight rearrangements in (2.2.40), we obtain that

$$\frac{g(b) - g(a)}{(b-a)} \leq \frac{g(c) - g(a)}{(c-a)}, \quad a < b < c, \qquad (2.2.41)$$

and this implies that

$$h^{-1}\{g(x+h) - g(x)\} \quad (h > 0) \quad \text{is nondecreasing in } x. \qquad (2.2.42)$$

Therefore, letting $EX = \theta$, we obtain, by using (2.2.42), that

$$g(x) - g(\theta) \geq g'_+(\theta)\{x - \theta\}, \quad x \geq \theta, \qquad (2.2.43)$$
$$g(x) - g(\theta) \geq g'_-(\theta)\{x - \theta\}, \quad x < \theta, \qquad (2.2.44)$$

where

$$g'_+(\theta) = \lim_{h \downarrow 0} h^{-1}\{g(\theta + h) - g(\theta)\},$$
$$g'_-(\theta) = \lim_{h \downarrow 0} h^{-1}\{g(\theta) - g(\theta - h)\},$$

and, by (2.2.42), $g'_+(\theta) \geq g'_-(\theta)$. Defining $g'(\theta) = g'_+(\theta) = g'_-(\theta)$ whenever $g'_+(\theta) = g'_-(\theta)$ and $g'(\theta)$ as some value in the interval $[g'_-(\theta), g'_+(\theta)]$ whenever $g'_-(\theta) \neq g'_+(\theta)$, we obtain from (2.2.43) and (2.2.44) that

$$g(x) - g(\theta) \geq g'(\theta)\{x - \theta\}, \quad x \in \mathbb{R}, \qquad (2.2.45)$$

and this implies that $Eg(X) - g(\theta) \geq g'(\theta)E\{X - \theta\} = 0$, i.e., $Eg(X) \geq g(\theta)$. Further, in (2.2.45), a strict equality sign holds for all x only when $g(x)$ is linear in x, and hence, for $g(\theta) = Eg(X)$, we need that $g(x)$ be linear a.e. ∎

Proposition 2.2.5: *Convergence in the rth mean, for some $r > 0$ implies convergence in the sth mean, for every s such that $0 \leq s \leq r$.*

Proof: Suppose that for some $r > 0$, $T_n - T$ converges in the rth mean to 0. We write $U_n = |T_n - T|^r$ and $V_n = |T_n - T|^s$ for some s, $0 < s < r$. Then $U_n = V_n^t$, $t = r/s > 1$, and we note that x^t is a convex function of x ($\in \mathbb{R}^+$), for every $t \geq 1$ (strictly convex, for $t > 1$). Therefore, by the Jensen Inequality (2.2.38), we obtain that

$$E|T_n - T|^r = EU_n \geq (EV_n)^{r/s} = \{E|T_n - T|^s\}^{r/s}, \quad r \geq s, \qquad (2.2.46)$$

and this implies that

$$\{E|T_n - T|^s\}^{1/s} \text{ is nondecreasing in } s \ (\geq 0). \qquad (2.2.47)$$

Therefore, we conclude that $\lim_{n \to \infty} E|T_n - T|^r = 0$, for some $r > 0$, implies that $\lim_{n \to \infty} E|T_n - T|^s = 0$, for every $s \leq r$, and this completes the proof.

∎

We may note that to show a.s. convergence, we may need to verify (2.2.7), and this may not be a simple task in all cases. In fact, the rth mean convergence, by itself, may not convey much information toward this verification. There is another mode of convergence (stronger than a.s. convergence) that is often used to simplify (2.2.7), and we consider this below.

Definition 2.2.15 (Complete convergence): *A sequence $\{T_n\}$ of random elements is said to converge completely to a (possibly degenerate) ran-*

dom element T, if for every $\varepsilon > 0$,

$$\sum_{n \geq 1} P\{d(T_n, T) > \varepsilon\} \quad converges; \tag{2.2.48}$$

this mode of convergence is denoted by $T_n - T \xrightarrow{c} 0$. The metric $d(\cdot)$ is defined as in the beginning of this section.

Note that the convergence of the series in expression (2.2.48) implies that the tail-sum $\sum_{n \geq n_0} P\{d(T_n, T) > \varepsilon\}$ converges to 0 as $n_0 \to \infty$; as such, noting that for arbitrary events $\{A_k\}$, $P\{\bigcup A_k\} \leq \sum_k P(A_k)$, we conclude from (2.2.7) and (2.2.48) that

$$T_n - T \xrightarrow{c} 0 \quad \Rightarrow \quad T_n - T \xrightarrow{\text{a.s.}} 0. \tag{2.2.49}$$

However, the converse may not be true. In practice, to verify (2.2.48), all we need is to verify that for every $\varepsilon > 0$, $\ell(n; \varepsilon) = P\{d(T_n, T) > \varepsilon\}$ converges to 0 as $n \to \infty$ at a rate such that the series $\sum_{n \geq n_0} \ell(n; \varepsilon)$ converges, and, in this context, suitable probability inequalities may be used. We shall discuss them in detail in the next section. To illustrate this point, however, we may consider again the binomial example, where $T_n = n^{-1} X_n$ and $T = \pi$. We may note then that for all $\pi \in [0, 1]$,

$$\begin{aligned}
E(T_n - \pi)^4 &= n^{-3} \pi (1 - \pi)[1 + 3\pi(1 - \pi)(n - 2)] \\
&\leq n^{-2} \frac{3}{16}. \tag{2.2.50}
\end{aligned}$$

Hence, using Chebyshev Inequality (2.2.31) with $U_n = (T_n - \pi)^4$, we obtain that for every $\varepsilon > 0$,

$$\begin{aligned}
P\{\|T_n - \pi\| > \varepsilon\} &= P\{(T_n - \pi)^4 > \varepsilon^4\} \\
&= P\left\{U_n > EU_n \frac{\varepsilon^4}{EU_n}\right\} \\
&\leq \frac{EU_n}{\varepsilon^4} \leq n^{-2} \frac{3}{16\varepsilon^4},
\end{aligned}$$

so that the series in (2.2.48) converges and, hence, $T_n \xrightarrow{c} \pi$.

Let us next consider the implications of the modes of convergence in Definitions 2.2.1, 2.2.2, 2.2.13 and 2.2.15 on the weak convergence in Definition 2.2.14:

Proposition 2.2.6: *Convergence in probability implies convergence in distribution.*

Proof: Suppose that $T_n - T \xrightarrow{P} 0$ and that T has the distribution function F; for simplicity, we deal with the case of real-valued r.v.'s only. Let x be a point of continuity of F, and let $\varepsilon > 0$ be an arbitrary small number. Then

$$
\begin{aligned}
F(x - \varepsilon) &= P\{T \le x - \varepsilon\} \\
&= P\{T \le x - \varepsilon, |T_n - T| \le \varepsilon\} + P\{T \le x - \varepsilon, |T_n - T| > \varepsilon\} \\
&\le P\{T_n \le x\} + P\{|T_n - T| > \varepsilon\} \quad\quad\quad (2.2.51)
\end{aligned}
$$

and

$$
F(x + \varepsilon) \ge P\{T_n \le x\} - P\{|T_n - T| > \varepsilon\}. \quad\quad\quad (2.2.52)
$$

Since the last term on the rhs of (2.2.51) and (2.2.52) converges to 0 as $n \to \infty$, we obtain from the above that

$$
F(x - \varepsilon) \le \liminf_{n \to \infty} F_n(x) \le \limsup_{n \to \infty} F_n(x) \le F(x + \varepsilon). \quad\quad (2.2.53)
$$

Since x is a point of continuity of F, $F(x + \varepsilon) - F(x - \varepsilon)$ can be made arbitrarily small by letting $\varepsilon \downarrow 0$ and, hence, we conclude that (2.2.27) holds. That is, $T_n \xrightarrow{D} T$. This completes the proof. ∎

Combining the results in the preceding propositions, we arrive at the implication diagram presented in Figure 2.2.2.

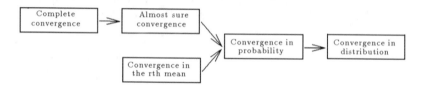

Figure 2.2.2. *Convergence implication diagram*

The picture is, however, a little incomplete in the sense that the implications of complete (or a.s.) convergence and the convergence in the rth mean $(r > 0)$ are not so straightforward.

We conclude this section with an important observation on the role of characteristic functions in weak convergence; this will also be useful in the study of stochastic convergence in the subsequent sections. For a distribution function F, defined on the real line $\mathbb{R}$, the characteristic function $\phi_F(t)$, for real t, has been defined in (1.4.55) as

$$
\phi_F(t) = \int_R e^{itx} dF(x), \quad t \in \mathbb{R}; \quad\quad\quad (2.2.54)
$$

in general, $\phi_F(t)$ is a complex-valued function. If the d.f. F is degenerate at the point c $(\in \mathbb{R})$, then, $\phi_F(t) = e^{itc}$, for every real t. The classical Lévy Theorem (see Section 1.4) asserts that if two d.f.'s F and G both have the same characteristic function $\phi_F(t)$, $t \in \mathbb{R}$, then F and G are the same,

and the converse is also true. A more useful result in this context, stated
without proof, is:

Theorem 2.2.2 (Lévy-Cramér): *Let $\{F_n, n \geq 1\}$ be a sequence of d.f.'s
(defined on $\mathbb{R}$) with corresponding characteristic functions $\{\phi_{F_n}(t), n \geq 1\}$.
Then, a necessary and sufficient condition for $F_n \xrightarrow{w} F$ [i.e., $F_n(x)$ converges weakly to a d.f. $F(x)$ at all points of continuity of F] is that, for
every real t, $\phi_{F_n}(t)$ converges to a limit $\phi_F(t)$ which is continuous at $t = 0$
and is the characteristic function of the d.f. F.*

A direct corollary to this theorem was stated in Proposition 2.2.1. This
proposition is quite useful in practice, and it provides the usual consistency result under minimal regularity conditions. To illustrate this point,
we consider a typical situation where we have a sequence $\{T_n\}$ of (real-
valued) estimators of a real parameter θ. As has been explained earlier,
T_n is said to be a (weakly) consistent estimator of θ if $T_n \xrightarrow{P} \theta$. We may
consider the Implication Diagram in Figure 2.2.2 and directly verify either
the convergence in probability criterion or any of the other three criteria
which imply convergence in probability. For any one of these, we may need
to verify certain regularity conditions. On the other hand, we may as well
use Proposition 2.2.1 and verify the same under (usually) less restrictive
conditions. As a specific case, consider $T_n = \overline{X}_n$, where the X_i are i.i.d.
r.v.'s with mean θ. By using the Levy-Cramér Theorem (2.2.2), it can easily be shown that $T_n \xrightarrow{D} T$, where T is degenerate at the point θ. (We
shall actually demonstrate this in Section 3.3.) Thus, by Proposition 2.2.1,
we may claim that $T_n \xrightarrow{P} \theta$, and this does not need any moment condition
on the d.f. of X_1 other than the finiteness of θ. On the other hand, using
the rth mean convergence (to ensure convergence in probability), we may
arrive at the same conclusion, but that generally needs the existence of the
moment of order r, for some $r > 1$; the simplest way of doing this is to
appeal to (2.2.31) on letting $U = (\overline{X}_n - \theta)^2$, so that $EU = n^{-1}\sigma^2 \downarrow 0$ as
$n \to \infty$, provided the second central moment (σ^2) of X_1 is finite. Evalu-
ation of $E|\overline{X}_n - \theta|^r$ for a non integer r (> 0) or for r other than 2 may
require more tedious manipulations and, hence, may not be generally advo-
cated. If, however, by any other means, it is possible to obtain a probability
inequality of the form

$$P\{|T_n - T| > \varepsilon\} \leq \psi_\varepsilon(n), \quad n \geq n_0, \quad \varepsilon > 0, \qquad (2.2.55)$$

where $\psi_\varepsilon(n) \downarrow 0$ as $n \to \infty$, then, of course, the desired result follows
directly. If $\sum_{m \geq n} \psi_\varepsilon(m) \to 0$ as $n \to \infty$, then the a.s. (or complete) convergence also follows from (2.2.55). Thus, refined probability inequalities
may provide information beyond convergence in probability (or a.s. con-

vergence). Alternatively, we do not need (2.2.55) with a specific rate for $\psi_\varepsilon(n)$, for the stochastic (or a.s.) convergence of a sequence of random elements; often, these may be achieved by the use of the so-called Laws of Large Numbers. Hence, before we consider some deeper results on the modes of convergence of random elements, we present some basic probability inequalities and laws of large numbers in the next section.

2.3 Probability inequalities and laws of large numbers

The Chebyshev Inequality (2.2.31) is the precursor of other forms of more general probability inequalities. Actually, if we have a sequence $\{T_n\}$ of random variables, and we put $\sigma_n^2 = E(T_n - \theta)^2$, then on letting $U = (T_n - \theta)^2$, we obtain from (2.2.31) the following form of the Chebyshev Inequality:

$$P\{|T_n - \theta| > t\sigma_n\} \le t^{-2}, \quad n \ge 1, \quad t \in \mathbb{R}^+. \tag{2.3.1}$$

Note that the inequality in (2.3.1) depends on the d.f. of T_n only through σ_n and, hence, may be termed a **distribution-free inequality**.

The simplicity of (2.3.1) is its main appeal. It follows that whenever σ_n^2 converges to 0 as $n \to \infty$, T_n converges in probability to θ. However, there are two points we should keep in mind here. First, for the stochastic convergence of T_n to θ, it is not necessary to assume that the mean square error (σ_n^2) exists and that it converges to 0 as $n \to \infty$. Second, if we assume that the mean square error exists, we may even obtain deeper convergence results in a more general setup. These will be explored throughout this section. We may also mention that (2.3.1), although quite simple, may not be generally very sharp. To illustrate this point, we may consider the classical case of i.i.d. r.v.'s $\{X_i\}$ having a normal distribution with mean θ and variance σ^2, and $T_n = \overline{X}_n$. Then, by (2.3.1), we obtain that

$$P\{|T_n - \theta| \ge t\sigma/\sqrt{n}\} \le t^{-2}, \quad t \in \mathbb{R}^+. \tag{2.3.2}$$

On the other hand, $n^{1/2}(\overline{X}_n - \theta)/\sigma$ has the standard normal distribution, and, hence,

$$P\{|T_n - \theta| > t\sigma/\sqrt{n}\} = 2[1 - \Phi(t)], \quad t \in \mathbb{R}^+, \tag{2.3.3}$$

where $\Phi(t)$, $t \in \mathbb{R}$, is the standard normal distribution function. Letting $t = 2$, we have, in (2.3.3), the value 0.0455, whereas the upper bound in (2.3.2) is 0.25 (quite higher). Similarly, for $t = 3$, (2.3.3) relates to 0.0029, whereas (2.3.2) to 1/9 (= 0.11111). Also, for $t^2 \le 1$, (2.3.2) is of no use, although (2.3.3) can be computed for every real t.

Better rates of convergence (or sharper bounds) can be obtained (at the price of assuming the existence of higher order moments) via the application of the following inequality.

Markov Inequality: *Let U be a non-negative r.v. with finite rth moment $\nu_r = EU^r$, for some $r > 0$. Then, for every $\varepsilon > 0$, we have*

$$P\{U > \varepsilon\} \le \varepsilon^{-r}\nu_r. \tag{2.3.4}$$

Proof: Put $V = U^r$, so that $EV = \nu_r$. Then, note that $[U > \varepsilon]$ is equivalent to $[V > \varepsilon^r]$, so that (2.3.4) follows directly from (2.2.31). ∎

To illustrate the utility of this inequality, we consider the binomial example treated in Section 2.2. Here $nT_n \sim \text{Bin}(n, \pi)$. From (2.2.35) and (2.3.2), we obtain that for every $\varepsilon > 0$,

$$P\{|T_n - \pi| > \varepsilon\} \le \frac{\pi(1 - \pi)}{n\varepsilon^2} = O(n^{-1}), \tag{2.3.5}$$

whereas, by (2.2.50) and (2.3.4), for $r = 4$, we have

$$P\{|T_n - \pi| > \varepsilon\} \le n^{-2}\frac{3}{16\varepsilon^4} = O(n^{-2}). \tag{2.3.6}$$

Thus, for large n, at least, we have a much better rate of convergence in (2.3.6) than in (2.3.5). In fact, using values of r even greater than 4, one may still obtain better rates of convergence. This is possible because T_n is a bounded r.v. and hence all moments of finite order exist. Taking the lead from this observation, we may go one step further and obtain a sharper form of a probability inequality when, in fact, moments of all finite order exist [in the sense that the moment generating function $M(t) = E(e^{tX})$ exists for t belonging to a suitable domain]. In this direction, we consider the following.

Bernstein Inequality: *Let U be a random variable such that $M_U(t) = E(e^{tU})$ exists for all $t \in [0, K]$, for some $K > 0$. Then, for every real u, we have*

$$P\{U \ge u\} \le \inf\left\{e^{-tu}M_U(t) : t \in [0, K]\right\}. \tag{2.3.7}$$

Proof: For any non-negative and nondecreasing function $g(x)$, $x \in \mathbb{R}$, using the Chebyshev Inequality (2.2.31), we obtain that

$$P\{U \ge u\} = P\{g(U) \ge g(u)\} \le Eg(U)/g(u); \tag{2.3.8}$$

so that letting $g(x) = \exp(tx)$, $t > 0$, $x \in \mathbb{R}$, we obtain from (2.3.8) that

$$P\{U > u\} \le \frac{E\left(e^{tU}\right)}{e^{tu}} = \frac{M_U(t)}{e^{tu}} = e^{-tu}M_U(t), \quad t > 0. \tag{2.3.9}$$

Since (2.3.9) holds for every $t \in [0, K]$, minimizing the rhs over t, we arrive at (2.3.7). ∎

Note that, in (2.3.7), U need not be a non-negative r.v.. Also, note that by the same arguments,

$$
\begin{aligned}
P\{U \le u\} &= P\{-U \ge -u\} \\
&\le \inf\{e^{tu}M_{(-U)}(t) : t \in [0, K']\} \\
&= \inf\{e^{tu}M_U(-t) : t \in [0, K^*]\},
\end{aligned}
$$

where K^* is a suitable positive number such that $M_U(-t)$ exists for every $t \le K^*$. Hence, we may write

$$
\begin{aligned}
P\{|U| \ge u\} &= P\{U \ge u\} + P\{U \le -u\} \\
&\le \inf\{e^{-tu}M_U(t) : t \in [0, K]\} \\
&\quad + \inf\{e^{-tu}M_U(-t) : t \in [0, K^*]\} \\
&\le \inf\{e^{-tu}(M_U(t) + M_U(-t)) : 0 < t \le K \wedge K^*\}.
\end{aligned}
$$
(2.3.10)

We illustrate the effectiveness of the Bernstein Inequality (2.3.7) with the same binomial example treated earlier. Note that $nT_n \sim \text{Bin}(n, \pi)$, so that the moment generating function of nT_n is $\sum_{r=0}^{n} \binom{n}{r}e^{tr}\pi^r(1-\pi)^{n-r} = \{e^t\pi + (1-\pi)\}^n$. Thus, taking $U = nT_n - n\pi$, we obtain from (2.3.10) that for all $\varepsilon > 0$ and $\pi \in (0, 1)$,

$$
\begin{aligned}
P\{|T_n - \pi| > \varepsilon\} &= P\{|nT_n - n\pi| > n\varepsilon\} \\
&= P\{U < -n\varepsilon\} + P\{U > n\varepsilon\} \\
&\le \inf_{t \ge 0}\left\{e^{-n(1-\pi)t - n\varepsilon t}[1 + (1-\pi)(e^t - 1)]^n\right\} \\
&\quad + \inf_{t > 0}\left\{e^{-n\pi t - n\varepsilon t}[1 + \pi(e^t - 1)]^n\right\} \\
&= [\rho(\varepsilon, \pi)]^n + [\rho(\varepsilon, 1-\pi)]^n,
\end{aligned}
$$
(2.3.11)

where, for $\pi \in (0, 1)$, $(\pi + \varepsilon) \in (0, 1)$,

$$
\rho(\varepsilon, \pi) = \inf_{t > 0}\left\{e^{-(\pi+\varepsilon)t}[1 + \pi(e^t - 1)]\right\}.
$$
(2.3.12)

Note that T_n lies between 0 and 1 (with probability 1), so that $\rho(\varepsilon, \pi) = 0$ for $\pi + \varepsilon > 1$. Next, to obtain an explicit expression for (2.3.12), we set $w = e^t$ and $g(w) = w^{-(\pi+\varepsilon)}[1 + \pi(w - 1)]$, so that $(\partial/\partial w)\log g(w) = -(\pi + \varepsilon)/w + \pi/[1 + \pi(w - 1)]$, and equating this to 0, we obtain that the point w^* at which $g(w)$ is an extremum is given by

$$
w^* = [(\pi + \varepsilon)(1 - \pi)]/[\pi(1 - \pi - \varepsilon)].
$$

Since

$$
\frac{\partial^2}{\partial w^2}\log g(w)\Big|_{w=w^*} = \left\{(\pi + \varepsilon)^{-1} - 1\right\}\pi^2/(1 - \pi)^2 > 0,
$$

it follows that w^* corresponds to a minimum. Therefore, we may write

$$\rho(\varepsilon, \pi) = g(w^*) = \left\{ \frac{(\pi + \varepsilon)(1 - \pi)}{\pi(1 - \pi - \varepsilon)} \right\}^{-(\pi + \varepsilon)} \left\{ \frac{1 + \pi\varepsilon}{\pi(1 - \pi - \varepsilon)} \right\}$$

$$= \left(\frac{\pi}{\pi + \varepsilon} \right)^{\pi + \varepsilon} \left(\frac{1 - \pi}{1 - \pi - \varepsilon} \right)^{1 - \pi - \varepsilon} \quad (> 0). \qquad (2.3.13)$$

Next, we note that by (2.3.13), $\rho(0, \pi) = 1$, for every $\pi \in (0, 1)$. Further,

$$\frac{\partial}{\partial \varepsilon} \log \rho(\varepsilon, \pi) = \log \left\{ \frac{\pi(1 - \pi - \varepsilon)}{(1 - \pi)(\pi + \varepsilon)} \right\} < 0, \quad \varepsilon > 0. \qquad (2.3.14)$$

Hence $\rho(\varepsilon, \pi) < 1$, for every (ε, π), such that $0 \le \pi + \varepsilon < 1$. Since both $\rho(\varepsilon, \pi)$ and $\rho(\varepsilon, 1 - \pi)$ are positive fractions less than 1, by (2.3.11) we may conclude that for every $\varepsilon > 0$, $P\{|T_n - \pi| > \varepsilon\}$ converges to 0 at an exponential rate as n increases. In fact, we may set

$$\rho^*(\varepsilon) = \sup \{\rho(\varepsilon, \pi) : 0 < \pi < 1 - \varepsilon\}, \quad \varepsilon > 0, \qquad (2.3.15)$$

and obtain from (2.3.11) and (2.3.15) that for every $\varepsilon > 0$ and $n \ge 1$,

$$P\{|T_n - \pi| > \varepsilon\} \le 2[\rho^*(\varepsilon)]^n, \quad \pi \in (0, 1). \qquad (2.3.16)$$

Interestingly, it may be noted that this powerful inequality is not confined to the binomial case alone. The same exponential rate of convergence appears in a wider class of problems involving a.s. bounded r.v.'s (for which the moment generating functions exist), and in this context, it may not be necessary to assume that these r.v.'s are identically distributed. We present the following elegant probability inequality due to Hoeffding (1963).

Hoeffding Inequality: *Let X_k, $k \ge 1$, be independent (but not necessarily identically distributed) random variables, such that $P\{0 \le X_k \le 1\} = 1$ for every $k \ge 1$. Let $\mu_k = EX_k$, $k \ge 1$, and $\overline{\mu}_n = n^{-1}\sum_{k=1}^{n} \mu_k$, $n \ge 1$. Also, let $\overline{X}_n = n^{-1}\sum_{k=1}^{n} X_k$, $n \ge 1$. Then, for every $\varepsilon > 0$,*

$$P\{|\overline{X}_n - \overline{\mu}_n| > \varepsilon\} \le [\rho(\varepsilon, \overline{\mu}_n)]^n + [\rho(\varepsilon, 1 - \overline{\mu}_n)]^n, \qquad (2.3.17)$$

where $\rho(\varepsilon, a)$ is defined as in (2.3.12). Thus, $\overline{X}_n - \overline{\mu}_n$ converges in probability to 0 at an exponential rate.

Proof: First, observe that for every $x \in [0, 1]$,

$$g_t(x) = e^{tx} \le 1 + x(e^t - 1), \quad t > 0$$

(see Figure 2.3.1). This ensures that for every k (as the $X_k \in [0, 1]$ with probability 1),

$$M_{X_k}(t) = E(e^{tX_k}) \le 1 + (e^t - 1)EX_k = 1 + (e^t - 1)\mu_k. \qquad (2.3.18)$$

Moreover, $X_1, \ldots, X_n$ are independent, so that for every real t,

$$E\left[\exp(t\{X_1 + \cdots + X_n\})\right] = \prod_{k=1}^{n} E(\exp\{tX_k\}) = \prod_{k=1}^{n} M_{X_k}(t). \quad (2.3.19)$$

Therefore, using (2.3.10) for $U = (X_1 + \cdots + X_n) - (\mu_1 + \cdots + \mu_n) = n(\overline{X}_n - \overline{\mu}_n)$, we obtain that for every $\varepsilon > 0$,

$$P\left\{|\overline{X}_n - \overline{\mu}_n| > \varepsilon\right\} = P\left\{|U| > n\varepsilon\right\}$$

$$\leq \inf_{t>0}\left\{\exp(-tn(\varepsilon + \overline{\mu}_n))\prod_{k=1}^{n} M_{X_k}(t)\right\}$$

$$+ \inf_{t>0}\left\{\exp(-tn(\varepsilon - \overline{\mu}_n))\prod_{k=1}^{n} M_{X_k}(-t)\right\}. \quad (2.3.20)$$

At this stage, we make use of (2.3.18) along with the AM/GM/HM In-

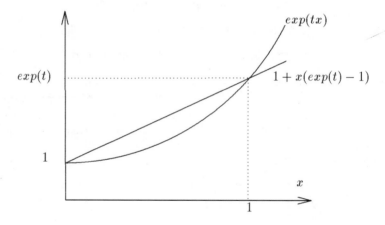

Figure 2.3.1.

equality (1.4.31) and obtain that

$$\prod_{k=1}^{n} M_{X_k}(t) \leq \left\{n^{-1}\sum_{k=1}^{n} M_{X_k}(t)\right\}^{n} \leq \left\{1 + (e^t - 1)\overline{\mu}_n\right\}^{n}. \quad (2.3.21)$$

As such, the first term on the rhs of (2.3.20) is bounded from above by the first term on the rhs of (2.3.11) where π has to be replaced by $\overline{\mu}_n$. A very similar inequality holds for the second term on the rhs of (2.3.20). As such, following (2.3.11)–(2.3.13), we conclude the proof. ∎

The above result may easily be extended to the case of r.v.'s belonging to an arbitrary compact interval $[a, b]$, for some $-\infty < a < b < +\infty$. In this context, we may note that if $P\{a \leq X_k \leq b\} = 1$ for every $k \geq 1$, we may set $Z_k = (X_k - a)/(b - a)$, so that $P\{0 \leq Z_k \leq 1\} = 1$, for every $k \geq 1$. Note that

$$(X_1 + \cdots + X_n) - (\mu_1 + \cdots + \mu_n)$$
$$= (b - a)[Z_1 + \cdots + Z_n - EZ_1 - \cdots - EZ_n], \qquad (2.3.22)$$

so that the only change in (2.3.17) would be to replace ε by $\varepsilon^* = \varepsilon/(b - a)$ and $\bar{\mu}_n$ by $(\bar{\mu}_n - a)/(b - a) = \bar{\mu}_n^*$, say. There are some other forms of probability inequalities for a.s. bounded r.v.'s, treated in detail in Hoeffding (1963), and, for brevity, we do not consider them here. Although they may be viewed as suitable reference materials. In passing, we may remark that by virtue of (2.3.14),

$$\frac{\partial}{\partial \varepsilon} \log \rho(\varepsilon, \pi)|_{\varepsilon=0} = 0$$

and

$$\frac{\partial^2}{\partial \varepsilon^2} \log \rho(\varepsilon, \pi) = -(1 - \pi - \varepsilon)^{-1} - (\pi + \varepsilon)^{-1}$$
$$= -\{(\pi + \varepsilon)(1 - \pi - \varepsilon)\}^{-1} \leq -4, \quad 0 \leq \pi + \varepsilon \leq 1.$$

As such, by a second order Taylor expansion, we have $\log \rho(\varepsilon, \pi) \leq -2\varepsilon^2$, so that $\rho(\varepsilon, \pi) \leq \exp(-2\varepsilon^2)$, for every π such that $0 \leq \pi + \varepsilon \leq 1$. Then, in (2.3.11) and (2.3.17), the rhs can be bounded from above by $2 \exp(-2n\varepsilon^2) = 2\{\exp(-2\varepsilon^2)\}^n$, and this specifies an upper bound for the exponential rate of convergence (to 0) of the tail probabilities in (2.3.11) or (2.3.17). We may also remark that for such an exponential rate of convergence, it may not be necessary to assume that the r.v.'s are a.s. bounded. Although for unbounded r.v.'s, (2.3.18) may not hold [see Figure 2.3.1], an exponential rate may often be achieved under the finiteness of the moment generating functions. We then present the following:

Corollary to the Bernstein Inequality: *Suppose that X_k, $k \geq 1$ are i.i.d. r.v.'s with $EX_k = 0$ (without any loss of generality) and $M(t) = E(e^{tX_1}) < \infty$, for t in some neighborhood of 0. For $\varepsilon > 0$ let*

$$\rho_+(\varepsilon) = \inf\{e^{-t\varepsilon} M(t) : t > 0\} \quad and \quad \rho_-(\varepsilon) = \inf\{e^{-t\varepsilon} M(-t) : t > 0\}.$$

Then, for every $\varepsilon > 0$ and $n \geq 1$,

$$P\{|\overline{X}_n| > \varepsilon\} \leq [\rho_+(\varepsilon)]^n + [\rho_-(\varepsilon)]^n. \qquad (2.3.23)$$

Proof: The proof is a direct consequence of (2.3.20) and the definitions of $\rho_+(\varepsilon)$ and $\rho_-(\varepsilon)$. It may also be noted that in this context, it may not be necessary to assume that the X_k are identically distributed. All that is needed is that $\prod_{k=1}^{n} M_{X_k}(t) \leq [\overline{M}_n(t)]^n$, where $\overline{M}_n(t) = n^{-1} \sum_{i=1}^{n} M_{X_i}(t)$ satisfies the condition ensuring (2.3.23) for every $n \geq n_0$. ∎

To illustrate this, we consider the following examples.

Example 2.3.1: Let $X_k \sim N(\mu, \sigma^2)$, $k \geq 1$, and let $T_n = \overline{X}_n - \mu$, $n \geq 1$. Then, for $X_k - \mu$, the moment generating function is $M(t) = \exp\{t^2\sigma^2/2\}$, t real. Thus, it is easy to verify that $\rho_+(\varepsilon) = \rho_-(\varepsilon) = \exp(-\varepsilon^2/2\sigma^2)$, so that along lines parallel to (2.3.23), we have here $P\{|\overline{X} - \mu| > \varepsilon\} \leq 2[\exp(-\varepsilon^2/2\sigma^2)]^n$, for every $\varepsilon > 0$. It is also possible to avoid the assumption of homogeneity of the means and the variances of these r.v.'s. For possibly different means, we need to take the average mean for μ in the definition of T_n, whereas as long as the variances are bounded from above by a finite γ^2, we have the same bound with σ^2 replaced by γ^2. ∎

Example 2.3.2: Let X_k be i.i.d. r.v.'s with the double exponential distribution [i.e., with a density function given by $f(x) = (1/2)\exp\{-|x|\}$, $x \in \mathbb{R}$], where, for simplicity, we take the mean equal to 0. Verify that for every $t : |t| < 1$, $M(t)$ is finite, and, hence or otherwise, verify (2.3.23). ∎

Example 2.3.3: Consider the simple logistic density function $f(x) = e^{-x}\{1 + e^{-x}\}^{-2}$, $x \in \mathbb{R}$. Verify that here $M(t)$ is finite for every t such that $|t| < 1$ and, hence, obtain (2.3.23) with explicit forms for $\rho_+(\varepsilon)$ and $\rho_-(\varepsilon)$. ∎

Example 2.3.4 (Counterexample): Consider the Student t-distribution with m degrees of freedom. Show that for this distribution, the moment generating function $M(t)$ is not finite for any real t, and, hence, comment on the rate of convergence for the t-statistic. Another counterexample is the classical Cauchy distribution for which the p.d.f. is given by $f(x) = \pi^{-1}\lambda\{\lambda^2 + (x - \theta)^2\}^{-1}$, $x \in \mathbb{R}$. Show that for this distribution, no positive moment of order 1 or more is finite, and, hence, none of the probability inequalities discussed so far apply. ∎

As has been stated earlier, these probability inequalities can be fruitfully incorporated in the proofs of various modes of convergence of suitable random elements. In this context, a very basic role is played by the **Borel-Cantelli Lemma**. In order to present this, we first consider the following.

Theorem 2.3.1: *Let $\{a_n\}$ be a sequence of real numbers, such that $0 < a_n < 1$, for all $n \geq 1$. Then $\sum_{n\geq1} a_n = +\infty$ implies $\prod_{k\geq1}(1 - a_k) = 0$.*

Proof: Let

$$K_n = \prod_{k=1}^{n}(1 - a_k) \text{ and } L_n = \sum_{k=1}^{n} a_k, n \geq 1. \qquad (2.3.24)$$

Since the a_k lie in the unit interval, we have for every $n \geq 1$,

$$
\begin{aligned}
\log K_n &= \sum_{k=1}^{n} \log(1 - a_k) \\
&= -\sum_{k=1}^{n}\left\{ a_k + \frac{1}{2}a_k^2 + \frac{1}{3}a_k^3 + \cdots \right\} \\
&\leq -\sum_{k=1}^{n} a_k = -L_n.
\end{aligned} \qquad (2.3.25)
$$

Allowing $n \to \infty$, it follows that $-L_n \to -\infty$ and, hence, $K_n \to 0$. ∎

Theorem 2.3.2 (Borel-Cantelli Lemma): *Let $\{A_n\}$ be a sequence of events and denote by $P(A_n)$ the probability that A_n occurs, $n \geq 1$. Also, let A denote the event that the A_n occur infinitely often (i.o.). Then*

$$\sum_{n \geq 1} P(A_n) < \infty \Rightarrow P(A) = 0, \qquad (2.3.26)$$

no matter whether the A_n are independent or not. If the A_n are independent, then

$$\sum_{n \geq 1} P(A_n) = +\infty \Rightarrow P(A) = 1. \qquad (2.3.27)$$

Proof: Let $B_n = \bigcup_{N \geq n} A_N$, $n \geq 1$. Then note that

$$A = \bigcap_{n \geq 1} \bigcup_{N \geq n} A_N = \bigcap_{n \geq 1} B_n \subseteq B_n, \quad n \geq 1. \qquad (2.3.28)$$

Thus, we have

$$
\begin{aligned}
P(A) &= P\left(\bigcap_{n \geq 1} B_n\right) \leq P(B_m) = P\left(\bigcup_{N \geq m} A_N\right) \\
&\leq \sum_{N \geq m} P(A_N), \quad m \geq 1.
\end{aligned} \qquad (2.3.29)
$$

In the last step, we use the basic inequality that $P(\bigcup A_r) \leq \sum_r P(A_r)$ without necessarily requiring that the A_r be independent. Thus, whenever $\sum_{n \geq 1} P(A_n)$ converges, by definition, $\sum_{n \geq m} P(A_n) \to 0$, and this proves

(2.3.26). Next, we note that by (2.3.28) $A^c = \bigcup_{n \geq 1} \bigcap_{N \geq n} A_N^c$, so that incorporating the independence of the A_n (or A_n^c, equivalently), we have

$$
\begin{aligned}
1 - P(A) &= P(A^c) \leq \sum_{n \geq 1} P\left(\bigcap_{N \geq n} A_N^c\right) \\
&= \sum_{n \geq 1} \prod_{N \geq n} P(A_N^c) = \sum_{n \geq 1} \prod_{N \geq n} \{1 - P(A_N)\}. \quad (2.3.30)
\end{aligned}
$$

By Theorem 2.3.1, the rhs of (2.3.30) is equal to 0 when $\sum_{n \geq 1} P(A_n) = +\infty$, and this completes the proof of (2.3.27).   ∎

The Borel-Cantelli Lemma (Theorem 2.3.2) occupies a central position in Probability Theory and has a variety of uses in Large Sample Theory. One of these uses relates to the verification of complete (and, hence, a.s.) convergence by using the Markov (or some other) Inequality (2.3.4) along with (2.3.26). Before we illustrate this point with the help of some examples, we consider the following:

Theorem 2.3.3: *For every $p > 1$, the series $\sum_{n \geq 1} n^{-p}$ converges.*

Proof. Note that

$$
\begin{aligned}
\sum_{n \geq 1} n^{-p} &= 1 + \sum_{k \geq 1} \left\{ \sum_{j=2^{k-1}+1}^{2^k} j^{-p} \right\} \\
&\leq 1 + \sum_{k \geq 1} (2^{-(k-1)})^p \left\{ 2^k - 2^{k-1} \right\} \\
&= 1 + \sum_{k \geq 1} 2^{-(k-1)(p-1)} \\
&= 1 + (1 - 2^{-(p-1)})^{-1}. \quad (2.3.31)
\end{aligned}
$$

If $p > 1$, it follows that $2^{-(p-1)} < 1$ and the rhs of (2.3.31) is finite, completing the proof.   ∎

Example 2.3.5: Let $X_k, k \geq 1$ be i.i.d. r.v.'s with $EX = \mu$, $E(X-\mu)^2 = \sigma^2$ and $E(X - \mu)^4 = \mu_4 < \infty$. Then let $T_n = \overline{X}_n$ and let $A_n = [|\overline{X}_n - \mu| > \varepsilon]$, for $n \geq 1$, where $\varepsilon > 0$ (arbitrary). Then, by the Markov Inequality (2.3.4) with $r = 4$, we have

$$
\begin{aligned}
P(A_n) &= P\left\{|\overline{X}_n - \mu| > \varepsilon\right\} \leq \varepsilon^{-4} E(\overline{X}_n - \mu)^4 \\
&= \frac{n^{-2}3\sigma^4 + n^{-3}(\mu_4 - 3\sigma^4)}{\varepsilon^4}, \quad n \geq 1. \quad (2.3.32)
\end{aligned}
$$

Consequently, using (2.3.31) for $p = 2$ and the first part of the Borel-

Cantelli Lemma [i.e., (2.3.26)], we immediately obtain that

$$P(A) = P(\bigcap_{n \geq 1} \bigcup_{N \geq n} A_N) = 0,$$

so that the A_n do not occur i.o., or $\overline{X}_n - \mu \xrightarrow{\text{a.s.}} 0$. In fact, in this case, $P(\bigcup_{N \geq n} A_N) \leq \sum_{N \geq n} P(A_N) \to 0$, as $n \to \infty$, and, hence, $\overline{X}_n \xrightarrow{c} \mu$. Note that the conclusions derived here remain true for the entire class of distributions for which the fourth moment is finite. Later on, we shall see that this fourth moment condition may also be relaxed. ∎

Example 2.3.6: Consider now the binomial example treated in (2.3.11)–(2.3.13). Here, letting $A_n = [|T_n - \pi| > \varepsilon]$, $n \geq 1$, $\varepsilon > 0$ (arbitrary), we have

$$\sum_{n \geq 1} P(A_n) \leq \sum_{n \geq 1} \{[\rho(\varepsilon, \pi)]^n + [\rho(\varepsilon, 1 - \pi)]^n\}$$

$$= \frac{\rho(\varepsilon, \pi)}{1 - \rho(\varepsilon, \pi)} + \frac{\rho(\varepsilon, 1 - \pi)}{1 - \rho(\varepsilon, 1 - \pi)} < \infty, \qquad (2.3.33)$$

and, hence, the Borel-Cantelli Lemma (Theorem 2.3.2) leads us to the conclusion that $T_n - \pi \xrightarrow{\text{a.s.}} 0$. We could have obtained the same result by using the previous example wherein we take $X_k = 1$ with probability π and $X_k = 0$, with probability $1 - \pi$. This result is known in the literature as the **Borel Strong Law of Large Numbers**. ∎

Essentially, the use of the Borel-Cantelli Lemma (2.3.2) to establish the a.s. or complete convergence is based on the specification of a sequence of the non-negative function $h(n)$, such that $P(A_n) = P\{|T_n - T| > \varepsilon\} \leq c_\varepsilon h(n)$, for all $n \geq n_0$ and $\sum_{n \geq n_0} h(n) < \infty$, and Theorem 2.3.3 provides a general class of such $h(n)$. In most of the practical problems, a choice of a suitable $h(n)$ is made by using the Markov (or some other) probability inequality (2.3.4), which generally deems the existence of positive order (absolute) moments of $T_n - T$ (up to some order ≥ 1). However, this is not necessary, and there are numerous examples where such a moment condition may not actually hold, but the desired order of $h(n)$ can be achieved by alternative methods. To illustrate this point, we consider the following example.

Example 2.3.7 (CMRR procedure): Consider the problem of estimating the number (N) of fishes in a lake. The so-called Capture-Mark-Release and Recapture (CMRR) procedure consists of (i) drawing an initial sample of n_1 fishes at random from the lake, marking them suitably (e.g., on the fin) and releasing them in the lake; (ii) drawing a second random sample of size n_2 fishes from the lake and counting the number (r_2) of marked fishes

in this sample, and (iii) estimating N by the maximum likelihood method
as

$$\widehat{N} = [n_1 n_2 / r_2],\qquad\qquad (2.3.34)$$

where $[s]$ denotes the integer part of $s\ (> 0)$. Note that given n_1 and n_2,
the probability law for r_2 is given by

$$P(r_2 = r \mid n_1, n_2, N) = \frac{\binom{n_1}{r}\binom{N-n_1}{n_2-r}}{\binom{N}{n_2}},\qquad r = 0, 1, \ldots, n_1 \wedge n_2,\qquad (2.3.35)$$

so that

$$\frac{P(r \mid N, n_1, n_2)}{P(r \mid N-1, n_1, n_2)} = \frac{(N-n_1)(N-n_2)}{N(N-n_1-n_2+r)} \geq (\leq) 1 \qquad (2.3.36)$$

according as $n_1 n_2 / r$ is $\geq (\leq) N$, and this characterizes the estimator in
(2.3.34). Since

$$P(0 \mid N, n_1, n_2) = \frac{(N-n_1)!(N-n_2)!}{N!(N-n_1-n_2)!} > 0,$$

the estimator $\widehat{N}$ in (2.3.34) may assume the value $+\infty$ with a positive
probability, and, hence, any positive order moment of its distribution fails
to be finite. Therefore, we are not in a position to use the probability
inequalities considered earlier which rest on the finiteness of the sth moment
for some $s > 0$. On the other hand, we may note that by (2.3.34)

$$N/\widehat{N} \simeq (r_2/n_2)/(n_1/N),\qquad\qquad (2.3.37)$$

where (r_2/n_2) converges in probability (or in the sth mean, for any $s > 0$) to
n_1/N, so that $N/\widehat{N}$ converges in the same mode to 1. Thus, $\widehat{N}/N$ converges
in probability to 1 although it does not converge in the rth mean, for any
$r > 0$.   ∎

   Motivated by this example, we may consider the following useful tool to
bypass such a moment existence problem through suitable transformations
of statistics.

Theorem 2.3.4: *Let $\{T_n\}$ be a sequence of r.v.'s, such that for some real θ,
T_n converges to θ, in probability, or a.s. or completely. Also, let $g : \mathbb{R} \to \mathbb{R}$,
be a function such that $g(x)$ is continuous at $x = \theta$. Then, $g(T_n)$ converges
to $g(\theta)$, in the same mode as T_n does (to θ).*

Proof: Note that by virtue of the assumed continuity of $g(\cdot)$ (at θ), for
every $\varepsilon > 0$, there exists a $\delta = \delta(\varepsilon) > 0$, such that

$$|t - \theta| < \varepsilon \Rightarrow |g(t) - g(\theta)| < \delta.\qquad\qquad (2.3.38)$$

Therefore, for every (fixed) n and $\varepsilon > 0$,

$$P\left\{|g(T_n) - g(\theta)| \leq \delta\right\} \;\geq\; P\left\{|T_n - \theta| \leq \varepsilon\right\};$$

$$P\left\{\bigcap_{N \geq n} |g(T_N) - g(\theta)| \leq \delta\right\} \;\geq\; P\left\{\bigcap_{N \geq n} |T_N - \theta| \leq \varepsilon\right\}.$$

$$(2.3.39)$$

Thus, if $T_n \xrightarrow{\mathrm{P}} \theta$, then $P\left\{|T_n - \theta| \leq \varepsilon\right\} \to 1$ as $n \to \infty$, and, hence, the result follows from the first inequality in (2.3.39); if $T_n \to \theta$ a.s. (or completely), then $P\left\{\bigcap_{N \geq n} |T_N - \theta| \leq \varepsilon\right\} \to 1$ (or $\sum_{N \geq n} P\left\{|T_N - \theta| > \varepsilon\right\} \to 0$), as $n \to \infty$, so that the result follows from the second inequality. ∎

Note that the theorem remains valid in the case of T_n and/or $g(\cdot)$ being vector valued; the proof is essentially the same, provided the appropriate distance function (considered earlier) is used. If, we want to maintain the same generality as in (2.2.1) or (2.2.5), i.e., replacing θ by a (possibly degenerate) r.v. T, then we may need to strengthen the continuity of $g(\cdot)$ to **uniform continuity** in order to derive a parallel result. This is presented below.

Theorem 2.3.5: *If $T_n - T \to 0$, in probability, a.s. or completely, and if $g(t)$ is uniformly continuous (a.e. T), then $g(T_n) - g(T) \to 0$ in the same mode, as $n \to \infty$.*

Proof: Note that if $g(t)$ is uniformly continuous (a.e. T), then (2.3.38) holds with θ and $g(\theta)$ replaced by T and $g(T)$, respectively, almost everywhere in T, and, hence, the rest of the proof follows as in the case of Theorem 2.3.4. Note that if $T = \theta$ with probability 1, then uniform continuity of $g(t)$ (a.e. T) is equivalent to the continuity of $g(t)$ at $t = \theta$, so that Theorem 2.3.4 is a special case of Theorem 2.3.5. ∎

We illustrate the use of the last two theorems with the following examples.

Example 2.3.8: With reference to Example 2.3.6 dealing with the binomial model, consider now the estimation of $\theta = \pi^{-1}$, so that $\theta \in (0, \infty)$. Define T_n as in that case so that $nT_n \sim \mathrm{Bin}(n, \pi)$, and $T_n \to \pi$, in probability or a.s. or completely as well as in the rth mean for any $r > 0$. A natural estimator of θ is $g(T_n) = T_n^{-1}$. Note that $P\{T_n = 0\} = (1-\pi)^n > 0$, so that T_n^{-1} assumes the value $+\infty$ with a positive probability (which converges to 0 exponentially in n), and, hence, $g(T_n) \to \pi^{-1} = \theta$, in probability or a.s. or completely, as $n \to \infty$ (although the rth mean convergence does not hold, for any $r > 0$).   ∎

Example 2.3.9: In the setup of Example 2.3.7, assume that N, n_1, n_2 are all large and $n_1/N \to \alpha$, $n_2/N \to \beta$, where $0 < \alpha, \beta < 1$. Let $T_{n_2} = r_2/n_2$, so that by (2.3.34), $\widehat{N}/N = (n_1/N)T_{n_2}^{-1} = g(T_{n_2})$, where $g(t) = (n_1/N)/t$, $t \geq 0$. Again, $g(t)$ is continuous, except at $t = 0$, whereas, by the Chebyshev Inequality (2.2.31), we have for all $\varepsilon > 0$,

$$P\left\{\left|T_{n_2} - \frac{n_1}{N}\right| > \varepsilon\right\} \leq (n_2\varepsilon^2)^{-1}n_1 \frac{(N - n_1)(N - n_2)}{N^2(N - 1)}, \qquad (2.3.40)$$

so that $T_{n_2} \to n_1/N$, in probability, as $N \to \infty$. Thus, by Theorem 2.3.4, we conclude that $\widehat{N}/N \to 1$, in probability, as N becomes large.   ∎

Example 2.3.10 (Ratio estimator): Let $(X_1, Y_1), \ldots, (X_n, Y_n)$ be n i.i.d. r.v.'s with mean vector (μ, ν) and dispersion matrix $\boldsymbol{\Sigma} = ((\sigma_{ij}))$, $i, j = 1, 2$; assume that all these parameters are finite and $\nu \neq 0$. Our concern is to estimate $\theta = \mu/\nu$. A natural estimator of θ is $\widehat{\theta}_n = \overline{X}_n/\overline{Y}_n$. By the use of the Chebyshev Inequality (2.2.31) or other inequalities, we may easily verify that $(\overline{X}_n, \overline{Y}_n)$ stochastically converges to (μ, ν), as $n \to \infty$. However, $\widehat{\theta}_n$ may not have any finite positive order moment, so that such a probability inequality may not be directly applicable to $\widehat{\theta}_n$. We note that $g(t_1, t_2) = t_1/t_2$ is a continuous function of (t_1, t_2), except on the line $t_2 = 0$, and, hence, whenever $\nu \neq 0$, we conclude that $(\overline{X}_n, \overline{Y}_n) \to (\mu, \nu)$, in probability or almost surely or completely, implies that $\widehat{\theta}_n \to \theta$, in the same mode.   ∎

Example 2.3.11 (ANOVA model): Let X_{ij}, $j = 1, \ldots, n$, be i.i.d. r.v.'s with mean μ_i and finite variance σ^2, for $i = 1, \ldots, k \ (\geq 2)$; all these k samples are assumed to be independent. Let $\overline{X}_i = n^{-1}\sum_{j=1}^{n} X_{ij}$, $i = 1, \ldots, k$, $\overline{X} = k^{-1}\sum_{i=1}^{k} \overline{X}_i$, $S_B^2 = \sum_{i=1}^{k}(\overline{X}_i - \overline{X})^2$, $S_n^2 = (k(n-1))^{-1}\sum_{i=1}^{k}(X_{ij} - \overline{X}_i)^2$. For testing the null hypothesis of homogeneity of $\mu_1, \ldots, \mu_k$, the usual analysis of variance test statistic is

$$F_{k,k(n-1)} = nS_B^2/(k-1)S_n^2. \qquad (2.3.41)$$

Side by side, we introduce the statistic $F_k^* = nS_B^2/\{(k-1)\sigma^2\}$. Computation of the actual central moments of $F_{k,k(n-1)}$ may be quite cumbersome (when the actual distribution of the X_{ij} is not known), and for possibly non-normal distributions, the mean, variance or other central moments of $F_{k,k(n-1)}$ may not exist. Thus, a direct use of the Chebyshev Inequality (2.3.31) or other moment-based probability inequality may not be that practicable. We shall see later on that whenever σ^2 is finite, $S_n^2 \xrightarrow{\text{a.s.}} \sigma^2$. Consequently, whenever $0 < \sigma^2 < \infty$, by using Theorem 2.3.5 we conclude that

$$F_{k,k(n-1)} - F_k^* \xrightarrow{\text{P}} 0. \tag{2.3.42}$$

For F_k^* , as we shall see in the next chapter, a suitable limiting distribution exists under quite general conditions, and, hence, we may use the same limiting distribution for $F_{k,k(n-1)}$ as well; in this context, we use the implications depicted in Figure 2.2.2.   ∎

In most of the cases dealing with simple statistical models, we encounter suitable statistics which can be expressed as averages over independent r.v.'s (or vectors) or as functions of such averages. In many other situations, by suitable expansions, we may also be able to express a statistic as an average over independent r.v.'s plus a remainder term which converges to 0 in a certain mode. As such, the study of the stochastic convergence of averages of independent r.v.'s occupies a central position in Large Sample Theory. Although the probability inequalities studied earlier (or some others to follow) may be used in this context, a more direct approach (requiring generally less restrictive regularity conditions) has been worked out; conventionally, such results are categorized under the denomination of Laws of Large Numbers. We may start with the following:

Theorem 2.3.6 (Khintchine Weak Law of Large Numbers): *Let* $\{X_k, k \geq 1\}$ *be a sequence of i.i.d. r.v.'s with* $\mathrm{E}X_1 = \theta$, *and let* $\overline{X}_n = n^{-1} \sum_{k=1}^n X_k$, $n \geq 1$. *Then*

$$\overline{X}_n \xrightarrow{\text{P}} \theta. \tag{2.3.43}$$

Proof. Let $F(x)$, $x \in \mathbb{R}$, be the distribution function of X_1, and let $\phi_F(t)$, $t \in \mathbb{R}$, be the characteristic function of X_1. Also, note that as $n \to \infty$,

$$\begin{aligned}
\phi_{\overline{X}_n}(t) &= \mathrm{E}\left\{ \prod_{k=1}^n e^{itX_k/n} \right\} = [\phi_F(t/n)]^n \\
&= \{1 + itn^{-1}\mathrm{E}X_1 + o(n^{-1})t\}^n \to e^{it\theta}. \tag{2.3.44}
\end{aligned}$$

Since $e^{it\theta}$ is the characteristic function of a r.v. Z which is degenerate at the point θ [i.e., $P(Z = \theta) = 1$], the proof of (2.3.43) can be completed by appealing to Proposition 2.2.1.   ∎

The simple treatment in (2.3.44) may not work out that well when the X_k are independent but not identically distributed; if we denote by $\phi_{F_k}(t)$ the characteristic function of X_k, then, of course, we have

$$\phi_{\overline{X}_n}(t) = \prod_{k=1}^n \phi_{F_k}(t/n) \Rightarrow \log \phi_{\overline{X}_n}(t) = \sum_{k=1}^n \log \phi_{F_k}(t/n), \tag{2.3.45}$$

but in order for the rhs to converge to a limiting characteristic function of a degenerate r.v., we may need additional (uniformity) conditions on the $\phi_{F_k}(t)$, and, these may, in turn, require higher order moment conditions on the X_k. In addition, if the X_k are not independent, then the factorization in (2.3.45) may not hold, and the treatment of the convergence of $\phi_{\overline{X}_n}(t)$ to a degenerate characteristic function may become still more complicated. Finally, if instead of the stochastic convergence in (2.3.43), we seek an a.s. convergence, then, even in the i.i.d. case, Proposition 2.2.1 fails to provide the desired tool, and a more involved proof is, therefore, in order. Keeping this spectrum in mind, we may appreciate the need for the different LLNs which address different degrees of dependence structure and/or more conditions on the underlying r.v.'s. Before we present some of these LLNs, we may note that for independent r.v.'s, if moments up to the order r exist for some $r \geq 2$, then by the use of the Chebyshev Inequality (2.2.31) or the Markov Inequality (2.3.4), the stochastic convergence of $\overline{X}_n - E\overline{X}_n$ to 0 can be easily established. On the other hand, for (2.3.43), in the i.i.d. case, the first moment suffices. Thus, there is a natural question: for random variables which are not necessarily identically distributed, do we still need the existence of the second moments for some LLN to hold, or we may relax this condition to a certain extent? Toward this, we consider the following example.

Example 2.3.12: Consider a sequence $\{X_k\}$ of independent r.v.'s, such that, for $k \geq 1$,

$$X_k = \begin{cases} -2^{k(1-\varepsilon)} & \text{with probability } 2^{-k-1} \\ 0 & \text{with probability } 1 - 2^{-k} \\ 2^{k(1-\varepsilon)} & \text{with probability } 2^{-k-1}, \end{cases} \quad (2.3.46)$$

where $0 < \varepsilon < 1$. Clearly, $EX_k = 0$ and $\text{Var}(X_k) = EX_k^2 = 2^{k(1-2\varepsilon)} < \infty$, for every $k \geq 1$. Thus, we have $E\overline{X}_n = 0$, for every $n \geq 1$, and

$$\text{Var}(\overline{X}_n) = \gamma_n^2 = n^{-2} \sum_{k=1}^{n} (2^{1-2\varepsilon})^k$$

$$= n^{-2} 2^{1-2\varepsilon} \left\{ \frac{2^{n(1-2\varepsilon)} - 1}{2^{(1-2\varepsilon)} - 1} \right\}. \quad (2.3.47)$$

Thus, for $\varepsilon > 1/2$, of course, $\gamma_n^2 \to 0$ as $n \to \infty$, and, hence, the Chebyshev Inequality (2.2.31) provides the access to the desired goal. A similar picture holds for the Markov Inequality (2.3.4) involving moment of order $r \geq 2$. However, none of these inequalities relates to necessary conditions, and the LLN may hold without either of them. This will be made clear from the following LLN. (We shall come back to this example later). ∎

Theorem 2.3.7 (Markov Weak Law of Large Numbers): *Let* X_k,

$k \geq 1$, *be independent r.v.'s, such that* $\mu_k = EX_k$, $k \geq 1$ *exist, and for some* δ *such that* $0 < \delta \leq 1$, *along with the existence of* $E|X_k - \mu_k|^{1+\delta}$, *suppose that*

$$n^{-1-\delta} \sum_{k=1}^{n} E|X_k - \mu_k|^{1+\delta} = \rho_n(\delta) \to 0, \quad \text{as } n \to \infty. \qquad (2.3.48)$$

Then $\overline{X}_n - E\overline{X}_n \xrightarrow{\text{p}} 0$.

Proof: The condition in (2.3.48) is known as the **Markov condition**. Note that without any loss of generality we may take $0 < \delta < 1$; as for $\delta \geq 1$, the Chebyshev Inequality (2.2.31) applies. We let $U_k = X_k - \mu_k$, so that $EU_k = 0$, $k \geq 1$. Then $\overline{X}_n - E\overline{X}_n = \overline{U}_n$, $n \geq 1$. We denote the characteristic functions of U_k and $\overline{U}_n$ by $\phi_{U_k}(t)$ and $\phi_{\overline{U}_n}(t)$, respectively. Since the U_k are independent, as in (2.3.45), we have for all $t \in \mathbb{R}$,

$$\phi_{\overline{U}_n}(t) = \prod_{k=1}^{n} \phi_{U_k}(t/n) \Rightarrow \log \phi_{\overline{U}_n}(t) = \sum_{k=1}^{n} \log \phi_{U_k}(t/n). \qquad (2.3.49)$$

At this stage, we recall Theorem 1.4.1 and note that under the assumed moment condition, for every $k \geq 1$ and real θ,

$$\phi_{U_k}(\theta) = 1 + i\theta EU_k + R_k(\theta), \qquad (2.3.50)$$

where

$$|R_k(\theta)| \leq c|\theta|^{1+\delta} \sum_{k=1}^{n} E|U_k|^{1+\delta}, \quad c < \infty.$$

Also, note that for every real t,

$$\max_{1 \leq k \leq n} |R_k(t/n)| \leq \sum_{k=1}^{n} |R_k(t/n)|$$

$$\leq cn^{-1-\delta}|t|^{1+\delta} \sum_{k=1}^{n} E|U_k|^{1+\delta}, \qquad (2.3.51)$$

where, by (2.3.48), the rhs of (2.3.51) converges to 0 as $n \to \infty$. Finally, note that $EU_k = 0$ for every $k \geq 1$, and if $a_{n1}, \ldots, a_{nn}$ are numbers (real or imaginary) such that $\sum_{k=1}^{n} |a_{nk}| \to 0$ as $n \to \infty$, then

$$\left| \sum_{k=1}^{n} \log(1 + a_{nk}) - \sum_{k=1}^{n} a_{nk} \right| \to 0 \quad \text{as} \quad n \to \infty.$$

Combining these with (2.3.49), we immediately obtain that

$$\left| \log \phi_{\overline{U}_n}(t) - \sum_{k=1}^{n} R_k(t/n) \right| \to 0 \quad \text{as } n \to \infty, \qquad (2.3.52)$$

so that, by (2.3.51) and (2.3.52), we obtain that under (2.3.48), for every real t, $\log \phi_{\overline{U}_n}(t) \to 0$ as $n \to \infty$, and, hence, $\phi_{\overline{U}_n}(t) \to 1$, as $n \to \infty$, for every real t. This ensures that $\overline{U}_n$ converges in law to a degenerate r.v. having the entire probability mass at the point 0, and the proof of the theorem follows by using Proposition 2.2.1.   ∎

We illustrate the use of the Markov LLN with Example 2.3.12. Note that

$$E|X_k|^{1+\delta} = 2^{-k+k(1-\varepsilon)(1+\delta)} \le 1 \quad \text{if} \quad (1+\delta)(1-\varepsilon) \le 1. \qquad (2.3.53)$$

In particular, if we choose $\delta = \varepsilon$, then $E|X_k|^{1+\delta} < 1$, for every $k \ge 1$, and, hence, (2.3.48) is bounded by $n^{-\varepsilon}$ ($\downarrow 0$, as $n \to \infty$). Thus, the weak LLN applies to the $\overline{X}_n$ under the Markov condition (2.3.48), for every $\varepsilon > 0$, although we have seen earlier that for $\varepsilon < 1/2$, the Chebyshev Inequality (2.2.31) or the Markov Inequality (2.3.4) fail to yield the desired result.

For the stochastic convergence of averages of independent (but not necessarily identically distributed) r.v.'s, the Markov weak LLN is the most general result. However, this is not enough to ensure a.s. convergence in the same setup. Such an a.s. convergence result can be established under a second moment condition on the U_k. In this context, we consider first a very useful probability inequality which has a variety of uses in Large Sample Theory.

Theorem 2.3.8 (Kolmogorov Maximal Inequality): *Let $X_1, \ldots, X_n$ be independent r.v.'s, such that $EX_i = \mu_i$ and $E(X_i - \mu_i)^2 = \sigma_i^2$ exist for every $i = 1, \ldots, n$. Then let $T_k = (X_1 - \mu_1) + \cdots + (X_k - \mu_k)$, for $k = 1, \ldots, n$, and, conventionally, we let $T_0 = 0$. Then, for every $t > 0$,*

$$P\left\{ \max_{0 \le k \le n} |T_k| > t \right\} \le t^{-2} \mathrm{Var}(T_n) = t^{-2} \sum_{k=1}^{n} \sigma_k^2. \qquad (2.3.54)$$

[Note the analogy with the Chebyshev Inequality in (2.3.1) where we have to take $T_n = \overline{X}_n$. For both, the rhs are the same, but compared to the marginal event that $|T_n| > t$, here we have the union of the events $|T_k| > t$, for $k = 1, \ldots, n$, and, hence, the statement on the lhs of (2.3.54) is much stronger than the one in (2.3.1).]

Proof: Without any loss of generality, we set $\mu_i = 0$, for $i = 1, \ldots, n$. Let us define the following mutually disjoint events

$$
\begin{aligned}
A_0 &= \{|T_k| \le t, \ k = 0, 1, \ldots, n\}, \\
A_k &= \{|T_j| \le t, \ j \le k - 1, \ |T_k| > t\}, \ k = 1, \ldots, n.
\end{aligned}
$$

$$(2.3.55)$$

Then, we may observe that

$$P\left\{\max_{1\le k\le n}|T_k| > t\right\} = P\left\{\bigcup_{k=1}^{n} A_k\right\}$$

$$= \sum_{k=1}^{n} P(A_k) = 1 - P(A_0). \qquad (2.3.56)$$

Also, denoting by I_A the indicator function of the set A, we have

$$\sum_{k=1}^{n} \sigma_k^2 = \mathrm{E}(T_n^2) = \sum_{k=0}^{n} \mathrm{E}(T_n^2 I_{A_k}) \ge \sum_{k=1}^{n} \mathrm{E}(T_n^2 I_{A_k}). \qquad (2.3.57)$$

Therefore, to establish (2.3.54), it suffices to show that for every $k = 1, \ldots, n$,

$$\mathrm{E}(T_n^2 I_{A_k}) \ge t^2 P(A_k). \qquad (2.3.58)$$

For this purpose, we write $T_n = T_k + (T_n - T_k)$ and note that $T_n - T_k$ is independent of $T_j, j \le k$, for every $k = 1, \ldots, n-1$. Moreover, the event A_k in (2.3.55) depends only on the $T_j, j \le k$, so that on the set A_k, $T_n - T_k$ has expectation 0. Thus, we have

$$\begin{aligned} \mathrm{E}(T_n^2 I_{A_k}) &= \mathrm{E}(T_k^2 I_{A_k}) + \mathrm{E}[(T_n - T_k)^2 I_{A_k}] + 2\mathrm{E}[T_k(T_n - T_k)I_{A_k}] \\ &= \mathrm{E}(T_k^2 I_{A_k}) + \mathrm{E}[(T_n - T_k)^2 I_{A_k}] + 0 \\ &\ge \mathrm{E}(T_k^2 I_{A_k}) \\ &\ge t^2 P(A_k), \end{aligned} \qquad (2.3.59)$$

where the last step follows from the fact that on A_k, $|T_k| > t$, for $k = 1, \ldots, n$. ∎

We shall consider various extensions of the Kolmogorov Maximal Inequality (2.3.54) for suitable sequences of dependent r.v.'s in the next section. However, in this section, we consider another extension of this inequality which plays a vital role in the subsequent developments.

Theorem 2.3.9 (Hájek-Rènyi Inequality): *Let $X_1, \ldots, X_n$ be independent r.v.'s with means $\mu_1, \ldots, \mu_n$ and finite variances $\sigma_1^2, \ldots, \sigma_n^2$, respectively. Let $T_k = \sum_{i=1}^{k}(X_i - \mu_i)$, $k = 1, \ldots, n$. Then, if $c_1 \ge c_2 \ge \cdots \ge c_n$ are positive constants, we have for every $t > 0$,*

$$P\left\{\max_{1\le k\le n} c_k|T_k| > t\right\} \le t^{-2} \sum_{k=1}^{n} c_k^2 \sigma_k^2. \qquad (2.3.60)$$

Proof: As in the proof of the Kolmogorov Inequality (2.3.54), we let $T_0 = 0$, and define

$$
\begin{aligned}
A_0 &= \{c_k|T_k| \le t, \, k = 0, \ldots, n\} \quad (c_0 = c_1), \\
A_k &= \{c_j|T_j| \le t, \, j \le k - 1; \, c_k|T_k| > t\}, \quad k = 1, \ldots, n.
\end{aligned}
$$
$$(2.3.61)$$

Then, by the definition of the A_k, we have

$$
\begin{aligned}
\sum_{k=1}^{n} \mathrm{E}[c_k^2 T_k^2 I_{A_k}] &\ge \sum_{k=1}^{n} \mathrm{E}[t^2 I_{A_k}] \\
&= t^2 \sum_{k=1}^{n} P(A_k) \\
&= t^2 P \left\{ \max_{1 \le k \le n} c_k|T_k| > t \right\}.
\end{aligned}
$$
$$(2.3.62)$$

Thus, to prove (2.3.60), it suffices to show that

$$
\sum_{k=1}^{n} c_k^2 \mathrm{E}(T_k^2 I_{A_k}) \le \sum_{k=1}^{n} c_k^2 \sigma_k^2. \tag{2.3.63}
$$

For this purpose, we let $B_0 = \Omega$ and $B_k = \left(\bigcup_{i=1}^{k} A_i\right)^c = \bigcap_{i=1}^{k} A_i^c$, for $k = 1, \ldots, n$. Then, noting that $I_{A_k} = I_{B_{k-1}} - I_{B_k}$ and $c_k \ge c_{k+1}$, $k = 1, \ldots$, we may write

$$
\begin{aligned}
\sum_{k=1}^{n} c_k^2 \mathrm{E}(T_n^2 I_{A_k}) &= \sum_{k=1}^{n} c_k^2 \left[\mathrm{E}(T_k^2 I_{B_{k-1}}) - \mathrm{E}(T_k^2 I_{B_k}) \right] \\
&\le c_1^2 \mathrm{E}(T_1^2) + \sum_{k=2}^{n} c_k^2 \mathrm{E}\left[(T_k^2 - T_{k-1}^2) I_{B_{k-1}} \right] - c_n^2 \mathrm{E}(T_n^2 I_{B_n}) \\
&\le c_1^2 \mathrm{E}(T_1^2) + \sum_{k=2}^{n} c_k^2 \left\{ \mathrm{E}\left[(X_k - \mu_k)^2 I_{B_{k-1}} \right] + \mathrm{E}\left[(X_k - \mu_k) T_{k-1} I_{B_{k-1}} \right] \right\} \\
&\le c_1^2 \sigma_1^2 + \sum_{k=2}^{n} c_k^2 \sigma_k^2 + 0 = \sum_{k=1}^{n} c_k^2 \sigma_k^2,
\end{aligned}
$$
$$(2.3.64)$$

where, in the penultimate step, we have made use of the fact that X_k is independent of B_{k-1}, for every $k = 1, \ldots, n$. ∎

We shall consider some extensions of this inequality for dependent r.v.'s in the next section. In passing, we may remark that in (2.3.54) and (2.3.60), the X_i are not necessarily identically distributed. As such, if we define $Y_1 = \cdots = Y_{M-1} = 0$ and $Y_k = T_k = \sum_{i=1}^{k}(X_i - \mu_i)$, for $k \ge M$, where the X_i are independent r.v.'s with means μ_i and finite variances σ_i^2, $i \ge 1$, we obtain, on letting $c_k = k^{-1}$, for $k \geq M$ and $n = N$, that for all $t > 0$,

$$P\left\{\max_{M \leq k \leq N} k^{-1}|T_k| > t\right\} \leq t^{-2}\left\{M^{-2}\mathrm{Var}(Y_M) + \sum_{k=M+1}^{N} k^{-2}\sigma_k^2\right\}$$

$$\leq t^{-2}\left\{M^{-2}\left(\sum_{i=1}^{M}\sigma_i^2\right) + \sum_{k=M+1}^{N} k^{-2}\sigma_k^2\right\}. \qquad (2.3.65)$$

The last inequality provides the key to the proof of the following:

Theorem 2.3.10 (Kolmogorov Strong Law of Large Numbers): *Let X_i, $i \geq 1$, be independent r.v.'s such that $\mathrm{E}X_i = \mu_i$ and $\mathrm{Var}(X_i) = \sigma_i^2$ exist for every $i \geq 1$. Also, let $\overline{\mu}_n = n^{-1}\sum_{i=1}^{n}\mu_i$, for $n \geq 1$. Then*

$$\sum_{k \geq 1} k^{-2}\sigma_k^2 < \infty \Rightarrow \overline{X}_n - \overline{\mu}_n \xrightarrow{a.s.} 0. \qquad (2.3.66)$$

Proof: Let $D_k = \sum_{n \geq k} n^{-2}\sigma_n^2$, for $k \geq 1$. Note that $D_1 < \infty$ ensures that D_k is nonincreasing in k and $\lim_{k \to \infty} D_k = 0$. Also, note that for every $M \geq 1$,

$$\frac{1}{M^2}\sum_{k \leq M}\sigma_k^2 = \frac{1}{M^2}\sum_{k \leq M}k^2[D_k - D_{k+1}]$$

$$\leq \frac{1}{M^2}\sum_{k=1}^{M}(2k-1)D_k, \qquad (2.3.67)$$

where the rhs converges to 0 as $M \to \infty$. Now, to prove that $\overline{X}_n - \overline{\mu}_n \xrightarrow{a.s.} 0$, we make use of (2.3.65) and obtain that for every $N > M$, $\varepsilon > 0$,

$$P\left\{\max_{M \leq k \leq N} |\overline{X}_n - \overline{\mu}_n| > \varepsilon\right\} = P\left\{\max_{M \leq k \leq N} k^{-1}|T_k| > \varepsilon\right\}$$

$$\leq \varepsilon^{-2}\left\{M^{-2}\left(\sum_{k=1}^{M}\sigma_k^2\right) + \sum_{k=M+1}^{N} k^{-2}\sigma_k^2\right\}. \qquad (2.3.68)$$

Thus, for any given M, first allowing $N \to \infty$, we obtain from (2.3.68) that

$$P\left\{\sup_{k \geq M} |\overline{X}_k - \overline{\mu}_k| > \varepsilon\right\} \leq \varepsilon^{-2}\left\{M^{-2}\sum_{k=1}^{M}\sigma_k^2 + D_{M+1}\right\}, \qquad (2.3.69)$$

and, hence, allowing M to go to $+\infty$, we obtain that the rhs of (2.3.69) converges to 0. ∎

In passing, we may remark that if the X_i are i.i.d. r.v.'s, then $\sigma_i^2 = \sigma^2$, $i \geq 1$, whereas $\sum_{k \geq 1} k^{-2} = \pi^2/6 < \infty$. Hence, $\overline{X}_n - \mathrm{E}\overline{X}_n \xrightarrow{a.s.} 0$, whenever

$\sigma^2 < \infty$. However, as we shall see later on, for the i.i.d. case we do not require the second moment condition, and the finiteness of the first moment suffices. The interesting point is that the Kolmogorov condition in (2.3.66) allows the σ_i^2 to be varying with i, but at a rate slower than i, and, in particular, if the σ_i^2 are all bounded (but not necessarily equal), then this condition holds automatically. In order to consider better results for the i.i.d. case, we first consider the following

Theorem 2.3.11: *Let X be a real-valued r.v. Then*

$$E|X| < \infty \Leftrightarrow \sum_{k \geq 1} kP\{k \leq |X| < k+1\} < \infty. \qquad (2.3.70)$$

Proof: First, note that for every $k \geq 0$,

$$kP\{k \leq |X| < k+1\} \quad \leq \quad \int_k^{k+1} x\, dP\{|X| \leq x\}$$

$$\leq \quad (k+1)P\{k \leq |X| < k+1\}. \qquad (2.3.71)$$

Hence, noting that $E|X| = \sum_{k>0}\int_k^{k+1} x\,dP\{|X| \leq x\}$, summing over k in (2.3.71), the first inequality yields that

$$E|X| < \infty \quad \text{implies} \quad \sum_{k \geq 1} kP\{k \leq |X| < k+1\} < \infty,$$

whereas the last inequality yields that

$$\sum_{k \geq 1} kP\{k \leq |X| < k+1\} < \infty \quad \text{implies} \quad E|X| < \infty.$$

$\blacksquare$

Theorem 2.3.12 (Khintchine Equivalence Lemma): *Let $\{X_n\}$ and $\{Y_n\}$ be two arbitrary sequences of r.v.'s. Then, if $\sum_{i \geq 1} P\{X_i \neq Y_i\} < \infty$, the Strong Law of Large Numbers (SLLN) holds for both sequences or for none.*

Proof: Note that

$$n^{-1}\sum_{i \leq n} X_i = n^{-1}\sum_{i \leq n} Y_i + n^{-1}\sum_{i \leq n}(X_i - Y_i), \quad n \geq 1.$$

If $\sum_{i \geq 1} P\{X_i \neq Y_i\} < \infty$, if follows from the Borel-Cantelli Lemma (Theorem 2.3.2) that only finitely many of the events $\{X_i \neq Y_i\}$ may occur and, hence, $n^{-1}\sum_{i \leq n}(X_i - Y_i) \xrightarrow{\text{a.s.}} 0$. Therefore, $n^{-1}\sum_{i \leq n} X_i$ and $n^{-1}\sum_{i \leq n} Y_i$ are **convergence equivalent**, that is, as $n \to \infty$, either both series converge or none.

$\blacksquare$

We are now in a position to prove the following SLLN.

Theorem 2.3.13 (Khintchine Strong Law of Large Numbers): *Let* X_i, $i \geq 1$ *be i.i.d. r.v.'s. Then* $\overline{X}_n \xrightarrow{\text{a.s.}} c$, *and only if* EX_1 *exists and* $c = EX_1$.

Proof. First assume that $\mu = EX_1$ exists, and let $U_i = X_i - \mu$, $i \geq 1$. Then, let us show that $\overline{U}_n \xrightarrow{\text{a.s.}} 0$. For this purpose, define

$$Y_i = U_i I_{\{|U_i| \leq i\}}, \quad i \geq 1. \tag{2.3.72}$$

Since the U_i are i.i.d. r.v.'s, we have

$$
\begin{aligned}
P\{Y_i \neq U_i\} &= P\{|U_i| > i\} = P\{|U_1| > i\} \\
&= \sum_{k \geq i} P\{k < |U_i| \leq k+1\}, \tag{2.3.73}
\end{aligned}
$$

for every $i \geq 1$, so that, by Theorem 2.3.11 and (2.3.73),

$$
\begin{aligned}
\sum_{i \geq 1} P\{Y_i \neq U_i\} &= \sum_{i \geq 1} \sum_{k \geq i} P\{k < |U_1| \leq k+1\} \\
&= \sum_{k \geq 1} k P\{k < |U_1| \leq k+1\} < \infty. \tag{2.3.74}
\end{aligned}
$$

Using the Khintchine Equivalence Lemma (Theorem 2.3.12) along with (2.3.74), it suffices to show that

$$\overline{Y}_n \xrightarrow{\text{a.s.}} 0. \tag{2.3.75}$$

Toward this, first, we observe that for $n \geq 1$,

$$
\begin{aligned}
E\overline{Y}_n &= n^{-1} \sum_{i=1}^{n} E\left(U_i I_{\{|U_i| \leq i\}}\right) = -n^{-1} \sum_{i=1}^{n} E\left(U_1 I_{\{|U_1| > i\}}\right) \\
&= -n^{-1} \sum_{i=1}^{n-1} i E\left(U_1 I_{\{i < |U_1| \leq i+1\}}\right) - E\left(U_1 I_{\{|U_1| > n\}}\right).
\end{aligned}
\tag{2.3.76}
$$

Now, EU_1 exists and, hence, $\left|E\left(U_1 I_{\{|U_1| > n\}}\right)\right| \to 0$ as $n \to \infty$. Similarly,

$$
\begin{aligned}
&\left| n^{-1} \sum_{i=1}^{n-1} i E\left(U_1 I_{\{i < |U_1| \leq i+1\}}\right) \right| \\
&\leq k^2 n^{-1} P\{|U_1| \leq k\} + \sum_{i=k+1}^{n-1} i P\{i < |U_1| \leq i+1\} \\
&\leq k^2 n^{-1} + E\left[|U_1| I_{\{|U_1| \geq k+1\}}\right], \quad k \in (1, n-1) \tag{2.3.77}
\end{aligned}
$$

Therefore, by (2.3.76) and (2.3.77), we conclude [on letting $k = k_n =$

$o(n^{1/2})]$ that

$$\overline{EY}_n \to 0 \text{ as } n \to \infty. \qquad (2.3.78)$$

Thus, to establish (2.3.75), it suffices to show that $\overline{Y}_n - \overline{EY}_n \xrightarrow{\text{a.s.}} 0$, and for this purpose, we may as well use the Kolmogorov SLLN (Theorem 2.3.10). For this, note that for all $i \geq 1$,

$$
\begin{aligned}
\text{Var}(Y_i) \leq EY_i^2 &= E\left(U_1^2 I_{\{|U_1| \leq i\}}\right) \\
&\leq \sum_{k=0}^{i-1} (k+1)^2 P\{k < |U_1| \leq k+1\}. \qquad (2.3.79)
\end{aligned}
$$

Consequently,

$$
\begin{aligned}
\sum_{k \geq 1} \frac{\text{Var}(Y_k)}{k^2} &\leq \sum_{k \geq 1} k^{-2} \sum_{i=0}^{k-1} (i+1)^2 P\{i < |U_1| \leq i+1\} \\
&\leq \sum_{i=0}^{\infty} (i+1)^2 P\{i < |U_1| \leq i+1\} \sum_{k=i+1}^{\infty} k^{-2} \\
&\leq P\{|U_1| \leq 1\} \sum_{i=1}^{\infty} i^{-2} \\
&\quad + \sum_{k=1}^{\infty} P\{k < |U_1| \leq k+1\} (k+1)^2 \sum_{i=k+1}^{\infty} i^{-2} \\
&\leq \left(\frac{\pi^2}{6}\right) P\{|U_1| \leq 1\} + 4 \sum_{k=1}^{\infty} kP\{k < |U_1| \leq k+1\}.
\end{aligned}
$$

$$(2.3.80)$$

Since $\sum_{i \geq 1} i^{-2} = \pi^2/6$ and for every $k \geq 1$,

$$(k+1)^2 \sum_{i \geq k+1} i^{-2} \leq 2(k+1)^2 \sum_{i \geq k+1} \{i(i+1)\}^{-1} = 2(k+1),$$

the rhs of (2.3.80) converges, and hence, by the Kolmogorov SLLN (Theorem 2.3.10), $\overline{Y}_n - \overline{EY}_n \xrightarrow{\text{a.s.}} 0$. Thus, if EX_1 exists, then $\overline{X}_n - EX_1 \xrightarrow{\text{a.s.}} 0$.

Alternatively, suppose now that $\overline{X}_n \xrightarrow{\text{a.s.}} c$, for some finite c. Then we show that $c = EX$. Let $U_i = X_i - c$, $i \geq 1$, so that by the definition of a.s. convergence, we claim that for every $\varepsilon > 0$,

$$P\left\{\bigcup_{N \geq n} (|\overline{U}_N| > \varepsilon)\right\} \to 0, \quad \text{as } n \to \infty. \qquad (2.3.81)$$

Note that $(N+1)\overline{U}_{n+1} = N\overline{U}_N + U_{N+1}$, for every N, so that $|\overline{U}_N| < \varepsilon$, $N \geq n$, implies that $|N^{-1}U_N| \leq 2\varepsilon$, for every $N \geq n$. Consequently, we have

$$P\left\{\bigcap_{N\geq n}\left(N^{-1}|U_N|\leq 2\varepsilon\right)\right\}\geq P\left\{\bigcap_{N\geq n}\left(|\overline{U}_N|\leq\varepsilon\right)\right\},\qquad(2.3.82)$$

which in conjunction with (2.3.81) implies that, for every positive ε, as $n\to\infty$,

$$P\left\{\bigcup_{N\geq n}\left(N^{-1}|U_N|>2\varepsilon\right)\right\}\leq P\left\{\bigcup_{N\geq n}\left(|\overline{U}_N|>\varepsilon\right)\right\}\to0.\qquad(2.3.83)$$

In view of the fact that the events $A_N=\left(N^{-1}|U_N|>\varepsilon\right)$, $N\geq1$, are independent and $\varepsilon>0$ is arbitrary, we obtain on letting $\varepsilon=1/2$ and using the second part of the Borel-Cantelli Lemma (Theorem 2.3.2) that (2.3.83) ensures that

$$\sum_{N\geq1}P\left\{N^{-1}|U_N|>1\right\}<\infty.\qquad(2.3.84)$$

On the other hand, the lhs of (2.3.84) is equal to

$$\sum_{N\geq1}\sum_{k\geq N}P\left\{k<|U_N|\leq k+1\right\}$$
$$=\sum_{N\geq1}\sum_{k\geq N}P\left\{k<|U_1|\leq k+1\right\}$$
$$=\sum_{k\geq1}kP\left\{k<|U_1|\leq k+1\right\}$$
$$\geq\mathrm{E}|U_1|-1,$$

by Theorem 2.3.11. Therefore, $\mathrm{E}U_1$ exists. As $\mathrm{E}U_1=\mathrm{E}X_1-c$, $\mathrm{E}X_1$ also exists. If $c\neq\mathrm{E}X_1$, then by the first part of the theorem, $\overline{X}_n\xrightarrow{\text{a.s.}}\mathrm{E}X_1=\mu$ which is a contradiction. Therefore, $c=\mathrm{E}X_1$.   ∎

The Khintchine SLLN (Theorem 2.3.13) provides the a.s. convergence for the average of i.i.d. r.v.'s under minimal conditions. The theorem extends directly to the vector case by treating each coordinate average separately. In fact, we may treat a somewhat more general result as follows. Suppose that $g(\mathbf{y})$ is a real-valued function of a vector $\mathbf{y}\in\mathbb{R}^p$, for some $p\geq1$. Suppose further that there exists a sequence $\{\mathbf{Z}_i,i\geq1\}$ of i.i.d. p-vectors, such that $\mathrm{E}\mathbf{Z}_1=\boldsymbol{\theta}$ exists. Finally, suppose that $g(\cdot)$ is continuous at $\boldsymbol{\theta}$. Then, $g(\overline{\mathbf{Z}}_n)\xrightarrow{\text{a.s.}}g(\boldsymbol{\theta})$. A parallel result holds when $g(\cdot)$ is itself a q-vector, for some $q\geq1$. We invite the reader to complete the proof of this result by using the definition of continuity of $g(\cdot)$ (in the multivariate case) and the Khintchine SLLN (Theorem 2.3.13) for the $\mathbf{Z}_i$. To illustrate the utility of the above results, we consider the following.

Example 2.3.13: Let $\{X_i, i \geq 1\}$ be i.i.d. r.v.'s such that $EX_1 = \theta$ and $\text{Var}(X_1) = \sigma^2$ where both θ and σ^2 are assumed to be finite. A natural estimator of σ^2 is the sample variance S_n^2. Note that

$$n^{-1}(n-1)S_n^2 = n^{-1}\sum_{i=1}^{n}(X_i - \theta)^2 - (\overline{X}_n - \theta)^2. \qquad (2.3.85)$$

Thus, if we let $Z_{1i} = (X_i - \theta)^2$, $Z_{2i} = (X_i - \theta)$ and $\mathbf{Z}_i = (Z_{1i}, Z_{2i})^t$, $i \geq 1$, then by the Khintchine SLLN (Theorem 2.3.13) (in the vector case), $\overline{\mathbf{Z}}_n \xrightarrow{\text{a.s.}} (\sigma^2, 0)^t$. On the other hand, $(n-1)/n \to 1$ as $n \to \infty$, whereas $g(t_1, t_2) = t_1 - t_2^2$ is a continuous function of (t_1, t_2). Hence, we conclude that

$$S_n^2 \xrightarrow{\text{a.s.}} \sigma^2 \quad \text{whenever } EX_1^2 < \infty. \qquad (2.3.86)$$

Note that for (2.3.86), the finiteness of the second moment suffices. If we had used the Kolmogorov SLLN (Theorem 2.3.10) for each component, then we would require that EX_1^4 be finite, and that would have been more restrictive than the second moment condition in (2.3.86). If, however, we use a slightly more general model where the X_i's have the same variance σ^2 (but possibly different distributions), then to establish the a.s. convergence result for S_n^2, we may not be able to use the Khintchine SLLN (Theorem 2.3.13), and some other conditions may be necessary.   ∎

2.4 Inequalities and laws of large numbers for some dependent variables

In the last section, we have considered some basic probability inequalities and laws of large numbers for sums (or averages of independent r.v.'s. In many statistical inference problems, although statistics are expressible as sums or averages of r.v.'s, these r.v.'s may not be all independent. Therefore, it is of natural interest to explore the scope of adaptability of the results of Section 2.3 to some dependent r.v.'s, and we shall consider this here.

A natural step to eliminate the assumption of independence is to bring in some weaker structure on conditional expectations. Fortunately, under such a formulation, some of the basic probability inequalities retain their validity, and we shall examine that carefully. In this context, we introduce first some basic definitions and illustrate them by suitable examples.

Definition 2.4.1 (Martingale): *Let $\{T_n; n \geq 1\}$ be a sequence of random variables. We assume that ET_n exists for every $n \geq 1$. If*

$$E[T_n \mid T_j, j \leq n-1] = T_{n-1} \quad \text{a.e.,} \quad n > 1, \qquad (2.4.1)$$

then $\{T_n\}$ is termed a martingale.

Definition 2.4.2 (Martingale array): *For each integer n (≥ 1), let $\{T_{n,k}, k \geq 1\}$ be a sequence of random variables such that $\mathrm{E}T_{n,k}$ exists for every n, k, and*

$$\mathrm{E}[T_{n,k} \mid T_{n,j}, \, j \leq k-1] = T_{n,k-1} \quad \text{a.e.,} \quad k \geq 1, \quad n \geq 1, \qquad (2.4.2)$$

then $\{T_{n,k}, k \geq 1, n \geq 1\}$ is termed a martingale array.

A more precise definition of martingales and martingale arrays may be given in terms of appropriate sub-sigma fields. However, for simplicity of presentation, we shall sacrifice some of these refinements. We consider the following examples.

Example 2.4.1: Let X_k, $k \geq 1$, be independent r.v.'s with means $\mu_k = \mathrm{E}X_k$, $k \geq 1$. Also, let $T_k = \sum_{j \leq k}\{X_j - \mu_j\}$, for $k \geq 1$. Then $T_n = T_{n-1} + (X_n - \mu_n)$, where X_n is independent of $X_1, \ldots, X_{n-1}$ (and hence of T_{n-1}). Therefore, (2.4.1) holds, so that $\{T_n\}$ is a martingale. Note that the X_k need not be identically distributed for (2.4.1) to hold. Similarly, if $\{X_{n,k}, k \geq 1, n \geq 1\}$ is a triangular scheme of (row-wise) independent r.v.'s, then (2.4.2) holds for $T_{n,k} = \sum_{j \leq k}\{X_{n,j} - \mathrm{E}X_{n,j}\}$, $k \geq 1$. Thus, the $T_{n,k}$ form a martingale array. ∎

Example 2.4.2 (Likelihood ratio statistics): Let $X_1, \ldots, X_n$ be a set of i.i.d. r.v.'s with the density function $f(x, \theta)$, $x \in \mathbb{R}$ and $\theta \in \Theta$. Then, the corresponding likelihood function is given by

$$L_n(\theta) = f(X_1, \theta) \cdots f(X_n, \theta) \quad n \geq 1, \quad \theta \in \Theta. \qquad (2.4.3)$$

For testing a null hypothesis $H_0 : \theta = \theta_0$ against an alternative $H_1 : \theta = \theta_1$, the likelihood ratio test statistic is given by

$$\lambda_n = \frac{L_n(\theta_1)}{L_n(\theta_0)} = \lambda_{n-1}\frac{f(X_n, \theta_1)}{f(X_n, \theta_0)} = \lambda_{n-1}h_n, \; n \geq 1, \qquad (2.4.4)$$

where $h_n = f(X_n, \theta_1)/f(X_n, \theta_0)$ and, conventionally, we take $\lambda_0 = 1$. Note that h_n is independent of $X_1, \ldots, X_{n-1}$ (and, hence, of $\lambda_j, \, j \leq n-1$), so that by (2.4.4), we have for all $n \geq 1$,

$$\begin{aligned}
\mathrm{E}[\lambda_n \mid \lambda_j \leq n-1] &= \lambda_{n-1}\mathrm{E}[h_n \mid \lambda_j, j \leq n-1] \\
&= \lambda_{n-1}\mathrm{E}(h_n). \qquad (2.4.5)
\end{aligned}$$

Also, by the definition of h_n,

$$\mathrm{E}_{\theta_0}[h_n] = \int h_n f(x_n, \theta_0)dx_n = \int f(x, \theta_1)dx = 1, \qquad (2.4.6)$$

so that from (2.4.5) and (2.4.6), we obtain that

$$\mathrm{E}_{\theta_0}[\lambda_n \mid \lambda_j, j \leq n-1] = \lambda_{n-1} \quad \text{a.e.,} \quad n \geq 1, \qquad (2.4.7)$$

and, hence, under H_0, $\{\lambda_n\}$ forms a martingale sequence. In this context, we may note that $\lambda_n - \lambda_{n-1} = \lambda_{n-1}\{h_n - 1\}$ is not necessarily independent of λ_{n-1}, so that the likelihood ratio statistics may not have independent summands, although they have the martingale structure under H_0. It may be noted further that if we consider the log-likelihood ratio statistics

$$W_n = \log \lambda_n = \sum_{i=1}^{n} \log h_i = \sum_{i=1}^{n} \log \left\{ \frac{f(X_i, \theta_1)}{f(X_i, \theta_0)} \right\}, \qquad (2.4.8)$$

then the summands are i.i.d. r.v.'s (regardless of whether H_0 holds), and, hence, when θ holds, $\{W_n - \mathrm{E}_\theta W_n, n \geq 1\}$ is a martingale, whenever $\mathrm{E}_\theta \log \{f(X_1, \theta_1)/f(X_1, \theta_0)\}$ exists. [Compare this with (2.4.7) where the martingale property holds for $\theta = \theta_0$.] Next, to illustrate the martingale array structure, we denote the ordered r.v.'s corresponding to $X_1, \ldots, X_n$ by $X_{n:1}, \ldots, X_{n:n}$, respectively. Note that for any $n \geq 1$, the joint density of $X_{n:1}, \ldots, X_{n:k}$, for $k \leq n$, is given by

$$L_{n,k}(\theta) = n(n-1)\cdots(n-k+1)f(X_{n:1}, \theta)\cdots f(X_{n:k}, \theta)[1 - F(X_{n:k}, \theta)]^{n-k}, \qquad (2.4.9)$$

where $F(x, \theta)$ is the distribution function corresponding to the density $f(x, \theta)$. Note that by (2.4.9), we have

$$L_{n,k+1}(\theta) = L_{n,k}(\theta) \left\{ (n-k)f(X_{n:k+1}, \theta) \frac{[1 - F(X_{n:k+1}, \theta)]^{n-k-1}}{[1 - F(X_{n:k}, \theta)]^{n-k}} \right\}, \qquad (2.4.10)$$

where defining the likelihood ratio statistics (based on $X_{n:1}, \ldots, X_{n:k}$) as in (2.4.4) and denoting it by $\lambda_{n,k}$, $k = 1, \ldots, n$ (and the corresponding h_n by $h_{n,k}$), we obtain on parallel lines that for $k = 1, \ldots, n-1$,

$$\mathrm{E}_\theta[h_{n,k+1} \mid L_{n,j}, j \leq k] = \mathrm{E}_\theta[h_{n,k+1} \mid X_{n:k}]$$

$$= \mathrm{E}_\theta \left[\frac{\lambda_{n,k+1}(\theta_1)}{\lambda_{n,k+1}(\theta_0)} \mid X_{n:k} \right]$$

$$= \int \lambda_{n,k+1}(\theta_1) \lambda_{n,k+1}(\theta) \{\lambda_{n,k+1}(\theta_0)\}^{-1} dx_{n:k+1}$$

$$= \int \lambda_{n,k+1}(\theta_1) dx_{n:k+1} = 1 \quad \text{whenever} \quad \theta = \theta_0, \quad (2.4.11)$$

so that under $\theta = \theta_0$, for every integer $n(\geq 1)$, $\{\lambda_{n,k}, k \leq n\}$ is a martingale array. ∎

A similar martingale array characterization holds for the partial sequence of log-likelihood ratio statistics [defined as in (2.4.8)] based on the order statistics. The likelihood function, as defined in (2.4.3) and (2.4.9), can be formulated in a more general manner and used for a wider class of dependent r.v.'s. In such a case too, the decomposition in (2.4.10) (in terms of the conditional density functions) or its natural extensions provide the martingale structure. In a large class of problems, as we shall see in later chapters, one may encounter the so called **likelihood ratio score statistics** (which are the derivatives of the log-likelihood function at the point θ_0) and we shall see later on that these statistics have also the martingale property. In passing, we may also remark that for the definition of the likelihood function in (2.4.3) we have taken $f(x, \theta)$ as the probability density function of the r.v. X_1 (under the assumption that the distribution function F admits a density). In case F is itself a discrete distribution (e.g., Poisson/binomial/hypergeometric etc.), we may as well work with the corresponding probability function and define the likelihood function accordingly. This will result in a summation (instead of an integral) in (2.4.7) or (2.4.11). We shall elaborate on this point in a later chapter.

Example 2.4.3: Consider a sequence $\{X_i; i \geq 1\}$ of i.i.d. r.v.'s, such that $P\{X_1 = 1\} = P\{X_1 = -1\} = 1/2$. Let $A_0 \, (> 0)$ be a positive number, and consider the model

$$Y_n = Y_{n-1} + X_n c_{n-1}, \quad n \geq 1; \quad Y_0 = A_0, \qquad (2.4.12)$$

where the bet $c_{n-1} = c_{n-1}(A_0, X_n; j \leq n-1)$ depends on the initial fortune A_0 as well as the outcome $X_j, \ j \leq n - 1$, up to the $(n - 1)$th stage, for $n \geq 1$. If $c_{n-1} = c, \ n \geq 1$, then $Y_n - Y_{n-1} \ (= cX_n)$ is independent of Y_{n-1}, and, hence, the characterization in Example 2.4.1 applies. In the general case, where the c_n are possibly r.v.'s, we may note that as the X_i are independent,

$$\begin{aligned} E[X_n c_{n-1} \mid X_j; j \leq n - 1] &= E\{E[X_n c_{n-1} \mid X_j; j \leq n - 1]\} \\ &= E\{E[X_n] c_{n-1}\} = 0, \quad n \geq 1. \end{aligned} \qquad (2.4.13)$$

Hence, we have $E[Y_n \mid Y_j, j \leq n - 1] = Y_{n-1}$ a.e., for every $n \geq 1$. Thus, $\{Y_n\}$ forms a martingale sequence, although the increments are not generally independent. Typically, in a gambling context where the model in (2.4.12) arises, c_{n-1} may be a proportion of the fortune (Y_{n-1}) at the $(n - 1)$th stage, whenever Y_{n-1} is positive, and 0 otherwise. ∎

Definition 2.4.3 (Submartingale): *A sequence $\{T_n\}$ of random variables is termed a submartingale (or supermartingale) if*

$$E[T_n \mid T_j, j \leq n - 1] \geq (\leq) T_{n-1} \quad \text{a.e.,} \quad n \geq 1. \qquad (2.4.14)$$

Before we present some examples of submartingales or supermartingales, we may note that if $\{Y_n\}$ is a martingale sequence and $g(y)$ is a convex function, then, for $T_n = g(Y_n)$, we may use a version of the Jensen Inequality

(2.2.38) adapted to the conditional expectations and obtain that

$$
\begin{aligned}
\mathrm{E}[T_n \mid T_j, j \leq n-1] \\
&= \mathrm{E}[g(Y_n) \mid g(Y_j), j \leq n-1] \\
&= \mathrm{E}[g(Y_n) \mid Y_1, \ldots, Y_{n-1}] \\
&\geq g(\mathrm{E}[Y_n \mid Y_j, j \leq n-1]) \\
&= g(Y_{n-1}) \quad \text{a.e., } n \geq 1.
\end{aligned}
\tag{2.4.15}
$$

Thus, for a martingale sequence $\{Y_n\}$, and for any convex function $g(\cdot)$, $\{T_n = g(Y_n)\}$ forms a submartingale sequence. Similarly, if $g(\cdot)$ is a concave function, we will have a supermartingale. In particular, if X_i, $i \geq 1$, are independent r.v.'s with zero means and finite pth order (absolute) moment, for some $p \geq 1$, then, on letting $Y_n = X_1 + \cdots + X_n$, $n \geq 1$, we obtain that $\{Y_n\}$ is a martingale. Let $T_n = |Y_n|^p$, $n \geq 1$. Then, $\{T_n\}$ forms a submartingale sequence. Specifically, in the case $p = 2$, for zero mean independent r.v.'s X_i, we get

$$
\mathrm{E}(Y_n^2 \mid Y_j, j \leq n-1) \geq Y_{n-1}^2 \quad \text{a.e.,} \quad n \geq 1; \quad Y_0 = 0.
\tag{2.4.16}
$$

Example 2.4.4: Let X_i, $i \geq 1$ be i.i.d. r.v.'s, and, for every $n \geq 1$, let $X_{n:n}$ (or $X_{n:1}$) denote the maximum (or minimum) of the X_j, $j \leq n$. Then, note that $X_{n:n}$ cannot be smaller than $X_{n-1:n-1}$, so that letting $T_n = X_{n:n}$, $n \geq 1$, we have $\mathrm{E}[T_n \mid T_1, \ldots, T_{n-1}] \geq T_{n-1}$ a.e., for every $n > 1$. Similarly, $\mathrm{E}[X_{n:1} \mid X_{j:1}, j \leq n-1] \leq X_{n-1:1}$, for every $n > 1$. It may be noted that in the above discussion we have tacitly assumed that $\mathrm{E}X_{n:n}$ exists (which may not always be true); if $\mathrm{E}X_{n:n}$ is not finite (we take that as $+\infty$), the inequality may still be interpreted in a meaningful way. A similar modification holds for $X_{n:1}$ where we take $\mathrm{E}X_{n:1} = -\infty$ whenever it does not exist.   ∎

Example 2.4.5: In Example 2.4.2, for the sequence $\{\lambda_n\}$ of likelihood ratio statistics in (2.4.4), define $T_{n,p} = \lambda_n^p$, for some $p > 0$. Note that by the Jensen Inequality (2.2.38),

$$
\mathrm{E}_{\theta_0}[h_n^p] \geq \{\mathrm{E}_{\theta_0}[h_n]\}^p, \quad p \geq 1,
\tag{2.4.17}
$$

and the opposite inequality holds for $p \in (0, 1)$. Therefore, proceeding as in (2.4.4) through (2.4.7), but using (2.4.17) instead of (2.4.6), we obtain that

i) for every (fixed) $p \geq 1$, $\{T_{n,p}\}$ is a submartingale,

ii) for $p \in (0, 1)$, $\{T_{n,p}\}$ is a supermartingale.   (2.4.18)

                                                                                    ∎

As in Definition 2.4.2, the definition of submartingales or supermartingales may also be extended to triangular schemes. However, we do not repeat the details. In passing, we may note that if $\{T_n\}$ is a submartingale, we may write

$$
\begin{aligned}
T_n &= \sum_{j=0}^{n-1} \{(T_{j+1} - \mathrm{E}[T_{j+1} \mid T_r, r \leq j]) + (\mathrm{E}[T_{j+1} \mid T_r, r \leq j] - T_j)\} \\
&= \sum_{j \leq n-1} Y_{j+1} + \sum_{j<n} U_j \\
&= T_{n0} + T_{n1}, \quad n \geq 1,
\end{aligned}
\tag{2.4.19}
$$

where $T_{n0} = \sum_{j \leq n-1} Y_{j+1}$ and $T_{n1} = \sum_{j<n} U_j$. By definition, T_{n0} forms a martingale sequence, whereas, by (2.4.14), the T_{n1} are non-negative and nondecreasing in n. Thus, we have the following.

Proposition 2.4.1: *A submartingale $\{T_n\}$ can be decomposed into a martingale $\{T_{n0}\}$ and a residual $\{T_{n1}\}$, where T_{n1} is a.s. nonnegative and nondecreasing in n. [Note that T_{n1} depends only on the T_j, $j \leq n - 1$.]*

Definition 2.4.4 (Reverse martingale): *A sequence $\{T_n\}$ of random variables (or vectors or more general elements) is called a reverse martingale (or reverse submartingale) if for every n*

$$
\mathrm{E}[T_n \mid T_{n+1}, T_{n+2}, \ldots] = (\geq)T_{n+1} \quad \text{a.e.}
\tag{2.4.20}
$$

Here also, by using the Jensen Inequality (2.2.38), we may conclude that if $\{T_n\}$ is a reverse martingale and if $g(\cdot)$ is a convex function, then $\{g(T_n)\}$ is a reverse submartingale. We consider the following illustrations.

Example 2.4.6: Let $\{X_i, i \geq 1\}$ be a sequence of i.i.d. r.v.'s with finite mean μ. Consider the sequence $T_n = \overline{X}_n$, $n \geq 1$. Then, for every $n \geq 1$,

$$
\mathrm{E}[\overline{X}_n \mid \overline{X}_{n+1}, \overline{X}_{n+2}, \ldots] = \mathrm{E}[\overline{X}_n \mid \sum_{j=1}^{n+1} X_j, X_{n+2}, X_{n+3}, \ldots].
\tag{2.4.21}
$$

For every n (≥ 1), let $X_{n:1}, \ldots, X_{n:n}$ be the ordered r.v.'s corresponding to $X_1, \ldots, X_n$; in this context, there is no need for these order statistics to be all distinct (so that we are not considering the case of continuous distributions only). Note that $\sum_{j=1}^{n+1} X_j = \sum_{j=1}^{n+1} X_{n+1:j}$ and is a function of the order statistics. Also, for all $j = 1, \ldots, n + 1$, we have

$$
P\left\{ X_k = X_{n+1:j} \mid \sum_{i=1}^{n+1} X_i, X_{n+2}, \ldots \right\}
$$

$$
= P\left\{ X_k = X_{n+1:j} \mid \sum_{i=1}^{n+1} X_{n+1:i}, X_{n+2}, \ldots \right\}
$$

$$= P\{X_k = X_{n+1:j} \mid X_{n+1:1}, \ldots, X_{n+1:n+1}\}$$
$$= (n+1)^{-1}, \quad k = 1, \ldots, n+1, \tag{2.4.22}$$

where the last step follows from the fact that given $(X_{n+1:1}, \ldots, X_{n+n:n+1})$, $(X_1, \ldots, X_{n+1})$ has the discrete uniform distribution over the $(n+1)!$ permutations of the order statistics. Therefore, by (2.4.21) and (2.4.22), we obtain that for all $n \geq 1$,

$$
\begin{aligned}
\mathrm{E}[\overline{X}_n \mid \overline{X}_{n+1}, \overline{X}_{n+2}, \ldots] &= n^{-1} \sum_{k=1}^{n} \mathrm{E}[X_k \mid \overline{X}_{n+1}, \overline{X}_{n+2}, \ldots] \\
&= \mathrm{E}\left[X_1 \mid \sum_{j=1}^{n+1} X_{n+1:j}, X_{n+2}, \ldots\right] \\
&= (n+1)^{-1} \sum_{j=1}^{n+1} X_{n+1:j} \\
&= (n+1)^{-1} \sum_{i=1}^{n+1} X_i = \overline{X}_{n+1} \text{ a.e..} \tag{2.4.23}
\end{aligned}
$$

Thus, $\{\overline{X}_n, n \geq 1\}$ is a reverse martingale. Note that the same treatment holds for $\overline{X}_n - a$, $n \geq 1$, for an arbitrary a, and, hence,

$$\{\overline{X} - \mu, \ n \geq 1\} \quad \text{is a reverse martingale.} \tag{2.4.24}$$

By (2.4.24) and the Jensen Inequality (2.3.38), we obtain that for every (fixed) $p \geq 1$,

$$\{|\overline{X}_n - \mu|^p, \ n \geq 1\} \quad \text{is a reverse submartingale.} \tag{2.4.25}$$

∎

Example 2.4.7 (U-statistics): Let X_i, $i \geq 1$, be i.i.d. r.v.'s with a d.f. F, defined on $\mathbb{R}^p$, for some $p \geq 1$. Consider a functional $\theta = \theta(F)$ defined by

$$\theta(F) = \int \cdots \int g(x_1, \ldots, x_m) dF(x_1) \cdots dF(x_m), \tag{2.4.26}$$

where $g(x_1, \ldots, x_m)$ is a symmetric function of its m (≥ 1) arguments. For example, if we let $g(x) = x$, i.e., $m = 1$, we have $\theta(F) = \mathrm{E}X_1 = \mu$. Similarly, if we take $m = 2$ and $g(x_1, x_2) = (x_1 - x_2)^2/2$, then $\theta(F) = \mathrm{E}(X_1 - X_2)^2/2 = \mathrm{E}[(X_1 - \mu) - (X_2 - \mu)]^2/2 = \sigma^2 = \mathrm{E}(X - \mu)^2$. In this way, a large class of parameters may be formulated as functionals of the underlying d.f. Note that (2.4.26) is equivalent to

$$\theta(F) = \mathrm{E}[g(X_1, \ldots, X_m)], \quad \text{for all } F \text{ belonging to a class F.} \tag{2.4.27}$$

In this form, $g(\cdot)$ is called a (symmetric) **kernel** and m the **degree** of the parameter. Corresponding to the n sample observations $X_1, \ldots, X_n$, we may take any subsample of size m (say, $X_{i_1}, \ldots, X_{i_m}$) and estimate $\theta(F)$ by $g(X_{i_1}, \ldots, X_{i_m})$ whenever $n \geq m$. Thus, a symmetric and unbiased estimator of $\theta(F)$ can be obtained by combining all these unbiased estimators. This is termed a **U-statistic** and is given by

$$U_n = \binom{n}{m}^{-1} \sum_{\{1 \leq i_1 < \cdots < i_m \leq n\}} g(X_{i_1}, \ldots, X_{i_m}). \tag{2.4.28}$$

For the case of $\theta(F) = \mu$, it is easy to verify that $U_n = \overline{X}_n$, $n \geq 1$. Also, for the case of $\theta(F) = \sigma^2$, we obtain, on using $g(a, b) = (a - b)^2/2$, that $U_n = S_n^2$.

As can be easily verified with the case of S_n^2 for $m \geq 2$, the summands in (2.4.28) are not all independent r.v.'s. As a result, the probability inequalities and the LLN developed in the earlier sections may not be directly applicable for U-statistics. However, we may note that, as in Example 2.4.6, we have here, for every $n \geq m$,

$$
\begin{aligned}
\mathrm{E}[U_n \mid U_{n+1}, U_{n+2}, \ldots] &= \mathrm{E}[g(X_1, \ldots, X_m) \mid U_{n+1}, U_{n+2}, \ldots] \\
&= \binom{n+1}{m}^{-1} \sum_{\{1 \leq i_1 < \cdots < i_m \leq n+1\}} g(X_{i_1}, \ldots, X_{i_m}) \\
&= U_{n+1} \quad \text{a.e.,} \tag{2.4.29}
\end{aligned}
$$

so that $\{U_n, n \geq m\}$ is a reverse martingale. Along the same lines of (2.4.24) and (2.4.25), we have for general U-statistics:

$$\{U_n - \theta(F), n \geq m\} \quad \text{is a reverse martingale,} \tag{2.4.30}$$

$$\{|U_n - \theta(F)|^p, n \geq m\} \quad \text{is a reverse submartingale for every } p \geq 1. \tag{2.4.31}$$

In passing, we may remark that the arguments used in (2.4.22) in favor of the order statistics easily extends to the vector case, where the order statistics are to be replaced by the collection of vectors, leading to the same permutation distribution of the indices $(s_{n+1,1}, \ldots, s_{n+1,n+1})$ over the set $\{1, \ldots, n+1\}$. Hence, the reverse martingale and submartingale results for U-statistics remain good even in the case where the X_i are random vectors and/or the kernel $g(\cdot)$ is a vector. ∎

Example 2.4.8 (Empirical d.f.): Let $X_1, \ldots, X_n$ be i.i.d. r.v.'s with the true d.f. F, defined on the real line $\mathbb{R}$ and F_n be the corresponding empirical d.f. defined in (2.2.10). Note that for all $x, y \in \mathbb{R}$,

$$\mathrm{E}F_n(x) = F(x) \quad \text{and} \quad \mathrm{E}\left\{I_{\{X_1 \leq x\}} I_{\{X_1 \leq y\}}\right\} = F(x \wedge y). \tag{2.4.32}$$

Letting $g_x(X_i) = I_{\{X_i \le x\}}$, for $x \in \mathbb{R}$, we obtain, as in (2.4.29), that for every (fixed) $x \in \mathbb{R}$,

$$E[F_n(x) \mid F_{n+1}(x), F_{n+2}(x), \ldots] = F_{n+1}(x) \quad \text{a.e.,} \quad n \ge 1. \qquad (2.4.33)$$

Also, we shall see in Chapter 4 that there is an one-to-one correspondence between the order statistics and the empirical d.f., so that the tail sequence $F_{n+1}, F_{n+2}, \ldots$ is equivalently defined in terms of the order statistics $X_{n+1:j}$, $j = 1, \ldots, n+1$, and by the X_{n+j}, $j \ge 2$. Consequently, the point-wise reverse martingale structure in (2.4.33) extends readily to the entire line, i.e. for all $n \ge 1$,

$$F[\{F_n(x), x \in \mathbb{R}\} \mid F_{n+1}, F_{n+2}, \ldots] = \{F_{n+1}(x), x \in \mathbb{R}\} \quad \text{a.e.} \quad (2.4.34)$$

Thus, the empirical d.f. is a reverse martingale (process). In fact, we have

$$\{F_n(x) - F(x), x \in \mathbb{R}\}, n \ge 1, \quad \text{is a reverse martingale process;}$$
$$(2.4.35)$$
$$\{|F_n(x) - F(x)|, x \in \mathbb{R}\}, n \ge 1, \quad \text{is a reverse submartingale process.}$$
$$(2.4.36)$$

Moreover, we have that $\sup\{f(x) : x \in \mathbb{R}\}$ and $\sup\{|f(x)| : x \in \mathbb{R}\}$ are both convex functions, and, hence, using (2.4.35) and the Jensen Inequality (2.3.38), we obtain that

$$\sup_{x \in R}[F_n(x) - F(x)] \text{ and } \sup_{x \in R}|F_n(x) - F(x)| \quad \text{are reverse submartingales.}$$
$$(2.4.37)$$

Similar characterizations hold for the multivariate r.v.'s also.   ∎

Example 2.4.9 (Finite population sampling): Let $A_N = \{a_1, \ldots, a_N\}$ stand for the vector of N observations in a finite population. A simple random sample of size n is drawn without replacement, and the sample observations are denoted by $X_1, \ldots, X_n$. Note that we may write $X_i = a_{R_i}$, $i = 1, \ldots, n$, where $R_1, \ldots, R_n$ can assume values in any subset of n integers out of the N natural numbers $1, \ldots, N$ with the common probability $(N^{[n]})^{-1}$, where $N^{[n]} = N \cdots (N - n + 1)$. Thus, we may formally write for $1 \le R_1 \ne \cdots \ne R_n \le N$,

$$P\{X_1 = a_{R_1}, \ldots, X_n = a_{R_n}\} = \{N \cdots (N - n + 1)\}^{-1}. \qquad (2.4.38)$$

Clearly, the X_i are not independent; they are, however, exchangeable r.v.'s. From the sample, we may compute the sample mean $\overline{X}_n$ or the variance S_n^2, and we may want to study their stochastic convergence properties. If we examine carefully (2.4.38), we shall see that it leads to the same conditional law in (2.4.24), and, as a result, in simple random sampling without replacement, U-statistics have the reverse martingale property (parallel to the case of i.i.d. r.v.'s). Thus, (2.4.29), (2.4.30) and (2.4.31) all remain valid in this case. If the sample units are drawn with replacement, then, of course, $X_1, \ldots, X_n$ are i.i.d. r.v.'s with $P\{X_i = a_k\} = N^{-1}$, for every k $(= 1, \ldots, N)$ and $i \geq 1$, and, hence, the characterization in (2.4.29)–(2.4.31) remains valid. ∎

There are other examples of reverse martingales and submartingales, and we shall consider some of them in later chapters. Next we consider the following.

Theorem 2.4.1 (Extension of the Kolmogorov Maximal Inequality): *Let* $\{T_n, n \geq 1\}$ *be a (zero mean) martingale such that* $\mathrm{E}(T_n^2)$ *exists, for every* $n \geq 1$. *Then, for every* $t > 0$, *we have*

$$P\left\{\max_{1 \leq k \leq n} |T_k| > t\right\} \leq t^{-2}\mathrm{E}(T_n^2). \qquad (2.4.39)$$

If $\{T_n, n \geq m\}$ *is a (zero mean) reverse martingale, then, for every* $N \geq n \geq m$,

$$P\left\{\max_{n \leq k \leq N} |T_k| > t\right\} \leq t^{-2}\mathrm{E}(T_n^2), \quad t > 0. \qquad (2.4.40)$$

Proof. First, consider the case of martingales. We follow the proof of the Kolmogorov Maximal Inequality (2.3.8) and we need to verify only that (2.3.58) holds for the martingales. For this, it suffices to show that $\mathrm{E}(T_n^2 \mid T_j, j \leq k) \geq T_k^2$ a.e., for every $k \leq n$, and this has already been proved in (2.4.16). Next, consider the case of reverse martingales. We write $S_1 = T_N, S_2 = T_{N-1}, \ldots, S_{N-n+1} = T_n$ and observe that the reverse martingale property of $\{T_n\}$ yields that the $\{S_k\}$ form a forward martingale. As such, (2.4.40) follows from (2.4.39) as adapted to the partial sequence $\{S_1, \ldots, S_{N-n+1}\}$. ∎

We may note that for the inequalities in (2.4.39) and (2.4.40), we do not need the $\{T_n\}$ to be a zero mean sequence. However, in the case of a zero mean sequence, we may as well replace $\mathrm{E}(T_n^2)$ by $\mathrm{Var}(T_n)$. In actual statistical applications, with suitable choice of t, the rhs of (2.4.39) or (2.4.40) can be made small when T_n has zero mean, but for non zero mean, the contribution of $[\mathrm{E}(T_n)/t]^2$ may make the bound of little utility. Further, it is not necessary to assume that $\{T_n\}$ is a martingale (or reverse martingale) sequence; we may as well take for $\{T_n\}$ a submartingale (or a reverse submartingale) property; this follows simply by noting that $[\max_{1 \leq k \leq n} |T_k| > t]$ is equivalent to $[\max_{1 \leq k \leq n} T_k^2 > t^2]$, and noting that the T_k^2 have the submartingale property. In fact, if $\{T_n\}$ is a submartingale and $g(\cdot)$ is any non-negative (nondecreasing) convex function such that $\mathrm{E}g(T_n)$ exists, then, by (2.4.15), $\{g(T_n)\}$ is a submartingale, and, hence, using Theorem 2.4.1 on these $g(T_k)$, we immediately obtain that for any positive t,

$$P\left\{\max_{1\leq k\leq n} T_k > t\right\} \leq \frac{\mathrm{E}\{g(T_n)\}}{g(t)}. \qquad (2.4.41)$$

In particular, if T_n has a finite moment generating function $M_n(\theta) = \mathrm{E}(e^{\theta T_n})$, for all $\theta \in (0, \theta_0)$, for some $\theta_0 > 0$, then, letting $g(x) = e^{\theta x}$, we obtain from (2.4.41) the following inequality [compare with the Bernstein Inequality (2.3.7)].

Bernstein Inequality for Submartingales: *Let $\{T_n\}$ be a submartingale such that $M_n(\theta) = \mathrm{E}(e^{\theta T_n})$ exists for all $\theta \in (0, \theta_0)$, for some $\theta_0 > 0$. Then, for every $t > 0$,*

$$P\left\{\max_{1\leq k\leq n} T_k > t\right\} \leq \inf_{\theta>0}\left\{e^{-\theta t} M_n(\theta)\right\}. \qquad (2.4.42)$$

If $\{T_n\}$ is a reverse submartingale such that $M_n(\theta)$ exists for all $\theta \in (0, \theta_0)$, then, for every $t > 0$,

$$P\left\{\sup_{N\geq n} T_N > t\right\} \leq \inf_{\theta>0}\left\{e^{-\theta t} M_n(\theta)\right\}. \qquad (2.4.43)$$

Example 2.4.10: Let X_i, $i \geq 1$ be i.i.d. r.v.'s with $\mathrm{E}X_1 = 0$ (without any loss of generality), and assume that $\mathrm{E}|X_1|^p < \infty$, for some $p \geq 1$. Then let $T_n = \overline{X}_n$, $n \geq 1$. Note that $\{\overline{X}_n, n \geq 1\}$ is a reverse martingale, so that $\{|\overline{X}_n|^p, n \geq 1\}$ is a reverse submartingale. Consequently, by a parallel version of (2.4.41) for reverse submartingales, we immediately obtain that for every $\varepsilon > 0$

$$P\left\{\sup_{N\geq n} |\overline{X}_n| > \varepsilon\right\} \leq \varepsilon^{-p}\mathrm{E}|\overline{X}_n|^p. \qquad (2.4.44)$$

On the other hand, $\mathrm{E}|X_1|^p < \infty$ implies $\mathrm{E}|\overline{X}_n|^p \to 0$ as $n \to \infty$, and, hence, the rhs of (2.4.44) converges to 0 as $n \to \infty$. Thus, $\overline{X}_n \xrightarrow{\text{a.s.}} 0$. In this context, we may note that for $p = 2$, $\mathrm{E}(\overline{X}_n^2) = \mathrm{Var}(\overline{X}_n) = n^{-1}\mathrm{Var}(X_1) = n^{-1}\mathrm{E}X_1^2$, whereas, for $p \in (1, 2)$, $\mathrm{E}|\overline{X}_n|^p = o(n^{1-p}) \to 0$, as $n \to \infty$; we may refer to Pyke and Root (1968) for some detailed treatment of these related L_p-convergence results.   ∎

Example 2.4.11: If the X_i are i.i.d. r.v.'s with $P(X_1 = 1) = 1 - P(X_1 = 0) = \pi$ $(0 < \pi < 1)$, then, for $T_n = \overline{X}_n$, we have noticed in Section 2.3 that the moment generating function exists, whereas the reverse martingale structure follows from the preceding example. Hence, using (2.4.43) we readily conclude that the Bernstein Inequality actually applies here to the entire tail, and, hence, $\overline{X}_n \xrightarrow{\text{a.s.}} \pi$ at an exponential rate.   ∎

Hájek-Rènyi-Chow Inequality: *Let $\{T_n, n \geq 1\}$ be a submartingale and let $\{c_n^*, n \geq 1\}$ be a nonincreasing sequence of positive numbers. Let $T_n^+ = \max\{T_n, 0\}$, $n \geq 1$, and assume that $\mathrm{E}T_n^+$ exists for every $n \geq 1$. Then, for every $\varepsilon > 0$,*

$$P\left\{\max_{1 \leq k \leq n} c_k^* T_k > \varepsilon\right\} \leq \varepsilon^{-1}\left[c_n^* \mathrm{E}T_1^+ + \sum_{k=2}^{n} c_k^* \mathrm{E}(T_k^+ - T_{k-1}^+)\right]. \quad (2.4.45)$$

If $\{T_n, n \geq 1\}$ is a reverse submartingale and the c_k^* are nondecreasing, we have for every $N \geq n$,

$$P\left\{\max_{n \leq k \leq N} c_k^* T_k > \varepsilon\right\} \leq \varepsilon^{-1}\left[c_n^* \mathrm{E}T_n^+ + \sum_{k=n+1}^{N} (c_k^* - c_{k-1}^*)\mathrm{E}T_k^+\right]. \quad (2.4.46)$$

[Note that if we take X_i, $i \geq 1$, as independent r.v.'s with zero mean, then $T_n = (X_1 + \cdots + X_n)^2$ is a non-negative submartingale (so that $T_n^+ = T_n$), and then (2.4.45) reduces to (2.3.60), with $c_k^* = c_k^2$, $k \geq 1$.] The proof of (2.4.45) consists in verifying (2.3.64) with the S_k^2 being replaced by T_k^+ and c_k^2 by c_k^*. We omit the details in view of the similarity and the basic definition in (2.4.14). The case of reverse submartingales can be treated as in the forward submartingale case after reversing the order of the index set (as we have done in the proof of Theorem 2.4.1).

In the case of independent r.v.'s, we have observed that the Kolmogorov SLLN (Theorem 2.3.10) provides the a.s. convergence result under a simple condition that $\sum_{k \geq 1} k^{-2}\sigma_k^2$ converges [see (2.3.66)]. Actually, the second moment condition in (2.3.66) can be relaxed and the independence assumption may as well be replaced by a martingale structure. Toward this, we present the following:

Theorem 2.4.2 (Kolmogorov SLLN for martingales): *First consider a martingale $\left\{T_n = \sum_{k \leq n} Y_k, n \geq 1\right\}$ such that for some p $(1 \leq p \leq 2)$, $\mathrm{E}|Y_k|^p$ exists for every $k \geq 1$, i.e., $\{T_n, n \geq 1\}$ is a zero mean L_p martingale. Also assume that there is a sequence $\{b_n, n \geq 1\}$ of increasing positive numbers such that $b_n \to \infty$ as $n \to \infty$, and*

$$\sum_{n \geq 2} b_n^{-p} \mathrm{E}[|Y_n|^p \mid Y_j, j < n] < \infty \quad \text{a.s.} \quad (2.4.47)$$

Then

$$b_n^{-1} T_n \xrightarrow{\text{a.s.}} 0. \quad (2.4.48)$$

We may note that if the Y_k are independent, then $\mathrm{E}[|Y_k|^p \mid Y_j, j < k] = \mathrm{E}[|Y_k|^p]$, for every $k \geq 2$, so that (2.4.47) reduces to (2.3.66) when we take $p = 2$ and $b_n = n$. For martingales, the conditional expectations are generally r.v.'s, and, hence, we need the series in (2.4.47) to converge a.s. It may also be noted that if the Y_k are independent or, more generally, if for some $p > 1$, $E[|Y_k|^p \mid Y_j < k] \leq c < \infty$, for every $k \geq 2$, then noting that $\sum_{n \geq 2} n^{-p} < \infty$, for every $p > 1$, we obtain that $n^{-1} T_n \xrightarrow{\text{a.s.}} 0$. Thus, the second moment condition in (2.3.66) is not that crucial for the SLLN to hold. For the i.i.d. case, a slightly more stringent condition, such as the one in (2.4.47), suffices.

Looking back at Example 2.4.10 and the reverse submartingale inequality, we may conclude that the following general result holds:

Theorem 2.4.3: *For a reverse submartingale $\{T_n, n \geq m\}$, there exists an (extended) r.v. T, such that $T_n - T \xrightarrow{\text{a.s.}} 0$. If $\{T_n\}$ is a reverse martingale, then $T_n - T \to 0$ a.s., as well as in the first mean, as $n \to \infty$.*

In simple random sampling (with or without replacement) from a finite population, or for i.i.d. r.v.'s, we have observed earlier that U-statistics form a reverse martingale. As such, we are in a position to use (2.4.46), i.e., the Hájek-Rènyi-Chow Inequality for U-statistics. In order to make use of the submartingale structure in (2.4.46), we may take $T_n = [U_n - EU_n]^2$, $n \geq m$, and $c_k^* = c_k^2$. We may, of course, use $T_n = |T_n - EU_n|^p$, for some $p \geq 1$. Thus, allowing in (2.4.46) N to be indefinitely large, we arrive at the a.s. convergence of U_n to $\theta = EU_n$. For $p > 1$, we may use the L_p convergence of U-statistics and claim that $E|U_n - \theta|^p = O(n^{-r(p)})$, where $r(p) = p - 1$ if $p \in (1, 2]$, and $p/2$ for $p \geq 2$, so that the rhs of (2.4.46) converges to 0 as $n \to \infty$. On the other hand, the treatment is a bit more delicate for $p = 1$. In this context, we first note that a tail event is an event whose probability remains the same when a finite number of the X_n are changed, i.e., the event belongs to the tail sigma-field of the X_n. Similarly, a tail function does not depend on any finite segment of the X_n. The celebrated Zero-One Law can be presented as follows.

Theorem 2.4.4 (Kolmogorov Zero-One Law): *Let $\{X_n\}$ be a sequence of independent r.v.'s. Then, the probability of any tail event is either 0 or 1, and any tail function is a constant with probability 1.*

To conceive of such tail functions, we may consider, for an arbitrary function $g(\cdot)$,

  i) $\limsup_{n \to \infty} g(X_n)$,

  ii) $\liminf_{n \to \infty} g(X_n)$,

  iii) $\limsup_{n \to \infty} n^{-1} \sum_{i=1}^n g(X_i)$,

etc. We have also noticed that the U-statistics are symmetric functions of the sample observations (i.e., they remain invariant under any permutation of the indices $1, \ldots, n$ of the r.v.'s $X_1, \ldots, X_n$). Such a function is termed **exchangeable**.

Theorem 2.4.5 (Hewitt-Savage Zero-One Law): *Let $\{X_i, i \geq 1\}$ be i.i.d. r.v.'s. Then every exchangeable event has probability either equal to 0 or 1.*

Coming back to the case of U-statistics, we observe that by virtue of Theorem 2.4.3, $U_n - U \xrightarrow{\text{a.s.}} 0$ where by the Hewitt-Savage Zero-One Law, we claim that $U = \theta$ with probability 1. Hence, $U_n \xrightarrow{\text{a.s.}} \theta$.

2.5 Some miscellaneous convergence results

Let us first consider the convergence of a series of r.v.'s. For a sequence $\{a_n\}$ of real numbers, we say that the series $\sum_{n \geq 1} a_n$ converges if for every $\varepsilon > 0$, there exists a positive integer $n_0 = n_0(\varepsilon)$, such that

$$| \sum_{n=m+1}^{m+N} a_n | \leq \varepsilon, \quad m \geq n_0, \ N \geq 1. \tag{2.5.1}$$

If instead of the sequence $\{a_n\}$ of real numbers, we have a sequence of r.v.'s, say $\{X_n\}$, and we define the partial sums as

$$T_n = X_1 + \cdots + X_n, \quad n \geq 1, \tag{2.5.2}$$

then a natural question may arise: Does T_n converge (in a meaningful sense) as $n \to \infty$? In this context, we may note that the T_n are themselves r.v.'s, and, hence, T_n may not converge to a nonstochastic limit as $n \to \infty$; rather, it may have some nondegenerate distribution. Thus, we may need to interpret the convergence of a series of r.v.'s in a different manner. It seems quite logical to incorporate the basic requirement in (2.5.1) in a stochastic sense and to define the desired convergence result as follows.

Definition 2.5.1: *A series T_n, $n \geq 1$, defined as in (2.5.2), converges almost surely if for every given positive numbers ε and η, there exists a positive integer $n_0 = n_0(\varepsilon, \eta)$, such that*

$$P \left\{ \max_{m+1 \leq n \leq m+N} | \sum_{i=m+1}^{n} X_i | > \varepsilon \right\} < \eta, \quad m \geq n_0, \ N \geq 1. \tag{2.5.3}$$

Note that this definition allows the T_n to retain their stochastic nature, but implies that the tail contribution becomes a.s. negligible as n is increased. Note that if the X_i are i.i.d. r.v.'s with a nondegenerate distribution F, then the series T_n does not converge a.s. in the sense of (2.5.3) (although

$n^{-1}T_n$ converges a.s. to EX whenever the latter exists). However, as we shall see later on, there are situations involving independent but possibly nonidentically distributed r.v.'s where this mode of convergence holds and has suitable applications in other problems too.

Theorem 2.5.1: *Let $\{X_i, i \geq 1\}$ be a sequence of independent r.v.'s, such that $EX_k = 0$ and $\mathrm{Var}(X_k) = \sigma_k^2 < \infty$, for all $k \geq 1$. If $\sum_{k \geq 1} \sigma_k^2 < \infty$, then $\sum_{n \geq 1} X_n$ converges a.s.*

Proof. It follows from the Kolmogorov Maximal Inequality in (2.3.54) that for every $n > 0$ and all $m \geq 1$, $N \geq 1$,

$$P\left\{ \max_{m+1 \leq n \leq m+N} \left| \sum_{k=m+1}^{m+N} X_k \right| > \eta \right\} \leq \eta^{-2} \sum_{k=m+1}^{m+N} \sigma_k^2 \leq \eta^{-2} \sum_{k > m} \sigma_k^2.$$
(2.5.4)

Since the series $\sum_{k \geq 1} \sigma_k^2$ converges, for every $\varepsilon > 0$, there exists an $n_0 = n_0(\varepsilon, \eta)$, such that $\sum_{i > n_0} \sigma_k^2 \leq \eta^2 \varepsilon$, and, hence, the result follows from (2.5.4). ∎

Example 2.5.1: Let Z_i, $i \geq 1$ be i.i.d. r.v.'s having the standard normal distribution and let $X_k = k^{-1} Z_k$, $k \geq 1$. Then $EX_k = 0$ and $EX_k^2 = \mathrm{Var}(X_k) = \sigma_k^2 = k^{-2}$, $k \geq 1$. Thus, $\sum_{k \geq 1} \sigma_k^2 = (\pi^2/6) < \infty$, and, hence, by Theorem 2.5.1 we may conclude that the series $\sum_{k \geq 1} X_k$ converges a.s. In fact, here $\sum_{k \geq 1} X_k$ has a normal distribution with zero mean and variance $\pi^2/6$. ∎

Before we proceed to consider the converse of Theorem 2.5.1, we present the following probability inequality; a proof of this may be found in most books in Probability Theory [it runs along the same lines as in the proof of (2.3.54)] and is omitted.

Kolmogorov Maximal Inequality (Lower Bound): *Let $X_1, \ldots, X_n$ be independent r.v.'s, such that for some finite positive c, $P\{|X_i| \leq c\} = 1$, for every $i \geq 1$. Let $T_k = \sum_{i \leq k} \{X_i - EX_i\}$, $k \geq 1$. Then, for every $\varepsilon > 0$,*

$$P\left\{ \max_{1 \leq k \leq n} |T_k| \geq \varepsilon \right\} \geq \frac{1 - (\varepsilon + 2c)^2}{\sum_{i=1}^{n} \sigma_i^2 + (\varepsilon + 2c)^2 - \varepsilon^2}.$$
(2.5.5)

It is of interest to compare (2.3.54) and (2.5.5). The former provided an upper bound, whereas the latter provides a lower bound under an additional condition that the r.v.'s are all bounded. We use (2.5.5) in the following:

Theorem 2.5.2: *Let $\{X_i, i \geq 1\}$ be a sequence of independent r.v.'s, such that $P\{|X_k| \leq c\} = 1$, for all k, for some finite c (> 0). Then the series*

$\sum_{k\geq 1} \{X_k - EX_k\}$ converges a.s., iff $\sum_{k\geq 1} \text{Var}(X_k) < \infty$.

Proof: Writing $Y_k = X_k - EX_k$, $k \geq 1$, the "if" part of the theorem follows directly from Theorem 2.5.1. For the "only if" part, we make use of (2.5.5) and have

$$P\left\{ \max_{m+1\leq n\leq m+N} \Big| \sum_{i=m+1}^{m+N} Y_i \Big| > \varepsilon \right\} \geq \frac{1 - (\varepsilon + 2c)^2}{\sum_{i=m+1}^{m+N} \sigma_i^2 + (\varepsilon + 2c)^2 - \varepsilon^2}. \quad (2.5.6)$$

If $\sum_{k\geq 1} \sigma_k^2 = +\infty$, (2.5.6) would imply (on letting $N \to \infty$) that

$$P\left\{ \max_{m<n\leq m+N} \Big| \sum_{i=m+1}^{m+N} Y_i \Big| > \varepsilon \right\} = 1,$$

contradicting the hypothesis of a.s. convergence of $\sum_{k>1} Y_k$. Therefore, we must have $\sum_{k\geq 1} \sigma_k^2 < \infty$. ∎

Theorem 2.5.3: Let $\{X_k; k \geq 1\}$ be independent r.v.'s, such that for some finite c (> 0), $P\{|X_k| \leq c\} = 1$, for all k. Then $\sum_{k\geq 1} X_k$ converges a.s. iff $\sum_{k\geq 1} EX_k < \infty$ and $\sum_{k\geq 1} \text{Var}(X_k) < \infty$.

Proof: If $\sum_{k\geq 1} \text{Var}(X_k) < \infty$, by Theorem 2.5.2., $\sum_{k\geq 1} \{X_k - EX_k\}$ converges a.s., so that the convergence of the series $\sum_{k\geq 1} EX_k$ ensures the a.s. convergence of $\sum_{k\geq 1} X_k$. Suppose next that $\sum_{k\geq 1} X_k$ converges a.s. Consider another sequence $\{X_k^*; k \geq 1\}$ of independent r.v.'s, such that for every $k \geq 1$, X_k^* has the same distribution as $(-1)X_k$. Let $Z_k = X_k + X_k^*$, $k \geq 1$; then $EZ_k = 0$ and $\text{Var}(Z_k) = 2\text{Var}(X_k)$, for $k \geq 1$. Since $\sum_{k\geq 1} Z_k| \leq |\sum_{k\geq 1} X_k| + |\sum_{k\geq 1} X_k^*|$, it follows that the series $\sum_{k\geq 1} Z_k$ converges a.s. whenever $\sum_{k\geq 1} X_k$ does so. Therefore by Theorem 2.5.2., we conclude that $\sum_{k\geq 1} \text{Var}(Z_k) = 2\sum_{k\geq 1} \text{Var}(X_k) < \infty$. Also, this, in turn [by Theorem 2.5.1], implies that $\sum_{k\geq 1} \{X_k - EX_k\}$ converges a.s., and the a.s. convergence of $\sum_{k\geq 1} X_k$ and $\sum_{k\geq 1} \{X_k - EX_k\}$ implies that the series $\sum_{k\geq 1} EX_k$ converges. ∎

We proceed now to formulate a basic theorem on the a.s. convergence of a series of r.v.'s. In this context, for every $c > 0$, we define

$$X_i^c = X_i I_{\{|X_i|\leq c\}}, \, i \geq 1. \quad (2.5.7)$$

Theorem 2.5.4 (Kolmogorov Three Series Criterion): Let X_i, $i \geq 1$ be independent r.v.'s. Then $\sum_{n=1}^{\infty} X_n$ converges a.s. if and only if the *following three conditions hold:*

$$\sum_{n\geq 1} P\{|X_n| > c\} < \infty, \tag{2.5.8}$$

$$\sum_{n\geq 1} \mathrm{E}(X_n^c) < \infty, \tag{2.5.9}$$

$$\sum_{n\geq 1} \mathrm{Var}(X_n^c) < \infty. \tag{2.5.10}$$

Proof: First assume that (2.5.8) and (2.5.9) hold. Since

$$P\{X_n \neq X_n^c\} = P\{|X_n| > c\},$$

by (2.5.8) and the Khintchine Equivalence Lemma (Theorem 2.3.12), it follows that $\sum_{n\geq 1} X_n$ and $\sum_{n\geq 1} X_n^c$ are convergence-equivalent. Since the X_n^c are bounded r.v.'s, by Theorem 2.5.3, we obtain that under (2.5.9) and (2.5.10), $\sum_{n\geq 1} X_n$ converges a.s. Next, suppose that $\sum_{n\geq 1} X_n$ converges a.s.; then $|\sum_{n=n_1+1}^{n_1+n_2} X_n| \to 0$ a.s., as $n_1 \to \infty$, and hence $X_n \xrightarrow{a.s.} 0$. Since the X_n are independent (so are the X_n^c), by the second part of the Borel-Cantelli Lemma (Theorem 2.3.2), we conclude that $X_n \xrightarrow{a.s.} 0$ implies $\sum_{n\geq 1} P\{|X_n| > c\} < \infty$, for all $c > 0$, i.e., (2.5.8) holds. Again (2.5.8) ensures the convergence-equivalence of $\sum_{n\geq 1} X_n$ and $\sum_{n\geq 1} X_n^c$, so that Theorem 2.5.3 leads to the satisfaction of (2.5.9) and (2.5.10). ∎

There are some other convergence theorems having fruitful applications in Large Sample Theory, among which the celebrated **Law of Iterated Logarithm** is the most notable one. The reader is referred to Chow and Teicher (1978) for details. We include in this chapter with a brief introduction to the situation in **Sequential Sampling Schemes** where the sample size is itself a non-negative integer-valued r.v. To this point, we consider first the following:

Example 2.5.2 (Inverse sampling): Often, to estimate the probability (π) of a rare event, instead of the binomial sampling scheme, one adopts the following alternative one. Units are drawn one by one (with replacement) until a specified number (say, m) of the event occurs. Thus, n, the total number of units drawn to yield exactly m occurrences, is itself a positive r.v. with $P\{n \geq m\} = 1$. For $\pi \in (0,1)$, the probability law for n is given by

$$P(n \mid m, \pi) = \binom{n-1}{m-1} \pi^m (1-\pi)^{n-m}, \quad n = m, m+1, \ldots. \tag{2.5.11}$$

An estimator of π is $T_n = m/n$, although here m is a given positive integer and n is stochastic. For the stochastic convergence of T_n to π, the results developed in this chapter may not be directly applicable. Further, in such a case, in view of the stochastic nature of n, we may like to formulate the stochastic convergence in terms of the nonstochastic variable m, i.e., we may pose the problem as follows: As m is made to increase, does T_n converges a.s. or in probability to π? The answer is, of course, affirmative, and this can be posed in a more general framework as follows.

Consider a sequence $\{N_m, m \geq 1\}$ of non-negative integer-valued r.v.'s, and let $T_{(m)} = T_{N_m}$, $m \geq 1$, be a sequence of (sequential) estimators of some parameter θ. Then note that if (i) $T_n \xrightarrow{\text{a.s.}} \theta$ and (ii) $N_m \to \infty$ a.s., as $m \to \infty$, we would have $T_{N_m} \to \theta$ a.s., as $m \to \infty$. Thus, under the additional condition (ii) the a.s. convergence of a sequential estimator follows from that of the classical version. On the other hand, suppose that there exists a sequence $\{a_m\}$ of positive numbers, such that

$$N_m/a_m \xrightarrow{\text{P}} 1, \quad \text{as } m \to \infty. \tag{2.5.12}$$

Then, for every positive ε and η, we have

$$P\{|T_{N_m} - \theta| > \varepsilon\}$$
$$= P\left\{|T_{N_m} - \theta| > \varepsilon, \left|\frac{N_m}{a_m} - 1\right| \leq \eta\right\}$$
$$+ P\left\{|T_{N_m} - \theta| > \varepsilon, \left|\frac{N_m}{a_m} - 1\right| > \eta\right\}$$
$$\leq P\left\{\max_{k:|k-a_m|\leq \eta a_m} |T_k - \theta| > \varepsilon\right\} + P\left\{\left|\frac{N_m}{a_m} - 1\right| > \eta\right\}. \tag{2.5.13}$$

By (2.5.12), the second term on the rhs of (2.5.13) converges to 0 as $m \to \infty$. On the other hand, the first term on the rhs of (2.5.13) is more stringent than the usual $P\{|T_{[a_m]} - \theta| > \varepsilon\} \to 0$ (but is less stringent than $T_n \xrightarrow{\text{a.s.}} \theta$). Fortunately, the Kolmogorov Maximal Inequality (Theorem 2.3.8) or its various extensions may often be used to verify that the first term on the rhs of (2.5.13) goes to 0 as $m \to \infty$. Thus, toward the proof of weak consistency of estimators based on stochastic sample sizes, the Kolmogorov or related Maximal inequalities may be conveniently used. In general, for the SLLN, the stochastic nature of the sample size makes no difference (provided the sample size goes to ∞ a.s.), but for the weak LLN, the stochastic convergence of the classical T_n may not be enough, and a little more stringent condition, such as the one in (2.5.13), is usually needed.

Example 2.5.3: Let X_i, $i \geq 1$, be i.i.d. r.v.'s with finite mean θ and variance $\sigma^2 < \infty$. Let $T_n = \overline{X}_n$, for $n \geq 1$ and N_m, $m \geq 1$ be a sequence of positive integer-valued r.v.'s, such that (2.5.12) holds. For example, letting

$S_n^2 = (n-1)^{-1} \sum_{i=1}^{n} (X_i - \overline{X}_n)^2$, $n \geq 2$, we may define

$$N_m = \min \left\{ n \geq 2 : \frac{S_n}{\sqrt{n}} \leq m^{-1/2} \right\}, \quad m > 0, \qquad (2.5.14)$$

then (2.5.12) holds with $a_m \sim m\sigma^2$ (use the a.s. convergence of S_n^2 to σ^2). Actually, here $N_m \to \infty$ a.s., as $m \to \infty$, whereas by the Khintchine SLLN (Theorem 2.3.13), $\overline{X}_n \xrightarrow{\text{a.s.}} \theta$. Consequently, $\overline{X}_{N_m} \to \theta$ a.s., as $m \to \infty$. Alternatively, we could have used the Kolmogorov Maximal Inequality (Theorem 2.3.8) for the $\overline{X}_k$ and verified that the first term on the rhs of (2.5.13) converges to 0 as $m \to \infty$. This yields the stochastic convergence of the sequential estimator $\overline{X}_{N_m}$ under the usual second moment condition.

∎

Example 2.5.4 (Renewal Theorem): Consider a sequence $\{X_i, \ i \geq 1\}$ of i.i.d. (usually non-negative) r.v.'s with $\mu = EX, 0 < \mu < \infty$, and let $T_n = \sum_{i \leq n} X_i$, $n \geq 1$, $T_0 = 0$. For $d > 0$, define $\tau_d = \inf \{n \geq 1 : T_n > d\}$. Then τ_d takes the role of N_m while $a_m = m/\mu$. Exercise 2.5.4 is set to verify the details.

We conclude this chapter with a brief introduction to other useful convergence concepts:

Definition 2.5.2 (Uniform integrability): *Let* $\{G_n, \ n \geq n_0\}$ *be a sequence of d.f.'s, defined on* $\mathbb{R}$, *and let* $h(y)$, $y \in \mathbb{R}$, *be a real-valued continuous function. If*

$$\sup_{n \geq n_0} \int_{\{|y| \geq a\}} |h(y)| dG_n(y) \to 0 \quad \text{as } a \to \infty, \qquad (2.5.15)$$

then $h(\cdot)$ *is called* **uniformly** *(in n)* **integrable** *relatively to* $\{G_n\}$. *This definition extends to* $\mathbb{R}^p$ *(for* G_n*) and/or* $\mathbb{R}^q$ *(for* $h(\cdot)$*), for* $p, q \geq 1$, *by replacing the norm* $|\cdot|$ *by the Euclidean norm* $\|\cdot\|$.

It may be recalled that if $G_n(y) = P\{T_n \leq y\}$ stands for the d.f. of a statistic T_n, such that $T_n \to \theta$, in probability (or a.s.), as $n \to \infty$, we are not automatically permitted to conclude that $T_n \to \theta$ in the first (or rth) mean (viz., Example 2.3.8). The uniform integrability condition (2.5.15) provides this access (and this is not true for Example 2.3.8). This concept is strengthened in the following.

Theorem 2.5.5 (Lebesgue dominated convergence theorem): *For a sequence* $\{X_n\}$ *of measurable functions, suppose that there exists an* Y, *such that*

$$|X_n| \leq Y \quad \text{a.e., where } EY < \infty, \qquad (2.5.16)$$

and either $X_n \to X$ a.e. or $X_n \xrightarrow{D} X$, for a suitable X. Then

$$E|X_n - X| \to 0, \quad \text{as } n \to \infty. \tag{2.5.17}$$

A related version of (2.5.17) wherein the role of the dominating r.v. Y is deemphasized is the following:

Let $\{X, X_n, n \geq n_0\}$ be a sequence of r.v.'s, such that $X_n - X \xrightarrow{P} 0$ and

$$E\left\{ \sup_{n \geq n_0} |X_n| \right\} < \infty. \tag{2.5.18}$$

Then (2.5.17) holds, and, hence, $EX_n \to EX$.

It may be remarked that the uniform integrability condition in (2.5.15) is implicit in (2.5.16) or (2.5.18), and, hence, the proof of the theorem follows by some standard steps; we may refer to Chow and Teicher (1978), for example, for details. For clarification of ideas, we set Exercises 2.5.2 and 2.5.3. Show that in Exercise 2.4.5 also, $\hat{\rho}_n \to \rho$ in the first mean.

2.6 Concluding notes

With due emphasis on the scope of applications, a variety of examples has been worked out throughout this chapter. However, a few of the basic results (viz., the Levy-Cramér Theorem, law of iterated logarithm, and lower bound to the Kolmogorov Maximal Inequality, among others) have been presented without derivation. Some supplementary reading of these left-out proofs is strongly recommended for a methodology-oriented reader. Nevertheless, for the rest of the book, this little omission will not create any impasse in the understanding of the outlined methodology as well as the contemplated applications. The probability inequalities and the laws of large numbers have primarily been developed for sums of independent r.v.'s and/or martingale/reverse martingale type of dependent sequences. Although these sequences cover a broad range of statistical models, there may be some other situations [viz., unequal probability sampling (without replacement) from a finite population] where, at best, one can approximate (in a meaningful way) the actual sequence of statistics by a (sub-)martingale or reverse (sub-)martingale. Further, the role of the sample size (n), in a large sample context, may be replaced by some other characteristic. Example 2.5.2 and the sequential scheme, following it, pertain to this extended mode. There are numerous other examples of this type, some of which will be considered in the subsequent chapters. For these reasons, at this stage, we present (below) only a selected few exercises (some of which are rather artificial) for further clarification of ideas. We will, of course, encourage applications-oriented readers to look into various applied problems and to see to what extent the current chapter provides insight.

2.7 Exercises

Exercise 2.1.1: Construct a sequence $\{f_n\}$ for which (2.1.1) holds, but f_n is not equal to f, for any finite n.

Exercise 2.1.2: Construct a sequence $\{f_n\}$ for which (2.1.1) does not hold, but there exists a subsequence $\{n_j\}$, such that (2.1.1) holds for $\{f_{n_j}\}$. Can different subsequences have different limits? Construct suitable $\{f_n\}$ to support your answer.

Exercise 2.2.1: Define F_n as in (2.2.10). Suppose that the true d.f. F is concentrated on three mass points a, b and c with respective probability masses p_a, p_b and p_c (where $p_a + p_b + p_c = 1$). Verify that $\|F_n - F\| \to 0$ a.s., as $n \to \infty$.

Exercise 2.2.2: Extend the result in the previous exercise to the case where F is discrete and (i) has finitely many mass points and (ii) has countably infinite number of mass points.

Exercise 2.2.3: Let $X_1, \ldots, X_n$ be n i.i.d. r.v. from the Unif$(0, \theta)$ d.f., $\theta > 0$. Let $T_n = \max\{X_1, \ldots, X_n\}$. Show that (2.2.55) holds with $\psi_\varepsilon(n) = O(n^{-r})$ for $r \geq 2$. Hence, or otherwise, show that $T_n \to \theta$ a.s. (as well as completely), as $n \to \infty$.

Exercise 2.2.4: Let $X_1, \ldots, X_n$ be n i.i.d. r.v. from the negative exponential d.f., defined in (1.4.45), and let $T_n = \min\{X_1, \ldots, X_n\}$. Show that (2.2.55) holds with $T = 0$ and $\psi_\varepsilon(n) = \exp(-n\varepsilon/\theta)$.

Exercise 2.2.5: A **system** has two components connected in series, so that it **fails** when at least one of the components fail. Suppose that each component has a negative exponential life distribution [defined by (1.4.45)]. What is the life distribution of the system? Suppose that there are n copies of this system, and let their lifetimes be denoted by $Y_1, \ldots, Y_n$ respectively. Show that $\overline{Y}_n \xrightarrow{\text{a.s.}} \theta/2$.

Exercise 2.2.6 (Continuation): If the system has two components in parallel, verify the a.s. convergence of $\overline{Y}_n$ and work out the limit.

Exercise 2.3.1: Let $nT_n \sim \text{Bin}(n, \theta)$, $0 < \theta < 1$. Compare (2.3.2) and (2.3.3) [assuming that the normality in (2.3.3) holds closely], when $n = 50$, $t = 1$ and $\theta = 1/2$, $1/4$ and $1/10$.

Exercise 2.3.2 (Continuation): For $\varepsilon = 0.05$ and $n = 50$, $\theta = 1/2 (= \pi)$, compare (2.3.11) with (2.3.2).

Exercise 2.3.3: Verify (2.3.14).

Exercise 2.3.4: For Example 2.3.2, verify (2.3.23).

Exercise 2.3.5: For Example 2.3.3, obtain explicit forms for $\rho_+(\varepsilon)$ and $\rho_-(\varepsilon)$.

Exercise 2.3.6: Consider the Pareto density f, given by $f(x; a, \nu) = \frac{1}{a}\nu(x/a)^{-\nu-1}I_{\{x \geq a\}}$, $a > 0$, $\nu > 0$. Show that for this density, the moment generating function $M(t)$ does not exist for any $t > 0$. For what range of values of ν, does EX^k exist, for a given $k > 0$? Hence, or otherwise, show that if $\nu > 1$, the Khintchine SLLN (Theorem 2.3.13) applies to samples from this distribution.

Exercise 2.3.7: For Examples 2.3.7 (CMRR procedure) and 2.3.9, consider the asymptotic situation where n_1/N and n_2/N both converge to 0 as $N \to \infty$, but $n_1 n_2/N \to \lambda$, for some $\lambda > 0$. Use (2.3.35) to verify that r_2 has closely a Poisson distribution with parameter λ. Hence, or otherwise, find out the asymptotic moment generating function of r_2 and show that $r_2/\lambda \xrightarrow{P} 1$, as $\lambda \to \infty$.

Exercise 2.3.8: Verify that the Khintchine SLLN [Theorem 2.3.13] holds when the r.v.'s X_i have a common distribution F, defined on $\mathbb{R}^p$, for some $p \geq 1$. Hence, or otherwise, show that the case of vector- or matrix-valued i.i.d. r.v.'s is covered by this extension.

Exercise 2.3.9: Let (X_i, Y_i), $i \geq 1$, be i.i.d. r.v.'s with a bivariate d.f. having finite moments up to the second order. Let r_n be the sample correlation coefficient (for a sample of size n) and ρ be the population counterpart. Show that $r_n \xrightarrow{a.s.} \rho$.

Exercise 2.4.1: Consider a **bundle** of n parallel filaments whose individual strengths are denoted by $X_1, \ldots, X_n$ (non-negative r.v.'s). Also, $X_{n:1} \leq \cdots \leq X_{n:n}$ be the ordered values of $X_1, \ldots, X_n$. The **bundle strength** is then defined by Daniels (1945) as

$$B_n = \max\left\{(n - i + 1)X_{n:i} : 1 \leq i \leq n\right\}.$$

Define F_n as in (2.2.10) and show that

$$Z_n = n^{-1}B_n = \sup\left\{x[1 - F_n(x)] : x \geq 0\right\}.$$

Hence, or otherwise, use (2.4.34), to verify that $\{Z_n\}$ is a non-negative reverse submartingale. Hence, use Theorem 2.4.3 to show that as $n \to \infty$

$$Z_n \xrightarrow{\text{a.s.}} \theta = \sup \{x[1 - F(x)] : x \geq 0\}.$$

Exercise 2.4.2 (Continuation): Show that as $n \to \infty$

$$n^{1/2}|\mathrm{E}(Z_n) - \theta| \to 0,$$

and $\mathrm{E}(Z_n) \geq \theta$, for all $n \geq 1$. [Sen, Bhattacharyya and Suh (1973)].

Exercise 2.4.4: For Example 2.4.9, show that for a U-statistic U_n and $n \in [m, N]$, $\mathrm{E}[U_n \mid U_N] = U_N$, and, hence, show that

$$\mathrm{E}[U_n - U_N]^2 = \mathrm{Var}(U_n) - \mathrm{Var}(U_N),$$

where the $\mathrm{Var}(\cdot)$ refers to the variance computed for i.i.d. r.v.'s.

Exercise 2.4.5: Consider a first order (stationary) Markov process: $X_k = \rho X_{k-1} + \varepsilon_k$, $k \geq 1$, where the ε_k are i.i.d. r.v. with zero mean and (finite) variance σ^2. ε_k is independent of X_{k-1}, for every $k \geq 1$. Set $U_n = \sum_{k=1}^{n-1} X_k^2$. Then set $\widehat{\rho}_n = U_n/V_n$. Show that $\{V_n(\widehat{\rho}_n - \rho), n \geq 2\}$ is a zero-mean martingale. Hence, or otherwise, verify that $\widehat{\rho}_n \xrightarrow{\text{a.s.}} \rho$.

Exercise 2.5.1: For the negative binomial law in (2.5.11), index the r.v. n by n_m, and rewrite m/n_m as T_m^*. Then show that T_m^* is a bounded r.v. for every $m \geq 1$. Moreover, using (2.5.11), compute the mean and variance of T_m^* and, hence, use the Chebyshev Inequality (2.3.1) to show that $T_m^* \xrightarrow{\text{P}} \pi$ as $m \to \infty$. Can you claim that $T_m^* \xrightarrow{\text{rth}} \pi$, as $m \to \infty$?

Exercise 2.5.2: Show that for every $m \geq 1$, n_m as defined in Exercise 2.5.1 can be expressed as $Z_1 + \cdots + Z_m$, where the Z_i are i.i.d. r.v.'s, each having a negative binomial law (with the index $m = 1$). Hence, show that (Theorem 2.3.13) $n_m/m \xrightarrow{\text{a.s.}} \mathrm{E}Z_1$ as $m \to \infty$, so that $T_m^* \xrightarrow{\text{a.s.}} \pi$ as $m \to \infty$.

Exercise 2.5.3: Define N_m as in (2.5.14). Show that

$$S_{N_m-1}^2 > \frac{1}{m}(N_m - 1) \geq S_{N_m}^2 - \frac{1}{m}, \quad m \geq 1.$$

Also verify that N_m is monotonically nondecreasing in m, and, for every fixed m, N_m is finite a.s. Further, by (2.3.86), $S_n^2 \to \sigma^2$ a.s., as $n \to \infty$. Thus, on letting $a_m = [m\sigma^2]$, verify that $N_m/a_m \to 1$ a.s., as $m \to \infty$.

Exercise 2.5.4 (Renewal Theorem): Let $\{X_i, i \geq 1\}$ be a sequence of i.i.d. r.v. with $0 < EX = \mu < \infty$, and set

$$T_n = \sum_{i=1}^{n} X_i, \quad n \geq 1, \quad \text{and} \quad \tau_d = \min\{n : T_n > d\}, \quad d > 0.$$

Note that T_d is nondecreasing in $d \ (> 0)$ and, further,

$$T_{\tau_d} > d \geq S_{\tau_d - 1}, \quad d > 0.$$

Moreover by Theorem 2.3.13, $n^{-1}T_n \to \mu$ a.s., as $n \to \infty$. Then, show that $d^{-1}\tau_d \to 1/\mu$ a.s., as $d \to \infty$. Also, $(E\tau_d)/d \to 1/\mu$, as $d \to \infty$.

CHAPTER 3

Weak Convergence and Central Limit Theorems

3.1 Introduction

The foundations of large sample theory are laid down by the concepts of **stochastic convergence** and **weak convergence**. Consistency is a minimal requirement in large sample theory, and the basic relationship between consistency and stochastic convergence has been thoroughly discussed in Chapter 2. Consistency (whether of an estimator or a test statistic) may, however, fail to convey the full statistical information contained in the data set at hand, and, hence, by itself, it may not provide an efficient statistical conclusion. To illustrate this point suppose that we have n independent random variables $X_1, \ldots, X_n$ drawn from an unknown distribution with mean μ and finite variance σ^2 (which are both unknown). The sample mean $\overline{X}_n$ is a natural estimator of μ, and from the results of Chapter 2, we may conclude that as n increases, $\overline{X}_n$ converges to μ in a well-defined manner (viz., in probability/almost surely/second mean). This certainly endows us with an increasing confidence on the estimator $(\overline{X}_n)$ with increasing sample sizes. Suppose now that we desire to set a **confidence interval** (L_n, U_n) for μ with a prescribed **coverage probability** or **confidence coefficient** $1 - \alpha$ (for some α, $0 < \alpha < 1$), i.e., we intend to determine suitable statistics L_n and U_n ($L_n \leq U_n$) based on a sample $X_1, \ldots, X_n$, such that

$$P\{L_n \leq \mu \leq U_n\} \geq 1 - \alpha. \tag{3.1.1}$$

If σ were known, we could have used the Chebyshev Inequality (2.3.1) to obtain that, for every $n \geq 1$,

$$P\{|\overline{X}_n - \mu| > t\} \leq \sigma^2/(nt^2), \tag{3.1.2}$$

so that, on setting $\alpha = \sigma^2/(nt^2)$, i.e., $t = \sigma/(n\alpha)^{1/2}$, we have

$$P\{L_n = \overline{X}_n - \sigma/(n\alpha)^{1/2} \leq \mu \leq \overline{X}_n + \sigma/(n\alpha)^{1/2} = U_n\} \geq 1 - \alpha. \tag{3.1.3}$$

We may even get sharper bounds (for L_n and U_n) by using the Markov Inequality (2.3.4) (assuming higher order moments) or the Bernstein In-

equality (2.3.7) (assuming a finite moment generating function) instead of
the Chebyshev Inequality. On the other hand, if X has a normal distribution with mean μ and variance σ^2, we have, for every real x and every
$n \geq 1$,

$$P\{n^{1/2}(\overline{X}_n - \mu)/\sigma \leq x\} = \Phi(x) = (2\pi)^{-1/2} \int_{-\infty}^{x} \exp(-t^2/2)dt. \quad (3.1.4)$$

Let us define τ_ε by $\Phi(\tau_\varepsilon) = 1 - \varepsilon$, $0 < \varepsilon \leq 1$. Then, we have from (3.1.4)
that, for every $n \geq 1$ and α $(0 < \alpha < 1)$,

$$P\{L_n^* = \overline{X}_n - n^{-1/2}\sigma\tau_{\alpha/2} \leq \mu \leq \overline{X}_n + n^{-1/2}\sigma\tau_{\alpha/2} = U_n^*\} = 1 - \alpha. \quad (3.1.5)$$

Let us compare the width of the two bounds in (3.1.3) and (3.1.5). Note
that

$$(U_n^* - L_n^*)/(U_n - L_n) = \alpha^{1/2}\tau_{\alpha/2}. \quad (3.1.6)$$

The rhs of (3.1.6) is strictly less than 1 for every α $(0 < \alpha < 1)$. For $\alpha=0.01$,
$0.025, 0.05$ and 0.10, it is equal to $0.258, 0.354, 0.438$ and 0.519, respectively.
A similar picture holds for the case of the Markov or Bernstein Inequality
based bounds. This simple example illustrates that the knowledge of the
actual sampling distribution of a statistic generally leads to more precise
confidence bounds than those obtained by using some of the probability
inequalities considered in Chapter 2. For a statistic T_n (viz., an estimator
of a parameter θ), more precise studies of its properties may be made when
its sampling distribution is known. A similar situation arises when one
wants to test for the null hypothesis (H_0) that μ is equal to some specified
value μ_0 (against one- or two-sided alternatives). In order that the test
has a specified margin (say, α, $0 < \alpha < 1$) for the Type I error, one may
use the probability inequalities from Chapter 2 for the demarcation of the
critical region. However, this generally entails some loss of the **power**
of the test. Knowledge of the distribution of the test statistic (under H_0)
provides a test which has generally better power properties.

    In small samples, a complete determination of the sampling distribution
of a statistic T_n may not only demand a knowledge of the functional form of
the underlying d.f. F but also cumbrous algebraic manipulations (especially
when T_n is not linear). The situation changes drastically as n increases.
Under fairly general regularity conditions, it is possible to approximate the
actual sampling distribution of the normalized version of a statistic T_n by
some simple ones [such as the normal law in (3.1.4)], and in, this way, one
has the vast flexibility to study suitable estimators and/or test statistics
which possess some optimality (or at least desirable) properties for large
sample sizes. To illustrate this point, let us go back to the same example
considered before, but assume now that both μ and σ are unknown (as is
generally the case). The sample variance S_n^2 is a consistent estimator of σ^2

(as has already been established in Chapter 2), and, hence, it may be quite tempting to use the normalized version [compared to (3.1.4)]:

$$t_{n-1} = n^{1/2}(\overline{X}_n - \mu)/S_n, \quad n \geq 2 \qquad (3.1.7)$$

(which is known as the Student t-statistic with $n - 1$ degrees of freedom). It is well-known that for $n > 30$, the actual sampling distribution of t_{n-1} can be very well approximated by the normal law in (3.1.4). Incidentally, to use some of the probability inequalities on t_{n-1} in (3.1.7) we need extra manipulations, and the simplicities of the Chebyshev/Markov/Bernstein inequalities are no longer attainable. Actually, even if the underlying d.f. F is not normal, but possesses a finite second order moment, (3.1.4) extends in a natural way to the following:

$$\lim_{n \to \infty} P\{n^{1/2}(\overline{X}_n - \mu)/\sigma \leq x\} = \Phi(x), \quad x \in \mathbb{R} \qquad (3.1.8)$$

and, in the literature, this is referred to as the classical **Central Limit Theorem** (CLT). Moreover, (3.1.8) holds even when σ is replaced by S_n, so that we may not have to presume the knowledge of the variance σ^2 in providing a confidence interval for μ or to test for a null hypothesis on μ. Presumably, one may have to pay a little penalty: the rate of convergence may be generally slower for the case in (3.1.7). Numerous other examples of this type will be considered in this chapter and in subsequent ones too.

Keeping in mind the case of $n^{1/2}(\overline{X}_n - \mu)/\sigma$ or $n^{1/2}(\overline{X}_n - \mu)/S_n$, we may gather that in a typical statistical inference problem (when the sample size is not small), we encounter a sequence $\{T_n; n \geq n_0\}$ of statistics, such that for suitable normalizing constants $\{a_n, b_n; n \geq n_0\}$, $(T_n - a_n)/b_n$ has a limiting distribution which is typically nondegenerate, and our goal is to formulate it in a simple manner. Let

$$F_n(t) = P\{(T_n - a_n)/b_n \leq t\}, \quad t \in \mathbb{R}, \quad n \geq n_0. \qquad (3.1.9)$$

Our contention is to provide, for large n, close approximation to F_n by some simple distribution F (viz., normal, Poisson, chi-squared) which has been extensively studied. Associated with such an F, we have a random variable say Z, such that $P\{Z \leq t\} = F(t)$, for all t. In that way, we say that $(T_n - a_n)/b_n$ converges in distribution (or law) to Z, and denote this by

$$(T_n - a_n)/b_n \xrightarrow{\mathcal{D}} Z. \qquad (3.1.10)$$

Some points of clarification are in order here. First, in (3.1.10), we really mean that $F_n(t) \to F(t)$ as $n \to \infty$, for almost all t. Now, the d.f. F_n may not be continuous everywhere (viz., the binomial case), nor necessarily is the d.f. F continuous everywhere (viz., the Poisson distribution). As such, if t is a jump point of F, $F_n(t)$ may not converge to $F(t)$. We eliminate this problem by saying that $F_n \to F$, weakly if at all points of continuity of F, $F_n(t) \rightarrow F(t)$, as $n \rightarrow \infty$. In symbols, we write

$$F_n \rightarrow F \quad \text{if} \quad F_n(t) \rightarrow F(t), \quad \forall t \in J, \quad \text{as} \quad n \rightarrow \infty, \tag{3.1.11}$$

where J is the set of continuity points t of F. This definition, we have already introduced in Section 2.2. Second, if we let $Z_n = (T_n - a_n)/b_n$, $n \geq n_0$, then Z_n and Z need not be defined on the same probability space, and $Z_n - Z$ (even if properly defined) may not converge to 0 in probability or in some other mode (as mentioned in Section 2.2) when n becomes large. For example, consider the binomial law where nT_n is $\text{Bin}(n, \pi)$, $0 < \pi < 1$, and let $Z_n = n^{1/2}(T_n - \pi)$, $n \geq 1$. Then, for every finite n, Z_n has a discrete distribution with positive probability masses attached to the points $n^{-1/2}(k - n\pi)$, $k = 0, 1, \ldots, n$, while we shall see later on that here (3.1.8) holds with $\sigma^2 = \pi(1 - \pi)$, so that $F(= \Phi)$ is an absolutely continuous d.f. In the same example, if $\pi = \pi_n$ is allowed to depend on n in such a way that as $n \rightarrow \infty$, $n\pi_n \rightarrow \lambda$, $0 < \lambda < \infty$, then the binomial law converges to a Poisson law with parameter λ. In this case, the Poisson distribution has jump points $0, 1, 2, \ldots$, at which the right- and left-hand side limits may not agree. Thus, we need to eliminate these jump points from the convergence criterion, and this reflects the necessity of the set J in (3.1.11). Finally, we shall attempt to provide a more general definition of the weak convergence in (3.1.11) covering the multidimensional case as well.

Whenever the $\mathbf{T}_n$ are random p-vectors, they have d.f.'s defined on $\mathbb{R}^p$, for some $p \geq 1$. In such a case, there are several possibilities. For example, we may consider a real-valued random element $Z_n = h(\mathbf{T}_n; n)$ and then apply the result in (3.1.10). A very common problem arising in this context is a linear combination of the elements of $\mathbf{T}_n$, in which case, we may have typically a normal asymptotic distribution. In a variety of other cases, we would also consider a quadratic form in the elements of $\mathbf{T}_n$, and these may relate to asymptotic (central/noncentral) chi-squared distributions. In a general multivariate setup, we consider a sequence of (normalized) stochastic vectors $\{\mathbf{T}_n, n \geq n_0\}$ and let $F_n(\mathbf{t}) = P\{\mathbf{T}_n \leq \mathbf{t}\}$, $\mathbf{t} \in \mathbb{R}^p$. Also, we conceive of a d.f. F defined on $\mathbb{R}^p$ and define the continuity-point set J ($\subset \mathbb{R}^p$) in a similar manner. Let F stand for the d.f. of a random vector $\mathbf{T}$. Then, we say that $\mathbf{T}_n$ converges in distribution (or law) to $\mathbf{T}$ ($\mathbf{T}_n \xrightarrow{\mathcal{D}} \mathbf{T}$) or F_n weakly converges to F, if

$$F_n(\mathbf{t}) \rightarrow F(\mathbf{t}), \quad \forall \mathbf{t} \in J \ (\subset \mathbb{R}^p), \quad \text{as} \quad n \rightarrow \infty. \tag{3.1.12}$$

The concept of weak convergence is not confined to the finite dimensional vector case. Its extension to the case of general probability spaces enables one to include infinite dimensional vectors or stochastic processes. However, to be consistent with our intended (intermediate) level of presentation, we find it difficult to include such a general treatment in this chapter.

The normal distribution (in univariate as well as multivariate setups) occupies a central position in asymptotic theory or large sample methods. The **De Moivre-Laplace Theorem** (on the convergence of the binomial to the normal law) is one of the classical examples of the weak convergence of $\{F_n\}$. For binomial, multinomial, hypergeometric and other discrete distributions, earlier work on weak convergence have exploited direct expansions of the associated probability functions (by the use of the **Stirling approximations** to factorials of natural integers), resulting in a mode of convergence stronger than in (3.1.12). However, in most of the cases, the algebraic manipulations are heavy and specifically dependent on the particular underlying model. Moreover, they are not that necessary for actual applications. For sums (or averages) of independent random variables, the weak convergence in (3.1.11) or (3.1.12) has been proved under increasing generality by a host of workers (Liapounov, Lindeberg, Feller and others). As we shall see later, for most of the statistical applications, this weak convergence suffices, and the other local limit theorems available for lattice distributions are not that necessary.

The normal (or Gaussian) distribution has many interesting properties. Among these, the structure of central moments has been widely used in weak convergence studies. Note that the $(2k)$th central moment of a standard normal distribution is $(2^k k!)^{-1}(2k)!$ and the $(2k+1)$th moment is 0, for every $k \geq 1$. This inspired a lot of workers to evaluate the first four moments of a statistic T_n and to show that these converge to the corresponding ones of a suitable normal distribution. From this moment convergence, they attempted to conclude that the distribution of T_n (for the normalized version) also converges to a normal one. Although this conclusion remains adaptable in some cases, it cannot be justified theoretically. If the convergence of the asymptotic distribution of T_n to a normal law has to be based on the convergence of moments, then one needs to consider moments of all finite orders (not just the first four), and, moreover, the limiting distribution must be uniquely defined by its moments. The evaluation of moments of all finite orders may indeed be a monumental task, and, therefore, this approach will not be stressed here too.

The characteristic functions introduced earlier play a fundamental role in the weak convergence studies to be pursued in this chapter. For example, for sums or averages or linear statistics, the approach based on characteristic functions yields general theorems which are much less model dependent, and these remain vastly adaptable in most of the statistical applications; they depend on some basic results that will be discussed in Section 3.2. We shall consider a general review of these broad results since they facilitate the extensions of the main theorems to triangular schemes of (row-wise independent) random variables and vectors and to multivariate central limit theorems.

The central limit theorems will be considered in Section 3.3. It is important to note that they are no longer confined to sums of independent random variables (vectors). During the past three decades, these central limit theorems have been successfully extended for various dependent random variables. In particular, **martingales, reverse martingales** and related sequences have been annexed in these developments, and they have some important applications too. We shall also review some of these developments in Section 3.3.

The Slutsky Theorem and some related **projection results** [mostly, due to Hoeffding (1948) and Hájek (1968)] are considered in Section 3.4; their applications are stressed too. Transformations of statistics and variables play an important role in large sample theory. In particular, the so-called **variance-stabilizing transformations** have interest on their own. Interestingly enough, such transformations often accelerate the rate of convergence to the limiting distributions. Related asymptotic theory is considered in the same section.

The **Cochran Theorem** (on quadratic forms) in a multinormal setup has many important uses in statistical inference. In this setup, too, it may not be necessary to assume multinormality of the underlying d.f., and incorporating the usual multidimensional central limit theorems, extensions of the Cochran Theorem will be considered in Section 3.4. We conclude this section with the following natural question: Is the weak convergence result in (3.1.10) accompanied by the convergence of the moments of Z_n (up to a certain order) to the corresponding ones of Z? Some additional regularity conditions pertain to this problem, and these will be briefly presented.

The last section in this chapter is essentially included for the sake of completeness and deals with the rates of convergence to normality. This is a very specialized topic and the reader may skip it without loss to the understanding of the rest of the book.

3.2 Some important tools

In this section we present some important results on weak convergence of distribution functions, which serve as the basis for many useful applications

Theorem 3.2.1 (Helly-Bray Lemma): *Let $\{F_n\}$ be a sequence of distribution functions; also, let g be a continuous function on $[a,b]$ where $-\infty < a < b < \infty$ are continuity points of a distribution function F.*

Then, if $F_n \xrightarrow{w} F$, *we have*

$$\int_a^b g(x)dF_n(x) \longrightarrow \int_a^b g(x)dF(x) \quad as \quad n \to \infty.$$

Proof: Let

$$I_n = \int_a^b g(x)dF_n(x) \quad and \quad I = \int_a^b g(x)dF(x);$$

now, given $\varepsilon > 0$, define a partition of $[a, b]$ by taking

$$a = x_0 < x_1 < \cdots < x_m < x_{m+1} = b,$$

where:

i) x_j, $0 \le j \le m+1$, are points of continuity of F;

ii) $x_{j+1} - x_j < \varepsilon$, $0 \le j \le m$.

Then, for $x_l \le x \le x_{l+1}$, $0 \le l < m$, define

$$g_m(x) = g[(x_l + x_{l+1})/2] = \sum_{l=0}^m g[(x_l + x_{l+1})/2]I_{(x_l,x_{l+1})}(x),$$

$$I_n(m) = \int_a^b g_m(x)dF_n(x),$$

$$I(m) = \int_a^b g_m(x)dF(x),$$

and observe that

$$|I_n - I| \le |I_n - I_n(m)| + |I_n(m) - I(m)| + |I(m) - I|.$$

Analyzing each term of this inequality, we have:

a) since g is continuous, for some $\delta = \delta(\varepsilon)$,

$$\begin{aligned} |I(m) - I| &\le \sup_{x \in [a,b]} |g_m(x) - g(x)| \int_a^b dF(x) \\ &\le \sup_{x \in [a,b]} |g_m(x) - g(x)| < \delta; \end{aligned}$$

b) since $F_n \xrightarrow{w} F$ and x_l, $0 \le l \le m+1$, are points of continuity of F,

$$I_n(m) - I(m) = \sum_{l=0}^m g\left(\frac{x_l + x_{l+1}}{2}\right)\left\{\Delta F_n(x_l) - \Delta F(x_l)\right\}$$

where $\Delta F_n(x_l) = F_n(x_{l+1}) - F_n(x_l)$ and $\Delta F(x_l) = F(x_{l+1}) - F(x_l)$ converges to 0 as $n \to \infty$;

c) since g is continuous, for some $\delta = \delta(\varepsilon)$,

$$|I_n - I_n(m)| \leq \sup_{x \in [a,b]} |g(x) - g_m(x)| \int_a^b dF_n(x)$$

$$\leq \sup_{x \in [a,b]} |g(x) - g_m(x)| < \delta.$$

Therefore, given $\delta > 0$ it is possible to choose a convenient $\varepsilon > 0$ and a sufficiently large n such that $|I_n - I| < 3\delta$, which completes the proof. ∎

Theorem 3.2.2 (Extension of the Helly-Bray Lemma): *Let $g(x)$ be a continuous and bounded function and let $\{F_n\}$ be a sequence of distribution functions such that $F_n \xrightarrow{w} F$ where F is a distribution function. Then*

$$\int_{-\infty}^{+\infty} g(x)dF_n(x) \to \int_{-\infty}^{+\infty} g(x)dF(x) \quad as \quad n \to \infty.$$

Proof: Given $\varepsilon' > 0$, let $-\infty < a < b < \infty$ be points of continuity of F such that $\int_{-\infty}^a dF(x) < \varepsilon'$ and $\int_b^\infty dF(x) < \varepsilon'$. Then let $\varepsilon = g^* \varepsilon'$ where $g^* = \sup_{x \in R} |g(x)| < \infty$ and observe that as $n \to \infty$ we have:

i) $\qquad |\int_{-\infty}^a g(x)dF_n(x)| \leq g^* F_n(a) \to g^* F(a) < g^* \varepsilon' = \varepsilon;$

ii) $\qquad |\int_b^\infty g(x)dF_n(x)| \leq g^*[1 - F_n(b)] \to g^*[1 - F(b)] < g^* \varepsilon' = \varepsilon.$

Now write $I_n = \int_{-\infty}^\infty g(x)dF_n(x)$ and $I = \int_{-\infty}^\infty g(x)dF(x)$ and note that

$$|I_n - I| \leq \left| \int_{-\infty}^a g(x)dF_n(x) - \int_{-\infty}^a g(x)dF(x) \right|$$

$$+ \left| \int_a^b g(x)dF_n(x) - \int_a^b g(x)dF(x) \right|$$

$$+ \left| \int_b^\infty g(x)dF_n(x) - \int_b^\infty g(x)dF(x) \right|. \qquad (3.2.1)$$

Applying the triangular inequality to each relation (i) and (ii), it follows that for sufficiently large n

$$\left| \int_{-\infty}^a g(x)dF_n(x) - \int_{-\infty}^a g(x)dF(x) \right| < 2\varepsilon, \qquad (3.2.2)$$

and

$$\left| \int_b^\infty g(x)dF_n(x) - \int_b^\infty g(x)dF(x) \right| < 2\varepsilon. \qquad (3.2.3)$$

Furthermore, an application of the Helly-Bray Lemma (3.2.1) yields

$$\left| \int_a^b g(x) dF_n(x) - \int_a^b g(x) dF(x) \right| < \varepsilon. \tag{3.2.4}$$

Then, from (3.2.1)–(3.2.4) we may conclude that for sufficiently large n, $|I_n - I| < 5\varepsilon$, which completes the proof. ∎

Note that the above theorem holds even in cases where g is not real valued. For example, let $g(x) = e^{itx} = \cos tx + i \sin tx$, $t \in \mathbb{R}$, and consider a sequence of distributions functions $\{F_n\}$ such that $F_n \xrightarrow{w} F$ where F is a distribution function. Applying Theorem 3.2.2 to $g_1(x) = \cos tx$ and $g_2(x) = \sin tx$ and using the triangular inequality, we have

$$\phi_{F_n}(t) = \int_{-\infty}^{\infty} e^{itx} dF_n(x) \to \int_{-\infty}^{\infty} e^{itx} dF(x) = \phi_F(t) \quad \text{as} \quad n \to \infty.$$

Actually, the converse proposition is also true, and together they constitute one of the most powerful tools in the field of weak convergence, the well-known Lévy-Cramér Theorem (2.2.2).

Example 3.2.1: Consider a sequence $\{R_n\}$ of random variable such that $R_n \sim \text{Bin}(n, \pi)$ and let $n \to 0$, $\pi \to 0$ in such a way that $n\pi \to \lambda < \infty$. We know that

$$\phi_{R_n}(t) = \text{E}(e^{itR_n}) = \{1 - \pi(1 - e^{it})\}^n = \{1 - \frac{\lambda}{n}(1 - e^{it})\}^n.$$

Then $\lim_{n \to \infty} \phi_{R_n}(t) = \exp\{-\lambda(1 - e^{it})\}$, which we recognize as the characteristic function of a random variable R following a Poisson(λ) distribution. Therefore, we may conclude that $R_n \xrightarrow{D} R$. ∎

The following result, parallel to that of Theorem 2.3.4, constitutes an important tool in the search for the asymptotic distribution of transformations of statistics with known asymptotic distributions.

Theorem 3.2.3 (Sverdrup): *Let $\{T_n\}$ be a sequence of random variables such that $T_n \xrightarrow{D} T$ and $g : \mathbb{R} \to \mathbb{R}$ be a continuous function. Then $g(T_n) \xrightarrow{D} g(T)$.*

Proof: For all $u \in \mathbb{R}$, let

$$\begin{aligned} \phi_{g(T_n)}(u) &= \text{E}\{\exp[iug(T_n)]\} \\ &= \int_{-\infty}^{+\infty} \cos[ug(t)] dF_{T_n}(u) + i \int_{-\infty}^{\infty} \sin[ug(t)] dF_{T_n}(u). \end{aligned}$$

Now, since $\cos x$, $\sin x$ and $g(x)$ are continuous and bounded functions, it follows from Theorem 3.2.2 that as $n \to \infty$

$$\phi_{g(T_n)}(u) \to \int_{-\infty}^{+\infty} \cos[ug(t)]dF_T(u) + i \int_{-\infty}^{+\infty} \sin[ug(t)]dF_T(u)$$
$$= \mathrm{E}\{\exp[iug(T)]\} = \phi_{g(T)}(u).$$

Thus, from the Lévy-Cramér Theorem (2.2.2) the result follows. ∎

Extensions to random vectors may be easily obtained via the following theorem which is the most important tool for the generalization of many univariate results to the multivariate case.

Theorem 3.2.4 (Cramér-Wold): *Let* $\mathbf{X}$, $\mathbf{X}_1$, $\mathbf{X}_2$, ... *be random vectors in* $\mathbb{R}^p$; *then* $\mathbf{X}_n \xrightarrow{D} \mathbf{X}$ *if and only if, for every fixed* $\boldsymbol{\lambda} \in \mathbb{R}^p$, *we have* $\boldsymbol{\lambda}^t\mathbf{X}_n \xrightarrow{D} \boldsymbol{\lambda}^t\mathbf{X}$.

Proof: Suppose that $\mathbf{X}_n \xrightarrow{D} \mathbf{X}$; then for every $\alpha \in \mathbb{R}$ and $\boldsymbol{\lambda} \in \mathbb{R}^p$ it follows from Theorem 3.2.2 that as $n \to \infty$

$$\begin{aligned}\phi_{\mathbf{X}_n}(\alpha\boldsymbol{\lambda}) &= \mathrm{E}\{\exp(i\alpha\boldsymbol{\lambda}^t\mathbf{X}_n)\} \\ &= \phi_{\boldsymbol{\lambda}^t\mathbf{X}_n}(\alpha) \to \phi_{\mathbf{X}}(\alpha\boldsymbol{\lambda}) \\ &= \mathrm{E}(i\alpha\boldsymbol{\lambda}^t\mathbf{X}) = \phi_{\boldsymbol{\lambda}^t\mathbf{X}}(\alpha).\end{aligned}$$

Using the Lévy-Cramér Theorem (2.2.2) we conclude that $\boldsymbol{\lambda}^t\mathbf{X}_n \xrightarrow{D} \boldsymbol{\lambda}^t\mathbf{X}$. The converse follows by similar arguments. ∎

Finally, another result which is continuously used in the remainder of the text is given by:

Theorem 3.2.5: *Let* $\{X_n\}$ *be a sequence of random variables such that* $X_n \xrightarrow{D} X$ *where* X *has a nondegenerate d.f.* F. *Then* $X_n = O_p(1)$.

Proof: From the definition of weak convergence it follows that given $\gamma > 0$, there exists $n_0 = n_0(\gamma)$ such that for all points of continuity of $F(x) = P\{X \leq x\}$, we have

$$|P\{X_n \leq x\} - P\{X \leq x\}| < \gamma, \quad \forall n \geq n_0. \tag{3.2.5}$$

Now, from the definition of $F(x)$, we know that given $\eta > 0$, there exists $M_1 = M_1(\eta)$ and $M_2 = M_2(\eta)$ such that $P\{X \leq x\} > 1 - \eta$, for every $x \geq M_1$ and $P\{X \leq x\} < \eta$, for every $x \leq M_2$; thus, taking $M = \max\{|M_1|, |M_2|\}$ we may conclude that $P\{X \leq x\} > 1 - \eta$, for every $x \geq M$ and $P\{X \leq x\} < \eta$, for every $x \leq -M$. Then, from (3.2.5) we have that for all $n \geq n_0$

$$P\{X_n \leq x\} > P\{X \leq x\} - \gamma \;\; \Rightarrow \;\; P\{X_n \leq M\} > 1 - \eta - \gamma$$
$$\Rightarrow \;\; P\{X_n > M\} < \eta + \gamma, \quad (3.2.6)$$

$$P\{X_n \leq x\} < P\{X \leq x\} + \gamma \Rightarrow P\{X_n \leq -M\} < \eta + \gamma. \quad (3.2.7)$$

Now, given $\varepsilon > 0$, it suffices to take $\eta = \gamma = \varepsilon/4$ and use (3.2.5)–(3.2.7) to show that for all $n \geq n_0$ we have

$$P\{|X_n| > M\} \leq P\{X_n > M\} + P\{X_n < -M\} < \varepsilon,$$

concluding the proof.   ∎

3.3 Central limit theorems

Here we essentially investigate the convergence of sequences of distribution functions $\{F_n\}$ of sums of random variables to the normal distribution function Φ. The theorems considered in this section may be classified under the denomination of **Central Limit Theorems** and although they constitute special cases of more general results, they are sufficient for most situations of practical interest. We start with the simplest form of such a theorem and proceed to extend the results to situations with less restrictive assumptions.

For the sake of simplicity, we will use the notation $X_n \xrightarrow{\mathcal{D}} N(\mu, \sigma^2)$ or $X_n \xrightarrow{\mathcal{D}} \chi_q^2$ to indicate that $X_n \xrightarrow{\mathcal{D}} X$ where $X \sim N(\mu, \sigma^2)$ or $X_n \xrightarrow{\mathcal{D}} X$ where $X \sim \chi_q^2$, respectively.

Theorem 3.3.1 (Classical Central Limit Theorem): *Let X_k, $k \geq 1$ be i.i.d. r.v.'s with mean μ and finite variance σ^2. Also let*

$$Z_n = (T_n - n\mu)/\sigma\sqrt{n}$$

where $T_n = X_1 + \cdots + X_n$. Then, $Z_n \xrightarrow{\mathcal{D}} N(0,1)$.

Proof: Let $\phi_{Z_n}(t) = \mathrm{E}\{\exp(itZ_n)\}$ and note that

$$
\begin{aligned}
\phi_{Z_n}(t) &= \mathrm{E}\left[\exp\left\{\frac{it}{\sigma\sqrt{n}}\left(\sum_{k=1}^{n} X_k - n\mu\right)\right\}\right] \\
&= \prod_{k=1}^{n} \mathrm{E}\left[\exp\left\{\frac{it}{\sigma\sqrt{n}}(X_k - \mu)\right\}\right] \\
&= \left[\mathrm{E}\left\{\exp\left(\frac{it}{\sigma\sqrt{n}}U\right)\right\}\right]^n = \left\{\phi_U\left(\frac{t}{\sqrt{n}}\right)\right\}^n,
\end{aligned}
$$

where $U = (X_1 - \mu)/\sigma$ and $\phi_U(t) = \mathrm{E}\{\exp(itU)\}$. Considering a Taylor expansion of $\phi_U(t/\sqrt{n})$ and observing that $EU = 0$ and $EU^2 = 1$, we may write

$$
\begin{aligned}
\phi_U\left(\frac{t}{\sqrt{n}}\right) &= 1 + iEU\frac{t}{\sqrt{n}} + i^2\frac{t^2}{2n}EU^2 + o\left(\frac{t^2}{n}\right) \\
&= 1 - \frac{t^2}{2n} + o\left(\frac{t^2}{n}\right).
\end{aligned}
$$

Therefore, as $n \to \infty$, we have

$$
\phi_{Z_n}(t) = \left\{1 - \frac{t^2}{2n} + o\left(\frac{t^2}{n}\right)\right\}^n \to \exp\left\{\frac{-t^2}{2}\right\}
$$

which is the characteristic function of the standard normal distribution, completing the proof.   ∎

Example 3.3.1: Let $T_n \sim \text{Bin}(n, \pi)$ and write $T_n = \sum_{k=1}^{n} X_k$ where

$$
X_k = \begin{cases} 1 & \text{with probability } \pi \\ 0 & \text{with probability } 1 - \pi. \end{cases}
$$

Since $EX_k = \pi$ and $\text{Var}X_k = \pi(1-\pi)$, a direct application of Theorem 3.3.1 implies that $(T_n - n\pi)/\{n\pi(1 - \pi)\}^{-1/2}$ is asymptotically distributed as $N(0, 1)$. This result is known as the **De Moivre-Laplace Theorem**.   ∎

An extension of Theorem 3.3.1 to cover the case of sums of independent but not identically distributed random variables is given by:

Theorem 3.3.2 (Liapounov): *Let X_k, $k \geq 1$, be independent random variables such that $EX_k = \mu_k$ and $\text{Var}X_k = \sigma_k^2$, and for some $0 < \delta\ (\leq 1)$,*

$$
\nu_{2+\delta}^{(k)} = E|X_k - \mu_k|^{2+\delta} < \infty, \quad k \geq 1.
$$

Also let $T_n = \sum_{k=1}^{n} X_k$, $\xi_n = ET_n = \sum_{k=1}^{n}\mu_k$, $s_n^2 = \text{Var}T_n = \sum_{k=1}^{n}\sigma_k^2$, $Z_n = (T_n - \xi_n)/s_n$ and $\rho_n = s_n^{-(2+\delta)}\sum_{k=1}^{n}\nu_{2+\delta}^{(k)}$. Then, if $\lim_{n\to\infty}\rho_n = 0$, we have $Z_n \xrightarrow{\mathcal{D}} N(0, 1)$.

Proof: First, for every $k \geq 1$ apply Jensen Inequality (2.2.38) to conclude that

$$
\begin{aligned}
\sigma_k^2 = E(X_k - \mu_k)^2 &\leq \{E|X_k - \mu_k|^{2+\delta}\}^{\frac{2}{2+\delta}} \\
&= \{\nu_{2+\delta}^{(k)}\}^{\frac{2}{2+\delta}}
\end{aligned}
$$

which implies

$$
\sigma_k^2/s_n^2 \leq \left\{\nu_{2+\delta}^{(k)}/s_n^{2+\delta}\right\}^{\frac{2}{2+\delta}}.
$$

Consequently, we have

$$\max_{1 \le k \le n} \frac{\sigma_k^2}{s_n^2} \le \max_{1 \le k \le n} \left\{ \frac{\nu_{2+\delta}^{(k)}}{s_n^{2+\delta}} \right\}^{\frac{2}{2+\delta}}$$

$$= \left[\max_{1 \le k \le n} \left\{ \frac{\nu_{2+\delta}^{(k)}}{s_n^{2+\delta}} \right\} \right]^{\frac{2}{2+\delta}}$$

$$\le \left\{ \sum_{k=1}^{n} \frac{\nu_{2+\delta}^{(k)}}{s_n^{2+\delta}} \right\}^{\frac{2}{2+\delta}}$$

$$= \rho_n^{\frac{2}{2+\delta}}.$$

Since $\lim_{n \to \infty} \rho_n = 0$, it follows that

$$\max_{1 \le k \le n} \frac{\sigma_k^2}{s_n^2} \to 0 \quad \text{as} \quad n \to \infty. \tag{3.3.1}$$

Next, let $U_k = (X_k - \mu_k)/\sigma_k$. Then $EU_k = 0$, $EU_k^2 = 1$ and $E|U_k|^{2+\delta} = \nu_{2+\delta}^{(k)}/\sigma_k^{2+\delta}$; furthermore, $Z_n = \sum_{k=1}^{n} \sigma_k U_k/s_n$. Now, writing $\phi_{Z_n}(t) = E\{\exp(itZ_n)\}$ and $\phi_{U_k}(t) = E\{\exp(itU_k)\}$ we have

$$\phi_{Z_n}(t) = E\left\{ \exp\left(it \sum_{k=1}^{n} \frac{\sigma_k}{s_n} U_k \right) \right\}$$

$$= \prod_{k=1}^{n} E\left\{ \exp\left(it \frac{\sigma_k}{s_n} U_k \right) \right\}$$

$$= \prod_{k=1}^{n} \phi_{U_k}\left(\frac{\sigma_k}{s_n} t \right)$$

which implies

$$\log \phi_{Z_n}(t) = \sum_{k=1}^{n} \log \phi_{U_k}\left(\frac{\sigma_k}{s_n} t \right). \tag{3.3.2}$$

Now, for $\left| 1 - \phi_{U_k}(\sigma_k t/s_n) \right| < 1$, consider the expansion

$$\log \phi_{U_k}\left(\frac{\sigma_k}{s_n} t \right) = \log\left\{ 1 - \left[1 - \phi_{U_k}\left(\frac{\sigma_k}{s_n} t \right) \right] \right\}$$

$$= -\left\{ 1 - \phi_{U_k}\left(\frac{\sigma_k}{s_n} t \right) \right\} - \sum_{r=2}^{\infty} \frac{1}{r} \left\{ 1 - \phi_{U_k}\left(\frac{\sigma_k}{s_n} t \right) \right\}^r \tag{3.3.3}$$

and let us examine the terms on the right-hand side; using the Taylor expansion

$$e^x = 1 + x + \frac{1}{2} x^2 e^{hx}, \quad 0 < h < 1,$$

we get

$$
\begin{aligned}
\left| 1 - \phi_{U_k}\left(\frac{\sigma_k}{s_n}t\right) \right| &= \left| \int_{-\infty}^{+\infty}\left\{ 1 - \exp\left(it\frac{\sigma_k}{s_n}u\right) \right\} dP\{U_k \le u\} \right| \\
&= \left| -it\frac{\sigma_k}{s_n}\int_{-\infty}^{+\infty} u\, dP\{U_k \le u\} \right. \\
&\quad \left. +\frac{1}{2}t^2\frac{\sigma_k^2}{s_n^2}\int_{-\infty}^{+\infty} u^2 \exp\left\{ it\frac{\sigma_k}{s_n}uh \right\} dP\{U_k \le u\} \right| \\
&\le \frac{1}{2}t^2\frac{\sigma_k^2}{s_n^2}\int_{-\infty}^{+\infty} u^2 \left| \exp\left\{ it\frac{\sigma_k}{s_n}uh \right\} \right| dP\{U_k \le u\} \\
&\le \frac{1}{2}t^2\frac{\sigma_k^2}{s_n^2}\int_{-\infty}^{+\infty} u^2\, dP\{U_k \le u\} = \frac{1}{2}t^2\frac{\sigma_k^2}{s_n^2}. \qquad (3.3.4)
\end{aligned}
$$

Therefore, using (3.3.1), it follows that

$$
\max_{1 \le k \le n}\left| 1 - \phi_{U_k}\left(\frac{\sigma_k}{s_n}t\right) \right| \le \frac{1}{2}t^2 \max_{1 \le k \le n}\frac{\sigma_k^2}{s_n^2} = o(1) \quad \text{as} \quad n \to \infty. \qquad (3.3.5)
$$

Furthermore, observe that

$$
\begin{aligned}
\left| \sum_{r=2}^{\infty}\left[-\frac{1}{r}\left\{ 1 - \phi_{U_k}\left(\frac{\sigma_k}{s_n}t\right) \right\}^r \right] \right| &\le \sum_{r=2}^{\infty}\frac{1}{r}\left| 1 - \phi_{U_k}\left(\frac{\sigma_k}{s_n}t\right) \right|^r \\
&\le \frac{1}{2}\sum_{r=2}^{\infty}\left| 1 - \phi_{U_k}\left(\frac{\sigma_k}{s_n}t\right) \right|^r = \frac{\frac{1}{2}|1 - \phi_{U_k}(\sigma_k t/s_n)|^2}{1 - |1 - \phi_{U_k}(t\sigma_k/s_n)|} \\
&\le \left| 1 - \phi_{U_k}\left(\frac{\sigma_k}{s_n}t\right) \right|^2, \quad n \ge n_0. \qquad (3.3.6)
\end{aligned}
$$

From (3.3.3)–(3.3.6) we obtain that, for $n \ge n_0$,

$$
\begin{aligned}
\sum_{k=1}^{n}\left| \log\phi_{U_k}\left(\frac{\sigma_k}{s_n}t\right) + \left\{ 1 - \phi_{U_k}\left(\frac{\sigma_k}{s_n}t\right) \right\} \right| &\le \sum_{k=1}^{n}\sum_{r=2}^{\infty}\frac{1}{r}\left| 1 - \phi_{U_k}\left(\frac{\sigma_k}{s_n}t\right) \right|^r \\
&\le \sum_{k=1}^{n}\left| 1 - \phi_{U_k}\left(\frac{\sigma_k}{s_n}t\right) \right|^2 \\
&\le \max_{1 \le k \le n}\left| 1 - \phi_{U_k}\left(\frac{\sigma_k}{s_n}t\right) \right| \sum_{k=1}^{n}\left| 1 - \phi_{U_k}\left(\frac{\sigma_k}{s_n}t\right) \right| \\
&\le o(1)\sum_{k=1}^{n}\frac{1}{2}t^2\frac{\sigma_k^2}{s_n^2} = o(1)\frac{t^2}{2} = o(1) \quad \text{as} \quad n \to \infty,
\end{aligned}
$$

which implies that as $n \to \infty$,

$$
\begin{aligned}
\log \phi_{Z_n}(t) &= \sum_{k=1}^{n} \log \phi_{U_k}\left(\frac{\sigma_k}{s_n}t\right) \\
&= -\sum_{k=1}^{n}\left\{1 - \phi_{U_k}\left(\frac{\sigma_k}{s_n}t\right)\right\} + o(1). \quad (3.3.7)
\end{aligned}
$$

Then, using Theorem 1.4.1, (3.3.7) enables us to write

$$
\log \phi_{Z_n}(t) = \sum_{k=1}^{n}\left\{-\frac{t^2}{2}\frac{\sigma_k^2}{s_n^2} + R_{2k}\left(\frac{\sigma_k}{s_n}t\right)\right\} + o(1) \quad \text{as} \quad n \to \infty, \quad (3.3.8)
$$

where

$$
|R_{2k}(t\sigma_k/s_n)| \leq c|t|^{2+\delta}\nu_{2+\delta}^{(k)}/s_n^{2+\delta} \quad \text{for some} \quad c > 0.
$$

Now, note that

$$
\begin{aligned}
\left|\sum_{k=1}^{n} R_{2k}\left(\frac{\sigma_k}{s_n}t\right)\right| &\leq \sum_{k=1}^{n}\left|R_{2k}\left(\frac{\sigma_k}{s_n}t\right)\right| \\
&\leq c|t|^{2+\delta}\sum_{k=1}^{n}\frac{\nu_{2+\delta}^{(k)}}{s_n^{2+\delta}} = c|t|^{2+\delta}\rho_n \to 0 \quad \text{as} \quad n \to \infty.
\end{aligned}
$$

Thus, from (3.3.8) we get $\log \phi_{Z_n}(t) = -t^2/2 + o(1)$, and it follows that as $n \to \infty$, $\phi_{Z_n}(t) \to \exp(-t^2/2)$ which is the characteristic function of the standard normal distribution. ∎

In the original formulation, Liapounov used $\delta = 1$, but even the existence of $\nu_{2+\delta}^{(k)}$, $0 < \delta \leq 1$ is not a necessary condition, as we may see from the following theorem.

Theorem 3.3.3 (Lindeberg-Feller): *Let X_k, $k \geq 1$, be independent random variables such that* $EX_k = \mu_k$ *and* $\mathrm{Var}X_k = \sigma_k^2$, $k \geq 1$; *also let* $T_n = \sum_{k=1}^{n} X_k$, $\xi_n = ET_n = \sum_{k=1}^{n}\mu_k$, $s_n^2 = \mathrm{Var}T_n = \sum_{k=1}^{n}\sigma_k^2$ *and* $Z_n = (T_n - \xi_n)/s_n = \sum_{k=1}^{n} Y_{nk}$ *where* $Y_{nk} = (X_k - \mu_k)/s_n$. *Consider the following conditions:*

A) Uniform asymptotic negligibility (UAN) condition*:*

$$
\max_{1 \leq k \leq n} \frac{\sigma_k^2}{s_n^2} \to 0 \quad \text{as} \quad n \to \infty.
$$

Note that this condition implies that the random variables Y_{nk} are **infinitesimal**, *that is,* $\max_{1 \leq k \leq n} P\{|Y_{nk}| > \varepsilon\} \to 0$ *as $n \to \infty$ for* *every $\varepsilon > 0$. To see this, observe that for every $\varepsilon > 0$*

$$
\begin{aligned}
0 \leq \max_{1 \leq k \leq n} P\{|Y_{nk}| > \varepsilon\} &= \max_{1 \leq k \leq n} \int_{\{|y| \geq \varepsilon\}} dP\{Y_{nk} \leq y\} \\
&\leq \frac{1}{\varepsilon^2} \max_{1 \leq k \leq n} \int_{\{|y| \geq \varepsilon\}} y^2 dP\{Y_{nk} \leq y\} \\
&\leq \frac{1}{\varepsilon^2} \max_{1 \leq k \leq n} \frac{\sigma_k^2}{s_n^2} \to 0 \quad as \quad n \to \infty.
\end{aligned}
$$

In other words, the UAN condition states that the random variables Y_{nk}, $1 \leq k \leq n$, are uniformly in k, asymptotically in n, negligible.

B) Asymptotic normality condition:

$$
P\{Z_n \leq z\} \to \frac{1}{\sqrt{2\pi}} \int_{-\infty}^{z} \exp\left(\frac{-t^2}{2}\right) dt = \Phi(z).
$$

C) Lindeberg-Feller condition (uniform integrability):

$$
\forall \varepsilon > 0, \quad \frac{1}{s_n^2} \sum_{k=1}^{n} E\left[(X_k - \mu_k)^2 I_{\{|X_k - \mu_k| > \varepsilon s_n\}}\right] \to 0 \quad as \quad n \to \infty.
$$

Then, (A) and (B) hold simultaneously if and only if (C) holds.

Proof: First suppose that (C) holds and let us show that this implies (A) and (B). Note that, for all $k \geq 1$,

$$
\begin{aligned}
\sigma_k^2 &= E\left[(X_k - \mu_k)^2 I_{\{|X_k - \mu_k| \leq \varepsilon s_n\}}\right] + E\left[(X_k - \mu_k)^2 I_{\{|X_k - \mu_k| > \varepsilon s_n\}}\right] \\
&\leq \varepsilon^2 s_n^2 + E\left[(X_k - \mu_k)^2 I_{\{|X_k - \mu_k| > \varepsilon s_n\}}\right].
\end{aligned}
$$

Thus, we have

$$
\begin{aligned}
\max_{1 \leq k \leq n} \frac{\sigma_k^2}{s_n^2} &\leq \varepsilon^2 + \max_{1 \leq k \leq n} \frac{1}{s_n^2} E\left[(X_k - \mu_k)^2 I_{\{|X_k - \mu_k| > \varepsilon s_n\}}\right] \\
&\leq \varepsilon^2 + \frac{1}{s_n^2} \sum_{k=1}^{n} E\left[(X_k - \mu_k)^2 I_{\{|X_k - \mu_k| > \varepsilon s_n\}}\right].
\end{aligned}
$$

From condition (C) and the fact that $\varepsilon > 0$ can be made arbitrarily small, it follows that (A) holds. Now let us show that (C) and (A) imply (B). In this direction we may follow the lines of the proof of Theorem 3.3.2 up to expression (3.3.7). Now, let

$$
R_{2k}(t) = \int_{-\infty}^{+\infty} \{e^{itu} - 1 - itu + t^2 u^2/2\} dP\{U_k \leq u\}
$$

and consider the expansion

$$
\begin{aligned}
\phi_{U_k}(t) &= 1 + it \int_{-\infty}^{+\infty} u\, dP\{U_k \le u\} - \frac{t^2}{2} \int_{-\infty}^{+\infty} u^2 dP\{U_k \le u\} + R_{2k}(t) \\
&= 1 - \frac{t^2}{2} + R_{2k}(t).
\end{aligned}
$$

Therefore, using (3.3.7) we may write

$$
\log \phi_{Z_n}(t) = -\frac{1}{2}t^2 + \sum_{k=1}^{n} R_{2k}\left(\frac{\sigma_k}{s_n}t\right) + o(1) \quad \text{as} \quad n \to \infty \qquad (3.3.9)
$$

and all we have to show is that $\sum_{k=1}^{n} |R_{2k}(t\sigma_k/s_n)| < \varepsilon$, for all $\varepsilon > 0$. Write

$$
g(u) = \exp\left(it\frac{\sigma_k}{s_n}u\right) - 1 - it\frac{\sigma_k}{s_n}u + \frac{t^2}{2}\frac{\sigma_k^2}{s_n^2}u^2
$$

and note that

$$
\begin{aligned}
\left| R_{2k}\left(\frac{\sigma_k}{s_n}t\right)\right| &\le \left| \int_{\{|u| \le \varepsilon s_n/\sigma_k\}} g(u)\, dP\{U_k \le u\}\right| \\
&\quad + \left| \int_{\{|u| > \varepsilon s_n/\sigma_k\}} g(u)\, dP\{U_k \le u\}\right|. \qquad (3.3.10)
\end{aligned}
$$

Using a third order Taylor expansion for $\exp(it\frac{\sigma_k}{s_n}u)$, we have

$$
g(u) = \frac{1}{3!}\left(\frac{itu\sigma_k}{s_n}\right)^3 \exp\left(\frac{it\sigma_k uh}{s_n}\right)
$$

for some $0 < h < 1$ and it follows that

$$
\begin{aligned}
&\left| \int_{\{|u| \le \varepsilon s_n/\sigma_k\}} g(u)\, dP\{U_k \le u\}\right| \\
&\le \int_{\{|u| \le \varepsilon s_n/\sigma_k\}} \frac{1}{3!}\left|\left(\frac{itu\sigma_k}{s_n}\right)^3\right| \left|\exp\left(\frac{it\sigma_k uh}{s_n}\right)\right| dP\{U_k \le u\} \\
&\le |t|^3\frac{\sigma_k^3}{s_n^3} \int_{\{|u| \le \varepsilon s_n/\sigma_k\}} |u|^3 dP\{U_k \le u\} \\
&\le \varepsilon|t|^3\frac{\sigma_k^2}{s_n^2} \int_{\{|u| \le \varepsilon s_n/\sigma_k\}} u^2 dP\{U_k \le u\} \\
&\le \varepsilon|t|^3\frac{\sigma_k^2}{s_n^2}. \qquad (3.3.11)
\end{aligned}
$$

Using a second order Taylor expansion for $\exp(it\frac{\sigma_k}{s_n}u)$, we have

$$
g(u) = \frac{1}{2}(tu\sigma_k/s_n)^2(1 - \exp(it\sigma_k uh/s_n))
$$

for some $0 < h < 1$ and it follows that

$$
\left| \int_{\{|u|>\varepsilon s_n/\sigma_k\}} g(u) dP\{U_k \leq u\} \right|
$$

$$
\leq \frac{1}{2}\left(\frac{t\sigma_k}{s_n}\right)^2 \int_{\{|u|>\varepsilon s_n/\sigma_k\}} u^2 |1 - e^{(it\sigma_k uh/s_n)}| dP\{U_k \leq u\}
$$

$$
\leq \frac{t^2}{s_n^2} \int_{\{|u|>\varepsilon s_n/\sigma_k\}} u^2 \sigma_k^2 dP\{U_k \leq u\}
$$

$$
= \frac{t^2}{s_n^2} \mathrm{E}\left[(X_k - \mu_k)^2 I_{\{|X_k - \mu_k|>\varepsilon s_n\}} \right]. \tag{3.3.12}
$$

From (3.3.10)–(3.3.12) we obtain

$$
\left| R_{2k}\left(\frac{\sigma_k}{s_n} t\right) \right| \leq \varepsilon |t|^3 \frac{\sigma_k^2}{s_n^2} + \frac{t^2}{s_n^2} \mathrm{E}\left[(X_k - \mu_k)^2 I_{\{|X_k - \mu_k|>\varepsilon s_n\}} \right],
$$

and summing over k we may write

$$
\sum_{k=1}^{n} \left| R_{2k}\left(\frac{\sigma_k}{s_n} t\right) \right| \leq \varepsilon |t|^3 + \frac{t^2}{s_n^2} \sum_{k=1}^{n} \mathrm{E}\left[(X_k - \mu_k)^2 I_{\{|X_k - \mu_k|>\varepsilon s_n\}} \right]
$$

Since the first term on the right-hand side can be made arbitrarily small by choosing $\varepsilon > 0$ sufficiently small and the second term converges to zero by condition (C), it follows from (3.3.9) that as $n \to \infty$, $\phi_{Z_n}(t) \to \exp\{-t^2/2\}$ which is characteristic function of a standard normal distribution and, therefore, (B) holds.

Now let us prove that (A) and (B) imply (C). First recall that condition (A) implies (3.3.7). Then we may write

$$
\log \phi_{Z_n}(t) = \ - \sum_{k=1}^{n} \mathrm{E}\left\{ 1 - \cos\left(\frac{t\sigma_k}{s_n} U_k\right) \right\}
$$

$$
+ \ i \sum_{k=1}^{n} \mathrm{E}\left\{ \sin\left(\frac{t\sigma_k}{s_n} U_k\right) \right\} + o(1) \quad \text{as} \quad n \to \infty.
$$

Since (B) holds, we may conclude that the coefficient of the imaginary term must converge to zero; furthermore, for the real-valued term we must have

$$
\sum_{k=1}^{n} \int_{-\infty}^{+\infty} \left\{ 1 - \cos\left(\frac{t\sigma_k}{s_n} u\right) \right\} dP\{U_k \leq u\} = \frac{t^2}{2} + o(1) \quad \text{as} \quad n \to \infty.
$$

Thus, as $n \to \infty$,

$$
\frac{t^2}{2} - \sum_{k=1}^{n} \int_{\{|u|\leq\varepsilon s_n/\sigma_k\}} \left\{ 1 - \cos\left(\frac{t\sigma_k}{s_n} u\right) \right\} dP\{U_k \leq u\}
$$

$$= \sum_{k=1}^{n} \int_{\{|u|>\varepsilon s_n/\sigma_k\}} \left\{1 - \cos\left(\frac{t\sigma_k}{s_n}u\right)\right\} dP\{U_k \le u\} + o(1).$$

$$(3.3.13)$$

Now observe that for $|y| < 1$ we have

$$\cos y = 1 - \frac{y^2}{2} + \frac{y^4}{4!} - \frac{y^6}{6!} + \cdots$$

which implies $1 - \cos y \le y^2/2$. Thus, for the left-hand side of (3.3.13) we get

$$\frac{t^2}{2} - \sum_{k=1}^{n} \int_{\{|u|\le\varepsilon s_n/\sigma_k\}} \left\{1 - \cos\left(\frac{t\sigma_k}{s_n}u\right)\right\} dP\{U_k \le u\}$$

$$\ge \frac{t^2}{2} - \frac{t^2}{2} \sum_{k=1}^{n} \frac{\sigma_k^2}{s_n^2} \int_{\{|u|\le\varepsilon s_n/\sigma_k\}} u^2 dP\{U_k \le u\}$$

$$= \frac{t^2}{2} \left\{1 - \frac{1}{s_n^2} \sum_{k=1}^{n} \mathrm{E}\left[(X_k - \mu_k)^2 I_{\{|X_k-\mu_k|\le\varepsilon s_n\}}\right]\right\}.$$

$$(3.3.14)$$

Observing that $1 - \cos y \le 2$ and that $u^2\sigma_k^2/\varepsilon^2 s_n^2 > 1$ in the region $\{|u| > \varepsilon s_n/\sigma_k\}$, for the term on the right-hand side of (3.3.13) we may write:

$$\sum_{k=1}^{n} \int_{\{|u|>\varepsilon s_n/\sigma_k\}} \left\{1 - \cos\left(\frac{t\sigma_k}{s_n}u\right)\right\} dP\{U_k \le u\}$$

$$\le 2 \sum_{k=1}^{n} \int_{\{|u|>\varepsilon s_n/\sigma_k\}} dP\{U_k \le u\}$$

$$\le \frac{2}{\varepsilon^2 s_n^2} \sum_{k=1}^{n} \sigma_k^2 \int_{\{|u|>\varepsilon s_n/\sigma_k\}} u^2 dP\{U_k \le u\}$$

$$= \frac{2}{\varepsilon^2 s_n^2} \sum_{k=1}^{n} \mathrm{E}\left[(X_k - \mu_k)^2 I_{\{|X_k-\mu_k|>\varepsilon s_n\}}\right]. \qquad (3.3.15)$$

Then, from (3.3.13)–(3.3.15) we have

$$\frac{t^2}{2} \left\{1 - \frac{1}{s_n^2} \sum_{k=1}^{n} \mathrm{E}[(X_k - \mu_k)^2 I_{\{|X_k-\mu_k|\le\varepsilon s_n\}}]\right\}$$

$$\le \frac{2}{\varepsilon^2 s_n^2} \sum_{k=1}^{n} \mathrm{E}\left[(X_k - \mu_k)^2 I_{\{|X_k-\mu_k|>\varepsilon s_n\}}\right] + o(1) \quad \text{as} \quad n \to \infty.$$

Dividing both members by $t^2/2$, it follows that

$$1 - \frac{1}{s_n^2} \sum_{k=1}^{n} \mathrm{E}\left[(X_k - \mu_k)^2 I_{\{|X_k - \mu_k| \leq \varepsilon s_n\}}\right]$$

$$\leq \frac{4}{\varepsilon^2 t^2 s_n^2} \sum_{k=1}^{n} \mathrm{E}\left[(X_k - \mu_k)^2 I_{\{|X_k - \mu_k| > \varepsilon s_n\}}\right] + o(t^{-2})$$

$$\leq \frac{4}{\varepsilon^2 t^2} \sum_{k=1}^{n} \frac{\sigma_k^2}{s_n^2} + o(t^{-2})$$

$$= \frac{4}{\varepsilon^2 t^2} + o(t^{-2}) \quad \text{as} \quad n \to \infty.$$

Since this must hold for all t, we may let $t \to \infty$ as well, to conclude that the right-hand side converges to zero; the left-hand side being positive must also converge to zero. Hence, we have

$$1 - \frac{1}{s_n^2} \sum_{k=1}^{n} \mathrm{E}\left[(X_k - \mu_k)^2 I_{\{|X_k - \mu_k| \leq \varepsilon s_n\}}\right]$$

$$= \frac{1}{s_n^2}\left\{ s_n^2 - \sum_{k=1}^{n} \mathrm{E}\left[(X_k - \mu_k)^2 I_{\{|X_k - \mu_k| \leq \varepsilon s_n\}}\right]\right\}$$

$$= \frac{1}{s_n^2} \sum_{k=1}^{n} \mathrm{E}\left[(X_k - \mu_k)^2 I_{\{|X_k - \mu_k| > \varepsilon s_n\}}\right] = o(1) \quad \text{as} \quad n \to \infty$$

and the proof is complete. ∎

The first part of the proof, (C) implies (A, B) is due to Lindeberg and the second part (A, B) implies (C) is due to Feller. It is interesting to note that one of (A) or (B) may hold if (C) does not hold; in this direction, consider the following example:

Example 3.3.1: Let Y_k, $k \geq 1$, be independent $N(0, 1)$ random variables and let $X_k = a^k Y_k$, $k \geq 1$, with $a \neq 1$. Then it follows that X_k, $k \geq 1$, are independent $N(0, a^{2k})$ random variables; also, denoting $\mathrm{Var} X_k = \sigma_k^2$, we have $s_n^2 = \sum_{k=1}^{n} \sigma_k^2 = a^2(a^{2n} - 1)/(a^2 - 1)$. Hence, for $a > 1$,

$$\max_{1 \leq k \leq n} \frac{\sigma_k^2}{s_n^2} = \frac{a^{2n}(a^2 - 1)}{a^2(a^{2n} - 1)} \to 1 - \frac{1}{a^2} > 0 \quad \text{as} \quad n \to \infty,$$

and for $a < 1$,

$$\max_{1 \leq k \leq n} \frac{\sigma_k^2}{s_n^2} = \frac{a^2(a^2 - 1)}{a^2(a^{2n} - 1)} \to 1 - a^2 > 0 \quad \text{as} \quad n \to \infty.$$

Thus, (A) does not hold and from the first part of the Lindeberg-Feller

Theorem (3.3.3) it follows that (C) does not hold. However, $\sum_{k=1}^{n} X_k \sim N(0, s_n^2)$ which implies that $s_n^{-1} \sum_{k=1}^{n} X_k \sim N(0, 1)$ and, consequently, that (B) holds. ∎

It is also noteworthy to note that the Liapounov Theorem (3.3.2) is a corollary of the Lindeberg-Feller Theorem (3.3.3); this follows directly from the fact that

$$\frac{1}{s_n^2} \sum_{k=1}^{n} \mathrm{E}\left[(X_k - \mu_k)^2 I_{\{|X_k - \mu_k| > \varepsilon s_n\}}\right]$$

$$\leq \frac{1}{\varepsilon^\delta s_n^{2+\delta}} \sum_{k=1}^{n} \mathrm{E}\left[|X_k - \mu_k|^{2+\delta} I_{\{|X_k - \mu_k| > \varepsilon s_n\}}\right]$$

$$\leq \frac{1}{\varepsilon^\delta s_n^{2+\delta}} \sum_{k=1}^{n} \mathrm{E}|X_k - \mu_k|^{2+\delta} = \frac{\rho_n}{\varepsilon^\delta}.$$

If the random variables under consideration are bounded, an even simpler result is valid:

Theorem 3.3.4: *Let X_k, $k \geq 1$, be independent random variables such that $P\{a \leq X_k \leq b\} = 1$ for some finite scalars $a < b$. Also let $\mathrm{E}X_k = \mu_k$, $\mathrm{Var}X_k = \sigma_k^2$, $T_n = \sum_{k=1}^{n} X_k$, $\xi_n = \sum_{k=1}^{n} \mu_k$ and $s_n^2 = \sum_{k=1}^{n} \sigma_k^2$. Then $Z_n = (T_n - \xi_n)/s_n \xrightarrow{D} N(0, 1)$ if and only if $s_n \to \infty$ as $n \to \infty$.*

Proof: First suppose that $s_n \to \infty$ as $n \to \infty$. Then note that

$$|X_k - \mu_k|^3 = |X_k - \mu_k|(X_k - \mu_k)^2 \leq (b - a)(X_k - \mu_k)^2$$

which implies

$$\mathrm{E}|X_k - \mu_k|^3 \leq (b - a)\sigma_k^2.$$

Therefore

$$\rho_n = \sum_{k=1}^{n} \mathrm{E}|X_k - \mu_k|^3 / s_n^3 \leq (b - a)/s_n \to 0 \quad \text{as} \quad n \to \infty$$

and the result follows from Liapounov Theorem (3.3.2). Now suppose that $s_n \to s < \infty$ as $n \to \infty$ and write

$$Z_n = (T_n - \xi_n)/s_n = (X_1 - \mu_1)/s_n + \sum_{k=2}^{n}(X_k - \mu_k)/s_n.$$

Then, if $Z_n \xrightarrow{D} N(0, 1)$ each of the terms on the right-hand side must also converge to $N(0, 1)$ random variables; this is absurd since $(X_1 - \mu_1)/s$ is a bounded random variable. Therefore, $s_n \to \infty$ as $n \to \infty$. ∎

Up to this point we have devoted attention to the weak convergence of sequences of statistics $\{T_n, n \geq 1\}$ constructed from **independent** underlying random variables $X_1, X_2, \ldots$ We consider now some extensions of the central limit theorems where such restriction may be relaxed. The first of such extensions holds for sequences of (possibly dependent) random variables which may be structured as a **double array** of the form

$$
\begin{matrix}
X_{11}, & X_{12}, & \cdots, & X_{1k_1} \\
X_{21}, & X_{22}, & \cdots, & X_{2k_2} \\
\vdots & \vdots & \ddots & \vdots \\
X_{n1}, & X_{n2}, & \cdots, & X_{nk_n}
\end{matrix}
$$

provided that the X_{nk}'s are row-wise independent. The case where $k_n = n$, $n \geq 1$, is usually termed a **triangular array** of random variables. As we shall see in Chapter 4, such a result is very useful in the field of order statistics.

Theorem 3.3.5: *Consider a double array of random variables $\{X_{nk}, 1 \leq k \leq k_n, n \geq 1\}$ where $k_n \to \infty$ as $n \to \infty$ and such that for each n, $\{X_{nk}, 1 \leq k \leq k_n\}$ are independent. Then*

  i) $\{X_{nk}, 1 \leq k \leq k_n, n \geq 1\}$ *is an infinitesimal system of random variables, that is, for every $\varepsilon > 0$, $\max_{1 \leq k \leq k_n} P\{|X_{nk}| > \varepsilon\} \to 0$ as $n \to \infty$ (the UAN condition), and*

 ii) $Z_n = \sum_{k=1}^{k_n} X_{nk} \xrightarrow{D} N(0, 1)$

hold simultaneously, if and only if, for every $\varepsilon > 0$, as $n \to \infty$

A) $\displaystyle \sum_{k=1}^{k_n} P\{|X_{nk}| > \varepsilon\} \to 0$

B) $\displaystyle \sum_{k=1}^{k_n} \left\{ \int_{\{|x| \leq \varepsilon\}} x^2 dP(X_{nk} \leq x) - \left[\int_{\{|x| \leq \varepsilon\}} x \, dP(X_{nk} \leq x) \right]^2 \right\} \to 1.$

Proof: Here we prove that (A) and (B) are sufficient conditions; the reader may refer to Tucker (1967, p. 197) for a proof of the necessity part. First note that $\max_{1 \leq k \leq k_n} P\{|X_{nk}| > \varepsilon\} \leq \sum_{k=1}^{k_n} P\{|X_{nk}| > \varepsilon\}$; thus (i) follows from (A). Then let $Y_{nk} = X_{nk} I_{\{|X_{nk}| \leq \varepsilon\}}$, $1 \leq k \leq k_n$, $n \geq 1$, and $Z_n^* = \sum_{k=1}^{k_n} Y_{nk}$. Now observe that

$$
\begin{aligned}
P\{Z_n \neq Z_n^*\} &\leq P\{X_{nk} \neq Y_{nk} \text{ for some } k = 1, \ldots, k_n\} \\
&\leq \sum_{k=1}^{k} P\{|X_{nk}| > \varepsilon\}.
\end{aligned}
$$

Therefore from (A), it follows that $P\{Z_n \neq Z_n^*\} \to 0$ as $n \to \infty$ and we may restrict ourselves to the limiting distribution of Z_n^*. Next note that since $P\{|Y_{nk}| \leq \varepsilon\} = 1$, we may write

$$\rho_n = \sum_{k=1}^{k_n} \frac{\mathrm{E}|Y_{nk} - \mathrm{E}Y_{nk}|^3}{\left\{\sum_{k=1}^{k_n} \mathrm{Var} Y_{nk}\right\}^{3/2}} \leq \frac{2\varepsilon \sum_{k=1}^{k_n} \mathrm{E}(Y_{nk} - \mathrm{E}Y_{nk})^2}{\left\{\sum_{k=1}^{k_n} \mathrm{Var} Y_{nk}\right\}^{3/2}}$$

$$= 2\varepsilon \left\{\sum_{k=1}^{k_n} \mathrm{Var} Y_{nk}\right\}^{-1/2}$$

Then, from the fact that ε can be made arbitrarily small and that (B) implies $\sum_{k=1}^{k_n} \mathrm{Var} Y_{nk} \to 1$ as $n \to \infty$ we may conclude that $\rho_n \to 0$ as $n \to \infty$. The result follows from Liapounov Theorem (3.3.2). ∎

An application of the above theorem which is especially useful in Regression Analysis is given by the following result:

Theorem 3.3.6 (Hájek-Šidak): *Let $\{Y_n\}$ be a sequence of i.i.d. r.v.'s with mean μ and finite variance σ^2; let $\{\mathbf{c}_n\}$ be a sequence of real vectors $\mathbf{c}_n = (c_{n1}, \ldots, c_{nn})^t$. Then if*

$$\max_{1 \leq i \leq n} c_{ni}^2 \Big/ \sum_{i=1}^{n} c_{ni}^2 \to 0 \quad as \quad n \to \infty,$$

it follows that

$$Z_n = \left\{\sum_{i=1}^{n} c_{ni}(Y_i - \mu)\right\} \Big/ \left\{\sigma^2 \sum_{i=1}^{n} c_{ni}^2\right\}^{1/2} \xrightarrow{\mathcal{D}} N(0,1).$$

Proof: Let $X_{ni} = c_{ni}Y_i$ and write $T_n = \sum_{i=1}^{n} X_{ni}$, $\mathrm{E}(T_n) = \mu \sum_{i=1}^{n} c_{ni}$. Without loss of generality, we take $\sum_{i=1}^{n} c_{ni}^2 = 1$ so that $s_n^2 = \mathrm{Var} T_n = \sigma^2$. Then note that for all $\varepsilon > 0$, as $n \to \infty$,

$$\frac{1}{s_n^2} \sum_{i=1}^{n} \mathrm{E}\left[(X_{ni} - c_{ni}\mu)^2 I_{\{|X_{ni} - c_{ni}\mu| > \varepsilon s_n\}}\right]$$

$$= \frac{1}{\sigma^2} \sum_{i=1}^{n} c_{ni}^2 \mathrm{E}\left[(Y_i - \mu)^2 I_{\{|Y_i - \mu| > \varepsilon \sigma / c_{ni}\}}\right]$$

$$\leq \frac{1}{\sigma^2} \mathrm{E}\left[(Y_1 - \mu)^2 I_{\{(Y_1 - \mu)^2 > \varepsilon^2 \sigma^2 / \max_{1 \leq i \leq n} c_{ni}^2\}}\right] \to 0 \quad (3.3.16)$$

since $\sum_{i=1}^{n} c_{ni}^2 / \max_{1 \leq i \leq n} c_{ni}^2 \to \infty$ as $n \to \infty$. All we have to do is to verify (A) and (B) of Theorem 3.3.5 which follows trivially from (3.3.16). ∎

It is still possible to relax further the independence assumption on the underlying random variables. The following theorems, stated without proofs, constitute examples of Central Limit Theorems for dependent random variables having a martingale (or reverse martingale) structure. For further details, the reader is referred to Brown (1971a), Dvoretzky (1971), Loynes (1970) or McLeish (1974).

Theorem 3.3.7 (Martingale Central Limit Theorem): *Consider a sequence* $\{X_k, k \geq 1\}$ *of random variables such that* $EX_k = 0$, $EX_k^2 = \sigma_k^2 < \infty$ *and* $E\{X_k \mid X_1, \ldots, X_{k-1}\} = 0$ $(X_0 = 0)$. *Also let* $T_n = \sum_{k=1}^{n} X_k$, $s_n^2 = \sum_{k=1}^{n} \sigma_k^2$, $v_k^2 = E\{X_k^2 \mid X_1, \ldots, X_{k-1}\}$ *and* $w_n^2 = \sum_{k=1}^{n} v_k^2$. *Then, if*

A) $w_n^2/s_n^2 \xrightarrow{P} 1$ *as* $n \to \infty$;

B) for every $\varepsilon > 0$, $s_n^{-2} \sum_{k=1}^{n} E\left[X_k^2 I_{\{|X_k| > \varepsilon s_n\}}\right] \to 0$ *as* $n \to \infty$

(*the Lindeberg-Feller condition*),

it follows that the sequence $\{X_k, k \geq 1\}$ *is infinitesimal and*

$$Z_n = T_n/s_n \xrightarrow{D} N(0,1).$$

It is of interest to remark that:

i) The v_k^2's are random variables (they depend on $X_1, \ldots, X_{k-1}$); condition (A) essentially states that all the information about the variability in th X_k's is contained in $X_1, \ldots, X_{k-1}$.

ii) $\{T_n, \quad n \geq 1\}$ is a **zero mean martingale** since

$$
\begin{aligned}
E(T_n \mid T_1, \ldots, T_{n-1}) &= E(T_n \mid X_1, \ldots, X_{n-1}) \\
&= E(T_{n-1} + X_n \mid X_1, \ldots, X_{n-1}) \\
&= T_{n-1}
\end{aligned}
$$

and $ET_n = 0$; furthermore, since for all $j > i$,

$$EX_i X_j = E\{X_i E(X_j \mid X_1, \ldots, X_i)\} = 0,$$

it follows that

$$ET_n^2 = E\left\{\sum_{k=1}^{n} X_k\right\}^2 = \sum_{k=1}^{n} EX_k^2 + 2 \sum_{1 \leq i \leq j \leq n} EX_i X_j = \sum_{k=1}^{n} \sigma_k^2 = s_n^2.$$

Theorem 3.3.8 (Reverse martingale Central Limit Theorem): *Consider a sequence* $\{T_k, k \geq 1\}$ *of random variables such that*

$$E\{T_n \mid T_{n+1}, T_{n+2}, \ldots\} = T_{n+1} \quad and \quad ET_n = 0,$$

i.e., $\{T_k, k \geq 1\}$ *is a* **zero mean reverse martingale**. *Assume that* $\mathrm{E}T_n^2 < \infty$ *and let* $Y_k = T_k - T_{k+1}, k \geq 1$, $v_k^2 = \mathrm{E}\{Y_k^2 \mid T_{k+1}, T_{k+2}, \ldots\}$ *and* $w_n^2 = \sum_{k=n}^{\infty} v_k^2$. *If*

A)
$$w_n^2/\mathrm{E}w_n^2 \xrightarrow{\text{a.s.}} 1$$

B)

$$w_n^{-2} \sum_{k=n}^{\infty} \mathrm{E}\left[Y_k^2 I_{\{|Y_k| > \varepsilon w_n\}} \middle| T_{k+1}, T_{k+2}, \ldots\right] \xrightarrow{\mathrm{P}} 0, \quad \varepsilon > 0$$

or
$$w_n^{-2} \sum_{k=n}^{\infty} Y_k^2 \xrightarrow{\text{a.s.}} 1,$$

then $T_n/\sqrt{\mathrm{E}w_n^2} \xrightarrow{\mathcal{D}} N(0,1)$.

Example 3.3.2: Let $\{X_k, k \geq 1\}$ be a sequence of i.i.d. r.v.'s such that $\mathrm{E}X_k = 0$ and $\mathrm{Var}X_k = \sigma^2 < \infty$. We know from previous results that $n^{1/2}\overline{X}_n/\sigma \xrightarrow{\mathcal{D}} N(0,1)$. We now indicate how the same conclusion may be reached by an application of the above theorem. First, recall from Example 2.4.6 that $\{\overline{X}_n, n \geq 1\}$ is a **zero mean reverse martingale**, i.e., $\mathrm{E}\{\overline{X}_k \mid \overline{X}_{k+1}, \overline{X}_{k+2}, \ldots\} = \overline{X}_{k+1}$, $\mathrm{E}\overline{X}_k = 0$. Then note that for every $k \leq n$ we have:

$$\begin{aligned}
\mathrm{E}\overline{X}_k\overline{X}_n &= \mathrm{E}\{\mathrm{E}(\overline{X}_k\overline{X}_n \mid \overline{X}_n, \overline{X}_{n+1}, \ldots)\} \\
&= \mathrm{E}\{\overline{X}_n\mathrm{E}(\overline{X}_k \mid \overline{X}_n, \overline{X}_{n+1}, \ldots)\} = \mathrm{E}\overline{X}_n^2.
\end{aligned}$$

Now, let $Y_k = \overline{X}_k - \overline{X}_{k+1}$; this implies $\mathrm{E}\{Y_k \mid \overline{X}_n, \overline{X}_{n+1}, \ldots\} = 0$, for every $k \leq n$. Furthermore, it follows that for every $j \geq 1$

$$\begin{aligned}
\mathrm{E}Y_kY_{k+j} &= \mathrm{E}(\overline{X}_k - \overline{X}_{k+1})(\overline{X}_{k+j} - \overline{X}_{k+j+1}) \\
&= \mathrm{E}\overline{X}_k\overline{X}_{k+j} - \mathrm{E}\overline{X}_k\overline{X}_{k+j+1} - \mathrm{E}\overline{X}_{k+1}\overline{X}_{k+j} + \mathrm{E}\overline{X}_{k+1}\overline{X}_{k+j+1} \\
&= \mathrm{E}\overline{X}_{k+j}^2 - \mathrm{E}\overline{X}_{k+j+1}^2 - \mathrm{E}\overline{X}_{k+j}^2 + \mathrm{E}\overline{X}_{k+j+1}^2 = 0.
\end{aligned}$$

Write

$$v_k^2 = \mathrm{E}\{Y_k^2 \mid \overline{X}_{k+1}, \overline{X}_{k+2}, \ldots\} = \mathrm{E}\{(\overline{X}_k - \overline{X}_{k+1})^2 \mid \overline{X}_{k+1}, \overline{X}_{k+2}, \ldots\}$$

and note that since $\overline{X}_k - \overline{X}_{k+1} = \frac{1}{k}(\overline{X}_{k+1} - X_{k+1})$, we have

$$\begin{aligned}
v_k^2 &= \frac{1}{k^2}\mathrm{E}\{(\overline{X}_{k+1}^2 - 2\overline{X}_{k+1}X_{k+1} + X_{k+1}^2) \mid \overline{X}_{k+1}, \overline{X}_{k+2}, \ldots\} \\
&= \frac{1}{k^2}\{\overline{X}_{k+1}^2 - 2\overline{X}_{k+1}\mathrm{E}(X_{k+1} \mid \overline{X}_{k+1}, \overline{X}_{k+2}, \ldots) \\
&\quad + \mathrm{E}(X_{k+1}^2 \mid \overline{X}_{k+1}, \overline{X}_{k+2}, \ldots)\} \\
&= \frac{1}{k^2}\{\overline{X}_{k+1}^2 - 2\overline{X}_{k+1}^2 + \mathrm{E}(X_{k+1}^2 \mid \overline{X}_{k+1}, \overline{X}_{k+2}, \ldots)\}. \quad (3.3.17)
\end{aligned}$$

Now observe that

$$
\begin{aligned}
& \mathrm{E}(X_{k+1} \mid \overline{X}_{k+1}, \overline{X}_{k+2}, \ldots) \\
&= \binom{k+1}{k}^{-1} \sum_{\{1 \leq \alpha_1 < \cdots < \alpha_k \leq k+1\}} \left\{ \sum_{j=1}^{k+1} X_j - \sum_{j=1}^{k} X_{\alpha_j} \right\} \\
&= \frac{1}{k+1} \left\{ (k+1) \sum_{j=1}^{k+1} X_j - \sum_{\{1 \leq \alpha_1 < \cdots < \alpha_k \leq k+1\}} \sum_{j=1}^{k} X_{\alpha_j} \right\} \\
&= \frac{1}{k+1} \left\{ (k+1) \sum_{j=1}^{k+1} X_j - k \sum_{j=1}^{k+1} X_j \right\} \\
&= \overline{X}_{k+1}.
\end{aligned}
\tag{3.3.18}
$$

Following a similar argument we have

$$
\begin{aligned}
& \mathrm{E}(X_{k+1}^2 \mid \overline{X}_{k+1}, \overline{X}_{k+2}, \ldots) \\
&= \binom{k+1}{k}^{-1} \sum_{\{1 \leq \alpha_1 < \cdots < \alpha_k \leq k+1\}} \left\{ \sum_{j=1}^{k+1} X_j^2 - \sum_{j=1}^{k} X_{\alpha_j}^2 \right\} \\
&= \frac{1}{k+1} \left\{ (k+1) \sum_{j=1}^{k+1} X_j^2 - k \sum_{j=1}^{k+1} X_j^2 \right\} \\
&= \frac{1}{k+1} \sum_{j=1}^{k+1} X_j^2.
\end{aligned}
\tag{3.3.19}
$$

From (3.3.17)–(3.3.19) it follows that

$$
\begin{aligned}
v_k^2 &= \frac{1}{k^2} \left\{ \frac{1}{k+1} \sum_{j=1}^{k+1} X_j^2 - \overline{X}_{k+1}^2 \right\} \\
&= \frac{1}{k^2} \left\{ \frac{1}{k+1} \sum_{j=1}^{k+1} (X_j - \overline{X}_{k+1})^2 \right\} \\
&= \frac{1}{k(k+1)} \left\{ \frac{1}{k} \sum_{j=1}^{k+1} (X_j - \overline{X}_{k+1})^2 \right\} \\
&= \left(\frac{1}{k} - \frac{1}{k+1} \right) s_{k+1}^2,
\end{aligned}
$$

where $s_{k+1}^2 = \frac{1}{k} \sum_{j=1}^{k+1} (X_j - \overline{X}_{k+1})^2$. Now, using (2.3.87), it follows that $s_{k+1}^2 \xrightarrow{\text{a.s.}} \sigma^2$ as $k \to \infty$. Then given $\varepsilon > 0$, there exists $n_0 = n_0(\varepsilon)$ such that for all $k \geq n_0$, $\sigma^2 - \varepsilon \leq s_{k+1}^2 \leq \sigma^2 + \varepsilon$. Thus, using the definition of

w_n^2, for all $n \geq n_0$ we have

$$nw_n^2 = n\sum_{k=n}^{\infty} v_k^2 = n\sum_{k=n}^{\infty}\left(\frac{1}{k} - \frac{1}{k+1}\right)s_{k+1}^2$$

$$\leq (\sigma^2 + \varepsilon)n\sum_{k=n}^{\infty}\left(\frac{1}{k} - \frac{1}{k+1}\right) = \sigma^2 + \varepsilon.$$

Similarly we may show that for all $n \geq n_0$, $nw_n^2 \geq \sigma^2 - \varepsilon$, and since ε is arbitrary, we may conclude that $nw_n^2 \xrightarrow{\text{a.s.}} \sigma^2$. Then, observing that $\mathrm{E}w_n^2 = \sum_{k=n}^{\infty}\left(\frac{1}{k} - \frac{1}{k+1}\right)\mathrm{E}s_{k+1}^2 = \sigma^2/n$, it follows that condition (A) of Theorem 3.3.8 holds.

The proof of condition (B) is somewhat more elaborate; the reader is referred to Loynes (1970) for details. ∎

The central limits theorems discussed above may be generalized to cover the multivariate case. Given a sequence $\{\mathbf{X_n}\}$ of random vectors in $\mathbb{R}^p$, with mean vectors $\boldsymbol{\mu}_n$ and covariance matrices $\boldsymbol{\Sigma}_n$, $n \geq 1$, to show that $n^{-1/2}\sum_{i=1}^{n}(\mathbf{X}_i - \boldsymbol{\mu}_i) \xrightarrow{D} \mathrm{N}_p(\mathbf{O}, \boldsymbol{\Sigma})$ with $\boldsymbol{\Sigma} = \lim_{n\to\infty} n^{-1}\sum_{i=1}^{n}\boldsymbol{\Sigma}_i$ (here we abuse notation as in the univariate case). One generally proceeds according to the following strategy:

i) Use one of the univariate central limit theorems to show that for every fixed $\boldsymbol{\lambda} \in \mathbb{R}^p$, $n^{-1}\sum_{i=1}^{n}\boldsymbol{\lambda}^t(\mathbf{X}_i - \boldsymbol{\mu}_i) \xrightarrow{D} N(0, \gamma)$ with $\gamma = \lim_{n\to\infty} n^{-1}\boldsymbol{\lambda}^t(\sum_{i=1}^{n}\boldsymbol{\Sigma}_i)\boldsymbol{\lambda}$.

ii) Use the Cramér-Wold Theorem (3.2.4) to complete the proof.

As an example consider the following result:

Theorem 3.3.9: Let $\{\mathbf{X}_n\}$ be a sequence of random vectors in $\mathbb{R}^p$ with mean vectors $\boldsymbol{\mu}_n$ and finite covariance matrices $\boldsymbol{\Sigma}_n$, $n \geq 1$, such that

$$\max_{1\leq i\leq n}\max_{1\leq j\leq p} \mathrm{E}|X_{ij} - \mu_{ij}|^{2+\delta} < \infty \quad \text{for some} \quad 0 < \delta < 1,$$

and

$$\boldsymbol{\Sigma} = \lim_{n\to\infty} n^{-1}\sum_{i=1}^{n}\boldsymbol{\Sigma}_i$$

exist. Then $n^{-1/2}\sum_{i=1}^{n}(\mathbf{X}_i - \boldsymbol{\mu}_i) \xrightarrow{D} \mathrm{N}_p(\mathbf{O}, \boldsymbol{\Sigma})$.

Proof: For every $\boldsymbol{\lambda} \in \mathbb{R}^p$ define $\gamma_n^2 = n^{-1}\sum_{i=1}^{n}\boldsymbol{\lambda}^t\boldsymbol{\Sigma}_i\boldsymbol{\lambda}$. Now, using the C_r Inequality (1.4.29) we have

$$\mathrm{E}|\boldsymbol{\lambda}^t(\mathbf{X}_i - \boldsymbol{\mu}_i)|^{2+\delta} \leq p^{1+\delta}\big(\max_{1\leq j\leq p} |\lambda_j|\big)\sum_{i=1}^{p} \mathrm{E}|X_{ij} - \mu_{ij}|^{2+\delta}.$$

Then, for all $\boldsymbol{\lambda} \in I\!\!R^p$ such that $\liminf \gamma_n > 0$, we have

$$\rho_n = \sum_{i=1}^{n} \frac{E|\boldsymbol{\lambda}^t(\mathbf{X}_i - \boldsymbol{\mu}_i)|^{2+\delta}}{(\sqrt{n}\gamma_n)^{2+\delta}}$$

$$\leq \frac{p^{1+\delta}}{\gamma_n^{2+\delta}} \Big(\max_{1 \leq j \leq p} |\lambda_j|\Big)\Big(\sum_{i=1}^{n}\sum_{j=1}^{p} \frac{E|X_{ij} - \mu_{ij}|^{2+\delta}}{n^{1+\delta/2}}\Big).$$

Since

$$\sum_{i=1}^{n}\sum_{j=1}^{p} n^{-(1+\delta/2)} E|X_{ij} - \mu_{ij}|^{2+\delta}$$

$$\leq np/n^{1+\delta/2} \max_{1 \leq i \leq n} \max_{1 \leq j \leq p} E|X_{ij} - \mu_{ij}|^{2+\delta} = O(n^{-\delta/2}),$$

it follows that $\rho_n \to 0$ as $n \to \infty$; thus, from Liapounov Theorem (3.3.2) we may conclude that

$$n^{-1/2}\gamma_n^{-1} \sum_{i=1}^{n}(\boldsymbol{\lambda}^t\mathbf{X}_i - \boldsymbol{\lambda}^t\boldsymbol{\mu}) \xrightarrow{\mathcal{D}} N(0,1).$$

Finally, a direct application of Cramér-Wold Theorem (3.2.4) completes the proof. ∎

To finalize this section we want to present the central limit theorem related to the elementary **Renewal Theorem** introduced via Example 2.5.4. Let $\{X_i, \ i \geq 1\}$ be a sequence of i.i.d. non-negative r.v.'s with mean $\mu \in (0, \infty)$ and variance σ^2, $0 < \sigma^2 < \infty$. For every $n \ (\geq 1)$ let

$$T_n = X_1 + \cdots + X_n, \quad T_0 = 0. \tag{3.3.20}$$

Also, for every $t > 0$, let

$$N_t = \max\{k : T_k \leq t\}. \tag{3.3.21}$$

Thus, $N_0 = 0$ and N_t is $\nearrow$ in t, $t \geq 0$. Note that, by definition,

$$T_{N_t} \leq t \leq T_{N_t+1}, \quad t \geq 0. \tag{3.3.22}$$

From every $a \in I\!\!R$, consider the event

$$\{t^{-1/2}(N_t - t/\mu) < a\} = \{N_t < t/\mu + t^{1/2}a\}. \tag{3.3.23}$$

Then, let $\{r_t; t \geq 0\}$ be a sequence of non-negative integers, defined by

$$r_t \leq t/\mu + t^{1/2}a < r_t + 1, \quad t > 0, \tag{3.3.24}$$

where a is held fixed. Recall that $0 < \mu < \infty$, so that for any fixed a,

$$\lim_{t \to \infty} r_t(t/\mu) = 1, \quad \text{i.e.,} \quad r_t \to \infty \text{ as } t \to \infty. \tag{3.3.25}$$

Consequently, by (3.3.22) and (3.3.23), we have

$$
\lim_{t \to \infty} P\{t^{-1/2}(N_t - t/\mu) < \alpha\} = \lim_{r_t \to \infty} P\{T_{r_t+1} > t\}
$$

$$
= \lim_{r_t \to \infty} P\{T_{r_t+1} - (r_t + 1)\mu > t - (r_t + 1)\mu\}
$$

$$
= \lim_{r_t \to \infty} P\left\{(r_t + 1)^{-1/2}[T_{r_t+1} - \mu(r_t + 1)] > \frac{t - (r_t + 1)\mu}{\sqrt{r_t + 1}}\right\}.
$$

$$(3.3.26)$$

Now, from (3.3.24), we obtain

$$
\lim_{t \to \infty} \frac{t - (r_t + 1)\mu}{\sqrt{r_t + 1}} = \lim_{t \to \infty} \frac{-a\mu\sqrt{t}}{\sqrt{t/\mu + \sqrt{t}/a}} = -a\mu^{3/2}. \qquad (3.3.27)
$$

On the other hand, by Theorem 3.3.1, as $n \to \infty$,

$$
n^{-1/2}(T_n - n\mu)/\sigma \xrightarrow{\mathcal{D}} N(0, 1) \qquad (3.3.28)
$$

so that by (3.3.25), (3.3.27) and (3.3.28), for $a \in \mathbb{R}$, the rhs of (3.3.26) reduces to

$$
P\{N(0, 1) > -a\mu^{3/2}/\sigma\} = 1 - \Phi(-a\mu^{3/2}/\sigma) = \Phi(a\mu^{3/2}/\sigma), \qquad (3.3.29)
$$

so that

$$
t^{-1/2}(N_t - t/\mu) \xrightarrow{\mathcal{D}} N(0, \sigma^2/\mu^3). \qquad (3.3.30)
$$

In this context, we may note that if $\{T_n - n\mu;\ n \geq 0\}$ is a zero mean martingale sequence, (3.3.28) follows from Theorem 3.3.7, so that for the renewal central limit theorem in (3.3.30), the independence of the X_i's may also be replaced by the aforesaid martingale structure.

3.4 Projection results and variance-stabilizing transformations

In this section we are mainly concerned with the asymptotic normality of statistics which may not be expressed as sums of independent random variables as in:

Example 3.4.1: Let $\{X_n\}$ be a sequence of i.i.d. r.v.'s with mean μ and variance σ^2; we are interested in verifying whether the sample variance S_n^2 follows an asymptotic normal distribution. ∎

A convenient strategy in that direction consists of decomposing the statistic of interest in two terms, one of which is a sum of i.i.d. r.v.'s and the other converges in probability to zero and then using one of the results considered below:

Theorem 3.4.1: *Consider the statistic $T_n = T(X_1, \ldots, X_n)$ where the X_i's are i.i.d. r.v.'s. Let $T_n = G_n + R_n$ where $G_n = \sum_{i=1}^{n} g(X_i)$ and $n^{-1/2} R_n \xrightarrow{P} 0$. Furthermore, let $\mathrm{E}g(X_i) = \xi$ and $\mathrm{Var}(X_i) = \nu^2 < \infty$ and suppose that $(G_n - n\xi)/\sqrt{n}\nu \xrightarrow{D} N(0, 1)$. Then $(T_n - n\xi)/\sqrt{n}\nu \xrightarrow{D} N(0, 1)$.*

Proof: First observe that given $\varepsilon > 0$:

$$
\begin{aligned}
P&\{(T_n - n\xi)/\nu\sqrt{n} \leq x\} \\
&= P\{(G_n - n\xi)/\sqrt{n}\nu + R_n/\sqrt{n}\nu \leq x\} \\
&= P\{(G_n - n\xi)/\nu\sqrt{n} + R_n/\sqrt{n}\nu \leq x; |R_n|/\nu\sqrt{n} \leq \varepsilon\} \\
&\quad + P\{(G_n - n\xi)/\nu\sqrt{n} + R_n/\sqrt{n}\nu \leq x; |R_n|/\nu\sqrt{n} > \varepsilon\} \\
&\leq P\{(G_n - n\xi)/\nu\sqrt{n} \leq x + \varepsilon; |R_n|/\nu\sqrt{n} \leq \varepsilon\} \\
&\quad + P\{|R_n|/\nu\sqrt{n} > \varepsilon\} \\
&\leq P\{(G_n - n\xi)/\nu\sqrt{n} \leq x + \varepsilon\} \\
&\quad + P\{|R_n|/\nu\sqrt{n} > \varepsilon\} \to \Phi(x + \varepsilon) \quad \text{as} \quad n \to \infty.
\end{aligned}
$$

Similarly,

$$
\begin{aligned}
P&\{(T_n - n\xi)/\nu\sqrt{n} \leq x\} \\
&= P\{(G_n - n\xi)/\nu\sqrt{n} + R_n/\sqrt{n}\nu \leq x\} \\
&\geq P\{(G_n - n\xi)/\nu\sqrt{n} \leq x - \varepsilon; |R_n|/\nu\sqrt{n} \leq \varepsilon\} \\
&\geq P\{(G_n - n\xi)/\nu\sqrt{n} \leq x - \varepsilon\} \\
&\quad - P\{(G_n - n\xi)/\nu\sqrt{n} \leq x - \varepsilon; |R_n|/\nu\sqrt{n} > \varepsilon\} \\
&\geq P\{(G_n - n\xi)/\nu\sqrt{n} \leq x - \varepsilon\} \\
&\quad - P\{|R_n|/\nu\sqrt{n} > \varepsilon\} \to \Phi(x - \varepsilon) \quad \text{as} \quad n \to \infty.
\end{aligned}
$$

From the two expressions above we have,

$$
\Phi(x - \varepsilon) \leq \lim_{n \to \infty} P\left\{\frac{T_n - n\xi}{\nu\sqrt{n}} \leq x\right\} \leq \Phi(x + \varepsilon).
$$

Now, since for all $\varepsilon > 0$,

$$
\Phi(x + \varepsilon) - \Phi(x - \varepsilon) = (2\pi)^{-1/2} \int_{x-\varepsilon}^{x+\varepsilon} \exp(-t^2/2) dt \leq 2\varepsilon/\sqrt{2\pi},
$$

it follows that $\lim_{n\to\infty} P\left\{(T_n - n\xi)/\nu\sqrt{n} \leq x\right\} = \Phi(x)$. ∎

Example 3.4.1 (Continued): Assume that $\mathrm{E}X_i^4 < \infty$ and write $\mathrm{E}(X_i -$

$\mu)^4 = \mu_4 < \infty$ and $\mathrm{Var}(X_i - \mu)^2 = \mu_4 - \sigma^4 = \gamma^2 > 0$. Then note that

$$\frac{\sqrt{n}}{\gamma}(S_n^2 - \sigma^2) = \frac{n}{n-1}\frac{1}{\sqrt{n}\gamma}\{T_n - (n-1)\sigma^2\} \qquad (3.4.1)$$

where

$$T_n = \sum_{i=1}^{n}(X_i - \overline{X}_n)^2 = \sum_{i=1}^{n}\left\{(X_i - \mu)^2 - \frac{\sigma^2}{n}\right\} - n(\overline{X}_n - \mu)^2 + \sigma^2$$
$$= G_n + R_n,$$

$G_n = \sum_{i=1}^{n} g(X_i)$ with $g(X_i) = (X_i - \mu)^2 - \sigma^2/n$ and $R_n = -n(\overline{X}_n - \mu)^2 + \sigma^2$. Note that

$$E|R_n/\sqrt{n}| = \sqrt{n}E(\overline{X}_n - \mu)^2 + \sigma^2/\sqrt{n} = 2\sigma^2/\sqrt{n} \to 0 \quad \text{as} \quad n \to \infty;$$

thus, from Proposition 2.2.3, it follows that $n^{-1/2}R_n \xrightarrow{P} 0$. Observing that $Eg(X_i) = (n - 1/n)\sigma^2$ and $\mathrm{Var}g(X_i) = \gamma^2$, a direct application of Theorem 3.4.1 in connection with (3.4.1) permits us to conclude that $\sqrt{n}(S_n^2 - \sigma^2)/\gamma \xrightarrow{D} N(0,1)$. ∎

This type of decomposition technique may be generalized further along the lines of the following well-known result:

Theorem 3.4.2 (Slutsky): *Let $\{X_n\}$ and $\{Y_n\}$ be sequences of random variables such that $X_n \xrightarrow{D} X$ and $Y_n \xrightarrow{P} c$ where c is a constant. Then, it follows that*

i) $X_n + Y_n \xrightarrow{D} X + c$,

ii) $Y_n X_n \xrightarrow{D} cX$,

iii) $X_n/Y_n \xrightarrow{D} X/c$ if $c \neq 0$.

Proof: Let F denote the distribution function of X; then, for every $\varepsilon > 0$ and every x, a continuity point of F, we may write

$$P\{X_n + Y_n \leq x\}$$
$$= P\{X_n + Y_n \leq x; |Y_n - c| \leq \varepsilon\} + P\{X_n + Y_n \leq x; |Y_n - c| > \varepsilon\}$$
$$\leq P\{X_n \leq x - c + \varepsilon; |Y_n - c| \leq \varepsilon\} + P\{|Y_n - c| > \varepsilon\}$$
$$\leq P\{X_n \leq x - c + \varepsilon\} + P\{|Y_n - c| > \varepsilon\}.$$

Therefore,

$$\limsup_{n\to\infty} P\{X_n + Y_n \leq x\}$$
$$\leq \lim_{n\to\infty} P\{X_n \leq x - c + \varepsilon\} + \lim_{n\to\infty} P\{|Y_n - c| > \varepsilon\}$$
$$= P\{X \leq x - c + \varepsilon\} = F(x - c + \varepsilon) \qquad (3.4.2)$$

Also,

$$P\{X_n + Y_n \leq x\}$$
$$\geq P\{X_n + Y_n \leq x; |Y_n - c| \leq \varepsilon\}$$
$$\geq P\{X_n \leq x - c - \varepsilon; |Y_n - c| \leq \varepsilon\}$$
$$= P\{X_n \leq x - c - \varepsilon\} - P\{X_n \leq x - c - \varepsilon; |Y_n - c| > \varepsilon\}$$
$$\geq P\{X_n \leq x - c - \varepsilon\} + P\{|Y_n - c| > \varepsilon\}.$$

Therefore

$$\lim_{n \to \infty} \inf P\{X_n + Y_n \leq x\}$$
$$\geq \lim_{n \to \infty} P\{X_n \leq x - c - \varepsilon\} + \lim_{n \to \infty} P\{|Y_n - c| > \varepsilon\}$$
$$\geq P\{X \leq x - c - \varepsilon\}$$
$$= F(x - c - \varepsilon). \tag{3.4.3}$$

From (3.4.2) and (3.4.3), we have

$$F(x - c - \varepsilon) \leq \lim_{n \to \infty} \inf P\{X_n + Y_n \leq x\}$$
$$\leq \lim_{n \to \infty} \sup P\{X_n + Y_n \leq x\}$$
$$\leq F(x - c + \varepsilon).$$

Letting $\varepsilon \to 0$, it follows that

$$\lim_{n \to \infty} P\{X_n + Y_n \leq x\} = F(x - c)$$

and the proof of (i) is completed. To prove (ii), note that, for $x \geq 0$, without loss of generality,

$$P\{X_n Y_n \leq x\}$$
$$= P\left\{X_n Y_n \leq x; \left|\frac{Y_n}{c} - 1\right| \leq \varepsilon\right\} + P\left\{X_n Y_n \leq x; \left|\frac{Y_n}{c} - 1\right| > \varepsilon\right\}$$
$$\leq P\left\{X_n \leq \frac{x}{c(1 - \varepsilon)}; \left|\frac{Y_n}{c} - 1\right| \leq \varepsilon\right\} + P\left\{\left|\frac{Y_n}{c} - 1\right| > \varepsilon\right\}$$
$$\leq P\left\{X_n \leq \frac{x}{c(1 - \varepsilon)}\right\} + P\left\{\left|\frac{Y_n}{c} - 1\right| > \varepsilon\right\}.$$

Therefore, for $x \geq 0$,

$$\lim_{n \to \infty} \sup P\{X_n Y_n \leq x\}$$
$$\leq \lim_{n \to \infty} P\left\{X_n \leq \frac{x}{c(1 - \varepsilon)}\right\} + \lim_{n \to \infty} P\left\{\left|\frac{Y_n}{c} - 1\right| > \varepsilon\right\}$$
$$= F\{x/c(1 - \varepsilon)\}. \tag{3.4.4}$$

Similarly, for $x \geq 0$,

$$P\{X_n Y_n \leq x\}$$

$$\geq P\left\{X_n \leq \frac{x}{c(1+\varepsilon)}; \left|\frac{Y_n}{c} - 1\right| \leq \varepsilon\right\}$$

$$= P\left\{X_n \leq \frac{x}{c(1+\varepsilon)}\right\} - P\left\{X_n \leq \frac{x}{c(1+\varepsilon)}; \left|\frac{Y_n}{c} - 1\right| > \varepsilon\right\}$$

$$\geq P\left\{X_n \leq \frac{x}{c(1+\varepsilon)}\right\} - P\left\{\left|\frac{Y_n}{c} - 1\right| > \varepsilon\right\}.$$

Therefore, for $x \geq 0$,

$$\liminf_{n \to \infty} P\{X_n Y_n \leq x\}$$

$$\geq \lim_{n \to \infty} P\left\{X_n \leq \frac{x}{c(1+\varepsilon)}\right\} - \lim_{n \to \infty} P\left\{\left|\frac{Y_n}{c} - 1\right| > \varepsilon\right\}$$

$$= F\{x/c(1+\varepsilon)\}. \tag{3.4.5}$$

From (3.4.4) and (3.4.5), we have, for $x \geq 0$,

$$F\{x/c(1+\varepsilon)\} \leq \liminf_{n \to \infty} P\{X_n Y_n \leq x\} \leq \limsup_{n \to \infty} P\{X_n Y_n \leq x\}$$

$$\leq F\{x/c(1-\varepsilon)\}.$$

Letting $\varepsilon \to 0$, it follows that $\lim_{n \to \infty} P\{X_n Y_n \leq x\} = F(x/c)$ and the proof of (ii) is complete. (iii) is proved along similar arguments. ∎

Example 3.4.2: Let $\{X_n\}$ be a sequence of i.i.d. r.v.'s with mean μ and variance $\sigma^2 < \infty$. We are interested in the asymptotic distribution of the statistic $t_n = \sqrt{n}(\overline{X}_n - \mu)/S_n$. First write

$$t_n = \frac{\sqrt{n}(\overline{X} - \mu)/\sigma}{\sqrt{S_n^2/\sigma^2}}$$

and note that

i) the Khintchine Weak Law of Large Numbers (Theorem 2.3.6) implies that

$$n^{-1} \sum_{i=1}^{n}(X_i - \mu)^2 \xrightarrow{P} E(X_i - \mu)^2 = \sigma^2 \quad \text{and} \quad \overline{X}_n \xrightarrow{P} \mu;$$

consequently, from the definition of convergence in probability it follows that

$$S_n^2 = \frac{n}{n-1}\left\{\frac{1}{n}\sum_{i=1}^{n}(X_i - \mu)^2 - (\overline{X}_n - \mu)^2\right\} \xrightarrow{P} \sigma^2$$

which, in turn, implies that $S_n \xrightarrow{P} \sigma$;

ii) $Z_n = \sqrt{n}(\overline{X}_n - \mu)/\sigma \xrightarrow{\mathcal{D}} N(0,1)$ by the Classical Central Limit Theorem 3.3.1.

From (i) and (ii), a direct application of the Slutsky Theorem (3.4.2) implies that $t_n \xrightarrow{\mathcal{D}} N(0,1)$. ∎

A multivariate version of the Slutsky Theorem may be stated as:

Theorem 3.4.3: *Let* $\{\mathbf{X}_n\}$ *and* $\{\mathbf{Y}_n\}$ *be sequences of random p-vectors such that* $\mathbf{X}_n \xrightarrow{\mathcal{D}} \mathbf{X}$ *and* $\mathbf{Y}_n \xrightarrow{P} \mathbf{0}$; *also let* $\{\mathbf{W}_n\}$ *be a sequence of random* $(w \times p)$ *matrices such that* $\mathrm{tr}\{(\mathbf{W}_n - \mathbf{W})^t(\mathbf{W}_n - \mathbf{W})\} \xrightarrow{P} 0$ *where* $\mathbf{W}$ *is a nonstochastic matrix. Then*

i) $\mathbf{X}_n \pm \mathbf{Y}_n \xrightarrow{\mathcal{D}} \mathbf{X}$,

ii) $\mathbf{W}_n\mathbf{X}_n \xrightarrow{\mathcal{D}} \mathbf{W}\mathbf{X}$.

Proof: Given $\boldsymbol{\lambda} \in \mathbb{R}^p$, arbitrary but fixed, we have for all $r \in \mathbb{R}$

$$
\begin{aligned}
\phi_{\boldsymbol{\lambda}^t\mathbf{X}_n}(r) &= \mathrm{E}\{\exp(ir\boldsymbol{\lambda}^t\mathbf{X}_n)\} \\
&= \phi_{\mathbf{X}_n}(r\boldsymbol{\lambda}) \to \phi_{\mathbf{X}}(r\boldsymbol{\lambda}) \\
&= \mathrm{E}\{\exp(ir\boldsymbol{\lambda}^t\mathbf{X})\} = \phi_{\boldsymbol{\lambda}^t\mathbf{X}}(r)
\end{aligned}
$$

as $n \to \infty$ by the Lévy-Cramér Theorem (2.2.2). Thus, $\boldsymbol{\lambda}^t\mathbf{X}_n \xrightarrow{\mathcal{D}} \boldsymbol{\lambda}^t\mathbf{X}$. Observing that $\boldsymbol{\lambda}^t\mathbf{Y}_n \xrightarrow{P} 0$, it follows from the Slutsky Theorem (3.4.2) that $\boldsymbol{\lambda}^t(\mathbf{X}_n \pm \mathbf{Y}_n) \xrightarrow{\mathcal{D}} \boldsymbol{\lambda}^t\mathbf{X}$. Since this is true for all $\boldsymbol{\lambda} \in \mathbb{R}^p$, (i) follows from the Cramér-Wold Theorem (3.2.4).

To prove (ii) first write

$$\mathbf{W}_n\mathbf{X}_n = \mathbf{W}\mathbf{X}_n + (\mathbf{W}_n - \mathbf{W})\mathbf{X}_n. \tag{3.4.6}$$

Then use an argument similar to the one above to show that for all $\boldsymbol{\lambda} \in \mathbb{R}^p$ (arbitrary, but fixed)

$$\boldsymbol{\lambda}^t\mathbf{W}\mathbf{X}_n \xrightarrow{\mathcal{D}} \boldsymbol{\lambda}^t\mathbf{W}\mathbf{X} \tag{3.4.7}$$

and note that

$$
\begin{aligned}
|\boldsymbol{\lambda}^t(\mathbf{W}_n - \mathbf{W})\mathbf{X}_n| &\leq \|\boldsymbol{\lambda}\|\|(\mathbf{W}_n - \mathbf{W})\mathbf{X}_n\| \\
&= \|\boldsymbol{\lambda}\||\mathbf{X}_n^t(\mathbf{W}_n - \mathbf{W})^t(\mathbf{W}_n - \mathbf{W})\mathbf{X}_n|^{1/2} \\
&\leq \|\boldsymbol{\lambda}\||\mathrm{ch}_1\{(\mathbf{W}_n - \mathbf{W})^t(\mathbf{W}_n - \mathbf{W})\}\mathbf{X}_n^t\mathbf{X}_n|^{1/2} \\
&\leq \|\boldsymbol{\lambda}\||\mathrm{tr}\{(\mathbf{W}_n - \mathbf{W})^t(\mathbf{W}_n - \mathbf{W})\}\mathbf{X}_n^t\mathbf{X}_n|^{1/2}.
\end{aligned}
$$

From Theorem 3.2.3 it follows that $\mathbf{X}_n^t\mathbf{X}_n \xrightarrow{\mathcal{D}} \mathbf{X}^t\mathbf{X}$ and an application of the Slutsky Theorem (3.4.2) yields

$$\boldsymbol{\lambda}^t(\mathbf{W}_n - \mathbf{W})\mathbf{X}_n \xrightarrow{P} \mathbf{0} \tag{3.4.8}$$

which in conjunction with (3.4.6) and (3.4.7) implies (ii) via a subsequent application of the Slutsky Theorem (3.4.2).   ∎

In Example 3.4.1 we have shown that the sample variance S_n^2 is such that $\sqrt{n}(S_n^2 - \sigma^2)/\gamma \xrightarrow{\mathcal{D}} N(0,1)$ using a decomposition result. Alternatively, we could have used the fact that $S_n^2 = \binom{n}{2}^{-1} \sum_{1 \leq i < j \leq n} \psi(X_i, X_j)$ with $\psi(a,b) = (a-b)^2/2$ is a **reverse martingale** to prove the same result via Theorem 3.3.8, although some rather tedious algebraic manipulation would have been necessary to show condition (A). Another way of obtaining such result relies on the so called **Projection Technique** due to Hoeffding (1948), which makes use of the Slutsky Theorem (3.4.2).

Let $\{X_k, k \geq 1\}$ be a sequence of i.i.d. r.v.'s and consider a statistic $T_n = T_n(X_1, \ldots, X_n)$; suppose we are interested in verifying whether $(T_n - ET_n)/\sqrt{\mathrm{Var} T_n}$ is asymptotically normal. First note that $T_{ni} = \mathrm{E}(T_n \mid X_i) = f_n(X_i)$, $i = 1, \ldots, n$ are i.i.d. r.v.'s such that $ET_{ni} = ET_n$. Then, letting $v_n = \sum_{i=1}^{n}(T_{ni} - ET_n)$ we may use the Central Limit Theorem for triangular schemes (3.3.5) to show the asymptotic normality of $\sqrt{n} v_n/\sqrt{\mathrm{Var} v_n}$; if we can show that $(T_n - ET_n)/\sqrt{\mathrm{Var} T_n} - v_n/\sqrt{\mathrm{Var} v_n} \xrightarrow{\mathrm{P}} 0$, then by the Slutsky Theorem (3.4.2) the asymptotic distribution of $(T_n - ET_n)/\sqrt{\mathrm{Var} T_n}$ will be the same as that of $v_n/\sqrt{\mathrm{Var} v_n}$. This technique is useful in the case of U-statistics and will be discussed in detail in Chapter 5.

In many practical situations one might be interested in obtaining the asymptotic distribution of statistics which may not be decomposed in a suitable form to apply Theorem 3.4.1 or the Slutsky Theorem (3.4.2) as in the case of the sample standard deviation. If, however, the statistic of interest is a well-behaved function of some other statistic known to be asymptotically normal, its asymptotic distribution may be determined via the following result known in the literature as the **Delta method**.

Theorem 3.4.4: *Suppose that* $\sqrt{n}(T_n - \theta)/\sigma \xrightarrow{\mathcal{D}} N(0,1)$ *and let g be a continuous function such that $g'(\theta)$ exists and $g'(\theta) \neq 0$. Then it follows that*

$$\sqrt{n}\{g(T_n) - g(\theta)\}/\sigma g'(\theta) \xrightarrow{\mathcal{D}} N(0,1).$$

Proof: Let $Z_n = \sqrt{n}(T_n - \theta)/\sigma$ and $G_n = \{g(T_n) - g(\theta)\}/(T_n - \theta)$ and observe that

$$\frac{\sqrt{n}}{\sigma}\{g(T_n) - g(\theta)\} = G_n Z_n. \tag{3.4.9}$$

Now, using the fact that $Z_n \xrightarrow{\mathcal{D}} Z$ and Theorem 3.2.5 we may conclude that $\sqrt{n}\sigma^{-1}(T_n - \theta) = O_p(1)$, which implies $(T_n - \theta) = O_p(1)O_p(n^{-1/2}) = O_p(n^{-1/2}) = o_p(1)$, that is, $T_n - \theta \xrightarrow{\mathrm{P}} 0$; consequently it follows that

$G_n \xrightarrow{\text{P}} g'(\theta)$. Then applying the Slutsky Theorem (3.4.2) to (3.4.9) we have $\sqrt{n}\sigma^{-1}\{g(T_n) - g(\theta)\} \xrightarrow{\mathcal{D}} g'(\theta)Z$ and the result follows.   ■

Example 3.4.3: Let $\{X_n\}$ be a sequence of i.i.d. r.v.'s with mean μ, variance σ^2 and such that $\mu_4 = \mathrm{E}(X-\mu)^4 < \infty$. We have seen in Example 3.4.1 that $\sqrt{n}(S_n^2 - \sigma^2) \xrightarrow{\mathcal{D}} N(0, \gamma^2)$. We are interested in the asymptotic distribution of $\sqrt{n}(s_n - \sigma)$. Note that $S_n = \{(n-1)^{-1}\sum_{i=1}^n (X_i - \overline{X})^2\}^{1/2}$ is not expressible as a sum of independent random variables as in the case of S_n^2. However, since $S_n = \sqrt{S_n^2}$, we may take $g(x) = \sqrt{x}$ [which implies $g'(x) = (2\sqrt{x})^{-1}$] and apply Theorem 3.4.4 to conclude that

$$\frac{\sqrt{n}(S_n - \sigma)}{(2\sigma)^{-1}\gamma} \xrightarrow{\mathcal{D}} N(0, 1).$$

■

Example 3.4.4: Let $R_n \sim \mathrm{Bin}(n, \pi)$ and write $T_n = R_n/n$. We have seen in Example 3.3.1 that $\sqrt{n}(T_n - \pi)/\{\pi(1-\pi)\}^{1/2} \xrightarrow{\mathcal{D}} N(0, 1)$. Suppose we are interested in the asymptotic distribution of $\sqrt{n}(G_n - \pi^{-1})$ where $G_n = T_n^{-1}$ is an estimate of π^{-1}. Taking $g(x) = 1/x$ [which implies $g'(x) = -1/x^2$] it follows from Theorem 3.4.4 that

$$\frac{\sqrt{n}(G_n - \pi^{-1})}{\pi^{-2}\sqrt{\pi(1-\pi)}} = \frac{\sqrt{n}(G_n - \pi^{-1})}{\sqrt{(1-\pi)/\pi^3}} \xrightarrow{\mathcal{D}} N(0, 1).$$

■

Example 3.4.5: Let $R_n \sim \mathrm{Bin}(n, \pi)$ and suppose we are interested in the asymptotic distribution of $\sqrt{n}\{G_n - \pi(1-\pi)\}$ where $G_n = R_n(n-R_n)/n^2$ is an estimator of $\pi(1-\pi)$. Observe that $G_n = g(R_n/n)$ where $g(x) = x(1-x)$ [which implies $g'(x) = 1 - 2x$] and apply Theorem 3.4.4 to see that, for $\pi \neq 1/2$,

$$\frac{\sqrt{n}\{G_n - \pi(1-\pi)\}}{(1-2\pi)\sqrt{\pi(1-\pi)}} \xrightarrow{\mathcal{D}} N(0, 1).$$

■

The generalization of the above results to the multivariate case is of great practical importance and will be considered next. Let $\{\mathbf{T}_n\}$ be a sequence of p-dimensional random vectors and suppose that $\sqrt{n}(\mathbf{T}_n - \boldsymbol{\theta}) \xrightarrow{\mathcal{D}} N_p(\mathbf{0}, \boldsymbol{\Sigma})$; essentially, we are interested in the asymptotic distribution of $\sqrt{n}\{g(\mathbf{T}_n) - g(\boldsymbol{\theta})\}$ where $g(\cdot)$ is a real-valued function of $\mathbf{T}_n$. This is the object of the following:

Theorem 3.4.5: *Let $\{\mathbf{T}_n\}$ be a sequence of random p-vectors such that $\sqrt{n}(\mathbf{T}_n - \boldsymbol{\theta}) \xrightarrow{\mathcal{D}} N(\mathbf{0}, \boldsymbol{\Sigma})$ and consider a real-valued function $g(\mathbf{T}_n)$ such* that $\dot{\mathbf{g}}(\boldsymbol{\theta}) = \partial g(\mathbf{x})/\partial \mathbf{x}|_{\mathbf{x}=\boldsymbol{\theta}}$ *is non-null and continuous in a neighborhood of* $\boldsymbol{\theta}$. *Then*

$$\sqrt{n}\{g(\mathbf{T}_n) - g(\boldsymbol{\theta})\} \xrightarrow{\mathcal{D}} N(0, \gamma^2) \quad \text{with} \quad \gamma^2 = [\dot{\mathbf{g}}(\boldsymbol{\theta})]^t \boldsymbol{\Sigma} [\dot{\mathbf{g}}(\boldsymbol{\theta})].$$

Proof: First note that for $\mathbf{x}, \mathbf{y} \in \mathbb{R}^p$ we may write

$$
\begin{aligned}
g(\mathbf{x}) - g(\mathbf{y}) &= g(x_1, \ldots, x_p) - g(y_1, \ldots y_p) \\
&= g(x_1, \ldots, x_p) - g(x_1, \ldots, x_{p-1}, y_p) \\
&\quad + g(x_1, \ldots, x_{p-1}, y_p) - g(x_1, \ldots, x_{p-2}, y_{p-1}, y_p) \\
&\quad + g(x_1, \ldots, x_{p-2}, y_{p-1}, y_p) \\
&\quad - g(x_1, \ldots, x_{p-3}, y_{p-2}, y_{p-1}, y_p) \\
&\quad + \cdots \\
&\quad + g(x_1, y_2, \ldots, y_p) - g(y_1, y_2, \ldots, y_p) \\
&= \sum_{i=1}^{p} (x_i - y_i) \tilde{g}_i(\mathbf{x}, \mathbf{y})
\end{aligned}
$$

$$(3.4.10)$$

where

$$\tilde{g}_i(\mathbf{x}, \mathbf{y}) = \{g(x_1, \ldots, x_i, y_{i+1}, \ldots, y_p) - g(x_1, \ldots, x_{i-1}, y_i, \ldots, y_p)/(x_i - y_i)\}$$

is such that

$$\lim_{x_i \to y_i} \tilde{g}_i(\mathbf{x}, \mathbf{y}) = \partial g(\mathbf{x})/\partial x_i|_{\mathbf{x}=\mathbf{y}} = \dot{g}_i(\mathbf{y}).$$

Now let

$$\tilde{\mathbf{g}}(T_n, \boldsymbol{\theta}) = [\tilde{g}_1(\mathbf{T}_n, \boldsymbol{\theta}), \ldots, \tilde{g}_p(\mathbf{T}_n, \boldsymbol{\theta})]^t$$

and observe that from (3.4.10) we may write

$$\sqrt{n}\{g(\mathbf{T}_n) - g(\boldsymbol{\theta})\} = \sqrt{n}(\mathbf{T}_n - \boldsymbol{\theta})^t \tilde{\mathbf{g}}(\mathbf{T}_n, \boldsymbol{\theta}). \qquad (3.4.11)$$

From a direct extension of Theorem 3.2.5 it follows that $\sqrt{n}(\mathbf{T}_n - \boldsymbol{\theta}) = O_p(1)$ which implies $(\mathbf{T}_n - \boldsymbol{\theta}) \xrightarrow{P} 0$ and, therefore, $\tilde{\mathbf{g}}(\mathbf{T}_n, \boldsymbol{\theta}) \xrightarrow{P} \dot{\mathbf{g}}(\boldsymbol{\theta})$. Then, applying the Slutsky Theorem (3.4.2) to (3.4.11) we have

$$\sqrt{n}(\mathbf{T}_n - \boldsymbol{\theta})^t \tilde{\mathbf{g}}_n \xrightarrow{\mathcal{D}} \mathbf{T}^t \dot{\mathbf{g}}(\boldsymbol{\theta}) = G$$

where $G \sim N(0, \gamma^2)$. ∎

Example 3.4.6: Let $\{X_n\}$ be a sequence of i.i.d. r.v.'s with mean μ, variance σ^2 and $EX_1^4 < \infty$. Let $\nu = \sigma/\mu$ denote the coefficient of variation and $v_n = S_n/\overline{X}_n$ the corresponding sample counterpart. We are interested in the asymptotic distribution of $\sqrt{n}(v_n - \nu)$. First let us show that the vector $\sqrt{n}\mathbf{T}_n$ where $\mathbf{T}_n = [\overline{X}_n - \mu, S_n^2 - \sigma^2]^t$ is asymptotically normally distributed. In this direction, observe that we may write

$$
\begin{aligned}
(S_n^2 - \sigma^2) &= \frac{n-1}{n}(S_n^2 - \sigma^2) + \frac{1}{n}(S_n^2 - \sigma^2) \\
&= \frac{1}{n}\sum_{i=1}^{n}\{(X_i - \mu)^2 - \sigma^2\} - (\overline{X}_n - \mu)^2 + \frac{\sigma^2}{n} + \frac{1}{n}(S_n^2 - \sigma^2) \\
&= \frac{1}{n}\sum_{i=1}^{n} Y_i + R_n,
\end{aligned}
$$

where $Y_i = (X_i - \mu)^2 - \sigma^2$ is such that $EY_i = 0$ and $\mathrm{Var}Y_i = \mu_4 - \sigma^4 = \gamma^2$ and $R_n = (S_n^2 - \sigma^2)/n + \sigma^2/n - (\overline{X}_n - \mu)^2$ is such that $\sqrt{n}R_n = o_p(1)$. Then, let $\lambda_1, \lambda_2 \in \mathbb{R}$ be arbitrary but fixed and let

$$
\begin{aligned}
Z_n &= \sqrt{n}\{\lambda_1(\overline{X}_n - \mu) + \lambda_2(S_n^2 - \sigma^2)\} \\
&= \sqrt{n}\left[\frac{1}{n}\sum_{i=1}^{n}\{\lambda_1(X_i - \mu) + \lambda_2 Y_i\}\right] + \sqrt{n}\lambda_2 R_n \\
&= \sqrt{n}\left\{\frac{1}{n}\sum_{i=1}^{n} W_i\right\} + o_p(1), \tag{3.4.12}
\end{aligned}
$$

where $W_i = \lambda_1(X_i - \mu) + \lambda_2 Y_i$, $i = 1, 2, \ldots, n$ are independent random variables such that $EW_i = 0$ and $\mathrm{Var}W_i = \lambda_1^2\sigma^2 + \lambda_2^2(\mu_4 - \sigma^4) + 2\mu_3\lambda_1\lambda_2$ with $\mu_4 = E(X_i - \mu)^4$ and $\mu_3 = E(X_i - \mu)^3$. Applying the Classical Central Limit Theorem 3.3.1 in conjunction with the Slutsky Theorem (3.4.2) to (3.4.12) we may conclude that Z_n is asymptotically normally distributed. Since λ_1 and λ_2 are arbitrary, it follows from the Cramér-Wold Theorem (3.2.4) that

$$
\sqrt{n}\mathbf{T}_n \xrightarrow{D} N(\mathbf{0}, \boldsymbol{\Sigma}) \quad \text{where} \quad \boldsymbol{\Sigma} = \begin{pmatrix} \sigma^2 & \mu_3 \\ \mu_3 & \mu_4 - \sigma^4 \end{pmatrix}.
$$

Now, let $\boldsymbol{\theta} = (\mu, \sigma^2)^t$ and $g(x, y) = \sqrt{y}/x$; thus, $g(\overline{X}_n, S_n^2) = S_n/\overline{X}_n = v_n$ and $g(\boldsymbol{\theta}) = \sigma/\mu = \nu$. Furthermore, $\dot{g}_1(\boldsymbol{\theta}) = \partial g(x, y)/\partial x|_{(x,y)=\boldsymbol{\theta}} = -\sigma/\mu^2$ and $\dot{g}_2(\boldsymbol{\theta}) = \partial g(x, y)/\partial y|_{(x,y)=\boldsymbol{\theta}} = 1/2\mu\sigma$ exist; therefore, applying Theorem 3.4.5 it follows that

$$
\sqrt{n}(v_n - \nu) \xrightarrow{D} G \quad \text{where} \quad G \sim N\left(0, \frac{\sigma^4}{\mu^4} - \frac{\mu_3}{\mu^3} + \frac{\mu_4}{4\mu^2\sigma^2} - \frac{\sigma^2}{4\mu^2}\right).
$$

∎

Example 3.4.7: Let $\{X_n, Y_n\}$ be a sequence of independent random variables following a bivariate distribution with mean vector $\boldsymbol{\mu} = (\mu_X, \mu_Y)^t$ and p.d. covariance matrix:

$$
\boldsymbol{\Sigma} = \begin{pmatrix} \sigma_X^2 & \rho\sigma_X\sigma_Y \\ \rho\sigma_X\sigma_Y & \sigma_Y^2 \end{pmatrix}.
$$

PROJECTIONS AND VARIANCE-STABILIZING TRANSFORMATIONS   135

We are interested in the asymptotic distribution of the sample coefficient of correlation

$$R_n = \sum_{i=1}^{n}(X_i - \overline{X}_n)(Y_i - \overline{Y}_n) \Big/ \left\{\sum_{i=1}^{n}(X_i - \overline{X}_n)^2 \sum_{i=1}^{n}(Y_i - \overline{Y}_n)^2\right\}^{1/2}.$$

First observe that

$$
\begin{aligned}
S_{nXY} &= n^{-1}\sum_{i=1}^{n}(X_i - \overline{X}_n)(Y_i - \overline{Y}_n) \\
&= n^{-1}\sum_{i=1}^{n}(X_i - \mu_X)(Y_i - \mu_Y) - (\overline{X}_n - \mu_X)(\overline{Y}_n - \mu_Y) \\
&= n^{-1}\sum_{i=1}^{n}(X_i - \mu_X)(Y_i - \mu_Y) + O_p(n^{-1}),
\end{aligned}
$$

since from Theorem 3.2.5 it follows that $\overline{X}_n - \mu_X = O_p(n^{-1/2})$. Similarly we may conclude that

$$S_{nX}^2 = n^{-1}\sum_{i=1}^{n}(X_i - \overline{X}_n)^2 = n^{-1}\sum_{i=1}^{n}(X_i - \mu_X)^2 + O_p(n^{-1}),$$

$$S_{nY}^2 = n^{-1}\sum_{i=1}^{n}(Y_i - \overline{Y}_n)^2 = n^{-1}\sum_{i=1}^{n}(Y_i - \mu_Y)^2 + O_p(n^{-1}).$$

Then, consider $\mathbf{t} = (t_1, t_2, t_3)^t \in \mathbb{R}^3$, fixed but arbitrary and write

$$
\begin{aligned}
&\sqrt{n}\{t_1(S_{nX}^2 - \sigma_X^2) + t_2(S_{nY}^2 - \sigma_Y^2) + t_3(S_{nXY} - \rho\sigma_X\sigma_Y)\} \\
&= \frac{1}{\sqrt{n}}\sum_{i=1}^{n}\Big[t_1\{(X_i - \mu_X)^2 - \sigma_X^2\} + t_2\{(Y_i - \mu_Y)^2 - \sigma_Y^2\} \\
&\quad + t_3\{(X_i - \mu_X)(Y_i - \mu_Y) - \rho\sigma_X\sigma_Y\}\Big] + O_p(n^{-1}) \\
&= \frac{1}{\sqrt{n}}\sum_{i=1}^{n}U_i + O_p(n^{-1}),
\end{aligned}
$$

where

$$
\begin{aligned}
U_i = t_1\{(X_i - \mu_X)^2 - \sigma_X^2\} &+ t_2\{(Y_i - \mu_Y)^2 - \sigma_Y^2\} \\
&+ t_3\{(X_i - \mu_X)(Y_i - \mu_Y) - \rho\sigma_X\sigma_Y\},
\end{aligned}
$$

$i = 1,\ldots,n$ are i.i.d. random variables with $EU_1 = 0$ and $\mathrm{Var}U_1 = EU_1^2 = \mathbf{t}^t\mathbf{\Gamma}\mathbf{t}$ with $\mathbf{\Gamma} = \mathrm{Var}(W_1, W_2, W_3)$, $W_1 = (X_1 - \mu_X)^2 - \sigma_X^2$, $W_2 = (Y_1 - \mu_Y)^2 - \sigma_Y^2$ and $W_3 = (X_1 - \mu_X)(Y_1 - \mu_Y) - \rho\sigma_X\sigma_Y$. Thus, from the Classical Central Limit Theorem 3.3.1 and the Cramér-Wold Theorem (3.2.4) it follows that

$$\sqrt{n}\{(S_{nX}^2, S_{nY}^2, S_{nXY})^t - (\sigma_X^2, \sigma_Y^2, \rho\sigma_X\sigma_Y)^t\} \xrightarrow{D} N(\mathbf{0}, \boldsymbol{\Gamma}).$$

Now, since $R_n = S_{nXY}/S_{nX}S_{nY}$, if we define $g : \mathbb{R}_+ \times \mathbb{R}_+ \times \mathbb{R} \to \mathbb{R}$ by $g(\mathbf{x}) = x_3/\sqrt{x_1 x_2}$, we may apply Theorem 3.4.5 to obtain the desired result. More specifically, letting $\boldsymbol{\theta} = (\sigma_X^2, \sigma_Y^2, \rho\sigma_X\sigma_Y)^t$ we have

$$\dot{g}_1(\mathbf{x}) = -x_3/(2x_1\sqrt{x_1 x_2}),$$

$$\dot{g}_2(\mathbf{x}) = -x_3/(2x_2\sqrt{x_1 x_2}),$$

$$\dot{g}_3(\mathbf{x}) = (x_1 x_2)^{-1}$$

and

$$\dot{\mathbf{g}}(\boldsymbol{\theta}) = \rho\left(-\frac{1}{2\sigma_X^2}, -\frac{1}{2\sigma_Y^2}, \frac{1}{\rho\sigma_X\sigma_Y}\right)^t$$

so that $\sqrt{n}(R_n - \rho) \xrightarrow{D} N(0, \gamma^2)$ with $\gamma^2 = \dot{\mathbf{g}}(\boldsymbol{\theta})^t \boldsymbol{\Gamma} \dot{\mathbf{g}}(\boldsymbol{\theta})$. ∎

Example 3.4.8: Consider the Pearsonian system of distributions, i.e., the family characterized by (at most) the first four moments, $\boldsymbol{\tau} = \boldsymbol{\tau}(\boldsymbol{\theta}) = [\mu_1(\boldsymbol{\theta}), \mu_2(\boldsymbol{\theta}), \mu_3(\boldsymbol{\theta}), \mu_4(\boldsymbol{\theta})]$ where $\boldsymbol{\theta} = [\theta_1, \theta_2, \theta_3, \theta_4]^t$ are unknown parameters. The Method of Moments estimator of $\boldsymbol{\theta}$ based on a random sample $X_1, \ldots, X_n$ is defined as a solution $\widehat{\boldsymbol{\theta}}_n$ to the equation $\mathbf{T}_n = \boldsymbol{\tau}(\boldsymbol{\theta})$ where $\mathbf{T}_n = [m_{n1}, m_{n2}, m_{n3}, m_{n4}]^t$ with $m_{nk} = n^{-1}\sum_{i=1}^n X_i^k$. In the case of the normal distribution, we have: $\boldsymbol{\theta} = [\mu, \sigma^2]^t$, $\boldsymbol{\tau} = [\mu, \mu^2 + \sigma^2]^t$, $\mathbf{T}_n = [\overline{X}_n, n^{-1}\sum_{i=1}^n (X_i - \overline{X}_n)^2]^t$ which implies $\widehat{\boldsymbol{\theta}}_n = [\widehat{\mu}_n, \widehat{\sigma}_n^2]^t$ with $\widehat{\mu}_n = \overline{X}_n$ and $\widehat{\sigma}_n^2 = n^{-1}\sum_{i=1}^n (X_i - \overline{X}_n)^2$. Using some version of the Central Limit Theorem we may show that $\sqrt{n}\{\mathbf{T}_n - \boldsymbol{\tau}(\boldsymbol{\theta})\} \xrightarrow{D} N(\mathbf{0}, \boldsymbol{\Sigma})$, provided the moment of order eight exists. The question is whether we can obtain the asymptotic distribution of $\sqrt{n}(\widehat{\boldsymbol{\theta}}_n - \boldsymbol{\theta})$. Now, since the correspondence between $\boldsymbol{\theta}$ and $\boldsymbol{\tau}$ is one-to-one, the inverse function $\boldsymbol{\tau}^{-1}$ exists and we may write $\sqrt{n}(\widehat{\boldsymbol{\theta}}_n - \boldsymbol{\theta}) = \boldsymbol{\tau}^{-1}\sqrt{n}\{\mathbf{T}_n - \boldsymbol{\tau}(\boldsymbol{\theta})\}$. The desired result is a consequence of the following generalization of the previous theorem to the case of vector-valued functions. ∎

Theorem 3.4.6: *Let $\{\mathbf{T}_n\}$ be a sequence of random p-vectors such that $\sqrt{n}(\mathbf{T}_n - \boldsymbol{\theta}) \xrightarrow{D} N(\mathbf{0}, \boldsymbol{\Sigma})$ and consider a vector-valued function $\mathbf{g} : \mathbb{R}^p \to \mathbb{R}^q$ such that $\dot{\mathbf{G}}(\boldsymbol{\theta}) = \frac{\partial}{\partial \mathbf{x}^t}\mathbf{g}(\mathbf{x})\big|_{\mathbf{x}=\boldsymbol{\theta}}$ exists. Then*

$$\sqrt{n}\{\mathbf{g}(\mathbf{T}_n) - \mathbf{g}(\boldsymbol{\theta})\} \xrightarrow{D} N\{\mathbf{0}, \dot{\mathbf{G}}(\boldsymbol{\theta})\boldsymbol{\Sigma}\dot{\mathbf{G}}(\boldsymbol{\theta})^t\}.$$

Proof: Let $\dot{\mathbf{G}}(\boldsymbol{\theta}) = [\dot{\mathbf{g}}_1(\boldsymbol{\theta}), \ldots, \dot{\mathbf{g}}_q(\boldsymbol{\theta})]$ and write for each coordinate j, $1 \le j \le q$,

$$\sqrt{n}\{g_j(\mathbf{T}_n) - g_j(\boldsymbol{\theta})\} = \sqrt{n}(\mathbf{T}_n - \boldsymbol{\theta})^t \widetilde{\mathbf{g}}_j(\mathbf{T}_n, \boldsymbol{\theta})$$

where $\widetilde{\mathbf{g}}_j(\mathbf{T}_n, \boldsymbol{\theta})$ as in the previous theorem is such that $\widetilde{\mathbf{g}}_j(\mathbf{T}_n, \boldsymbol{\theta}) \xrightarrow{\text{P}} \dot{\mathbf{g}}_j(\boldsymbol{\theta})$. Then, let $\boldsymbol{\lambda} \in I\!\!R^q$ be arbitrary and write

$$G_n(\boldsymbol{\lambda}) = \sqrt{n}\boldsymbol{\lambda}^t\{\mathbf{g}(\mathbf{T}_n) - \mathbf{g}(\boldsymbol{\theta})\} = \sqrt{n}(\mathbf{T}_n - \boldsymbol{\theta})^t \boldsymbol{\lambda}^t \widetilde{\mathbf{G}}(\mathbf{T}_n, \boldsymbol{\theta})$$

where $\widetilde{\mathbf{G}}(\mathbf{T}_n, \boldsymbol{\theta}) = [\widetilde{\mathbf{g}}_1(\mathbf{T}_n, \boldsymbol{\theta}), \ldots, \widetilde{\mathbf{g}}_q(\mathbf{T}_n, \boldsymbol{\theta})]$. Now apply Theorem 3.4.5 to conclude that $G_n(\boldsymbol{\lambda}) \xrightarrow{\mathcal{D}} N\{0, \boldsymbol{\lambda}^t \dot{\mathbf{G}}(\boldsymbol{\theta})\boldsymbol{\Sigma}\dot{\mathbf{G}}^t(\boldsymbol{\theta})\boldsymbol{\lambda}\}$. Finally, since $\boldsymbol{\lambda}$ is arbitrary, the result follows from the Cramér-Wold Theorem (3.2.4).

Consider now a sequence of random p-vectors $\{\mathbf{T}_n\}$ such that $\sqrt{n}(\mathbf{T}_n - \boldsymbol{\theta}) \xrightarrow{\mathcal{D}} N(\mathbf{0}, \boldsymbol{\Sigma})$. A problem of great practical interest concerns the determination of the asymptotic distribution of quadratic forms of the type $Q_n = n(\mathbf{T}_n - \boldsymbol{\theta})^t \mathbf{A}_n(\mathbf{T}_n - \boldsymbol{\theta})$ where $\{\mathbf{A}_n\}$ is a sequence of positive semidefinite matrices. Theorem 3.4.5 may not be applied here, since $\partial Q_n(\mathbf{x})/\partial \mathbf{x}|_{\mathbf{x}=\boldsymbol{\theta}} = \mathbf{0}$; however, the following results may be considered instead.

Theorem 3.4.7 (Cochran): *Let $\{\mathbf{T}_n\}$ be a sequence of random p-vectors such that $\sqrt{n}(\mathbf{T}_n - \boldsymbol{\theta}) \xrightarrow{\mathcal{D}} N(\mathbf{0}, \boldsymbol{\Sigma})$, $\mathrm{rank}(\boldsymbol{\Sigma}) = q \le p$, and $\{\mathbf{A}_n\}$ be a sequence of p.s.d. nonstochastic matrices. Then*

$$Q_n = n(\mathbf{T}_n - \boldsymbol{\theta})^t \mathbf{A}_n(\mathbf{T}_n - \boldsymbol{\theta}) \xrightarrow{\mathcal{D}} \chi_q^2$$

iff $\mathbf{A}_n$ converges to some generalized inverse of $\boldsymbol{\Sigma}$.

Proof: Since $\boldsymbol{\Sigma}$ is symmetric p.s.d., there exists a $(p \times q)$ matrix $\mathbf{B}$ of rank q such that $\boldsymbol{\Sigma} = \mathbf{B}\mathbf{B}^t$. Now, let $\mathbf{Z}_n = (\mathbf{B}^t\mathbf{B})^{-1}\mathbf{B}^t\sqrt{n}(\mathbf{T}_n - \boldsymbol{\theta})$ and note that $\mathbf{Z}_n \xrightarrow{\mathcal{D}} N(\mathbf{0}, \mathbf{I}_q)$. Then take $\sqrt{n}(\mathbf{T}_n - \boldsymbol{\theta}) = \mathbf{B}\mathbf{Z}_n$ and observe that

$$Q_n = n\mathbf{Z}_n^t\mathbf{B}^t\mathbf{A}_n\mathbf{B}\mathbf{Z}_n \xrightarrow{\mathcal{D}} \mathbf{Z}^t\mathbf{B}^t\mathbf{A}\mathbf{B}\mathbf{Z}$$

where $\mathbf{A} = \lim_{n\to\infty} \mathbf{A}_n$. A necessary and sufficient condition for $\mathbf{Z}^t\mathbf{B}^t\mathbf{A}\mathbf{B}\mathbf{Z}$ to be distributed as χ_q^2 is that $\mathbf{B}^t\mathbf{A}\mathbf{B}$ be idempotent, i.e.,

$$\mathbf{B}^t\mathbf{A}\mathbf{B}\mathbf{B}^t\mathbf{A}\mathbf{B} = \mathbf{B}^t\mathbf{A}\boldsymbol{\Sigma}\mathbf{A}\mathbf{B} = \mathbf{B}^t\mathbf{A}\mathbf{B}$$

which occurs if and only if $\mathbf{A}$ is a generalized inverse of $\boldsymbol{\Sigma}$, i.e., $\mathbf{A}\boldsymbol{\Sigma}\mathbf{A} = \mathbf{A}$. Note that if $\mathbf{T} \sim N(\boldsymbol{\mu}, \boldsymbol{\Sigma})$ with $\boldsymbol{\mu} \ne \mathbf{0}$, Q_n will follow an asymptotic noncentral χ_q^2 distribution with noncentrality parameter $\boldsymbol{\mu}^t\boldsymbol{\Sigma}\boldsymbol{\mu}$, denoted by $\chi_q^2(\boldsymbol{\mu}^t\boldsymbol{\Sigma}\boldsymbol{\mu})$. ∎

Theorem 3.4.8: *Let $\{\mathbf{T}_n\}$ be a sequence of random p-vectors such that $\sqrt{n}(\mathbf{T}_n - \boldsymbol{\theta}) \xrightarrow{\mathcal{D}} N(\mathbf{0}, \boldsymbol{\Sigma})$, $\mathrm{rank}(\boldsymbol{\Sigma}) = q \le p$ and $\{\mathbf{A}_n\}$ be a sequence of* *random matrices such that* $\mathbf{A}_n \xrightarrow{\text{P}} \mathbf{A}$ *where* $\mathbf{A}\boldsymbol{\Sigma}\mathbf{A} = \mathbf{A}$. *Then*

$$Q_n = n(\mathbf{T}_n - \boldsymbol{\theta})^t \mathbf{A}_n (\mathbf{T}_n - \boldsymbol{\theta}) \xrightarrow{\mathcal{D}} \chi_q^2.$$

Proof: We shall prove the result for $\text{rank}(\boldsymbol{\Sigma}) = p$. First, note that since $\mathbf{A}_n \xrightarrow{\text{P}} \mathbf{A}$ it follows that $\mathbf{x}^t \mathbf{A}_n \mathbf{x}/\mathbf{x}^t \mathbf{A}\mathbf{x} \xrightarrow{\text{P}} 1$, for all $\mathbf{x} \in I\!\!R^p$. Therefore, given $\varepsilon > 0$ we may write

$$P\left\{ 1 - \varepsilon < \frac{\mathbf{x}^t \mathbf{A}_n \mathbf{x}}{\mathbf{x}^t \mathbf{A}\mathbf{x}} < 1 + \varepsilon \right\} \rightarrow 1 \quad \text{as} \quad n \rightarrow \infty. \tag{3.4.13}$$

Then, let $\lambda_{np} = \sup_{\mathbf{x} \neq \mathbf{0}} \mathbf{x}^t \mathbf{A}_n \mathbf{x}/\mathbf{x}^t \mathbf{A}\mathbf{x}$ and observe that given $\varepsilon > 0$, there exists $\mathbf{x}_0 \in I\!\!R^p$ such that

i) $\mathbf{x}_0^t \mathbf{A}_n \mathbf{x}_0/\mathbf{x}_0^t \mathbf{A}\mathbf{x}_0 > \lambda_{np} - \varepsilon \Rightarrow \lambda_{np} < \mathbf{x}_0^t \mathbf{A}_n \mathbf{x}_0/\mathbf{x}_0^t \mathbf{A}\mathbf{x}_0 + \varepsilon$,

ii) $\mathbf{x}_0^t \mathbf{A}_n \mathbf{x}_0/\mathbf{x}_0^t \mathbf{A}\mathbf{x}_0 \leq \lambda_{np} \Rightarrow \lambda_{np} \geq \mathbf{x}_0^t \mathbf{A}_n \mathbf{x}_0/\mathbf{x}_0^t \mathbf{A}\mathbf{x}_0 - \varepsilon$.

Thus, given $\varepsilon > 0$, it follows that

$$\left\{ 1 - \varepsilon < \frac{\mathbf{x}_0^t \mathbf{A}_n \mathbf{x}_0}{\mathbf{x}_0^t \mathbf{A}\mathbf{x}_0} < 1 + \varepsilon \right\} \subset \{ 1 - 2\varepsilon < \lambda_{np} < 1 + 2\varepsilon \}$$

which implies that $\lambda_{np} \xrightarrow{\text{P}} 1$ by (3.4.13). Using a similar argument we may show that $\lambda_{n1} = \inf_{\mathbf{x} \neq \mathbf{0}} \mathbf{x}^t \mathbf{A}_n \mathbf{x}/\mathbf{x}^t \mathbf{A}\mathbf{x} \xrightarrow{\text{P}} 1$. Now use the previous theorem to conclude that

$$Q_n^* = n(\mathbf{T}_n - \boldsymbol{\theta})^t \mathbf{A} (\mathbf{T}_n - \boldsymbol{\theta}) \xrightarrow{\mathcal{D}} \chi_p^2,$$

and let $Q_n = Q_n^* W_n$ where

$$W_n = (\mathbf{T}_n - \boldsymbol{\theta})^t \mathbf{A}_n (\mathbf{T}_n - \boldsymbol{\theta})/(\mathbf{T}_n - \boldsymbol{\theta})^t \mathbf{A} (\mathbf{T}_n - \boldsymbol{\theta}).$$

Since $\lambda_{n1} \leq W_n \leq \lambda_{np}$ we have $W_n \xrightarrow{\text{P}} 1$ and the result follows from a direct application of the Slutsky Theorem (3.4.2).   ■

Example 3.4.9: Let $\{\mathbf{X}_n\}$ be a sequence of i.i.d. random p-vectors with means $\boldsymbol{\mu}$ and p.d. covariance matrices $\boldsymbol{\Sigma}$. We know from the Central Limit Theorem that $\sqrt{n}(\overline{\mathbf{X}}_n - \boldsymbol{\mu}) \xrightarrow{\mathcal{D}} N(\mathbf{0}, \boldsymbol{\Sigma})$; we also know that

$$\mathbf{S}_n = (n-1)^{-1} \sum_{i=1}^{n} (\mathbf{X}_i - \overline{\mathbf{X}}_n)(\mathbf{X}_i - \overline{\mathbf{X}}_n)^t \xrightarrow{\text{P}} \boldsymbol{\Sigma}.$$

Thus, from the above result it follows that Hotelling's T^2-statistic, namely,

$$T_n^2 = n(\overline{\mathbf{X}}_n - \boldsymbol{\mu}_0)^t \mathbf{S}_n^{-1} (\overline{\mathbf{X}}_n - \boldsymbol{\mu}_0),$$

is such that $T_n^2 \xrightarrow{\mathcal{D}} \chi_p^2$ under the hypothesis that $\boldsymbol{\mu} = \boldsymbol{\mu}_0$.   ■

The results on the asymptotic distribution of functions of asymptotically normal statistics are of special interest for the so-called **variance-stabilizing transformations**. Consider, for example, the problem of obtaining an asymptotic confidence interval for σ^2 in the setup of Example 3.4.1. Since

$$\sqrt{n}(S_n^2 - \sigma^2) \xrightarrow{\mathcal{D}} N(0, (\beta_2 - 1)\sigma^4), \ \beta_2 = \mu_4/\sigma^4,$$

we have that as $n \to \infty$,

$$P\{S_n^2 - z_{\alpha/2}\sqrt{(\mu_4 - \sigma^4)/n} \le \sigma^2 \le S_n^2 + z_{\alpha/2}\sqrt{(\mu_4 - \sigma^4)/n}\} \to 1 - \alpha,$$

where $z_{\alpha/2}$ represents the $(100\alpha/2)$th percentile of the $N(0,1)$ distribution. Now, since the asymptotic variance depends on the unknown parameter σ^2 this statement is of no practical use. The question is whether we may consider a suitable transformation of S_n^2 for which the asymptotic variance does not depend on the unknown parameter. More generally, let

$$\sqrt{n}(T_n - \theta) \xrightarrow{\mathcal{D}} N(0, h(\theta)).$$

We want to consider a transformation $g(T_n)$ such that

$$\sqrt{n}\{g(T_n) - g(\theta)\} \xrightarrow{\mathcal{D}} N(0, c^2),$$

c independent of θ. Note that when such **variance-stabilizing transformations** exists we may write

$$P\{g(T_n) - z_{\alpha/2}c/\sqrt{n} \le g(\theta) \le g(T_n) + z_{\alpha/2}c/\sqrt{n}\} \to 1 - \alpha \quad \text{as} \quad n \to \infty.$$

Furthermore, if g is monotone, the above statement may be "inverted" to produce an asymptotic confidence interval for θ, i.e. as $n \to \infty$,

$$P\{g^{-1}[g(T_n) - z_{\alpha/2}c/\sqrt{n}] \le \theta \le g^{-1}[g(T_n) + z_{\alpha/2}c/\sqrt{n}]\} \to 1 - \alpha.$$

In the above example, letting $g(x) = \log x$, we have $g'(x) = x^{-1}$ and $g'(\sigma^2) = 1/\sigma^2 > 0$; thus, by Theorem 3.4.4. it follows that

$$\sqrt{n}\{\log S_n^2 - \log \sigma^2\} \xrightarrow{\mathcal{D}} N(0, c^2)$$

with $c^2 = \sigma^{-4}(\beta_2 - 1)\sigma^4 = \beta_2 - 1$. Thus, as $n \to \infty$

$$P\{\log S_n^2 - z_{\alpha/2}\sqrt{(\beta_2 - 1)/n} \le \log \sigma^2 \le \log S_n^2 + z_{\alpha/2}\sqrt{(\beta_2 - 1)/n}\}$$

converges to $1 - \alpha$ which implies that as $n \to \infty$

$$P\{\exp[\log S_n^2 - z_{\alpha/2}\sqrt{(\beta_2 - 1)/n}] \le \sigma^2 \le \exp[\log S_n^2 + z_{\alpha/2}\sqrt{(\beta_2 - 1)/n}]\}$$

converges to $1-\alpha$. Now observe that if g is continuous and $0 < |g'(\theta)| < \infty$, it follows from Theorem 3.4.4 that

$$\sqrt{n}\{g(T_n) - g(\theta)\} \xrightarrow{\mathcal{D}} N(0, [g'(\theta)]^2 h(\theta)).$$

Thus, to obtain a variance-stabilizing transformation we need to define g such that $[g'(\theta)]^2 h(\theta) = c^2$ which implies

$$g'(\theta) = \frac{c}{\sqrt{h(\theta)}}$$

so that

$$g(\theta) = \int g'(\theta)d\theta = \int \frac{c}{\sqrt{h(\theta)}}d\theta.$$

Returning to the setup of Example 3.4.1, letting $\theta = \sigma^2$, we have $h(\theta) = (\beta_2 - 1)\theta^2$; then $g(\theta) = c(\beta_2 - 1)^{-1/2} \int \theta^{-1} d\theta = c^* \log\theta$, where $c^* = c/\sqrt{\beta_2 - 1}$. Choosing $c^* = 1$ we get $g(T_n) = \log T_n$ as the appropriate transformation.

Example 3.4.10: Let $R_n \sim \text{Bin}(n, \pi)$ and consider the statistic $T_n = R_n/n$; we have seen that $\sqrt{n}(T_n - \pi) \to Z$ where $Z \sim N(0, \pi(1 - \pi))$. Here $h(\pi) = \pi(1 - \pi)$ and the variance-stabilizing transformation may be obtained from $g(\pi) = c \int \{\pi(1 - \pi)\}^{-1/2}d\pi$. Letting $\pi = \sin^2 x$, we have $1 - \pi = \cos^2 x$ and $d\pi = 2 \sin x \cos x dx$ which yields

$$g(\pi) = c \int \frac{2 \sin x \cos x}{\sin x \cos x}dx = 2cx = 2c \sin^{-1}\sqrt{\pi}.$$

Choosing $c = 1/2$, we get $g(T_n) = \sin^{-1}(\sqrt{T_n})$, $0 < T_n < 1$. More explicitly we may write

$$\sqrt{n}\{\sin^{-1}\sqrt{T_n} - \sin^{-1}\sqrt{\pi}\} \xrightarrow{\mathcal{D}} N(0, 1/4) \quad \text{for} \quad 0 < \pi < 1.$$

An approximate $100(1 - \alpha)\%$ confidence interval for π has lower and upper limits respectively given by $L_n = \sin^2\{\sin^{-1}\sqrt{T_n} - z_{\alpha/2}/2\sqrt{n}\}$ and $U_n = \sin^2\{\sin^{-1}(\sqrt{T_n}) + z_{\alpha/2}/2\sqrt{n}\}$, where $z_{\alpha/2}$ is the $(100\alpha/2)$th percentile of the $N(0, 1)$ distribution. It is interesting to note the approximation is fairly good, even for $n \cong 10$. Anscombe (1948) suggested the alternative transformation $g(T_n) = \sin^{-1}\sqrt{(R_n + 3/8)/(n + 3/4)}$, which achieves a better approximation for moderate values of n.   ∎

Example 3.4.11: Let $\{X_\theta\}$ be a sequence of independent random variables with a Poisson(θ) distribution. Then, $\theta^{-1/2}(\overline{X}_\theta - \theta) \xrightarrow{\mathcal{D}} N(0, \theta)$. Letting $h(\theta) = \theta$ we get $g(\theta) = c \int d\theta/\sqrt{\theta} = 2c\sqrt{\theta}$ and choosing $c = 1/2$ we conclude that the appropriate variance-stabilizing transformation is $g(X_\theta) = \sqrt{X_\theta}$. More explicitly we may write

$$\theta^{-1/2}(\sqrt{X_\theta} - \sqrt{\theta}) \xrightarrow{\mathcal{D}} N(0, 1/4).$$

Anscombe (1948) suggested an alternative transformation where $g(X_\theta) = \sqrt{X_\theta + 3/8}$, which works out better for moderate θ.   ∎

Example 3.4.12: Consider the setup of Example 3.4.7 and suppose that (X_n, Y_n), $n \geq 1$, have bivariate normal distribution. In this case it follows that

$$\gamma^2 = \frac{\rho^2}{4} \left\{ E\left(\frac{X_1 - \mu_X}{\sigma_X}\right)^4 + E\left(\frac{Y_1 - \mu_Y}{\sigma_Y}\right)^4 \right\}$$
$$+ \rho^2 \left(\frac{1}{\rho^2} + \frac{1}{2}\right) E\left(\frac{X_1 - \mu_X}{\sigma_X}\right)^2 \left(\frac{Y_1 - \mu_Y}{\sigma_Y}\right)^2$$
$$- \rho \left\{ E\left(\frac{X_1 - \mu_X}{\sigma_X}\right)^3 \left(\frac{Y_1 - \mu_Y}{\sigma_Y}\right) + E\left(\frac{X_1 - \mu_X}{\sigma_X}\right) \left(\frac{Y_1 - \mu_Y}{\sigma_Y}\right)^3 \right\}.$$

$$(3.4.14)$$

Since $E\{(X_1 - \mu_X)/\sigma_X\}^2 \sim \chi_1^2$, we have

$$E\left(\frac{X_1 - \mu_X}{\sigma_X}\right)^4 = \mathrm{Var}\left(\frac{X_1 - \mu_X}{\sigma_X}\right)^2 + E^2\left(\frac{X_1 - \mu_X}{\sigma_X}\right)^2 = 3.$$

Similarly, it follows that $E\{(Y_1 - \mu_Y)/\sigma_Y\}^4 = 3$. On the other hand, we know that the conditional distribution of Y_1 given $X_1 = x$ is

$$N\left(\mu_Y + \rho \frac{\sigma_Y}{\sigma_X}(x - \mu_X), \sigma_Y^2(1 - \rho^2)\right);$$

therefore,

$$E\left(\frac{X_1 - \mu_X}{\sigma_X}\right)^2 \left(\frac{Y_1 - \mu_Y}{\sigma_Y}\right)^2 = E\left\{ E\left[\left(\frac{X_1 - \mu_X}{\sigma_X}\right)^2 \left(\frac{Y_1 - \mu_Y}{\sigma_Y}\right)^2 \Big| X_1\right]\right\}$$
$$= E\left\{ \left(\frac{X_1 - \mu_X}{\sigma_X}\right)^2 \left[(1 - \rho^2) + \rho^2 \left(\frac{X_1 - \mu_X}{\sigma_X}\right)^2\right]\right\}$$
$$= (1 - \rho^2) + \rho^2 E\left(\frac{X_1 - \mu_X}{\sigma_X}\right)^4$$
$$= 1 - \rho^2 + 3\rho^2$$
$$= 1 + 2\rho^2$$

and also,

$$E\left(\frac{Y_1 - \mu_Y}{\sigma_Y}\right)\left(\frac{X_1 - \mu_X}{\sigma_X}\right)^3 = E\left\{ E\left[\left(\frac{Y_1 - \mu_Y}{\sigma_Y}\right)\left(\frac{X_1 - \mu_X}{\sigma_X}\right)^3 \Big| X_1\right]\right\}$$
$$= E\left\{ \left(\frac{X_1 - \mu_X}{\sigma_X}\right)^3 \rho\left(\frac{X_1 - \mu_X}{\sigma_X}\right)\right\}$$
$$= 3\rho.$$

Similarly we have $E\{(Y_1 - \mu_Y)/\sigma_Y\}^3\{(X_1 - \mu_X)/\sigma_X\} = 3\rho$; substituting these results in (3.4.14) we obtain $\gamma^2 = (1 - \rho^2)^2$, so that

$$\sqrt{n}(R_n - \rho) \xrightarrow{\mathcal{D}} N(0, (1 - \rho^2)^2).$$

To obtain a variance stabilizing transformation, we let $h(p) = (1 - \rho^2)^2$ and

$$
\begin{aligned}
g(\rho) = c \int \frac{d\rho}{1 - \rho^2} &= c \int \left\{ \frac{1}{1 - \rho} + \frac{1}{1 + \rho} \right\} d\rho \\
&= \frac{c}{2} \int \frac{d\rho}{1 - \rho} + \frac{c}{2} \int \frac{d\rho}{1 + \rho} \\
&= \frac{c}{2} \log \frac{1 + \rho}{1 - \rho}.
\end{aligned}
$$

Choosing $c = 1$ and recalling that $\tanh(x) = (e^x - e^{-x})/(e^x + e^{-x})$, it follows that $g(R_n) = \tanh^{-1}(R_n)$. Therefore, we may write $\sqrt{n}\{\tanh^{-1}(R_n) - \tanh^{-1}(\rho)\} \xrightarrow{\mathcal{D}} N(0, 1)$. Gayen (1951) showed that a better approximation is obtained if we replace $\sqrt{n}$ by $\sqrt{n-3}$ in the above expression; this produces reasonable approximations for $n \geq 10$. ∎

A final but important topic in this section is that of the **convergence of moments** of functions of statistics which are asymptotically normal. In the setup of Example 3.4.6 we have seen that the sample correlation coefficient $v_n = S_n/\overline{X}_n$ is such that $\sqrt{n}(v_n - \nu) \xrightarrow{\mathcal{D}} N(0, \gamma^2)$; the question is whether we may approximate the moments of $\sqrt{n}(v_n - \nu)$ by appropriate functions of the moments of $\overline{X}_n$ and S_n^2. More specifically, let $\{T_n\}$ be a sequence of statistics such that $\sqrt{n}(T_n - \theta) \xrightarrow{\mathcal{D}} N(0, \sigma^2)$ and let g be a function such that $g'(\theta) \neq 0$ exists. We have seen above that

$$\sqrt{n}\{g(T_n) - g(\theta)\} \xrightarrow{\mathcal{D}} N(0, [g'(\theta)]^2 \sigma^2).$$

Furthermore, we also know that $(T_n - \theta) = O_p(1/\sqrt{n})$ and that by Taylor's expansion formula $\sqrt{n}\{g(T_n) - g(\theta)\} = g'(\theta)\sqrt{n}(T_n - \theta) + o_p(1)$. Then, if r is a positive integer, we have

$$
\begin{aligned}
[\sqrt{n}\{g(T_n) - g(\theta)\}]^r &= \sum_{i=0}^{r} \binom{r}{i} \{g'(\theta)\sqrt{n}(T_n - \theta)\}^i \{o_p(1)\}^{r-i} \\
&= \{g'(\theta)\}^r \{\sqrt{n}(T_n - \theta)\}^r + \sum_{i=0}^{r-1} \binom{r}{i} \{O_p(1)\}^i \{o_p(1)\}^{r-i} \\
&= \{g'(\theta)\}^r \{\sqrt{n}(T_n - \theta)\}^r + o_p(1).
\end{aligned}
$$

Thus, we may write

$$E[\sqrt{n}\{g(T_n) - g(\theta)\}]^r = \{g'(\theta)\}^r E\{\sqrt{n}(T_n - \theta)\}^r + Eo_p(1).$$

The question is whether we may claim that $\mathrm{E}o_p(1) = o(1)$. In general, the answer is negative, but under some further conditions on g, the result holds. In this direction we consider the following.

Theorem 3.4.9 (Cramér): *Let $\{T_n\}$ be a sequence of random variables with mean θ, r a positive integer and g a function satisfying the following conditions:*

i) $|g(T_n)| \leq cn^p$ *where c and p are finite constant;*

ii) *there exists $k > r(p+1)$ and $n_0 = n_0(k)$ such that for every $n \geq n_0$*

$$\mu_{2k} = \mathrm{E}\{\sqrt{n}(T_n - \theta)\}^{2k} < \infty;$$

iii) $g''(x)$ *is continuous and bounded in some neighborhood of θ.*

Then

$$\mathrm{E}[\sqrt{n}\{g(T_n) - g(\theta)\}]^r = \{g'(\theta)\}^r \mathrm{E}\{\sqrt{n}(T_n - \theta)\}^r + O(1/\sqrt{n})$$

or equivalently

$$\mathrm{E}\{g(T_n) - g(\theta)\}^r = \{g'(\theta)\}^r \mathrm{E}(T_n - \theta)^r + O(n^{-(r+1)/2}).$$

Proof: Let $Z_n = \sqrt{n}\{g(T_n) - g(\theta)\}$ and $U_n = \sqrt{n}(T_n - \theta)$. Then, for all $\varepsilon > 0$

$$\mathrm{E}Z_n^r = \mathrm{E}[Z_n^r I_{\{|U_n| < \varepsilon\sqrt{n}\}}] + \mathrm{E}[Z_n^r I_{\{|U_n| \geq \varepsilon\sqrt{n}\}}]. \tag{3.4.15}$$

Now, note that

$$
\begin{aligned}
|Z_n| \leq \sqrt{n}\{|g(T_n)| + |g(\theta)|\} &\leq cn^{p+1/2} + n^{1/2}|g(\theta)| \\
&< n^{p+1/2}\{c + |g(\theta)|\} \\
&= C_1 n^{p+1/2}
\end{aligned}
$$

where $C_1 = c + |g(\theta)|$. Then considering the second term on the right-hand side of (3.4.15) we have

$$
\begin{aligned}
|\mathrm{E}[Z_n^r I_{\{|U_n| \geq \varepsilon\sqrt{n}\}}]| &\leq \mathrm{E}[|Z_n^r| I_{\{|U_n| \geq \varepsilon\sqrt{n}\}}] \\
&< C_1^r n^{r(p+1/2)} P\{|U_n| \geq \varepsilon\sqrt{n}\} \\
&\leq C_1^r n^{r(p+1/2)} \frac{\mathrm{E}|T_n - \theta|^{2k}}{\varepsilon^{2k}} \\
&= C_1^r n^{r(p+1/2)} \frac{\mu_{2k}}{\varepsilon^{2k} n^k} \\
&= C_2 n^{r(p+1/2) - k}
\end{aligned}
$$

where $C_2 = C_1 \mu_{2k}/\varepsilon^{2k}$. Now, for all $n \geq n_0$ we have from (ii)

$$r(p + 1/2) - k < r(p + 1/2) - r(p + 1) = -r/2 \leq -1/2$$

so it follows that, for all $n \geq n_0$

$$|\mathrm{E}[Z_n^r I_{\{|U_n| \geq \varepsilon \sqrt{n}\}}]| \leq C_2 n^{-1/2} = O(n^{-1/2}). \tag{3.4.16}$$

In order to deal with the first term on the right-hand side of (3.4.15), note that

$$
\begin{aligned}
Z_n^r I_{\{|U_n| < \varepsilon \sqrt{n}\}} &= n^{r/2} \{g(T_n) - g(\theta)\}^r I_{\{|U_n| < \varepsilon \sqrt{n}\}} \\
&= n^{r/2} \left\{ g'(\theta)(T_n - \theta) + \frac{g''(\xi_n)}{2}(T_n - \theta)^2 \right\}^r I_{\{|U_n| < \varepsilon \sqrt{n}\}}
\end{aligned}
$$

where $\theta - \varepsilon < \xi_n < \theta + \varepsilon$. Then

$$
\begin{aligned}
Z_n^r I_{\{|U_n| < \varepsilon \sqrt{n}\}} &= \left\{ g'(\theta)U_n + \frac{g''(\xi_n)}{2\sqrt{n}} U_n^2 \right\}^r I_{\{|U_n| < \varepsilon \sqrt{n}\}} \\
&= \sum_{i=0}^{r} \binom{r}{i} \left\{ \frac{1}{2\sqrt{n}} g''(\xi_n) U_n^2 \right\}^i \{g'(\theta)U_n\}^{r-i} I_{\{|U_n| < \varepsilon \sqrt{n}\}} \\
&= \{g'(\theta)\}^r U_n^r I_{\{|T_n - \theta| < \varepsilon \sqrt{n}\}} \\
&\quad + \sum_{i=1}^{r} \binom{r}{i} \left(\frac{1}{2\sqrt{n}} \right)^i \{g'(\theta)\}^{r-i} \{g''(\xi_n)\}^i U_n^{r+i} I_{\{|T_n - \theta| < \varepsilon \sqrt{n}\}}.
\end{aligned}
$$

Therefore,

$$
\begin{aligned}
\mathrm{E}[Z_n^r &I_{\{|U_n| < \varepsilon \sqrt{n}\}}] \\
&= \{g'(\theta)\}^r [\mathrm{E}\{\sqrt{n}(T_n - \theta)\}^r - \mathrm{E}\{\sqrt{n}(T_n - \theta)\}^r I_{\{|U_n| \geq \varepsilon \sqrt{n}\}}] \\
&\quad + \sum_{i=1}^{r} \binom{r}{i} \left(\frac{1}{2\sqrt{n}} \right)^i \{g'(\theta)\}^{r-i} \mathrm{E}[\{g''(\xi_n)\}^i U_n^{r+i} I_{\{|U_n| \geq \varepsilon \sqrt{n}\}}].
\end{aligned}
\tag{3.4.17}
$$

Now observe that

$$
\begin{aligned}
|\{\mathrm{E}\sqrt{n}(T_n - \theta)\}^r &I_{\{|U_n| \geq \varepsilon \sqrt{n}\}}| \\
&\leq \mathrm{E}\{[\sqrt{n}(T_n - \theta)]^{2r}\}^{1/2} \{P(|U_n| \geq \varepsilon \sqrt{n})\}^{1/2} \\
&\leq \mathrm{E}\{n^r(T_n - \theta)^{2r}\}^{1/2} \left\{ \frac{\mathrm{E}|T_n - \theta|^{2k}}{\varepsilon^k} \right\}^{1/2} \\
&= \mu_{2r}^{1/2} \frac{\mu_{2k}^{1/2}}{\varepsilon^k n^{k/2}} \\
&= C_3 n^{-k/2} < C_3 n^{-1/2} = O(n^{-1/2})
\end{aligned}
\tag{3.4.18}
$$

where $C_3 = (\mu_{2r}\mu_{2k}/\varepsilon^{2k})^{1/2}$.

Let A_n denote the second term on the right-hand side of (3.4.17) and note that

$$|A_n| \leq \sum_{i=1}^{r} \binom{r}{i} \left(\frac{1}{2\sqrt{n}}\right)^i |g'(\theta)|^{r-i} \mathrm{E}[|g''(\xi_n)|^i |U_n|^{r+i} I_{\{|U_n|<\varepsilon\sqrt{n}\}}].$$

Using condition (iii), it follows that there exists $\varepsilon > 0$ such that $A = \sup\{|g''(X)| : \theta - \varepsilon \leq X \leq \theta + \varepsilon\} < \infty$; then we have

$$
\begin{aligned}
|A_n| &\leq \sum_{i=1}^{r} \binom{r}{i} \left(\frac{1}{2\sqrt{n}}\right)^i A^i |g'(\theta)|^{r-i} \mathrm{E}[|U_n^{r+i}| I_{\{|U_n|<\varepsilon\sqrt{n}\}}] \\
&\leq \sum_{i=1}^{r} \binom{r}{i} \left(\frac{1}{2\sqrt{n}}\right)^i A^i |g'(\theta)|^{r-i} \mathrm{E}|U_n|^{r+i} \\
&= \sum_{i=1}^{r} \binom{r}{i} \left(\frac{1}{2\sqrt{n}}\right)^i A^i |g'(\theta)|^{r-i} \mu_{r+i} \\
&\leq \frac{1}{\sqrt{n}} \left\{ \sum_{i=1}^{r} \binom{r}{i} A^i |g'(\theta)|^{r-i} \mu_{r+i} \right\} = O(n^{-1/2}). \qquad (3.4.19)
\end{aligned}
$$

Thus, from (3.4.17)–(3.4.19) it follows that

$$\mathrm{E}[Z_n^r I_{\{|U_n|<\varepsilon\sqrt{n}\}}] = \{g'(\theta)\}^r \mathrm{E}\{\sqrt{n}(T_n - \theta)\}^r + O(n^{-1/2})$$

which in connection with (3.4.15) completes the proof.   ∎

A useful generalization of the above result is given by:

Theorem 3.4.10 (Cramér): *Let $\{\mathbf{T}_n\}$ be a sequence of random q-vectors with mean vector $\boldsymbol{\theta}$, r a positive integer and $g : \mathbb{R}^q \to \mathbb{R}$ satisfying the following conditions:*

i) *$|g(\mathbf{T}_n)| \leq cn^p$ where c and p are finite positive constants;*

ii) *there exists $k > r(p+1)$ and $n_0 = n_0(k)$ such that for all $n \geq n_0$,*

$$\mu_{2k}^{(j)} = \mathrm{E}\{\sqrt{n}(T_{nj} - \theta_j)\}^{2k} < \infty, \; j = 1, \ldots, q.$$

iii) *$g_{ij}''(\mathbf{x}) = (\partial^2/\partial x_i \partial x_j)g(\mathbf{x})$ is a continuous and bounded in some neighborhood of $\boldsymbol{\theta}$, $i, j = 1, \ldots, q$.*

Then,

$$\mathrm{E}[\sqrt{n}\{g(\mathbf{T}_n) - g(\boldsymbol{\theta})\}]^r = \mathrm{E}\{\sqrt{n} \sum_{j=1}^{q} g_j'(\boldsymbol{\theta})(T_{nj} - \theta_j)\}^r + O(n^{-1/2}).$$

Example 3.4.13: Consider the setup of Example 3.4.6. and let $v_n =$

$S_n/\overline{X}_n = g(\overline{X}_n, S_n^2)$. Now note that

$$v_n^2 = \frac{\frac{1}{n-1}\left\{\sum_{i=1}^n X_i^2 - n\overline{X}_n^2\right\}}{\overline{X}_n^2} \leq \frac{\frac{1}{n-1}\left\{\sum_{i=1}^n X_i^2\right\}}{\frac{1}{n}\left\{\sum_{i=1}^n X_i\right\}^2}$$

$$\leq \frac{n}{n-1}\sum_{i=1}^n \left\{X_i \Big/ \sum_{i=1}^n X_i\right\}^2 \leq n$$

and then $|v_n| = |g(\overline{X}_n, S_n^2)| \leq n^{1/2}$ which implies that (i) in Theorem 3.4.10 holds with $c = 1$ and $p = 1/2$. If $\mu > 0$ and $\sigma^2 < \infty$, then (iii) is also satisfied. Therefore, if we want to obtain an approximation for the second moment of $\sqrt{n}(v_n - \nu)$, we must have $\mu_{2k}^{(j)} < \infty$ for some $k > r(p+1) = 3$. This implies that $E(S_n^2 - \sigma^2)^8 < \infty$, for all $n \geq n_0$ and consequently that $E|X_1|^{16} < \infty$.

Suppose, for example that the distribution function of X_1 is given by $F(x) = 1 - x^{-5}$ if $x \geq 1$ and $F(x) = 0$ for $x < 1$. Then

$$EX^4 = \int_1^\infty x^4 5x^{-6} dx = 5 \int_1^\infty x^{-2} dx = 5\left[-x^{-1}\right]_1^\infty = 5 < \infty.$$

Therefore, from Example 3.4.6. it follows that

$$\sqrt{n}(v_n - \nu) \xrightarrow{D} N(0, \gamma^2).$$

However, $EX^{16} = \int_1^\infty x^{16} 5x^{-6} dx = 5 \int_1^\infty x^{10} dx = \infty$, so we may not use the previous theorem to approximate the moments of $\sqrt{n}(v_n - \nu)$. ∎

Finally we point that although the assumptions required by the Cramér Theorem (3.4.9) are sufficient, they are not necessary. If, for example, $g(T_n) = T_n = \overline{X}_n$, (i) may not hold (i.e., when the d.f. F has infinite support). Nevertheless, we may still be able to obtain convenient moment convergence, provided some extra regularity conditions are met. In this direction we state without proof the following result due to von Bahr (1965).

Theorem 3.4.11: Let $X_1, \ldots, X_n$ be i.i.d. r.v.'s with zero means and unit variances. If $E|X_i|^k < \infty$ for an integer $k \geq 3$, we have

$$E(\sqrt{n}\,\overline{X}_n)^k = \int_{-\infty}^\infty x^k d\Phi(x) + O(n^{-1/2}).$$

3.5 Rates of convergence to normality

Let $\{X_n\}$ be a sequence of random variables such that $\mathrm{E}X_i = \mu_i$ and $\mathrm{Var}X_i = \sigma_i^2$, $i \geq 1$. Let $T_n = \sum_{i=1}^{n} X_i$, $\xi_n = \sum_{i=1}^{n} \mu_i$, $s_n^2 = \sum_{i=1}^{n} \sigma_i^2$ and suppose that the conditions required by some version of the Central Limit Theorem are satisfied. Then, for every $x \in \mathbb{R}$ we may write

$$F_n(x) = P\left\{\frac{T_n - \xi_n}{s_n} \leq x\right\} \to \int_{-\infty}^{x} \frac{1}{\sqrt{2\pi}} e^{-t^2/2} dt = \Phi(x) \quad \text{as} \quad n \to \infty.$$

A question of both theoretical and practical interest concerns the speed with which the above convergence takes place. In other words, how large should n be so that $\Phi(x)$ may be considered as a good approximation for $F_n(x)$. Although there are no simple answers to this question, the following results may serve as general guidelines in that direction.

Theorem 3.5.1 (Polya): *Let $\{F_n\}$ be a sequence of distribution functions weakly converging to a continuous distribution function F. Then*

$$\lim_{n \to \infty} \sup_{x \in \mathbb{R}} |F_n(x) - F(x)| = 0.$$

Proof: Since F is continuous, given $\varepsilon > 0$, there exists $n_0 = n_0(\varepsilon)$ such that for all $x \in \mathbb{R}$, $|F_n(x) - F(x)| < \varepsilon$. Therefore, $\sup_{x \in \mathbb{R}} |F_n(x) - F(x)| \leq \varepsilon$ and the result follows from the definition of limit.   ∎

Theorem 3.5.2 (Berry-Esséen): *Let $\{X_n\}$ be a sequence of i.i.d. r.v.'s with $\mathrm{E}X_1 = \mu$, $\mathrm{Var}X_1 = \sigma^2$ and suppose that $\mathrm{E}|X_1 - \mu|^{2+\delta} = \nu_{2+\delta} < \infty$ for some $0 < \delta \leq 1$. Also let $T_n = \sum_{i=1}^{n} X_i$, $F_n(x) = P\left\{\frac{T_n - n\mu}{\sigma\sqrt{n}} \leq x\right\}$, $x \in \mathbb{R}$. Then there exists a constant C such that*

$$\Delta_n = \sup_{x \in \mathbb{R}} |F_n(x) - \Phi(x)| \leq C \frac{\nu_{2+\delta} n^{-\delta/2}}{\sigma^{2+\delta}}.$$

The proof is out of the scope of this text and the reader is referred to Feller (1971) for details. Berry (1941) proved the result for $\delta = 1$ and sharper limits for the constant C have been obtained by Zolotarev (1967) and Van Beeck (1972) (actually $C=0.41$). The usefulness of the theorem, however, is limited, since the rates of convergence attained are not very sharp. Suppose, for example, that $\delta = 1$, $n = 100$ and $\gamma_3 = \nu_3/\sigma^3 = 1.5$; then $\Delta_n \leq C\gamma_3 n^{-1/2} = 0.41 \times 1.5 \times 1/10 = 0.0615$.

Alternatively, for the rates of convergence of the sequence of d.f.'s F_n to Φ or of the p.d.f.'s f_n (when they exist) to φ, the p.d.f. of the standard normal distribution may be assessed by **Gram-Charlier** or **Edgeworth**

expansions.

Essentially, the idea behind the Gram-Charlier expansions is based on the fact that any probability density function g may be formally expressed as

$$g(x) = \sum_{j=0}^{\infty} \frac{c_j}{j!} \varphi^{(j)}(x) \quad = \quad \frac{1}{\sqrt{2\pi}} \sum_{j=0}^{\infty} \frac{c_j}{j!} \frac{d^j}{dx^j} e^{-x^2/2}$$

$$= \quad \varphi(x) \sum_{j=0}^{\infty} (-1)^j \frac{c_j}{j!} H_j(x) \qquad (3.5.1)$$

where

$$H_j(x) = (-1)^j e^{-x^2/2} d^j / dx^j e^{-x^2/2}, \; j = 0, 1, \dots$$

are orthogonal polynomials associated with the normal distribution, known in the literature as **Hermite polynomials**, and ϕ is the p.d.f. for the normal distribution. In particular, they satisfy the relations

$$\frac{1}{\sqrt{2\pi}} \int_{-\infty}^{+\infty} H_m(x) H_n(x) e^{-x^2/2} dx = \begin{cases} m! & \text{if } m = n \\ 0 & \text{if } m \neq n. \end{cases}$$

Now, from (3.5.1) it follows that for $m = 0, 1, \dots$ we have

$$\int_{-\infty}^{+\infty} H_m(x) g(x) dx \quad = \quad \sum_{j=0}^{\infty} (-1)^j \frac{c_j}{j!} \frac{1}{\sqrt{2\pi}} \int_{-\infty}^{+\infty} H_m(x) H_j(x) e^{-x^2/2} dx$$

$$= \quad (-1)^m c_m$$

which implies that

$$c_m = (-1)^m \int_{-\infty}^{+\infty} g(x) H_m(x) dx. \qquad (3.5.2)$$

For example, if X is a random variable with $EX = 0$, $\text{Var}X = 1$ and probability density function g we may use (3.5.2) to obtain

$$c_0 \quad = \quad 1,$$

$$c_1 \quad = \quad -\int_{-\infty}^{+\infty} x g(x) dx = 0,$$

$$c_2 \quad = \quad (-1)^2 \int_{-\infty}^{+\infty} (x^2 - 1) g(x) dx = 1 - 1 = 0,$$

$$c_3 \quad = \quad (-1)^3 \int_{-\infty}^{+\infty} (x^3 - 3x) g(x) dx = -\mu_3,$$

$$\vdots$$

so that from (3.5.1) we may write

$$g(x) = \varphi(x)\left\{1 + \sum_{j=3}^{\infty}(-1)^j \frac{c_j}{j!} H_j(x)\right\}. \qquad (3.5.3)$$

Note that we are not really interested in the convergence of the series in (3.5.3); our main objective is to know whether a small number of terms (say, one or two) are sufficient to guarantee a good approximation to $g(x)$. In particular, we may consider

$$g(x) = \varphi(x)\{1 + \mu_3(x^3 - 3x)/6\} + r(x)$$

and the problem is to find out whether $r(x)$ is small enough.

Recalling that from the definition of the Hermite polynomials,

$$\Phi^{(j)}(x) = \frac{1}{\sqrt{2\pi}}(-1)^{j-1} H_{j-1}(x) e^{-x^2/2}, \quad j \geq 1,$$

a formal integration of (3.5.3) yields

$$\int_{-\infty}^{x} g(t)dt = \int_{\infty}^{x} \varphi(t)dt + \sum_{j=3}^{\infty}(-1)^j \frac{c_j}{j!} \int_{-\infty}^{x} \frac{1}{\sqrt{2\pi}} e^{-t^2/2} H_j(t)dt. \quad (3.5.4)$$

But since

$$\int_{-\infty}^{x} \frac{1}{\sqrt{2\pi}} e^{-t^2/2} H_j(t)dt = \int_{-\infty}^{x} (-1)^j \Phi^{(j+1)}(t)dt$$

$$= (-1)^j \Phi^{(j)}(t)\Big|_{-\infty}^{x} = (-1)^j \Phi^{(j)}(x),$$

it follows that

$$G(x) = \Phi(x)\left\{1 + \sum_{j=3}^{\infty} \frac{c_j}{j!} \Phi^{(j)}(x)\right\}$$

$$= \Phi(x) - \varphi(x) \sum_{j=3}^{\infty}(-1)^j \frac{c_j}{j!} H_{j-1}(x) \qquad (3.5.5)$$

where G is the distribution function of X. In particular we may write

$$G(x) = \Phi(x) + \varphi(x)\mu_3(x^2 - 1)/6 + R(x).$$

In order to verify how the above results may be applied in the context of large samples, let $\{X_n\}$ denote a sequence of i.i.d. r.v.'s with $EX_1 = 0$, $\mathrm{Var}X_1 = \sigma^2 < \infty$ and characteristic function ϕ. Let $Z_n = \sum_{i=1}^{n} X_i/\sigma\sqrt{n}$ and suppose that the corresponding probability density function f_n exists; also let F_n and ϕ_n respectively denote the distribution function and the characteristic function associated with f_n. In order to specify an expansion of the form (3.5.3) for f_n we must obtain the coefficients c_j, $j \geq 1$. In this direction we first consider the following formal Taylor series expansions:

$$\phi(t) = \sum_{j=0}^{\infty} \phi^{(j)}(0) \frac{t^j}{j!} = 1 + \sum_{j=1}^{\infty} \frac{\mu_j}{j!} (it)^j \qquad (3.5.6)$$

where $\mu_j = i^{-j} \phi^{(j)}(0)$, $j \geq 1$, are the moments of X_1 and

$$\log \phi(t) = \sum_{j=0}^{\infty} [\log \phi(t)]_{t=0}^{(j)} \frac{t^j}{j!} = \sum_{j=0}^{\infty} \frac{\lambda_j}{j!} (it)^j \qquad (3.5.7)$$

where $\lambda_j = i^{-j} [\log \phi(t)]_{t=0}^{(j)}$, $j \geq 1$, are the cumulants of X_1. From (3.5.6) and (3.5.7) it follows that

$$\log \phi(t) = \log \left\{ 1 + \sum_{j=1}^{\infty} \frac{\mu_j}{j!} (it)^j \right\} = \sum_{j=0}^{\infty} \frac{\lambda_j}{j!} (it)^j$$

which implies

$$\sum_{h=1}^{\infty} \frac{(-1)^{h+1}}{h} \left\{ \sum_{j=1}^{\infty} \frac{\mu_j}{j!} (it)^j \right\}^h = \sum_{j=0}^{\infty} \frac{\lambda_j}{j!} (it)^j. \qquad (3.5.8)$$

Expanding both members of (3.5.8) in powers of (it) and equating the coefficients of terms with the same power, we have $\lambda_1 = \mu_1$, $\lambda_2 = \mu_2 - \mu_1^2$, $\lambda_3 = \mu_3 - 3\mu_1\mu_2 + 2\mu_1^3$, $\lambda_4 = \mu_4 - 3\mu_2^2 - 4\mu_1\mu_3 + 12\mu_1^2\mu_2 - 6\mu_1^4$, ... and, consequently, $\mu_1 = \lambda_1$, $\mu_2 = \lambda_2 - \lambda_1^2$, $\mu_3 = \lambda_3 + 3\lambda_1\lambda_2 + \lambda_1^3$, $\mu_4 = \lambda_4 + 3\lambda_2^2 + 4\lambda_1\lambda_3 + 6\lambda_1^2\lambda_2 + \lambda_1^4 \cdots$.

Now observe that

$$\phi_n(t) = \{\phi(t/\sigma\sqrt{n})\}^n \rightarrow \log \phi_n(t) = n \log \phi(t/\sigma\sqrt{n});$$

expanding both members via (3.5.7) we obtain

$$\sum_{j=1}^{\infty} \lambda_j^{(n)} \frac{(it)^j}{j!} = \sum_{j=1}^{\infty} n\lambda_j \frac{(it/\sigma\sqrt{n})^j}{j!} = \sum_{j=1}^{\infty} \frac{\lambda_j}{j!\sigma^j n^{3/2-1}} (it)^j$$

where $\lambda_j^{(n)}$, $j \geq 1$, are the cumulants of Z_n; equating the coefficients of the terms with the same power of (it) we arrive at

$$\begin{aligned}
\lambda_1^{(n)} &= \lambda_1 = \mu_1 = 0, \\
\lambda_2^{(n)} &= \lambda_2/\sigma^2 = \sigma^2/\sigma^2 = 1, \\
\lambda_3^{(n)} &= \lambda_3/\sigma^3\sqrt{n} = \mu_3/\sigma^3\sqrt{n}, \\
\lambda_4^{(n)} &= \lambda_4/\sigma^4 n = (\mu_4 - 3\sigma^4)/\sigma^4 n,
\end{aligned}$$

$$\vdots$$

From the relation between cumulants and moments we get

$$
\begin{aligned}
\mu_1^{(n)} &= 0, \\
\mu_2^{(n)} &= 1, \\
\mu_3^{(n)} &= \mu_3/\sigma^3\sqrt{n}, \\
\mu_4^{(n)} &= \frac{\mu_4 - 3\sigma^4}{\sigma^4 n} + 3,
\end{aligned}
$$

$$\vdots$$

Finally, from (3.5.2) we may conclude that

$$
\begin{aligned}
c_1 &= c_2 = 0, \\
c_3 &= -\int_{-\infty}^{+\infty} (x^3 - 3x)f_n(x)dx = -\mu_3^{(n)} + 3\mu_1^{(n)} = -\mu_3/\sigma^3\sqrt{n}, \\
c_4 &= \int_{-\infty}^{+\infty} (x^4 - 6x^2 + 3)f_n(x)dx = \mu_4^{(n)} - 6\mu_2^{(n)} + 3 = \frac{\mu_4 - 3\sigma^4}{\sigma^4 n},
\end{aligned}
$$

$$\vdots$$

so that the Gram-Charlier expansion (3.5.3) is

$$f_n(x) = \varphi(x)\left\{1 + \frac{\mu_3}{6\sigma^3\sqrt{n}}(x^3 - 3x) + \frac{\mu_4 - 3\sigma^4}{24\sigma^4 n}(x^4 - 6x^2 + 3) + \cdots\right\}. \quad (3.5.9)$$

Analogously we may obtain an expansion for the distribution function:

$$F_n(x) = \Phi(x) + \varphi(x)\left\{\frac{\mu_3}{6\sigma^3\sqrt{n}}(1 - x^2) + \frac{\mu_4 - 3\sigma^4}{24\sigma^4 n}(3x - x^3) + \cdots\right\}. \quad (3.5.10)$$

Although the above expansions offer a better insight to the problem of evaluation of the rates of convergence to normality than that provided by Berry-Esséen Theorem (3.5.2), they have the disadvantage of requiring the knowledge of the moments of the parent distribution. Others alternatives, like Edgeworth's expansions are available, but they are out of the scope of this text. The reader is referred to Cramér (1946) for a detailed description and other references.

3.6 Concluding notes

As we have seen, the normal distribution occupies a central position in large sample theory. In the first place, it may be considered as an approximation to the distributions of many statistics (conveniently standardized) commonly used in practice. In many other cases, the corresponding asymptotic distributions (such as the central or noncentral chi-squared or f distributions) are basically related to suitable functions of normal variables. There are some problems, however, where normality, albeit being embedded, may not provide the appropriate results. A notable example of this kind is the U-statistic U_n defined by (2.4.28) when the parameter $\theta(F)$, given by (2.4.26), is stationary of order 1, in the sense that

$$\zeta_1(F) = \mathrm{E}_F g^2(X_1) - \theta^2(F) = 0 < \zeta_2(F) = \mathrm{E}_F g^2(X_1, X_2) - \theta^2(F)$$

where

$$g_c(x_1, \ldots, x_c) = \mathrm{E}\{g(X_1, \ldots, X_m) \mid X_1 = x_1, \ldots, X_c = x_c\}.$$

In such a case $n\{U_n - \theta(F)\} \xrightarrow{D} \sum_{j \geq 1} \lambda_j (Z_j^2 - 1)$, where the λ_j's are unknown coefficients and the Z_j's are i.i.d. r.v's following the $N(0,1)$ law. A similar problem will be encountered in Chapter 4, when dealing with Kolmogorov-Smirnov type statistics. The techniques developed in this chapter do not suffice to handle these situations and a survey of some modern probability tools more appropriate for such nonregular cases will be presented in Chapter 8.

Another related problem deals with the convergence of the binomial law (1.4.34) to the Poisson distribution when $n \to \infty$, $\pi \to \infty$ but $n\pi \to \lambda$ (> 0); this convergence to a Poisson law may also arise in a variety of other problems (see Exercise 2.3.8). For various reasons, we shall not enter into details of this Poisson convergence results. We would like to refer to Tucker (1967), Chow and Teicher (1978) and other standard texts in Probability Theory for some nice accounts of these developments. For applications of martingale central limit theorems in nonparametrics, the reader is referred to Sen (1981). Further results are posed in the form of exercises.

3.7 Exercises

Exercise 3.1.1: Let X_{ni}, $i \geq 1$, be independent r.v.'s such that $P\{X_{ni} = 1\} = 1 - P\{X_{ni} = 0\} = \pi_{ni}$, $i \geq 1$. If the π_{ni} are strictly positive and as $n \to \infty$, (i) $\sum_{i=1}^{n} \pi_{ni} = \pi_n \to \lambda$ $(0 < \lambda < \infty)$ and (ii) $\max_{i \leq i \leq n}(\pi_n^{-1} \pi_{ni}) \to 0$, show that $T_n = \sum_{i=1}^{n} X_{ni}$ has asymptotically a Poisson (λ) distribution.

Exercise 3.1.2 (Continued): Show that as λ increases, $(T_n - \lambda)/\sqrt{\lambda}$ converges in law to a normal variate.
Hint: For both Exercises 3.1.1 and 3.1.2 you may use characteristic functions.

Exercise 3.1.3: Consider the setup of Exercise 3.1.1. Show that $\mathrm{Var}\, T_n = \pi_n(1 - \bar{\pi}_n) - \sum_{i=1}^{n}(\pi_{ni} - \bar{\pi}_n)^2$, where $\bar{\pi}_n = n^{-1}\pi_n$. Hence, or otherwise, show that $\mathrm{Var}\, T_n \to \lambda$. Also, show that $\mathrm{Var} T_n / \{n\bar{\pi}_n(1 - \bar{\pi}_n)\} \leq 1$, where the upper bound is attained when $\sum_{i=1}^{n}(\pi_{ni} - \bar{\pi}_n)^2/n\bar{\pi}_n(1 - \bar{\pi}_n) \to 0$.

Exercise 3.3.1: In the same setup of Exercise 3.1.1 show that the divergence of the series $\sum_{i=1}^{n} \pi_{ni}(1 - \pi_{ni})$ is a necessary and sufficient condition for the asymptotic normality of $(T_n - \pi_n)/\sqrt{n}$.

Exercise 3.3.2: Let T_n be a r.v. having a noncentral chi-squared distribution with n degrees of freedom and noncentrality parameter Δ_n. Show $T_n \overset{\mathcal{D}}{=} X_1 + \cdots + X_n$, for all $n \geq 1$, where the X_j's are i.i.d. r.v.'s following a noncentral chi-squared distribution with 1 degree of freedom and noncentrality parameter $\delta_n = n^{-1}\Delta_n$. Hence, use Theorem 3.3.5 and show that $n^{1/2}(T_n - n - \Delta_n)$ is asymptotically normal. In particular, for $\Delta_n = 0$, verify that $(2n)^{-1/2}(T_n - n) \overset{\mathcal{D}}{\longrightarrow} N(0,1)$.

Exercise 3.3.3: Let $X_1, \ldots, X_n$ be i.i.d. r.v.'s with a continuous d.f. F and let $c_1, \ldots, c_n$ be n given constants, not all equal to zero. Let $a_n(i) = \mathrm{E}\psi(U_{ni})$, $1 \leq i \leq n$, where $\psi(u)$, $u \in (0,1)$ is square integrable and $U_{n1} < \cdots < U_{nn}$ are the ordered r.v.'s of a sample of size n from the uniform $(0,1)$ distribution. Consider the linear rank statistic $L_n = \sum_{i=1}^{n}(c_i - \bar{c}_n)a_n(R_{ni})$ where $\bar{c}_n = n^{-1}\sum_{i=1}^{n} c_i$ and R_{ni} is the rank of X_i among $X_1, \ldots, X_n$, for $i = 1, \ldots n$.

i) Show that $\{L_n; n \geq 1\}$ is a zero mean martingale sequence.

ii) Apply Theorem 3.3.7 to show that $C^{-1}L_n \overset{\mathcal{D}}{\longrightarrow} N(0,\gamma^2)$ where $C_n^2 = \sum_{i=1}^{n}(c_i - \bar{c}_n)^2$ and $\gamma^2 = \int_0^1 \psi^2(u)\,du - \bar{\psi}^2$ with $\bar{\psi} = \int_0^1 \psi(u)\,du$.

iii) Let $L_n^0 = \sum_{i=1}^{n}(c_i - \bar{c}_n)\psi\{F(X_i)\}$, so that $\mathrm{E}(L_n^0 \mid R_{n1}, \ldots, R_{nn}) = L_n$ and $C_n^{-2}\mathrm{E}(L_n^0 - L_n)^2 \to \infty$. Use Theorem 3.3.6 to verify that $C_n^{-1}L_n^0 \overset{\mathcal{D}}{\longrightarrow} N(0,\gamma^2)$, and, hence, use the Projection Technique discussed in Section 3.4 to obtain the asymptotic normality of $C_n^{-1}L_n$ from that of $C_n^{-1}L_n^0$.

Exercise 3.3.4: Let X_i, $i \geq 1$, be i.i.d. r.v.'s with a d.f. F symmetric about 0 [i.e., $F(x) + F(-x) = 1, \forall x$]. Also let $\psi(u)$ be skew-symmetric [i.e., $\psi(1/2 + u) + \psi(1/2 - u) = 0, \forall u \in (0, 1/2)$] and $\psi^+(u) = \psi\{(1 + u)/2\}$, $0 < u < 1$. Furthermore, define $T_n = \sum_{i=1}^{n} \mathrm{sign}\,(X_i)a_n^+(R_{ni}^+)$, $n \geq 1$, where $a_n^+(i) = \mathrm{E}\psi^+(U_{ni})$, $i = 1, \ldots, n$, $U_{n1} < \cdots < U_{nn}$ are ordered r.v.'s of a sample of size n from the Uniform$(0,1)$ distribution and R_{ni}^+ denotes the rank of $|X_i|$ among $|X_1|, \ldots, |X_n|$, for $i = 1, \ldots, n$.

i) Show that $\{T_n; n \geq 1\}$ is a zero mean martingale sequence. Hence, use Theorem 3.3.7 to verify the asymptotic normality of T_n (conveniently standardized).

ii) Define $T_n^0 = \sum_{i=1}^{n} \psi\{F(X_i)\}$, $n \geq 1$, and show that

$$\mathrm{E}\{T_n^0 \mid \mathrm{sign}\,(X_i), R_{ni}^+, 1 \leq i \leq n\} = T_n$$

and, hence, use Theorem 3.3.6 (on T_n^0) and the Projection Technique discussed in Section 3.4 to verify the asymptotic normality of T_n in an alternative manner.

Exercise 3.3.5: For the CMRR procedure in Example 2.3.7, let $n_1/N \to \alpha_1$ and $n_2/N \to \alpha_2$ $(0 < \alpha_1, \alpha_2 < 1)$ as $n_i \to \infty$, $i = 1, 2$ and $N \to \infty$. Then show that $N^{-1/2}(\widehat{N} - N) \xrightarrow{D} N(0, \sigma^2(\alpha_1, \alpha_2))$ where $\sigma^2(\alpha_1, \alpha_2) = (1 - \alpha_1)(1 - \alpha_2)/\alpha_1\alpha_2$. Verify that for a given α $(= \alpha_1 + \alpha_2)$, $\sigma^2(\alpha_1, \alpha_2)$ is minimized when $\alpha_1 = \alpha_2 = \alpha/2$.

Exercise 3.4.1: Let X be a random variable with finite moments; let μ_i denote the corresponding central moments and $\beta_1 = \mu_3^2/\mu_2^3$ and $\beta_2 = \mu_4/\mu_2^2$ be the conventional measures of **skewness** and **kurtosis**, respectively. Consider their sample counterparts $\widehat{\beta}_{1n} = m_{n3}^2/m_{n2}^3$ and $\widehat{\beta}_{2n} = m_{n4}/m_{n2}^2$ where $m_{nj} = n^{-1}\sum_{i=1}^{n}(X_i - \overline{X}_n)^j$, $j \geq 1$. Obtain the asymptotic normality of $\sqrt{n}(\widehat{\beta}_{jn} - \beta_j)$, $j = 1, 2$, as well as the joint one. (Use Theorems 3.4.1 and 3.4.4.)

Exercise 3.4.2: Let $\mathbf{X}_n \sim N_p(\mathbf{0}, \boldsymbol{\Sigma})$ and consider the quadratic form $A_n = \mathbf{X}_n^t \mathbf{A}_n \mathbf{X}_n$, where $\mathbf{A}_n$ $(p \times p)$ is p.s.d. Let $\alpha_0 = \lim_{n \to \infty} \mathrm{ch}_p(\mathbf{A}_n\boldsymbol{\Sigma})$ and $\alpha^0 = \lim_{n \to \infty} \mathrm{ch}_1(\mathbf{A}_n\boldsymbol{\Sigma})$. Show that $\alpha_0\mathbf{X}_n^t\boldsymbol{\Sigma}^{-1}\mathbf{X}_n \leq Q_n \leq \alpha^0\mathbf{X}_n^t\boldsymbol{\Sigma}^{-1}\mathbf{X}_n$. Hence, or otherwise, verify that for every real t (≥ 0) we have

$$P\{\chi_p^2 \leq t/\alpha^0\} \leq \lim_{n \to \infty} P\{Q_n \leq t\} \leq P\{\chi_p^2 \leq t/\alpha_0\}.$$

Use these inequalities to show that $\alpha_0 = \alpha^0$ implies $\lim_{n \to \infty} P\{Q_n \leq t\} = P\{\chi_p^2 \leq t/\alpha_0\}$.

Exercise 3.4.3: Extend the results in Exercise 3.4.2 to the case where $\mathbf{A}_n$ is a stochastic matrix, by making $\alpha_{0n} = \mathrm{ch}_p(\mathbf{A}_n\boldsymbol{\Sigma}_n) \xrightarrow{P} \alpha_0$ and $\alpha_{1n} = \mathrm{ch}_1(\mathbf{A}_n\boldsymbol{\Sigma}_n) \xrightarrow{P} \alpha^0$ $(\geq \alpha_0)$.

CHAPTER 4

Large Sample Behavior of Empirical Distributions and Order Statistics

4.1 Introduction

In this chapter we apply the techniques considered so far to study the asymptotic properties of some basic statistics like empirical distributions, sample quantiles and order statistics. Most of the results we discuss are intrinsically important because of their usefulness in nonparametric inference, reliability and life-testing; many are also important because they serve as paradigms to similar results developed for other types of statistics, like U-statistics, or M-, L- and R-statistics.

Let X be a real-valued r.v. having a d.f. F $(= \{F(x): x \in \mathbb{R}\})$. Consider a sample of n i.i.d. r.v.'s $\{X_1, \ldots, X_n\}$ drawn from the d.f. F. Then, as in Section 2.2, we may define the **sample** or **empirical distribution** by

$$F_n(x) = n^{-1} \sum_{i=1}^{n} I_{\{X_i \le x\}}, \quad x \in \mathbb{R}. \tag{4.1.1}$$

For each fixed sample, F_n is a distribution function when considered as a function of x. For every fixed $x \in \mathbb{R}$, when considered as a function of $X_1, \ldots, X_n$, $F_n(x)$ is a random variable; in this context, since the $I_{\{X_i \le x\}}, i = 1, \ldots, n$, are i.i.d. zero-one valued random variables, we have

$$\mathrm{E}F_n(x) = \frac{1}{n} \sum_{i=1}^{n} \mathrm{E}I_{\{X_i \le x\}} = \frac{1}{n} \sum_{i=1}^{n} P\{X_i \le x\} = F(x)$$

and

$$\mathrm{Var}F_n(x) = \frac{1}{n^2} \sum_{i=1}^{n} \mathrm{Var}I_{\{X_i \le x\}} = \frac{F(x)[1 - F(x)]}{n} \le \frac{1}{4n}.$$

Observing that $\mathrm{Var}F_n(x) \to 0$ as $n \to \infty$ and using the Central Limit Theorem 3.3.1 we may conclude that for each fixed x

$$\sqrt{n}\{F_n(x) - F(x)\} \xrightarrow{D} N(0, F(x)[1 - F(x)]). \tag{4.1.2}$$

It is also easy to verify that for every $x \leq y$

$$E[I_{\{X_i \leq x\}} I_{\{X_i \leq y\}}] = F(x) \tag{4.1.3}$$

so that for all $x, y \in \mathbb{R}$

$$
\begin{aligned}
E\{[F_n(x) &- F(x)]\}\{F_n(y) - F(y)\} \\
&= \text{Cov}\{F_n(x), F_n(y)\} \\
&= n^{-1} F(x \wedge y)\{1 - F(x \vee y)\}.
\end{aligned} \tag{4.1.4}
$$

On the other hand, $F_n - F = \{F_n(x) - F(x) : x \in \mathbb{R}\}$ is a **random function** defined on $\mathbb{R}$, and, hence, to study its various properties we may need more than (4.1.2); this topic will be considered partly in the present chapter and partly in Chapter 8.

For the time being, in order to simplify the analysis, let us assume that F is continuous a.e., so that

$$P\{X_i = X_j \text{ for some } i \neq j, \ i, j = 1, \ldots, n\} = 0. \tag{4.1.5}$$

Let us arrange the observations $X_1, \ldots, X_n$ is ascending order of magnitude and denote these ordered r.v.'s by

$$X_{n:1} \leq X_{n:2} \leq \cdots \leq X_{n:k} \leq \cdots \leq X_{n:n}. \tag{4.1.6}$$

By virtue of (4.1.5), ties among the X_i and (hence, the $X_{n:k}$) can be neglected with probability 1, so that in (4.1.6) we may as well replace all the "$\leq$" signs by strict "$<$" signs. Note that $X_{n:k}$ is the kth smallest observation in the sample and $X_{n:n-k+1}$, is the kth largest observation, for $k = 1, \ldots, n$. Also,

$$\mathbf{t}_n = (X_{n:1}, \ldots, X_{n:n})^t \tag{4.1.7}$$

is termed the vector of sample **order statistics**. Depending on whether k/n is close to 0 (or 1) or converging to some $p, 0 < p < 1$, $X_{n:k}$ will be termed an **extreme value** or a **sample quantile**, and we shall make this point clear later. Let

$$R_i = \sum_{j=i}^{n} I_{\{X_j \leq X_i\}} \tag{4.1.8}$$

be the **rank** of X_i among $X_1, \ldots, X_n$, for $i = 1, \ldots, n$, so that $R_1, \ldots, R_n$ represent the natural integers, permuted in some (random) manner. We let

$$\mathbf{R} = (R_1, \ldots, R_n)^t, \tag{4.1.9}$$

so that $\mathbf{R}$ takes on the permutations of $(1, \ldots, n)$, there being $n!$ possible realizations of $\mathbf{R}$. Then note that by (4.1.6) and (4.1.8),

$$X_i = X_{n:R_i}, \quad i = 1, \ldots, n. \tag{4.1.10}$$

By reversing the order of the index, it is also possible to write

$$X_{n:i} = X_{S_i}, \quad i = 1, \ldots, n, \tag{4.1.11}$$

where S_i is termed the **anti-rank**, $i = 1, \ldots, n$. For the last two equations, we have, therefore,

$$R_{S_i} = i = S_{R_i}, \quad i = 1, \ldots, n, \tag{4.1.12}$$

so that $\mathbf{S} = (S_1, \ldots, S_n)^t$ also takes on the permutations of $(1, \ldots, n)$. Although the X_i's are assumed to be i.i.d. r.v.'s, the $X_{n:k}$ are neither independent nor identically distributed. For this reason, the study of (exact or asymptotic) properties of $\mathbf{t}_n$ (or any subset of it) generally requires a somewhat different approach (relative to earlier chapters). Moreover, by (4.1.1) and (4.1.6), we have, for $1 \leq k \leq n - 1$,

$$F_n(x) = \begin{cases} 0, & x < X_{n:1} \\ k/n, & X_{n:k} \leq x \leq X_{n:k+1} \\ 1, & x \geq X_{n:n}. \end{cases} \tag{4.1.13}$$

Thus, F_n and $\mathbf{t}_n$ are very much interrelated, and this makes it possible to formulate a common approach for the study of asymptotic properties of both the empirical d.f. and the order statistics; this is considered in this chapter.

In Section 4.2, along with some preliminary notions, the one-to-one relationship between F_n and $\mathbf{t}_n$ [i.e., (4.1.13)] is considered in a relatively more general framework, and the same is incorporated in the derivation of the exact distribution of $X_{n:k}$ when F may not be continuous a.e. Section 4.3 deals with the large sample properties of sample quantiles, whereas the case of sample extreme values is treated in Section 4.4. Section 4.5 is devoted to the study of almost sure convergence of the empirical d.f. F_n. The concluding section deals with some applications in a motivated but nontechnical fashion. More technical versions of these results will be considered in later chapters.

4.2 Preliminary notions

We introduce first the basic notations and interrelationships. First, the empirical d.f. F_n may generally be defined in a multivariate setup as follows. Let $\mathbf{X}_1, \ldots, \mathbf{X}_n$ be p-vectors, for some $p \geq 1$, and let $\mathbf{a} \leq \mathbf{b}$ stand for the simultaneous inequalities $a_j \leq b_j$, $j = 1, \ldots, p$. Then, (4.1.1) extends to the p-variate case as

$$F_n(\mathbf{x}) = n^{-1} \sum_{i=1}^{n} I_{\{\mathbf{X}_i \leq \mathbf{x}\}}, \quad \mathbf{x} \in \mathbb{R}^p. \tag{4.2.1}$$

In this setup, (4.1.2)–(4.1.4) all go through in the vector case when we let

$$\mathbf{a} \wedge \mathbf{b} = (a_1 \wedge b_1, \ldots, a_p \wedge b_p)^t, \quad \mathbf{a} \vee \mathbf{b} = (a_1 \vee b_1, \ldots, a_p \vee b_p)^t. \quad (4.2.2)$$

An empirical distributional process is defined as $W_n = \{W_n(\mathbf{x}); \mathbf{x} \in \mathbb{R}^p\}$ where

$$W_n(\mathbf{x}) = \sqrt{n}\{F_n(\mathbf{x}) - F(\mathbf{x})\}, \quad \mathbf{x} \in \mathbb{R}^p. \quad (4.2.3)$$

The univariate case (i.e., $p = 1$) is the most used one and will be considered in detail; multivariate cases will be treated only briefly.

In the univariate case, to be more general, we drop the assumption that F is continuous a.e. In that case, (4.1.5) may not hold, so that the **ties** among the X_i may not be negligible, with probability 1. Suppose that in the sample of size n there are m ($=m_n$) distinct values (arranged in ascending order)

$$X_{n:1}^* < \cdots < X_{n:n}^*, \quad (4.2.4)$$

and that $X_{n:k}^*$ occurs with the frequency $f_k^*(\geq 1)$, $k = 1, \ldots, m$, so that $f_1^* + \cdots + f_m^* = n$. In this setup, the vectors $\mathbf{t}_n$ of sample order statistics can equivalently be expressed as

$$\mathbf{t}_n = (X_{n:1}^*, f_1^*; \ldots, X_{n:k}^*, f_k^*; \ldots; X_{n:m}, f_m^*), \quad (4.2.5)$$

where $m(= m_n)$ is itself a positive integer-valued r.v., such that $P\{1 \leq m \leq n\} = 1$. Let $F_k^* = f_1^* + \cdots + f_k^*$, $1 \leq k \leq m$, stand for the cumulative frequencies. With these notations, for $1 \leq k \leq m - 1$, (4.1.13) extends to

$$F_n(x) = \begin{cases} 0, & x < X_{n:1}^* \\ n^{-1}F_k^*, & X_{n:k}^* \leq x < X_{n:k+1}^* \\ 1 & x \geq X_{n:m}^*. \end{cases} \quad (4.2.6)$$

Thus, F_n is a **step function** having m **jump-points** $X_{n:1}^* < \cdots < X_{n:m}^*$ and the jump at $X_{n:k}^*$ is $n^{-1}f_k^*$, $k = 1, \ldots, m$. In the particular case of F being continuous a.e., we have $m = n$, $X_{n:k}^* = X_{n:k}$ and $f_k^* = 1$, $1 \leq k \leq n$, with probability 1, so that F_n has n jumps of equal magnitude (i.e., n^{-1}) at each of the sample order statistics $X_{n:1}, \ldots, X_{n:n}$. In any case, there is a one-to-one correspondence between F_n and $\mathbf{t}_n$, as defined above. This forms a basis for subsequent analysis in this chapter. Consider next a fixed positive number p ($0 < p < 1$). If for an order statistic $X_{n:k}, k/n$ converges to p in a suitable manner, then $X_{n:k}$ is termed a **sample p-quantile**. Although there are various ways of defining such a k, the following are the most adopted ones:

$$k = k_p = [np] + 1, \quad (4.2.7)$$

$$k = k_p = [(n + 1)p]. \quad (4.2.8)$$

Perhaps, it will be in order to define the population counterpart in an unambiguous manner. If the equation

$$F(x) = p \qquad (4.2.9)$$

admits a unique solution ξ_p [i.e., $F(\xi_p) = p$ is the unique root of (4.2.9)], then ξ_p is termed the **population p -quantile**. Such a definition entails that F is strictly monotone at ξ_p, so that no multiple roots exist (see Figure 4.2.1.a). In Figure 4.2.1.b, a is a jump-point for F, so that $F(a-) < p < F(a)$, and, hence, (4.2.9) is not properly defined. On the other hand, in Figure 4.2.1.c, $F(x) = p$ for $a \leq x \leq b$, so that there are multiple roots for (4.2.9). To avoid such technical difficulties, often one defines

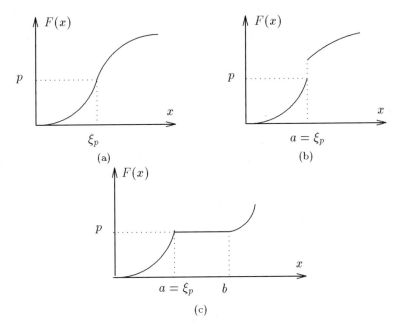

Figure 4.2.1. *Population Quantiles*

$$\xi_p = F^{-1}(p) = \inf\{x \colon F(x) \geq p\}. \qquad (4.2.10)$$

In the same manner, we may define the sample p-quantile as

$$\hat{\xi}_{p,n} = F_n^{-1}(p) = \inf\{x \colon F_n(x) \geq p\}. \qquad (4.2.11)$$

If F is continuous and strictly monotone at ξ_p, then, of course, ξ_p as defined by (4.2.9) and (4.2.10) are the same. But, (4.2.11) may not be in agreement with some other conventional definitions of sample quantiles. To clarify this point, let us consider the case of $p = 1/2$ for which $\xi_{0.5}$ is also termed the

median. If n is an odd number ($= 2m + 1$, say), then both (4.2.7) and (4.2.8) lead to $k_p = m + 1$. In the case of all n distinct order statistics, (4.2.11) also leads to $\hat{\xi}_{p,n} = X_{n:m+1}$. On the other hand, if n is even ($= 2m$, say), (4.2.7) yields $k_p = m + 1$, (4.2.8) leads to $k_p = m$ and (4.2.11) leads to $\hat{\xi}_{p,n} = X_{n:m}$. In this case, a more conventional definition of the sample median is $\tilde{X}_n = \frac{1}{2}(X_{n:m} + X_{n:m+1})$. In view of this, we may as well let

$$\hat{\xi}_{p,n} = \frac{1}{2}(\hat{\xi}_{p,n}^{(1)} + \hat{\xi}_{p,n}^{(2)}) \qquad (4.2.12)$$

$$\hat{\xi}_{p,n}^{(1)} = \sup\{x \colon F_n(x) < p\}, \qquad \hat{\xi}_{p,n}^{(2)} = \inf\{x \colon F_n(x) > p\}. \qquad (4.2.13)$$

[Note that by definition F_n is right-continuous, and hence, in (4.2.13), the possible equality sign is coupled with $>$ sign.] In other words, $\hat{\xi}_{p,n}$ is taken as the mid-point of the interval for which the F_n value is equal to p (whenever such an interval exists). However, as we shall see later that for large n, this modification is of minor importance, and any one of (4.2.7), (4.2.8) or (4.2.11), (4.2.12) will work out well.

Note that $X_{n:1}$ ($X_{n:n}$) is the sample smallest (largest) observation, and, in general, for any k (≥ 1), $X_{n:k}$ ($X_{n:n-k+1}$) is the kth smallest (largest) observation. In an asymptotic setup (where n is taken to be large), whenever $k/n \to 0$, these order statistics are termed **extreme values**. Their populations counterparts may be introduced as follows. Suppose that there exists $\xi_0(> -\infty)$, such that

$$F(x) > 0, \quad x > \xi_0 \qquad \text{and} \qquad F(x) = 0, \quad x \leq \xi_0. \qquad (4.2.14)$$

Then ξ_0 is termed a **lower end point** of the d.f. F. Similarly, if there exists ξ_1 ($< \infty$), such that

$$F(x) < 1, \quad x < \xi_1 \qquad \text{and} \qquad F(x) = 1, \quad x \geq \xi_1, \qquad (4.2.15)$$

then ξ_1 is an **upper end point** of the d.f. F. If $\xi_0 = -\infty$ ($\xi_1 = +\infty$), the d.f. F is said to have an **infinite lower (upper) end point**. This can be further clarified through the following example.

Example 4.2.1: If X is a r.v. following a Exp(θ) distribution, then $\xi_0 = 0$ and $\xi_1 = \infty$ as depicted in Figure 4.2.2.a. On the other hand, if X is a r.v. following a Unif$[\mu - \theta/2, \mu + \theta/2]$, distribution, then $\xi_0 = \mu - \theta/2$ and $\xi_1 = \mu + \theta/2$ as shown in Figure 4.2.2.b. ∎

The behavior of the sample extreme values depends heavily on whether the population end points are finite or not, and also on how the d.f. F behaves in its tails. These results will be considered in Section 4.4.

We conclude this section with some results on the d.f. of sample order statistics which may or may not require stringent regularity conditions on

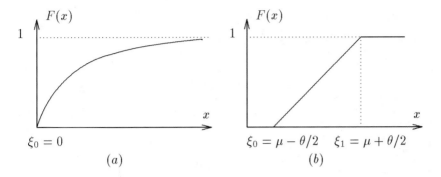

Figure 4.2.2. *End points of Exponential and Uniform distributions*

the d.f. F. First, consider the kth order statistic $X_{n:k}$ and let $Q_{n,k}(x) = P\{X_{n:k} \le x\}$ be the actual d.f. of $X_{n:k}$ when the original d.f. (of the X_i) is F.

Theorem 4.2.1: *Under the setup described above, for every $n \ge k \ge 1$ and F defined on $\mathbb{R}$, we have for all $x \in \mathbb{R}$*

$$Q_{n,k}(x) = \frac{\Gamma(n+1)}{\Gamma(k)\Gamma(n-k+1)} \int_0^{F(x)} t^{k-1}(1-t)^{n-k}\,dt. \qquad (4.2.16)$$

Proof: Note that by definition

$$
\begin{aligned}
Q_{n,k}(x) &= P\{X_{n:k} \le x\} \\
&= P\{k \text{ or more of } (X_1, \ldots, X_n) \text{ are } \le x\} \\
&= P\left\{ \sum_{i=1}^{n} I_{\{X_i \le x\}} \ge k \right\} \\
&= P\left\{ \frac{1}{n} \sum_{i=1}^{n} I_{\{X_i \le x\}} \ge \frac{k}{n} \right\} \\
&= P\left\{ F_n(x) \ge \frac{k}{n} \right\} \\
&= P\{nF_n(x) \ge k\}.
\end{aligned}
$$

Since $nF_n(x) \sim \text{Bin}(n, F(x))$, it follows that

$$Q_{n,k(x)}(x) = \sum_{r=k}^{n} \binom{n}{r} \{F(x)\}^r \{1 - F(x)\}^{n-r}. \qquad (4.2.17)$$

Now, recalling the definition of the incomplete Beta function we have for

$0 < u < 1$

$$
\begin{aligned}
B_{n,k}(u) &= \frac{\Gamma(n+1)}{\Gamma(k)\Gamma(n-k+1)} \int_0^u t^{k-1}(1-t)^{n-k}\,dt \\
&= \frac{n!}{(k-1)!(n-k)!} \int_0^u t^{k-1}(1-t)^{n-k}\,dt \\
&= \frac{n!}{k!(n-k)!} \int_0^u (1-t)^{n-k}\,dt^k. \qquad (4.2.18)
\end{aligned}
$$

Integrating by parts, we get

$$
\begin{aligned}
B_{n,k}(u) &= \binom{n}{k}\left[t^k(1-t)^{n-k} \right]_0^u + \binom{n}{k}\int_0^u t^k(n-k)(1-t)^{n-k-1}\,dt \\
&= \binom{n}{k} u^k(1-u)^{n-k} + \frac{n!}{k!(n-k-1)!}\int_0^u t^k(1-t)^{n-k-1}\,dt.
\end{aligned}
$$

Repeating this process $(n-k+2)$ times we obtain

$$
B_{n,k}(u) = \sum_{r=k}^{n} \binom{n}{r} u^r(1-u)^{n-r}. \qquad (4.2.19)
$$

From (4.2.17)–(4.2.19) it follows that (4.2.16) holds, completing the proof.

∎

Now suppose that F is absolutely continuous with density function f; then the density function of $X_{n:k}$ is given by

$$
\begin{aligned}
q_{n,k}(x) &= \frac{d}{dx}Q_{n,k}(x) \\
&= \frac{d}{dF(x)}Q_{n,k}(x)\frac{dF(x)}{dx} \\
&= \frac{\Gamma(n+1)}{\Gamma(k)\Gamma(n-k+1)}\{F(x)\}^{k-1}\{1-F(x)\}^{n-k}f(x) \\
&= k\binom{n}{k}\{F(x)\}^{k-1}\{1-F(x)\}^{n-k}f(x). \qquad (4.2.20)
\end{aligned}
$$

Obviously, the use of (4.2.16) and (4.2.20) is restricted to situations where the form of the underlying distribution is known. In general, this is not the case and we must rely on the asymptotic properties of the sample quantiles, which are considered in the subsequent section.

Next consider the problem of existence of moments of order statistics; this issue is important in view of the fact that linear functions of order statistics, like trimmed or Winsorized means, constitute appealing alternatives to estimating parameters.

Theorem 4.2.2: *Let $X_1, \ldots, X_n$ be a random sample corresponding to a* *random variable X with distribution function F and suppose that $\mathrm{E}|X|^a < \infty$ for some $a > 0$. Then, if n, k and r satisfy $r \leq a \min(k, n - k + 1)$, it follows that $\mathrm{E}|X_{n:k}|^r < \infty$.*

Proof: First note that $\mathrm{E}|X|^a < \infty$ implies $\mathrm{E}|X|^r < \infty$ for all $r \leq a$. Next observe that

i) as $x \to \infty$,

$$\infty > \mathrm{E}|X|^r \geq \bar{c}_r(x) = \int_x^\infty |t|^r dP\{X \leq t\} \to 0 \qquad (4.2.21)$$

ii)

$$\bar{c}_r(x) \geq |x|^r P\{X \geq x\} = |x|^r \{1 - F(x)\}$$

which implies

$$0 = \lim_{x \to \infty} \bar{c}_r(x) \;\; \geq \;\; \lim_{x \to \infty} |x|^r \{1 - F(x)\}$$
$$\geq \;\; \lim_{x \to \infty} |x|^r \{1 - F(x)\} F(x).$$

Since $|x|^r F(x)\{1 - F(x)\} \geq 0$, it follows that

$$\lim_{x \to \infty} |x|^r \{1 - F(x)\} = \lim_{x \to \infty} |x|^r \{1 - F(x)\} F(x) = 0. \qquad (4.2.22)$$

Furthermore, note that

$$\infty > \mathrm{E}|X|^a \;\; \geq \;\; \int_0^\infty x^a dP\{X \leq x\}$$
$$= \;\; \int_0^\infty x^a d[1 - P\{X \geq x\}]$$
$$= \;\; -\int_0^\infty x^a dP\{X \geq x\}.$$

Integrating by parts we obtain

$$-\int_0^\infty x^a dP\{X \geq x\}$$
$$= \;\; -x^{a-1} P\{X \geq x\}\Big|_0^\infty + a \int_0^\infty x^{a-1} P\{X \geq x\} dx$$
$$= \;\; a \int_0^\infty x^{a-1} \{1 - F(x)\} dx < \infty \qquad (4.2.23)$$

in view of (4.2.22). Then putting $P\{X_{n:k} \leq x\} = Q_{n,k}(x)$ we may write

$$\mathrm{E}|X_{n:k}|^r = \int_{-\infty}^0 |x|^r dQ_{n,k}(x) + \int_0^\infty x^r dQ_{n,k}(x). \qquad (4.2.24)$$

Using an argument similar to that considered in (4.2.23) and recalling (4.2.17) we get

$$\int_0^\infty x^r \, dQ_{n,k}(x) = r \int_0^\infty x^{r-1} P\{X_{n:k} \geq x\} dx$$

$$= r \sum_{i=0}^{k-1} \binom{n}{i} \int_0^\infty x^{r-1} \{F(x)\}^i \{1 - F(x)\}^{n-i} dx$$

$$= r \sum_{i=0}^{k-1} \binom{n}{i} \int_0^\infty x^{r-a} \{1 - F(x)\}^{n-i-1} \{F(x)\}^i x^{a-1} \{1 - F(x)\} dx$$

$$= r \sum_{i=0}^{k-1} \binom{n}{i} \int_0^\infty [x^a \{1 - F(x)\}]^{(r-a)/a} \{1 - F(x)\}^{n-i-r/a}$$

$$\times \{F(x)\}^i x^{a-1} \{1 - F(x)\} dx. \qquad (4.2.25)$$

From (4.2.21) it follows that $[x^a \{1 - F(x)\}]^{(r-a)/a}$ is finite; note also that $\{F(x)\}^i$ is bounded and that $\{1 - F(x)\}^{n-i-r/a}$ is bounded if $n - i - r/a \geq 0$ for $i = 0, \ldots, k-1$ or equivalently if $r \leq a(n-k+1)$. Thus, if $r \leq a(n-k-1)$ it follows from (4.2.25) and (4.2.23) that

$$\int_0^\infty x^r \, dQ_{n,k}(x) \leq M \int_0^\infty x^{a-1} \{1 - F(x)\} dx < \infty \qquad (4.2.26)$$

Using similar arguments we may show that

$$\int_{-\infty}^0 |x|^r \, dQ_{n,k}(x) < \infty \quad \text{if} \quad r \leq ak. \qquad (4.2.27)$$

Then, from (4.2.26) and (4.2.27) if $r \leq a \min(k, n - k + 1)$ it follows that

$$E|X_{n:k}|^r = \int_{-\infty}^0 |x|^r \, dQ_{n,k}(x) + \int_0^\infty x^r \, dQ_{n,k}(x) < \infty$$

completing the proof.   ∎

Example 4.2.2: Let X be a random variable with the Cauchy distribution. First observe that, since

$$E|X| = \frac{2}{\pi} \int_0^\infty \frac{x \, dx}{1 + x^2} = \frac{1}{\pi} \int_0^\infty \frac{d(1 + x^2)}{1 + x^2} = \frac{1}{\pi} \log x \bigg|_1^\infty = \infty,$$

the mean $\overline{X}_n$ of a random sample $X_1, \ldots, X_n$ from that distribution is such that $E|\overline{X}_n| = \infty$ for all $r \geq 1$. However, note that, for all $\varepsilon > 0$,

$$E|X|^{1-\varepsilon} = \frac{1}{\pi} \int_0^\infty x^{-\varepsilon} \frac{2x \, dx}{1 + x^2}$$

which upon integration by parts yields

$$
\begin{aligned}
\mathrm{E}|X|^{1-\varepsilon} &= \frac{1}{\pi}\left\{\left.\frac{\log(1+x^2)}{x^\varepsilon}\right|_0^\infty + \varepsilon \int_0^\infty x^{-1-\varepsilon}\log(1+x^2)dx\right\} \\
&= \frac{\varepsilon}{\pi}\int_0^\infty x^{-1-\varepsilon}\log(1+x^2)dx \\
&= \frac{\varepsilon}{\pi}\int_0^\infty x^{-1-\varepsilon/2}\frac{\log(1+x^2)}{x^{\varepsilon/2}}dx.
\end{aligned} \tag{4.2.28}
$$

Observing that $\lim_{x\to\infty} x^{-\varepsilon/2}\log(1+x^2) = 0$, it follows that there exist $x_0 > 0$ and $M > 0$ such that for all $x > x_0$ we have $x^{-\varepsilon/2}\log(1+x^2) < M$. Therefore, from (4.2.28) we may write

$$
\mathrm{E}|X|^{1-\varepsilon} = \frac{\varepsilon}{\pi}\left\{\int_0^{x_0} x^{-1-\varepsilon}\log(1+x^2)dx + M\int_{x_0}^\infty x^{-1-\varepsilon/2}dx\right\} < \infty.
$$

Suppose that $n = 3$ and consider the sample median (i.e., take $k = 2$); using Theorem 4.2.2 it follows that $\mathrm{E}|X_{3:2}|^r$ exists for $r \le (1-\varepsilon)\min(2,2) = 2 - 2\varepsilon$, which implies that the sample median from a sample of size 3 from a Cauchy distribution will have finite mean, but infinite variance.

Now consider the sample quantile $X_{n:k}$ where $k = [np] + 1$, $0 < p < 1$. Then

$$
\begin{aligned}
r &\le (1-\varepsilon)\min(k, n-k+1) \\
&= (1-\varepsilon)\min\{[np]+1, n-[np]\} = (1-\varepsilon)O(np).
\end{aligned}
$$

If, for example, $n = 50$ and $p = 1/2$, we may choose ε conveniently so that $r \le 24$; in other words, the sample median from a sample of size 50 from a Cauchy distribution has finite moments of order up to 24.

The same approach may be employed to verify the existence of moments of the trimmed mean

$$
T_{n,k} = (n-2k)^{-1}\sum_{i=k+1}^{n-k} X_{n:i}
$$

or of the Winsorized mean

$$
W_{n,k} = n^{-1}\left\{(k+1)X_{n:k+1} + \sum_{j=k+2}^{n-k+1} X_{n:j} + (k+1)X_{n:n-k}\right\}.
$$

If we take $k = 2$, both the trimmed and the Winsorized means from a sample of size $n \ge 5$ from a Cauchy distribution will have finite variance. ∎

Note that if $X_1, \ldots, X_n$ is a random sample from a random variable X, $\mathrm{E}|\overline{X}_n|^r < \infty$ iff $\mathrm{E}|X|^r < \infty$; however, we may have $\mathrm{E}|X_{n:k}|^r < \infty$ even if

$E|X|^r = \infty$ as in the example above.

To conclude this topic, we note that under the assumptions of Theorem 4.2.2 and some further regularity conditions on the probability density function f we have, as $n \to \infty$,

$$E|X_{n:k}|^r \to \left| F^{-1} \left(\frac{k}{n+1} \right) \right|^r \quad \text{with} \quad \frac{k}{n+1} \to p, \quad 0 < p < 1. \quad (4.2.29)$$

We may refer to Blom (1958), Sen (1959) and van Zwet (1965) for some details.

There are situations in which the interest lies on functions of the sample quantiles, or order statistics, like the sample range $X_{n:n} - X_{n:1}$. Statistical inference in such cases generally relies on the joint distributions of order statistics, which we derive below for the bivariate case.

Let $\{X_1, \ldots, X_n\}$ be a random sample corresponding to a random variable X with absolutely continuous distribution function F; we are interested in obtaining the exact joint density function of $X_{n:s}$, $X_{n:l}$, $1 \leq s < l \leq n$. In this direction assume that Δt and Δu are sufficiently small so that the probabilities of more than one observation in the interval $(t, t+\Delta t)$ or $(u, u + \Delta u)$ are negligible and consider the multinomial distribution to verify that

$$P\{t \leq X_{n:s} \leq t + \Delta t, u \leq X_{n:l} \leq u + \Delta u\}$$
$$= \frac{n!}{(s-1)!1!(l-s-1)!1!(n-l)!}$$
$$\times [F(t)]^{s-1}[F(t+\Delta t) - F(t)][F(u) - F(t+\Delta t)]^{l-s-1}$$
$$\times [F(u+\Delta u) - F(u)][1 - F(u+\Delta u)]^{n-l}.$$

Dividing both members by $\Delta t \Delta u$ and considering the limit for $\Delta t \to 0$ and $\Delta u \to 0$ we obtain the joint density function

$$q_{n,s,l}(t, u) = \frac{n!}{(s-1)!(l-s-1)!(n-l)!}[F(t)]^{s-1}[F(u) - F(t)]^{l-s-1}$$
$$\times [1 - F(u)]^{n-l}f(t)f(u) \quad (4.2.30)$$

where f is the density function corresponding to F. The extension to more than two quantiles is straightforward.

4.3 Sample quantiles

In this section we concentrate on some fundamental results regarding the stochastic convergence of sample quantiles; we emphasize some of the techniques used in the proofs, even when this approach might lead to redundancies, as is the case with the following theorem:

Theorem 4.3.1: *Let $\{X_1, \ldots, X_n\}$ be a random sample corresponding to a random variable with distribution function F and assume that its pth quantile, ξ_p, $0 < p < 1$, is uniquely defined, i.e., for all $\eta > 0$, $F(\xi_p - \eta) < F(\xi_p) = p < F(\xi_p + \eta)$. Then, for the pth sample quantile $X_{n:k}$, with $k = k_p$ such that (4.2.7) or (4.2.8) holds, we have $X_{n:k} \xrightarrow{P} \xi_p (X_{n:k} \xrightarrow{a.s.} \xi_p)$.*

Proof: For every $\varepsilon > 0$, let $p_\varepsilon = F(\xi_p + \varepsilon)$ and note that $p_\varepsilon > p$. Then observe that

$$
\begin{aligned}
Q_{n,k}(\xi_p + \varepsilon) &= P\{X_{n:k} \leq \xi_p + \varepsilon\} \\
&= P\{k \text{ or more among } (X_1, \ldots, X_n) \text{ are } \leq \xi_p + \varepsilon\} \\
&= P\left\{ \sum_{i=1}^{n} I_{\{X_i \leq \xi_p + \varepsilon\}} \geq k \right\} \\
&= P\left\{ \frac{1}{n} \sum_{i=1}^{n} I_{\{X_i \leq \xi_p + \varepsilon\}} - p_\varepsilon \geq \frac{k}{n} - p_\varepsilon \right\} \\
&= P\{U_n \geq a_n\} \qquad\qquad\qquad (4.3.1)
\end{aligned}
$$

where $U_n = \frac{1}{n} \sum_{i=1}^{n} I_{\{X_i \leq \xi_p + \varepsilon\}} - p_\varepsilon$ is such that $U_n \xrightarrow{P} 0$ by Khintchine (Borel) Weak Law of Large Numbers (Theorem 2.3.6), and $a_n = k/n - p_\varepsilon$ is such that $a_n \to a = p - p_\varepsilon < 0$ as $n \to \infty$. Thus, given $\eta > 0$, there exists $n_0 = n_0(\eta)$ such that as $n \to \infty$

$$
\begin{aligned}
P\{U_n \geq a_n\} &= P\{U_n - a \geq a_n - a\} \\
&\geq P\{U_n - a \geq -\eta\} \\
&= P\{U_n \geq a - \eta\} \\
&\geq P\{a - \eta \leq U_n \leq -a + \eta\} \to 1. \qquad (4.3.2)
\end{aligned}
$$

From (4.3.1) and (4.3.2) we obtain

$$
Q_{n,k}(\xi_p + \varepsilon) = P\{X_{n:k} - \xi_p \leq \varepsilon\} \to 1 \quad \text{as} \quad n \to \infty. \qquad (4.3.3)
$$

Now, letting $p_\varepsilon^* = F(\xi_p - \varepsilon)$ we have $p_\varepsilon^* < p$ and proceeding analogously we may show that

$$
Q_{n,k}(\xi_p - \varepsilon) = P\{X_{n:k} - \xi_p \leq -\varepsilon\} \to 0 \quad \text{as} \quad n \to \infty. \qquad (4.3.4)
$$

From (4.3.3) and (4.3.4) it follows that $X_{n:k} \xrightarrow{P} \xi_p$.

To show strong consistency, first let

$$
\begin{aligned}
A_n = \{X_{n:k} \geq \xi_p + \varepsilon\} &= \left\{ \sum_{i=1}^{n} I_{\{X_i \leq \xi_p + \varepsilon\}} \leq k \right\} \\
&= \{n F_n(\xi_p + \varepsilon) \leq k\}, \qquad (4.3.5)
\end{aligned}
$$

$$B_n = \{X_{n:k} \leq \xi_p - \varepsilon\} \quad = \quad \left\{\sum_{i=1}^{n} I_{\{X_i \leq \xi_p - \varepsilon\}} \geq k\right\}$$

$$= \quad \{nF_n(\xi_p - \varepsilon) \geq k\} \qquad (4.3.6)$$

and note that since $nF_n(\xi_p \pm \varepsilon) \sim \text{Bin}(n, F(\xi_p \pm \varepsilon))$, we may use the Bernstein Inequality (2.3.16) for binomial events to see that

$$P\{A_n\} \leq \{\rho[\varepsilon, F(\xi_p)]\}^n$$

and

$$P\{B_n\} \leq \{\rho^*[\varepsilon, F(\xi_p)]\}^n$$

where $0 \leq \rho[\varepsilon, F(\xi_p)]$, $\rho^*[\varepsilon, F(\xi_p)] < 1$. Therefore, we have $\sum_{i=1}^{\infty} P\{A_n\} < \infty$ and $\sum_{n=1}^{\infty} P\{B_n\} < \infty$ and using the Borel-Cantelli Lemma (Theorem 2.3.2), it follows that both $P\{\cup_{n \geq m} A_n\} \to 0$ and $P\{\cup_{n \geq m} B_n\} \to 0$ as $m \to \infty$. The desired result is then a consequence of the fact that

$$P[\cup_{n \geq m}\{|X_{n:k_p} - \xi_p| \geq \varepsilon\}] = P\{\cup_{n \geq m} A_n\} + P\{\cup_{n \geq m} B_n\}. \qquad (4.3.7)$$

∎

It is noteworthy to remark that by the very use of the Bernstein Inequality we are able to achieve an exponential rate of convergence. Also, note that the above theorem holds for any p, $0 < p < 1$.

We now focus our attention on the asymptotic normality of the sample quantiles. The first authors to deal with this topic, like Cramér (1946) or Mosteller (1946), showed the convergence of the density function (4.2.20) to the normal density function. Their proofs were essentially based on taking

$$x = \xi_p + n^{-1/2}t, \ |t| \leq K, \ K > 0,$$

and using Stirling's approximation

$$n! = \sqrt{2\pi}n^{n+1/2}e^{-n+1/12n}$$

in (4.2.20). We shall consider a weaker result, which is sufficient for most statistical applications, however.

Theorem 4.3.2: *Let* $\{X_1, \ldots, X_n\}$ *be a random sample corresponding to a random variable with distribution function F and density function f, continuous at ξ_p, $0 < p < 1$, and such that $f(\xi_p) > 0$; define k in such a way that $k = np + o(n^{1/2})$. Then*

$$\sqrt{n}(X_{n:k} - \xi_p) \xrightarrow{D} N(0, p(1-p)/f^2(\xi_p)). \qquad (4.3.8)$$

Proof: Let $Z_n = \sqrt{n}(X_{n:k} - \xi_p)$ and note that

$$
\begin{aligned}
P\{Z_n \le x\} &= P\left\{X_{n:k} \le \xi_p + n^{-1/2}x\right\} \\
&= P\left\{\sum_{i=1}^{n} I_{\{X_i \le \xi_p + n^{-1/2}x\}} \ge k\right\} \\
&= P\left\{\frac{1}{\sqrt{n}}\sum_{i=1}^{n} Y_{ni} \ge t_n\right\}
\end{aligned} \tag{4.3.9}
$$

where

$$
Y_{ni} = I_{\{X_i \le \xi_p + n^{-1/2}x\}} - F(\xi_p + n^{-1/2}x), \quad i = 1, \ldots, n,
$$

and

$$
t_n = \{k - nF(\xi_p + n^{-1/2}x)\}/\sqrt{n}.
$$

Now, since $Y_{ni} = 1 - F(\xi_p + n^{-1/2}x)$ with probability $F(\xi_p + n^{-1/2}x)$ and $Y_{ni} = -F(\xi_p + n^{-1/2}x)$ with probability $1 - F(\xi_p + n^{-1/2}x)$ and $F(\xi_p + n^{-1/2}x) = F(\xi_p) + n^{-1/2}xf(\xi_p) + o(n^{-1/2})$, it follows that $EY_{ni} = 0$, $EY_{ni}^2 = p(1-p) + O(n^{-1/2})$ and that

$$
t_n = \frac{1}{\sqrt{n}}\{k - nF(\xi_p) - n^{1/2}xf(\xi_p) + o(n^{-1/2})\} = -xf(\xi_p) + o(1).
$$

Therefore, using the Central Limit Theorem 3.3.4 and the Slutsky Theorem (3.4.2) in (4.3.9), we may conclude that as $n \to \infty$

$$
\begin{aligned}
P\{Z_n \le x\} &= P\left\{\frac{1}{\sqrt{np(1-p)}}\sum_{i=1}^{n} Y_{ni} \ge \frac{t_n}{\sqrt{p(1-p)}}\right\} \\
&\to 1 - \Phi\left\{\frac{-xf(\xi_p)}{\sqrt{p(1-p)}}\right\} = \Phi\left\{\frac{xf(\xi_p)}{\sqrt{p(1-p)}}\right\}
\end{aligned} \tag{4.3.10}
$$

which implies that as $n \to \infty$

$$
P\left\{\frac{f(\xi_p)}{\sqrt{p(1-p)}}\sqrt{n}(X_{n:k} - \xi_p) \le x\right\} \to \Phi(x), \tag{4.3.11}
$$

which is equivalent to (4.3.8). ∎

The next theorem generalizes the preceding result by examining the asymptotic joint distribution of two sample quantiles.

Theorem 4.3.3: Let $\{X_1, \ldots, X_n\}$ be a random sample corresponding to a random variable with distribution function F and density function f continuous at $\xi_{p_j}, j = 1, 2, 0 < p_1 < p_2 < 1$, and such that $f(\xi_{p_j}) > 0, j = 1, 2$;

define k_j in such a way that $k_j = np_j + o(n^{1/2})$, $j = 1, 2$. Then

$$\sqrt{n}(X_{n:k_1} - \xi_{p_1}, X_{n:k_2} - \xi_{p_2})^t \xrightarrow{D} N_2(0, \boldsymbol{\Sigma})$$

where $\boldsymbol{\Sigma} = ((\sigma_{ij}))$ with $\sigma_{11} = p_1(1 - p_1)/f^2(\xi_{p_1})$, $\sigma_{12} = \sigma_{21} = p_1(1 - p_2)/f(\xi_{p_1})f(\xi_{p_2})$ and $\sigma_{22} = p_2(1 - p_2)/f^2(\xi_{p_2})$.

Proof: Let $\mathbf{Z}_n = (Z_{n1}, Z_{n2})^t$ where $Z_{nj} = \sqrt{n}(X_{n:k_j} - \xi_{p_j})$, $j = 1, 2$, and $\mathbf{x} = (x_1, x_2)^t$, and use arguments similar to the ones considered in the previous theorem to see that

$$P\{\mathbf{Z}_n \leq \mathbf{x}\} = P\left\{ \frac{1}{\sqrt{n}} \sum_{i=1}^{n} \mathbf{Y}_{ni} \geq \mathbf{t}_n \right\} \tag{4.3.12}$$

where $\mathbf{Y}_{ni} = (Y_{ni1}, Y_{ni2})^t$ with

$$Y_{nij} = I_{\{X_i \leq \xi_{p_j} + n^{-1/2}x_j\}} - F(\xi_{p_j} + n^{-1/2}x_j), \quad j = 1, 2,$$

and $\mathbf{t}_n = (t_{n1}, t_{n2})^t$ with

$$t_{nj} = \{k_j - nF(\xi_{p_j} + n^{-1/2}x_j)\}/\sqrt{n}, \quad j = 1, 2.$$

Also, note that $EY_{nij} = 0$ and $EY_{nij}^2 = p_j(1 - p_j) + O(n^{-1/2})$, $j = 1, 2$. Furthermore, since we are assuming $p_1 < p_2$, it follows that

$$
\begin{aligned}
EY_{ni1}Y_{ni2} &= E\{I_{\{X_i \leq \xi_{p_1} + n^{-1/2}x_1\}} I_{\{X_i \leq \xi_{p_2} + n^{-1/2}x_2\}}\} \\
&\quad - F(\xi_{p_1} + n^{-1/2}x_1)F(\xi_{p_2} + n^{-1/2}x_2) \\
&= F(\xi_{p_1} + n^{-1/2}x_1) - F(\xi_{p_1} + n^{-1/2}x_1)F(\xi_{p_2} + n^{-1/2}x_2) \\
&= p_1(1 - p_2) + O(n^{-1/2}).
\end{aligned}
$$

Thus, putting

$$\boldsymbol{\Gamma} = \begin{pmatrix} p_1(1 - p_1) & p_1(1 - p_2) \\ p_1(1 - p_2) & p_2(1 - p_2) \end{pmatrix}$$

we may write $\text{Var}\mathbf{Y}_{ni} = \boldsymbol{\Gamma} + \boldsymbol{\Delta}_n$ where $\boldsymbol{\Delta}_n = O(n^{-1/2})$. Finally, observing that

$$\mathbf{t}_n = -\mathbf{D}\mathbf{x} + o(1)$$

where $\mathbf{D} = \text{diag}\{f(\xi_{p_1}), f(\xi_{p_2})\}$, we may apply the multivariate versions of the Central Limit Theorem and the Slutsky Theorem (3.4.2) to (4.3.12), to conclude that as $n \to \infty$

$$
\begin{aligned}
P\{\mathbf{Z}_n \leq \mathbf{x}\} &= P\left\{ \boldsymbol{\Gamma}^{-1/2} \sum_{i=1}^{n} \mathbf{Y}_{ni} \geq \boldsymbol{\Gamma}^{-1/2}\mathbf{t}_n \right\} \\
&\to 1 - \Phi_2\{-\boldsymbol{\Gamma}^{-1/2}\mathbf{D}\mathbf{x}\} = \Phi_2\{\boldsymbol{\Gamma}^{-1/2}\mathbf{D}\mathbf{x}\}
\end{aligned}
$$

where Φ_2 denotes the bivariate standard normal distribution function. Therefore, we may conclude that

$$\mathbf{Z}_n \xrightarrow{\mathcal{D}} N_2(\mathbf{0}, \mathbf{D}^{-1}\boldsymbol{\Gamma}\mathbf{D}^{-1}), \qquad (4.3.13)$$

and since $\boldsymbol{\Sigma} = \mathbf{D}^{-1}\boldsymbol{\Gamma}\mathbf{D}^{-1}$, the proof is completed. ∎

Note that this result may be easily extended to cover the joint asymptotic distribution of $q \geq 2$ sample quantiles. In this direction, let $0 < p_1 < \cdots < p_q < 1$, $\mathbf{Z}_n = (Z_{n1}, \ldots, Z_{nq})^t$ where $Z_{nj} = \sqrt{n}(X_{n:k_1} - \xi_{p_j})$, $j = 1, \ldots, q$ and $\mathbf{x} = (x_1, \ldots, x_q)^t$. Then (4.3.13) holds with $\mathbf{D} = \operatorname{diag}\{f(\xi_{p_1}), \ldots, f(\xi_{p_q})\}$ and $\boldsymbol{\Gamma} = ((\gamma_{ij}))$ with $\gamma_{ij} = p_i \wedge p_j - p_i p_j$, $i, j = 1, \ldots, q$. An extension to the case of sample quantiles of multivariate distributions was considered by Mood (1941) through elaborate analysis; the method of proof outlined in Theorem 4.3.2 is much simpler and goes through without much change. Without loss of generality we state the result for bivariate distributions.

Theorem 4.3.4: *Let* $\{X_1, \ldots, X_n\}$ *be a random sample corresponding to a bivariate random variable* $\mathbf{X} = (X^{(1)}, X^{(2)})^t$; *let* F, F_1 *and* F_2 *respectively denote the joint and marginal distribution functions of* $\mathbf{X}$ *and assume that the corresponding joint and marginal density functions,* f, f_1 *and* f_2 *are such that* f_j *is continuous at* $\xi_{p_j}^{(j)}$ *and* $f_j(\xi_{p_j}^{(j)}) > 0$, $0 < p_j < 1$, $j = 1, 2$. *Define* k_j *in such a way that* $k_{p_j} = np_j + o(n^{1/2})$, $j = 1, 2$. *Then*

$$\sqrt{n}(X_{n:k_1}^{(1)} - \xi_{p_1}^{(1)}, \quad X_{n:k_2}^{(2)} - \xi_{p_2}^{(2)})^t \xrightarrow{\mathcal{D}} N(\mathbf{0}, \boldsymbol{\Sigma})$$

where $\boldsymbol{\Sigma} = ((\sigma_{ij}))$ *with* $\sigma_{11} = p_1(1 - p_1)/f_1^2(\xi_{p_1}^{(1)})$, $\sigma_{22} = p_2(1 - p_2)/f_2^2(\xi_{p_2}^{(2)})$ *and* $\sigma_{12} = \{F(\xi_{p_1}^{(1)}, \xi_{p_2}^{(2)}) - p_1 p_2\}/f_1(\xi_{p_1}^{(1)})f_2(\xi_{p_2}^{(2)})$.

Proof: Let $\mathbf{Z}_n = (Z_{n1}, Z_{n2})^t$ with $Z_{nj} = \sqrt{n}(X_{n:k_j}^{(j)} - \xi_{p_j}^{(j)})$, $j = 1, 2$, and note that

$$P\{\mathbf{Z}_n \leq \mathbf{x}\} = P\left\{\frac{1}{\sqrt{n}}\sum_{i=1}^n \mathbf{Y}_{ni} \geq \mathbf{t}_n\right\}$$

where $\mathbf{Y}_{ni} = (Y_{ni1}, Y_{ni2})^t$, with

$$Y_{nij} = I_{\{X_i^{(j)} \leq \xi_{p_j}^{(j)} + n^{-1/2}x_j\}} - F_j(\xi_{p_j}^{(j)} + n^{-1/2}x_j), \quad j = 1, 2,$$

and $\mathbf{t}_n = (t_{n1}, t_{n2})^t$ with

$$t_{nj} = n^{-1/2}\{k_j - nF_j(\xi_{p_j}^{(j)} + n^{-1/2}x_j)\}, \quad j = 1, 2.$$

Now, recalling that for $a, b \in \mathbb{R}$

$$E[I_{\{X_i^{(1)} \leq a\}} I_{\{X_i^{(2)} \leq b\}}] = P\{X_i^{(1)} \leq a, X_i^{(2)} \leq b\} = F(a, b),$$

it follows that

$$
\begin{aligned}
EY_{ni1}Y_{ni2} &= E\{I_{\{X_i^{(1)} \le \xi_{p_1}^{(1)} + n^{-1/2}x_1\}} I_{\{X_i^{(2)} \le \xi_{p_2} + n^{-1/2}x_2\}}\} \\
&\quad F_1(\xi_{p_1}^{(1)} + n^{-1/2}x_1) F_2(\xi_{p_2}^{(2)} + n^{-1/2}x_2) \\
&= F(\xi_{p_1}^{(1)}, \xi_{p_2}^{(2)}) - p_1 p_2 + O(n^{-1/2}).
\end{aligned}
$$

The rest of the proof follows the same arguments as those in Theorem 4.3.3.

∎

A multivariate generalization of this theorem is straightforward.

An interesting application of the above results is in the estimation of the parameters of the so-called **location-scale** families of distributions. Consider a random sample $X_1, \ldots, X_n$ corresponding to a random variable X with distribution function F and density function f such that $f(y) = \lambda^{-1} f_0\{(y - \mu)/\lambda\}$ where $\mu \in \mathbb{R}$ and $\lambda > 0$ are unknown location and scale parameters, respectively, and f_0 is a known density function, free of μ and λ and F_0 is the corresponding d.f. In many cases the solution to the likelihood equations is difficult to obtain and alternative methods to estimate μ and λ must be employed. In this direction let $Y_i = (X_i - \mu)/\lambda$, $i = 1, \ldots, n$, and note that the density function of X_i is f_0. Now consider the order statistics $X_{n:1} \le X_{n:2} \le \cdots \le X_{n:n}$ which imply $X_{n:i} = \mu + \lambda Y_{n:i}$, $i = 1, \ldots, n$ and note that for large n, $EX_{n:i} = \xi_{p_i}$ may be approximated by $F_0^{-1}(p_i)$, $p_i = i/(n+1)$, $i = 1, \ldots, n$, and $n\text{Cov}(X_{n:i}, X_{n:j}) = \gamma_{ij}$ which may be approximated by $p_i(1 - p_j)/f_0(\xi_{p_i}) f_0(\xi_{p_j})$, $i \le j, i, j = 1, \ldots, n$. Also note that the ξ_{p_i}'s and γ_{ij}'s do not depend on μ and λ and, in many cases, may be obtained from appropriate tables. Letting $\mathbf{X}_n = (X_{n:1}, \ldots, X_{n:n})^t$, $\boldsymbol{\xi}_n = (\xi_{p_1}, \ldots, \xi_{p_n})^t$, $\boldsymbol{\Gamma}_n = ((\gamma_{ij}))$ and $\boldsymbol{\theta} = (\mu, \lambda)^t$ we may write

$$
E\mathbf{X}_n = \mu \mathbf{1}_n + \lambda \boldsymbol{\xi}_n = \mathbf{W}_n \boldsymbol{\theta} \tag{4.3.14}
$$

where $\mathbf{1}_n$ denotes a n-vector with all elements equal to 1 and $\mathbf{W}_n = (\mathbf{1}_n, \boldsymbol{\xi}_n)$. Then a generalized least-squares estimator of $\boldsymbol{\theta}$ may be obtained by minimizing

$$
Q_n(\boldsymbol{\theta}) = (\mathbf{X}_n - \mathbf{W}_n \boldsymbol{\theta})^t \boldsymbol{\Gamma}_n^{-1}(\mathbf{X}_n - \mathbf{W}_n \boldsymbol{\theta}), \tag{4.3.15}
$$

yielding

$$
\hat{\boldsymbol{\theta}}_n = (\mathbf{W}_n^t \boldsymbol{\Gamma}_n^{-1} \mathbf{W}_n)^{-1} \mathbf{W}_n^t \boldsymbol{\Gamma}_n^{-1} \mathbf{X}_n \tag{4.3.16}
$$

with covariance matrix

$$
\text{Var}\hat{\boldsymbol{\theta}}_n = \lambda^2 (\mathbf{W}_n^t \boldsymbol{\Gamma}_n^{-1} \mathbf{W}_n)^{-1}. \tag{4.3.17}
$$

Now, since

$$
\mathbf{W}_n^t \boldsymbol{\Gamma}_n^{-1} \mathbf{W}_n = \begin{pmatrix} \mathbf{1}_n^t \boldsymbol{\Gamma}_n^{-1} \mathbf{1}_n & \mathbf{1}_n^t \boldsymbol{\Gamma}_n^{-1} \boldsymbol{\xi}_n \\ \boldsymbol{\xi}_n^t \boldsymbol{\Gamma}_n^{-1} \mathbf{1}_n & \boldsymbol{\xi}_n^t \boldsymbol{\Gamma}_n^{-1} \boldsymbol{\xi}_n \end{pmatrix} \tag{4.3.18}
$$

it follows from (4.3.16) that $\hat{\mu}_n = -\boldsymbol{\xi}_n^t \mathbf{A}_n \mathbf{X}_n$ and $\hat{\lambda}_n = \mathbf{1}_n^t \mathbf{A}_n \mathbf{X}_n$ where $\mathbf{A}_n = \Delta_n^{-1}\{\boldsymbol{\Gamma}_n^{-1}(\mathbf{1}_n \boldsymbol{\xi}_n^t - \boldsymbol{\xi}_n \mathbf{1}_n^T)\boldsymbol{\Gamma}_n^{-1}\}$ and $\Delta_n = \det(\mathbf{W}_n^t \boldsymbol{\Gamma}_n^{-1} \mathbf{W}_n)$, or, in other words, that $\hat{\mu}_n$ and $\hat{\lambda}_n$ are linear combinations of the order statistics $\mathbf{X}_n$, with coefficients depending only on the known function f_0. Also, from the generalized version of the Gauss-Markov Theorem, $\hat{\mu}_n$ and $\hat{\lambda}_n$ are (nearly) **best** (nearly) **unbiased linear estimators** (BLUE) of μ and λ, respectively; from (4.3.17) and (4.3.18) it follows that

$$\mathrm{Var}\hat{\mu}_n = \lambda^2 \Delta_n^{-1} \boldsymbol{\xi}_n^t \boldsymbol{\Gamma}_n^{-1} \boldsymbol{\xi}_n,$$

$$\mathrm{Var}\hat{\lambda}_n = \lambda^2 \Delta_n^{-1} \mathbf{1}_n^t \boldsymbol{\Gamma}_n^{-1} \mathbf{1}_n$$

and

$$\mathrm{Cov}(\hat{\mu}_n, \hat{\lambda}_n) = -\lambda^2 \Delta_n^{-1} \mathbf{1}_n^t \boldsymbol{\Gamma}_n^{-1} \boldsymbol{\xi}_n.$$

For further details, the reader is referred to Section 7.2.

The same approach may be employed with censored data, i.e., when only a subset of the order statistics is available. Suppose that the only order statistics available correspond to $\mathbf{X}_n(J) = (X_{n:j_1}, \ldots, X_{n:j_m})^t$ where $J = \{j_1, \ldots, j_m\} \subseteq \{1, \ldots, n\}$; then, defining $\boldsymbol{\xi}_n(J) = (\xi_{p_{j_1}}, \ldots, \xi_{p_{j_m}})^t$ and $\boldsymbol{\Gamma}_n(J) = ((\gamma_{j_k j_l}))$ with $\gamma_{j_k j_l} = n\mathrm{Cov}(X_{n:j_k}, X_{n:j_l})$, $j_k, j_l \in J$, the procedure consists of minimizing

$$Q_n(\boldsymbol{\theta}, J) = \{\mathbf{X}_n(J) - \mathbf{W}_n(J)\boldsymbol{\theta}\}^t \boldsymbol{\Gamma}_n^{-1}(J)\{\mathbf{X}_n(J) - \mathbf{W}_n(J)\boldsymbol{\theta}\}$$

where $\mathbf{W}_n(J) = (\mathbf{1}_{\#J}, \boldsymbol{\xi}_n(J))$ and $\#J$ denotes the cardinality of J. A nice account of these topics is given in Sahran and Greenberg (1962) or David (1970), for example. We shall consider some related results in Section 4.6.

4.4 Extreme order statistics

In the last section we focused our attention on the asymptotic properties of the sample order statistics by letting both the sample size n and the order k_p tend to infinity holding the ratio k_p/n constant. Here we concentrate on the limiting behavior of the sample extreme values (minimum and maximum), i.e., taking $k_p = 1$ (or n) and letting the sample size increase. Such statistics have applications in many fields, of which the study of natural disasters, failure of equipment or air pollution are some examples. We start with the stochastic convergence of such extreme values.

Theorem 4.4.1: Let $\{X_1, \ldots, X_n\}$ be a random sample corresponding to a random variable with distribution function F. If F has a finite upper- (lower-) end point $\xi_1(\xi_0)$, then $X_{n:n} \xrightarrow{\text{p,a.s.}} \xi_1$ $(X_{n:1} \xrightarrow{\text{p,a.s.}} \xi_0)$.

Proof: First note that if ξ_1 is a finite upper-end point of F, then given

$\varepsilon > 0$, there exists $0 < \eta = \eta(\varepsilon) < 1$ such that $F(\xi_1 - \varepsilon) = 1 - \eta$. Therefore, given $\varepsilon > 0$,

$$
\begin{aligned}
P\{X_{n:n} < \xi_1 - \varepsilon\} &= P\{\text{ all of } (X_1, \dots, X_n) < \xi_1 - \varepsilon\} \\
&\leq \{F(\xi_1 - \varepsilon)\}^n \\
&= (1 - \eta)^n.
\end{aligned}
$$

Thus

$$
\sum_{n=1}^{\infty} P\{X_{n:n} < \xi_1 - \varepsilon\} \leq \sum_{n=1}^{\infty}(1 - \eta)^n = (1 - \eta)/\eta,
$$

and by the Borel-Cantelli Lemma (Theorem 2.3.2), it follows that

$$
P[\lim_{n \to \infty}\{X_{n:n} < \xi_1 - \varepsilon\}] = 0. \tag{4.4.1}
$$

Now observe that

$$
P\{X_{n:n} > \xi_1 + \varepsilon\} = 1 - P\{X_{n:n} \leq \xi_1 + \varepsilon\} = 1 - F(\xi_1 + \varepsilon) = 0.
$$

Thus $\sum_{n=1}^{\infty} P\{X_{n:n} > \xi_1 + \varepsilon\} = 0$, and by the Borel-Cantelli Lemma (Theorem 2.3.2), we obtain

$$
P[\lim_{n \to \infty}\{X_{n:n} > \xi_1 + \varepsilon\}] = 0. \tag{4.4.2}
$$

Now, since $\{|X_{n:n} - \xi_1| > \varepsilon\} = \{X_{n:n} < \xi_1 - \varepsilon\} \cup \{X_{n:n} > \xi_1 + \varepsilon\}$, it follows from (4.4.1) and (4.4.2) that $X_{n:n} \xrightarrow{\text{a.s.}} \xi_1$; the fact that $X_{n:n} \xrightarrow{\text{P}} \xi_1$ follows then as a direct consequence of Proposition 2.2.2.

In order to show that $X_{n:1} \to \xi_0$, in probability or almost surely, observe that given $\varepsilon > 0$, there exists $0 < \eta = \eta(\varepsilon) < 1$ such that $F(\xi_0 + \varepsilon) = \eta$; therefore, given $\varepsilon > 0$, we have

$$
\begin{aligned}
P\{X_{n:1} > \xi_0 + \varepsilon\} &= P\{\text{ all of } (X_1, \dots, X_n) > \xi_0 + \varepsilon\} \\
&= \{1 - F(\xi_0 + \varepsilon)\}^n \\
&= (1 - \eta)^n.
\end{aligned}
$$

Proceeding as in the first part of the proof, the result follows. ∎

Now let us consider the case of infinite end points. In this direction let us introduce the following definition conventionally used in extreme value theory.

Definition 4.4.1: *Let X be a random variable with distribution function F. The **characteristic largest (smallest) observation** of F is defined as a solution $\xi_{1n}^*(\xi_{0n}^*)$ to $F(\xi) = 1 - 1/n\ (= 1/n)$.*

Given a random sample $\{X_1, \dots, X_n\}$ corresponding to F we know that since F_n is left-continuous, $F_n(X_{n:n}^-) = 1 - 1/n$; therefore, it seems reasonable to expect that $X_{n:n} - \xi_{1n}^* \xrightarrow{P} 0$; similarly, since $F_n(X_{n:1}^+) = 1/n$, we expect that $X_{n:1} - \xi_{0n}^* \xrightarrow{P} 0$. As we shall see through some examples, this is not always true.

Note that if x_n is such that $F(x_n) = 1 - cn^{-1}\log n$ for some $c > 1$, we have $x_n \to \infty$ as $n \to \infty$; furthermore, observe that

$$P\{X_{n:n} < x_n\} = (1 - cn^{-1}\log n)^n \approx n^{-c},$$

and, hence, by the Borel-Cantelli Lemma (Theorem 2.3.2), $P\{X_{n:n} < x_n \text{ infinitely often }\} \to 0$ as $n \to \infty$. Therefore $X_{n:n} \xrightarrow{a.s.} \infty$. In order to show that $X_{n:n} - \xi_{1n}^* \xrightarrow{P} 0$ we must prove that for all $\varepsilon > 0$

$$P\{X_{n:n} \le \xi_{1n}^* + \varepsilon\} \to 1 \quad \text{and} \quad P\{X_{n:n} \le \xi_{1n}^* - \varepsilon\} \to 0 \quad \text{as} \quad n \to \infty$$

which is equivalent to

$$\{F(\xi_{1n}^* + \varepsilon)\}^n \to 1 \quad \text{and} \quad \{F(\xi_{1n}^* - \varepsilon)\}^n \to 0 \quad \text{as} \quad n \to \infty$$

which, in turn, can be equivalently stated as

$$n \log F(\xi_{1n}^* + \varepsilon) \to 0 \quad \text{and} \quad n \log F(\xi_{1n}^* - \varepsilon) \to -\infty \quad \text{as} \quad n \to \infty. \quad (4.4.3)$$

Analogous conditions are required to show that $X_{n:1} - \xi_{0n}^* \xrightarrow{P} 0$.

Example 4.4.1: Let X be a random variable with the Logistic distribution, i.e., $F(x) = \{1 + \exp(-x)\}^{-1}$, $x \in \mathbb{R}$. Then

$$F(\xi_{1n}^*) = \{1 + \exp(-\xi_{1n}^*)\}^{-1} = \frac{n-1}{n}$$

so that

$$n\exp(\xi_{1n}^*) = n\exp(\xi_{1n}^*) - \exp(\xi_{1n}^*) + n - 1$$

which implies $\exp(\xi_{1n}^*) = n - 1$; thus $\xi_{1n}^* = \log(n-1)$. Now note that

$$F(\xi_{1n}^* + \varepsilon) = \{1 + e^{-\xi_{1n}^* - \varepsilon}\}^{-1} = \{1 + e^{-\varepsilon}(n-1)^{-1}\}^{-1}.$$

Thus, as $n \to \infty$

$$\begin{aligned} n \log F(\xi_n^* + \varepsilon) &= n \log\left\{1 + \frac{e^{-\varepsilon}}{n-1}\right\}^{-1} \\ &= -n \log\left\{1 + \frac{e^{-\varepsilon}}{n-1}\right\} \\ &= -n\left\{\frac{e^{-\varepsilon}}{n-1} - \frac{e^{-2\varepsilon}}{2(n-1)^2} + o(n^{-2})\right\} \to -e^{-\varepsilon} \end{aligned}$$

and in view of (4.4.3) we may not conclude that $X_{n:n} - \xi_{1n}^* \xrightarrow{P} 0$. Along similar lines we may show that $\xi_{0n}^* = -\log(n-1)$ and that $n\log\{1 - F(\xi_{0n}^* + \varepsilon)\} \to -e^{\varepsilon}$ as $n \to \infty$ so we cannot conclude that $X_{n:1} - \xi_{0n}^* \xrightarrow{P} 0$. ∎

Example 4.4.2: Let X be a random variable with a $N(0,1)$ distribution, i.e., with d.f.

$$\Phi(x) = \int_{-\infty}^{x} \frac{1}{\sqrt{2\pi}} e^{-\frac{1}{2}t^2} dt, \quad x \in \mathbb{R}.$$

Writing $1 - \Phi(x) = \int_{x}^{\infty} \frac{1}{\sqrt{2\pi}} \exp(-t^2/2)\,dt$, and integrating by parts and letting $\varphi = \Phi'$, we can show that as $x \to \infty$

$$1 - \Phi(x) = \frac{\varphi(x)}{x}\{1 + O(x^{-2})\}.$$

Therefore, the corresponding characteristic largest observation ξ_{1n}^* is defined as the solution to

$$1 - \Phi(\xi_{1n}^*) = \frac{1}{\xi_{1n}^* \sqrt{2\pi}} \exp(-\frac{1}{2}\xi_{1n}^{*2})\{1 + o(1)\} = \frac{1}{n}. \tag{4.4.4}$$

Re-expressing (4.4.4) as

$$e^{-\frac{1}{2}\xi_{1n}^{*2}} = \sqrt{2\pi}\,\xi_{1n}^* n^{-1}\{1 + o(1)\}$$

and taking logarithms of both sides we obtain

$$-\frac{1}{2}\xi_{1n}^{*2} = \log\sqrt{2\pi} + \log\xi_{1n}^* - \log n + \log\{1 + o(1)\}$$

which implies

$$\begin{aligned}
\xi_{1n}^{*\,2} &= 2\log n - 2\log\sqrt{2\pi} - 2\log\xi_{1n}^* + o(1) \\
&= 2\log n\left\{1 - \frac{\log\sqrt{2\pi}}{\log n} - \frac{\log\xi_{1n}^*}{\log n} + o\left(\frac{1}{\log n}\right)\right\}. \tag{4.4.5}
\end{aligned}$$

Now, from (4.4.4), for sufficiently large n we have

$$1 - \Phi(\xi_{1n}^*) = \frac{1}{n} < \frac{1}{\sqrt{2\pi}\xi_{1n}^*} e^{-\frac{1}{2}\xi_{1n}^{*2}} < e^{-\frac{1}{2}\xi_{1n}^{*2}}$$

and then we may write

$$-\log n < -\frac{1}{2}\xi_{1n}^{*\,2} \Rightarrow 2\log n > \xi_{1n}^{*\,2} \Rightarrow \log(2\log n) > 2\log\xi_{1n}^*$$

which, in turn, implies that

$$\log\xi_{1n}^* < \frac{\log(2\log n)}{2} = \frac{\log(\log n^2)}{2}.$$

Thus,

$$0 < \sqrt{\log n}\,\frac{\log\xi_{1n}^*}{\log n} = \frac{\log\xi_{1n}^*}{\sqrt{\log n}} < \frac{\log(\log n^2)}{2\sqrt{\log n}}. \tag{4.4.6}$$

Applying l'Hôpital's rule, we may show that

$$\lim_{n\to\infty} \log(\log n^2)/2\sqrt{\log n} = 0$$

and it follows from (4.4.6) that

$$\log \xi_{1n}^* / \log n = o\{(\log n)^{-1/2}\}.$$

Using this fact in (4.4.5) we get

$$\xi_{1n}^{*2} = 2 \log n \left\{ 1 + o\left(\frac{1}{\sqrt{\log n}}\right) \right\}$$

which implies that

$$\xi_{1n}^* = \sqrt{2 \log n} \left\{ 1 + o\left(\frac{1}{\sqrt{\log n}}\right) \right\}^{1/2}.$$

Finally, observing that for $x \geq -1$,

$$(1 + x)^{1/2} = 1 + x/2 + o(x),$$

we have

$$\left\{ 1 + o\left(\frac{1}{\sqrt{\log n}}\right) \right\}^{1/2} = 1 + o\left(\frac{1}{\sqrt{\log n}}\right)$$

and then

$$\xi_{1n}^* = \sqrt{2 \log n} \left\{ 1 + o\left(\frac{1}{\sqrt{\log n}}\right) \right\} = \sqrt{2 \log n} + o(1).$$

Then observe that

$$n \log \Phi(\xi_{1n}^* + \varepsilon) = n \log \left\{ 1 - [1 - \Phi(\xi_{1n}^* + \varepsilon)] \right\}$$

$$= n \log \left\{ 1 - \frac{1}{\sqrt{2\pi}(\xi_{1n}^* + \varepsilon)} e^{-\frac{1}{2}(\xi_{1n}^* + \varepsilon)^2} [1 + o(1)] \right\}$$

$$= n \log \left\{ 1 - \frac{\xi_{1n}^*}{\xi_{1n}^* + \varepsilon} \frac{1}{\sqrt{2\pi}\xi_{1n}^*} e^{-\frac{1}{2}\xi_{1n}^{*2} - \xi_{1n}^*\varepsilon - \frac{1}{2}\varepsilon^2} [1 + o(1)] \right\}$$

$$= n \log \left\{ 1 - \frac{\xi_{1n}^*}{\xi_{1n}^* + \varepsilon} \frac{1}{n} e^{-\xi_{1n}^*\varepsilon - \frac{1}{2}\varepsilon^2} [1 + o(1)] \right\}$$

$$= -n \left\{ \frac{\xi_{1n}^*}{\xi_{1n}^* + \varepsilon} \frac{1}{n} e^{-\xi_{1n}^*\varepsilon - \frac{1}{2}\varepsilon^2} + o(n^{-2}) \right\}$$

$$= -\frac{\xi_{1n}^*}{\xi_{1n}^* + \varepsilon} e^{-\xi_{1n}^*\varepsilon - \frac{1}{2}\varepsilon^2} + o(n^{-1}) \to 0 \quad \text{as} \quad n \to \infty.$$

Along the same lines we may show that

$$n \log \Phi(\xi_{1n}^* - \varepsilon) = -\frac{\xi_{1n}^*}{\xi_{1n}^* - \varepsilon} e^{\xi_{1n}^*\varepsilon - \frac{1}{2}\varepsilon^2} + o(n^{-1}) \to 0 \quad \text{as} \quad n \to \infty.$$

Thus, in view of (4.4.5) we may conclude that $X_{n:n} - \xi_{1n}^* \xrightarrow{P} 0$. Similarly, one can prove that $\xi_{0n}^* = -\sqrt{2 \log n} + o(1)$ and that $X_{n:n} - \xi_{0n}^* \xrightarrow{P} 0$. ∎

The basis for the asymptotic distribution theory is laid down by the following:

Theorem 4.4.2: *Let* $\{X_1, \ldots, X_n\}$ *be a random sample corresponding to a random variable with a continuous distribution function* F *and let* $X_{n:1} < \cdots < X_{n:n}$ *denote the set of associated order statistics. Also let* $V_n = n\{1 - F(X_{n:n})\}$. *Then* $V_n \xrightarrow{D} V$ *where* $V \sim \text{Exp}(1)$.

Proof: Since $Y_i = F(X_i) \sim \text{Unif}(0, 1)$, it follows that for $v \geq 0$

$$
\begin{aligned}
G_n(v) &= P\{V_n \leq v\} = P\{n[1 - F(X_{n:n})] \leq v\} = P\{1 - Y_{n:n} \leq v/n\} \\
&= P\{Y_{n:n} \geq 1 - v/n\} = 1 - P\{Y_{n:n} \leq 1 - v/n\} \\
&= 1 - (1 - v/n)^n \to 1 - e^{-v} \quad \text{as} \quad n \to \infty.
\end{aligned}
$$

■

Our next task is to determine conditions on the underlying distribution function F under which $X_{n:n}$ can be normalized by sequences of constants $\{a_n\}$ and $\{b_n\}$ so that $(X_{n:n} - a_n)/b_n$ converges to some nondegenerate distribution. In this direction we first consider the following definitions:

Definition 4.4.2: *Let* X *be a random variable with distribution function* F *and a finite upper end point* ξ_1. *We say that* F *has a* **terminal contact of order** m *at* ξ_1 *if* $1 - F(x)$ *and the left-hand derivates* $F^{(j)}(x)$, $j = 1, \ldots, m$, *vanish at* ξ_1, *whereas* $F^{(m+1)}(\xi_1) \neq 0$.

Definition 4.4.3: *Let* X *be a random variable with distribution function* F *with an infinite upper end point and such that* $F^{(j)}(x)$, $j = 1, 2, \ldots$, *exists. We say that* F *is of the* **Exponential type** *if for large* x

$$
-\frac{F^{(1)}(x)}{1 - F(x)} \cong \frac{F^{(2)}(x)}{F^{(1)}(x)} \cong \frac{F^{(3)}(x)}{F^{(2)}(x)} \cong \cdots .
$$

Definition 4.4.4: *Let* X *be a random variable with distribution function* F. *We say that* F *is of the* **Cauchy type** *if for some* $k > 0$ *and* $c > 0$, *we have*

$$
x^k\{1 - F(x)\} \to c \quad \text{as} \quad x \to \infty.
$$

Distributions of the Exponential type have finite moments of all orders and include those commonly employed in statistical methods such as the normal, Exponential or Gamma distributions. Cauchy type distributions, on the other hand, have no finite moments of order $\geq k$ and are named after the typical member, the Cauchy distribution for which $k = 1$. As we shall see in the sequel, the asymptotic behavior of sample minima or maxima depends on the type of the underlying distribution.

Theorem 4.4.3: *Let $\{X_1, \ldots, X_n\}$ be a random sample corresponding to a random variable X with distribution function F having an mth order terminal contact at the upper end point ξ_1. Then there exist sequences of constants $\{a_n\}$ and $\{b_n\}$ such that, as $n \to \infty$,*

$$P\{X_{n:n} - \xi_1)/b_n \le t\} \to \begin{cases} \exp\{-(-t)^{m+1}\}, & t \le 0 \\ 1, & t > 0. \end{cases}$$

Proof: Consider the following Taylor expansion:

$$\begin{aligned} F(\xi_1 - s) &= F(\xi_1) - sF^{(1)}(\xi_1) + \ldots + \frac{(-1)^m}{m!} s^m F^{(m)}(\xi_1) \\ &\quad + \frac{(-1)^{m+1}}{(m+1)!} s^{m+1} F^{(m+1)}(\xi_1 - \theta s) \\ &= F(\xi_1) + \frac{(-1)^{m+1}}{(m+1)!} s^{m+1} F^{(m+1)}(\xi_1 - \theta s) \qquad (4.4.7) \end{aligned}$$

for some $0 < \theta < 1$. Then note that

$$\begin{aligned} V_n = n\{1 - F(X_{n:n})\} &= n\{1 - F(\xi_1)\} + n\{F(\xi_1) - F(X_{n:n})\} \\ &= n\{F(\xi_1) - F(X_{n:n})\} \end{aligned}$$

and let $s = \xi_1 - X_{n:n}$ in (4.4.7) to see that

$$\begin{aligned} V_n &= \frac{(-1)^m}{(m+1)!} nF^{(m+1)}(\xi_1)(\xi_1 - X_{n:n})^{m+1} \frac{F^{(m+1)}\{(1-\theta)\xi_1 + \theta X_{n:n}\}}{F^{(m+1)}(\xi_1)} \\ &= \left(\frac{\xi_1 - X_{n:n}}{b_n}\right)^{m+1} W_n \\ &= \left\{-\left(\frac{X_{n:n} - \xi_1}{b_n}\right)\right\}^{m+1} W_n \qquad (4.4.8) \end{aligned}$$

where

$$b_n = \left\{(-1)^m(m+1)!/nF^{(m+1)}(\xi_1)\right\}^{\frac{1}{m+1}}$$

and

$$W_n = F^{(m+1)}\{(1-\theta)\xi_1 + \theta X_{n:n}\}/F^{(m+1)}(\xi_1) \xrightarrow{P} 1.$$

Therefore, from (4.4.8) and the Slutsky Theorem (3.4.2), it follows that the asymptotic distribution of $\{-[(X_{n:n} - \xi_1)/b_n]\}^{m+1}$ is the same as that of V_n. Finally, using Theorem 4.4.2 we may conclude that, as $n \to \infty$

$$P\left\{\frac{X_{n:n} - \xi_1}{b_n} \le t\right\} = P\left\{\left[-\frac{X_{n:n} - \xi_1}{b_n}\right]^{m+1} \ge (-t)^{m+1}\right\}$$

which converges to

$$\exp\{-(-t)^{m+1}\} \quad \text{if} \quad t \le 0 \quad \text{or} \quad 1 \quad \text{if} \quad t > 0. \qquad (4.4.9)$$

∎

Note that (4.4.9) is known as the **extreme value distribution of the first type.**

Example 4.4.3: Let $\{X_1, \ldots, X_n\}$ be a random sample corresponding to a random variable with the Uniform $(0, \theta)$ distribution. Note that since $F(x) = x/\theta$, $0 \le x \le \theta$, and $F^{(1)}(x) = \theta^{-1}, 0 \le x \le \theta$, it follows that F has terminal contact of order $m = 0$ and in view of Theorem 4.4.3 we may write

$$P\left\{\frac{n}{\theta}(X_{n:n} - \theta) \le t\right\} \to e^t, \quad t \le 0 \quad \text{as} \quad n \to \infty.$$

It can be shown that:

i) $EX_{n:n} = n\theta/(n+1)$,

ii) $\text{Var}X_{n:n} = \theta^2 n/\{(n+2)(n+1)^2\}$.

Although the sample maximum $X_{n:n}$ is a biased estimate of θ, its variance is smaller than that of the sample median, $X_{n:k}$ with $k = [n/2] + 1$, an alternative estimate. In this direction recall that from Theorem 4.3.2 we have

$$\sqrt{n}\left(X_{n:k} - \frac{\theta}{2}\right) \xrightarrow{D} N(0, \theta^2/4)$$

which implies that for large n, $n\text{Var}\,X_{n:k} \sim \theta^2/4n$. ∎

Example 4.4.4: Let $\{X_1, \ldots, X_n\}$ be a random sample corresponding to a random variable with the Triangular$(0, \theta)$ distribution. Then

$$F(x) = \begin{cases} 2x^2/\theta^2, & 0 \le x \le \theta/2 \\ 4x/\theta - 2x^2/\theta^2 - 1, & \theta/2 \le x \le \theta, \end{cases}$$

$$F^{(1)}(x) = \begin{cases} 4x/\theta^2, & 0 \le x \le \theta/2 \\ 4/\theta - 4x/\theta^2, & \theta/2 \le x \le \theta \end{cases}$$

which imply $F^{(1)}(\theta) = 0$ and $F^{(2)}(\theta) = -4/\theta^2$ so that F has terminal contact of order $m = 1$. In this case $b_n = \theta/\sqrt{2n}$ and $\text{Var}X_{n:n} = \theta^2/2n$ which is of the same order as that of the variance of the corresponding sample median. We note that for $m \ge 2$ the rate of convergence of the distribution of the (standardized) sample maximum becomes slow and in such cases this statistic is of little practical interest. ∎

Theorem 4.4.4: *Let* $\{X_1, \ldots, X_n\}$ *be a random sample corresponding to a random variable with distribution function* F *of the Cauchy type; i.e., such*

*that as $x \to \infty$, $x^k \{1 - F(x)\} \to c$, for some $k > 0$ and $c > 0$; also let ξ_{1n}^*
denote the characteristic largest observation of F. Then as $n \to \infty$*

$$P\left\{\frac{X_{n:n}}{\xi_{1n}^*} \le t\right\} \to \exp\{-t^{-k}\}, \quad t \ge 0.$$

Proof: Since $1 - F(\xi_{1n}^*) = n^{-1}$ we may write

$$
\begin{aligned}
V_n &= n\{1 - F(X_{n:n})\} \\
&= \frac{1 - F(X_{n:n})}{1 - F(\xi_{1n}^*)} \\
&= \left(\frac{\xi_{1n}^*}{X_{n:n}}\right)^k \left\{\frac{X_{n:n}^k [1 - F(X_{n:n})]}{\xi_{1n}^{*\ k}[1 - F(\xi_{1n}^*)]}\right\}.
\end{aligned}
$$

Observing that the term within $\{\ \}$ converges in probability to 1 and using
the Slutsky Theorem (3.4.2), it follows that the asymptotic distribution
of $(\xi_{1n}^*/X_{n:n})^k$ is the same as that of V_n. Thus, from Theorem 4.4.2, as
$n \to \infty$,

$$P\left\{\frac{X_{n:n}}{\xi_{1n}^*} \le t\right\} = P\left\{\left(\frac{\xi_{1n}^*}{X_{n:n}}\right)^k \ge t^{-k}\right\} \to \exp(-t^{-k}), \quad t \ge 0. \quad (4.4.10)$$

$\blacksquare$

We note that (4.4.10) is known as the **extreme value distribution of
the second type**.

Theorem 4.4.5: *Let $\{X_1, \ldots, X_n\}$ be a random sample corresponding to
random variable with distribution function F of the Exponential type; also
let ξ_{1n}^* denote the characteristic largest observation of F. Then there exists
a sequence of constants $\{b_n\}$ such that, as $n \to \infty$,*

$$P\left\{\frac{X_{n:n} - \xi_{1n}^*}{b_n} \le t\right\} \to \exp\{-\exp(-t)\}, \quad t \in \mathbb{R}. \quad (4.4.11)$$

Proof: Let $f(x) = F^{(1)}(x)$, $\gamma_n = nf(\xi_{1n}^*)$ and $X_{n:n} = \xi_{1n}^* + h/\gamma_n$, for some
$h \in \mathbb{R}$. Then observe that

$$
\begin{aligned}
V_n &= n\{1 - F(X_{n:n})\} \\
&= n\{1 - F(\xi_{1n}^*)\} - n\{F(X_{n:n}) - F(\xi_{1n}^*)\} \\
&= 1 - n\{F(X_{n:n}) - F(\xi_{1n}^*)\} \quad (4.4.12)
\end{aligned}
$$

and consider the following Taylor expansion

$$F(X_{n:n}) = F(\xi_{1n}^*) + (X_{n:n} - \xi_{1n}^*)f(\xi_{1n}^*)$$

$$+ \ \frac{1}{2!}(X_{n:n} - \xi_{1n}^*)^2 F^{(2)}(\xi_{1n}^*) + \cdots . \quad (4.4.13)$$

Substituting (4.4.12) into (4.4.13) we obtain

$$V_n = 1 - n(X_{n:n} - \xi_{1n}^*)f(\xi_{1n}^*) - \frac{n}{2!}(X_{n:n} - \xi_{1n}^*)^2 F^{(2)}(\xi_{1n}^*) - \cdots . \quad (4.4.14)$$

Recalling that $1 - F(\xi_{1n}^*) = n^{-1}$, the typical term of (4.4.14) is given by

$$\frac{n}{k!}(X_{n:n} - \xi_{1n}^*)^k F^{(k-1)}(\xi_{1n}^+) = \frac{n}{k!}\frac{h^k}{\gamma_n^k}F^{(k-1)}(\xi_{1n}^*)$$

$$= \frac{h^k}{k!}\frac{n}{[nf(\xi_{1n}^*)]^k}\frac{F^{(k-1)}(\xi_{1n}^*)}{F^{(k-2)}(\xi_{1n}^*)}\frac{F^{(k-2)}(\xi_{1n}^*)}{F^{(k-3)}(\xi_{1n}^*)}\cdots\frac{F^2(\xi_{1n}^*)}{f(\xi_{1n}^*)}f(\xi_{1n}^*)$$

$$= \frac{h^k}{k!}\left\{\left[\frac{1-F(\xi_{1n}^*)}{f(\xi_{1n}^*)}\right]^k\right.$$

$$\left.\frac{F^{(k-1)}(\xi_{1n}^*)}{F^{(k-2)}(\xi_{1n}^*)}\frac{F^{(k-2)}(\xi_{1n}^*)}{F^{(k-3)}(\xi_{1n}^*)}\cdots\frac{F^2(\xi_{1n}^*)}{f(\xi_{1n}^*)}\frac{f(\xi_{1n}^*)}{[1-F(\xi_{1n}^*)]}\right\}$$

$$\cong (-1)^k \frac{h^k}{k!}$$

since the assumption that F is of the Exponential type implies that we may approximate the term within { } by $(-1)^k$ for large n. Then, from (4.4.14) we get

$$V_n = 1 - h + \frac{h^2}{2!} - \frac{h^3}{3!} + \ldots = \exp(-h) = \exp\{-\gamma_n(X_{n:n} - \xi_{1n}^*)\}$$

and taking $b_n = \gamma_n^{-1}$ we may write, for all $t \in \mathbb{R}$,

$$P\left\{\frac{X_{n:n} - \xi_{1n}^*}{b_n} \le t\right\} = P\{\exp[-\gamma_n(X_{n:n} - \xi_{1n}^*)] \ge \exp(-t)\}$$

$$\cong P\{V_n \ge \exp(-t)\} \to e^{\{-e^{(-t)}\}} \quad (4.4.15)$$

as $n \to \infty$ [by Theorem 4.4.2.]   ∎

Note that (4.4.15) is known as the **extreme value distribution of the third type.**

The above results extend directly to the case of the sample minimum $X_{n:1}$. More specifically, consider a random sample $\{X_1, \ldots, X_n\}$ corresponding to a random variable X having distribution function F; then

i) if F has an mth order terminal contact at the lower end point ξ_0, it follows that, as $n \to \infty$,

$$P\{(X_{n:1} - \xi_0)/b_n \le t\} \to \begin{cases} 1 - \exp(-t)^{m+1}, & t \ge 0 \\ 0, & t < 0 \end{cases}$$

where $b_n = \{(m+1)!/nF^{(m+1)}(\xi_0)\}^{\frac{1}{m+1}}$;

ii) if F is of the Cauchy type, it follows that , as $n \to \infty$,

$$P\{X_{n:1}/\xi_{0n}^* \le t\} \to 1 - \exp\{-t^{-k}\}, \quad t \in \mathbb{R}.$$

iii) if F is of the Exponential type, it follows that, as $n \to \infty$,

$$P\{(X_{n:1} - \xi_{0n}^*)/b_n \le t\} \to 1 - \exp\{-\exp(t)\}, \quad t \in \mathbb{R},$$

where $b_n = \{nf(\xi_{0n}^*)\}^{-1}$.

Gnedenko (1943) showed that the limiting distributions presented above are the only possible limiting distributions (domains of attraction) for the sample extreme values.

The class of Exponential type distributions is particularly important since its members are commonly employed for practical purposes. If the interest lies in approximating the largest (smallest) characteristic observation based on a (large) sample of size n, Theorem 4.4.5 allows us to obtain limits I_α and U_α such that

$$P\left\{\frac{I_\alpha}{\gamma_n} \le X_{n:n} - \xi_{1n}^* \le \frac{U_\alpha}{\gamma_n}\right\} \cong 1 - \alpha \qquad (4.4.16)$$

where $\gamma_n = nf(\xi_{1n}^*)$ is known as the **extremal intensity function** (extremal failure rate). Note that for practical applications, (4.4.16) is useful only if $\gamma_n \to \infty$ as $n \to \infty$, since, in such a case, $X_{n:n} - \xi_{1n}^* \xrightarrow{P} 0$. In view of this fact, the members of the Exponential class of distributions may be classified into three categories according to the behavior of the extremal intensity function:

i) **convex Exponential type**, when $\gamma_n \to \infty$ as $n \to \infty$ (e.g., normal distribution);

ii) **simple Exponential type**, when $\gamma_n \to c > 0$ as $n \to \infty$ (e.g., Exponential or Double Exponential distributions);

iii) **concave Exponential type**, when $\gamma_n \to 0$ as $n \to \infty$ (e.g., Laplace or Logistic distributions).

For simple or concave Exponential type distributions, the extreme observations do not provide much information for estimation of the extreme characteristic observations and may be discarded with little loss.

Example 4.4.5: Consider the standard normal distribution and note that

$$\gamma_n = n\varphi(\xi_{1n}^*) = \varphi(\xi_{1n}^*)/\{1 - \Phi(\xi_{1n}^*)\} \approx \xi_{1n}^* \approx \sqrt{2\log n}$$

for large n, in view of Example 4.2.3. Now, using Theorems 4.4.3 and 3.4.5, it follows that $\gamma_n(X_{n:n} - \xi_{1n}^*) = O_p(1)$ which, in turn, implies that

$X_{n:n} - \xi_{1n}^* = O_p\{(\log n)^{-1/2}\}$. Although the convergence is slow when compared to the pth-quantile case $(0 < p < 1)$, where $X_{n:k_p} - \xi_p = O_p(n^{-1/2})$, we may still use the information contained in the extreme observations. For the Exp(θ) distribution, we have $\gamma_n = f(\xi_{1n}^*)/\{1 - F(\xi_{1n}^*)\} = \theta$; following the same argument as above, we may only conclude that $X_{n:n} - \xi_{1n}^* = O_p(1)$, indicating that here the largest observation is of no utility for approximating the largest characteristic observation.   ∎

Extensions to the case of the lth largest (or sth smallest) order statistics have been considered by many authors and the reader is referred to Gumbel (1958), among others, for details. In particular, to obtain the asymptotic joint distribution of the sth smallest and the lth largest order statistics of samples of size n corresponding to an absolutely continuous distribution function F, first consider $V_n = nF(X_{n:s})$ and $W_n = n\{1 - F(X_{n:l})\}$ and note that from (4.2.30), the exact joint density function for V_n and W_n is given by

$$g_{n,l,s}(v, w) = \frac{n!}{n^2(s-1)!(n-l-s)!(l-1)!}$$
$$\times \left(\frac{v}{n}\right)^{s-1} \left(1 - \frac{w}{n} - \frac{v}{n}\right)^{n-l-s} \left(\frac{w}{n}\right)^{l-1} \quad (4.4.17)$$

since the Jacobian of the transformation

$$(X_{n:s}, X_{n:n-l}) \rightarrow (V_n, W_n) \quad (4.4.18)$$

is $\{n^2 f(t)f(u)\}^{-1}$. Then, if we let $n \rightarrow \infty$ while maintaining s and l fixed, we obtain the asymptotic joint density function

$$g_{l,s} = (v, w) = \frac{v^{s-1}e^{-v}}{\Gamma(s)} \frac{w^{l-1}e^{-w}}{\Gamma(l)} \quad (4.4.19)$$

which implies that V_n and W_n are asymptotically independent. If we impose the restriction that F is monotonically increasing, then (4.4.18) defines a one-to-one transformation and, therefore, we may conclude also that $X_{n:s}$ and $X_{n:n-l}$ are asymptotically independent. In particular, for $s = 1$ and $l = n$, it follows that the sample minimum and maximum are asymptotically independent.

4.5 Empirical distributions

As we have mentioned in Section 4.1, the so-called Kolmogorov distance, $\sup_x |F_n(x) - F(x)|$, is an important tool to evaluate the use of the empirical distribution function F_n as an estimator of a distribution function F. In this section we prove a basic result concerning the almost sure convergence of the Kolmogorov distance; the proof is essentially based on the reduction of the operation of considering the supremum over a compact set to that of taking the maximum over a finite number of points, a technique which may be successfully employed in many other applications.

Theorem 4.5.1 (Glivenko-Cantelli): *Let $\{X_1, \ldots, X_n\}$ be a random sample corresponding to a random variable X with distribution function F. Then, if F_n denotes the associated empirical distribution function, we have*

$$\sup_{x \in R} |F_n(x) - F(x)| \xrightarrow{\text{a.s.}} 0.$$

Proof: Assume that F is continuous. Then $Y_i = F(X_i) \sim \text{Unif}(0, 1)$, which implies $G(t) = P\{Y_i \leq t\} = t$, $0 \leq t \leq 1$. Let $x = F^{-1}(t)$; then $F(x) = t$ and

$$
\begin{aligned}
F_n(x) &= \frac{1}{n} \sum_{i=1}^n I_{\{X_i \leq F^{-1}(t)\}} = \frac{1}{n} \sum_{i=1}^n I_{\{F(X_i) \leq t\}} \\
&= \frac{1}{n} \sum_{i=1}^n I_{\{Y_i \leq t\}} = G_n(t).
\end{aligned}
\tag{4.5.1}
$$

(G_n is known as the reduced empirical d.f.) Consequently,

$$\sup_{x \in R} |F_n(x) - F(x)| = \sup_{0 \leq t \leq 1} |G_n(t) - t|. \tag{4.5.2}$$

Now note that G_n is nondecreasing; therefore, given $m > 0$, it follows that for all $t \in [(k-1)/m, \ k/m]$, $k = 1, \ldots, m$, we have

$$G_n\left(\frac{k-1}{m}\right) \leq G_n(t) \leq G_n\left(\frac{k}{m}\right)$$

$$\Rightarrow G_n\left(\frac{k-1}{m}\right) - \frac{k}{m} \leq G_n(t) - t \leq G_n\left(\frac{k}{m}\right) - \frac{k-1}{m}$$

$$\Rightarrow G_n\left(\frac{k-1}{m}\right) - \frac{k-1}{m} - \frac{1}{m} \leq G_n(t) - t \leq G_n\left(\frac{k}{m}\right) - \frac{k}{m} + \frac{1}{m}$$

$$\Rightarrow |G_n(t) - t| \leq \max_{s=k-1,k} \left| G_n\left(\frac{s}{m}\right) - \frac{s}{m} \right| + \frac{1}{m}$$

$$\Rightarrow \sup_{0 \leq t \leq 1} |G_n(t) - t| \leq \max_{1 \leq k \leq m} \sup_{\frac{k-1}{m} \leq t \leq \frac{k}{m}} |G_n(t) - t|$$

$$\leq \max_{1 \leq k \leq m} \left| G_n\left(\frac{k}{m}\right) - \frac{k}{m} \right| + \frac{1}{m}.$$

Given $\varepsilon > 0$, there exists m such that $m^{-1} < \varepsilon/2$, and then

$$\sup_{0 \leq t \leq 1} |G_n(t) - t| \leq \max_{1 \leq k \leq m} \left| G_n\left(\frac{k}{m}\right) - \frac{k}{m} \right| + \frac{\varepsilon}{2}. \tag{4.5.3}$$

Now recall that $nG_n(k/m) = \sum_{i=1}^{n} I_{\{Y_i \leq k/m\}} \sim \text{Bin}(n, k/m)$, and apply the Bernstein Inequality for binomial events (2.3.16) to see that, given $\eta > 0$,

$$P\left\{ \left| G_n\left(\frac{k}{m}\right) - \frac{k}{m} \right| > \eta \right\} \leq 2\{\rho(\eta)\}^n, \quad \text{where} \quad 0 < \rho(\eta) < 1.$$

Choose $\eta = \varepsilon/2$ and note that

$$P\left\{ \max_{0 \leq k \leq m} \left| G_n\left(\frac{k}{m}\right) - \frac{k}{m} \right| > \frac{\varepsilon}{2} \right\} \leq \sum_{k=1}^{m-1} P\left\{ \left| G_n\left(\frac{k}{m}\right) - \frac{k}{m} \right| > \frac{\varepsilon}{2} \right\}$$

$$\leq 2(m-1)\{\rho(\eta)\}^n$$

so that

$$P\left\{ \bigcup_{N \geq n} \max_{0 \leq k \leq m} \left| G_N\left(\frac{k}{m}\right) - \frac{k}{m} \right| > \frac{\varepsilon}{2} \right\}$$

$$\leq \sum_{N \geq n} P\left\{ \max_{0 \leq k \leq m} \left| G_N\left(\frac{k}{m}\right) - \frac{k}{m} \right| > \frac{\varepsilon}{2} \right\}$$

$$\leq 2(m-1) \sum_{N \geq n} \left\{ \rho\left(\frac{\varepsilon}{2}\right) \right\}^N$$

$$= \frac{2(m-1)\{\rho(\varepsilon/2)\}^n}{1 - \rho(\varepsilon/2)} \to 0 \quad \text{as} \quad n \to \infty,$$

implying that

$$\max_{1 \leq k \leq m} \left| G_n\left(\frac{k}{m}\right) - \frac{k}{m} \right| \xrightarrow{\text{a.s.}} 0.$$

In view of (4.5.2)–(4.5.3) the result is proved for F continuous.

Now suppose that F has a finite number of jump-points at $a_1 < a_2 < \cdots < a_M$. Then, given $\varepsilon > 0$, let M_ε denote the number of jump-points with jumps $> \varepsilon$. Excluding the neighborhoods of the jump-points, the first part of the proof holds. For the jump-point at a_j, observe that $nF_n(a_j) \sim \text{Bin}(n, F_n(a_j))$ and then apply the Bernstein Inequality for binomial events (2.3.16) to see that

$$P\{|F_n(a_j) - F(a_j)| > \varepsilon\} \leq 2\{\rho(\varepsilon)\}^n \quad \text{where} \quad 0 < \rho(\varepsilon) < 1.$$

Finally note that since there are only finitely many jumps of magnitude $> \varepsilon > 0$, we obtain

$$P\left\{ \bigcup_{N \geq n} \max_{1 \leq j \leq M} |F_N(a_j) - F(a_j)| > \varepsilon \right\} \to 0 \quad \text{as} \quad n \to \infty$$

and the proof follows.   ∎

To extend the above result to the bivariate case, consider a random sample $\{(X_1, Y_1)^t, \ldots, (X_n, Y_n)^t\}$ from random vector $(X, Y)^t$ with d.f. F and let

$$F_n(x, y) = \frac{1}{n} \sum_{i=1}^{n} I_{\{X \leq x\}} I_{\{Y \leq y\}}, \quad x, y \in \mathbb{R}.$$

Assuming F continuous, let $W_i = F(X_i, \infty)$ and $Z_i = F(\infty, Y_i)$ then write $X_i = F^{-1}(W_i, \infty)$ and $Y_i = F^{-1}(\infty, Z_i)$, so that

$$
\begin{aligned}
F_n(x, y) &= \frac{1}{n} \sum_{i=1}^{n} I_{\{X_i \leq F^{-1}(w, \infty)\}} I_{\{Y_i \leq F^{-1}(\infty, z)\}} \\
&= \frac{1}{n} \sum_{i=1}^{n} I_{\{F(X_i), \infty) \leq w\}} I_{\{F(\infty, Y_i) \leq z\}} \\
&= \frac{1}{n} \sum_{i=1}^{n} I_{\{W_i \leq w\}} I_{\{Z_i \leq z\}} = G_n(w, z).
\end{aligned}
$$

So, letting $G(w, z) = P\{W_i \leq w, Z_i \leq z\}$, we obtain

$$\sup_{(x,y)^t \in \mathbb{R}^2} |F_n(x, y) - F(x, y)| = \sup_{(w,z)^t \in [0,1] \times [0,1]} |G_n(w, z) - G(w, z)|.$$

Now, since G_n and G are nondecreasing functions, it follows that

$$
\begin{aligned}
G_n\left(\frac{k-1}{m}, \frac{l-1}{m}\right) &- G\left(\frac{k}{m}, \frac{l}{m}\right) \\
&\leq G_n(w, z) - G(w, z) \\
&\leq G_n\left(\frac{k}{m}, \frac{l}{m}\right) - G\left(\frac{k-1}{m}, \frac{l-1}{m}\right)
\end{aligned}
\tag{4.5.4}
$$

for all $(w, z) \in \left[\frac{k-1}{m}, \frac{k}{m}\right) \times \left[\frac{l-1}{m}, \frac{l}{m}\right)$, $k, l = 1, \ldots, m$, where m is a positive integer. Then, given $\varepsilon > 0$, we may choose $m^{-1} < \varepsilon/4$ so that

$$
\begin{aligned}
G_n&\left(\frac{k}{m}, \frac{l}{m}\right) - G\left(\frac{k-1}{m}, \frac{l-1}{m}\right) \\
&= \left\{G_n\left(\frac{k}{m}, \frac{l}{m}\right) - G\left(\frac{k}{m}, \frac{l}{m}\right)\right\} + \left\{G\left(\frac{k}{m}, \frac{l}{m}\right) - G\left(\frac{k-1}{m}, \frac{l-1}{m}\right)\right\} \\
&= G_n\left(\frac{k}{m}, \frac{l}{m}\right) - G\left(\frac{k}{m}, \frac{l}{m}\right) + \left\{G\left(\frac{k}{m}, \frac{l}{m}\right) - G\left(\frac{k}{m}, \frac{l-1}{m}\right)\right\} \\
&\quad + \left\{G\left(\frac{k}{m}, \frac{l-1}{m}\right) - G\left(\frac{k-1}{m}, \frac{l-1}{m}\right)\right\}
\end{aligned}
$$

$$\leq \; G_n\left(\frac{k}{m},\frac{l}{m}\right) - G\left(\frac{k}{m},\frac{l}{m}\right) + \left\{G\left(1,\frac{l}{m}\right) - G\left(1,\frac{l-1}{m}\right)\right\}$$

$$+\left\{G\left(\frac{k}{m},1\right) - G\left(\frac{k-1}{m},1\right)\right\}$$

$$\leq \; \left\{G_n\left(\frac{k}{m},\frac{l}{m}\right) - G\left(\frac{k}{m},\frac{l}{m}\right)\right\} + \frac{\varepsilon}{4} + \frac{\varepsilon}{4}, \quad k,l = 1,\ldots,m. \quad (4.5.5)$$

Using a similar argument we may show that

$$G_n\left(\frac{k-1}{m},\frac{l-1}{m}\right) - G\left(\frac{k}{m},\frac{l}{m}\right)$$

$$\leq \left\{G_n\left(\frac{k-1}{m},\frac{l-1}{m}\right) - G\left(\frac{k-1}{m},\frac{l-1}{m}\right)\right\} + \frac{\varepsilon}{2} \quad (4.5.6)$$

so that from (4.5.4)–(4.5.6) we may write

$$\sup_{(w,z)^t \in [0,1] \times [0,1]} \left|G_n(w,z) - G(w,z)\right|$$

$$\leq \max_{k,l=1,\ldots,m} \left|G_n\left(\frac{k}{m},\frac{l}{m}\right) - G\left(\frac{k}{m},\frac{l}{m}\right)\right| + \frac{\varepsilon}{2}.$$

Recalling that $nG_n\left(\frac{k}{m},\frac{l}{m}\right) \sim \mathrm{Bin}\{n, G\left(\frac{k}{m},\frac{l}{m}\right)\}$ and following the same steps as in Theorem 4.5.1 we may readily complete the proof.   ∎

Although the Central Limit Theorems introduced in Chapter 3 in conjunction with the results presented in this chapter constitute potent tools to study the asymptotic properties of the empirical distribution function F_n for each fixed x, they fail to provide sufficient power to deal with some statistics that depend on the stochastic process $\{F_n(x), x \in \mathbb{R}\}$, like the Cramér-von Mises statistic $C_n = n\int_{-\infty}^{+\infty}[F_n(x) - F(x)]^2 dF(x)$ and some other statistics which will be intuitively introduced in the next section; a relatively more technical presentation is outlined in Chapter 8.

4.6 Functions of order statistics and empirical distributions

In Section 4.3, we have discussed briefly the BLUE based on the vector of order statistics $\mathbf{t}_n$ or a subset of its coordinates. Since such estimators are linear functions of $\mathbf{t}_n$, they are often called L-estimators. In this context, perhaps, it will be more convenient to introduce the following type of L-estimator:

$$L_{n1} = d_{n1}X_{n:k_1} + \ldots + d_{nq}X_{n:k_q} \quad (4.6.1)$$

where the d_{nj} are suitable (nonstochastic) constants, q is a fixed positive integer, and $k_j \sim np_j$, $1 \leq j \leq q$, with $0 < p_1 < \cdots < p_q < 1$. Thus, L_{n1} is a linear combination of a selected number of sample quantiles. Theorem 4.2.2 and (4.2.29) may directly be adapted to study stochastic convergence of L_{n1}, and, thereby, we need a minimal set of regularity conditions on F. Further, a generalized version Theorem 4.4.1 may be incorporated in a direct verification of the asymptotic normality of the standardized form of L_{n1}. For this reason, we shall not discuss further properties of L_{n1} in what follows. Consider next the second type of L-estimator:

$$L_{n2} = \sum_{i=1}^{n} c_{ni} X_{n:i} \qquad (4.6.2)$$

where the c_{ni} are real numbers depending on (i, n), $1 \leq i \leq n$. For example, looking at (4.3.15) and its affinity to the classical weighted least-squares method, we may conclude that the two normal equations leading to the BLUE of (μ, λ) conform to L-estimators of the form (4.6.2) where the c_{ni} depend on $\mathbf{W}_n$ and $\boldsymbol{\Gamma}_n$, which are known. Further, we may use, for large n, the approximations for $\mathbf{W}_n$ and $\boldsymbol{\Gamma}_n$ discussed before (4.3.15) and obtain the so-called ABLUE estimators. This adaptation generally leads to a simplification:

$$c_{ni} = n^{-1} J_n(i/n), \quad 1 \leq i \leq n, \qquad (4.6.3)$$

where $J_n(u) = J_n(n^{-1}[nu])$, $u \in (0, 1)$, converges to some "smooth" function $J(u)$, $u \in (0, 1)$, and $J(\cdot)$ satisfies additional regularity conditions. For this reason, L_{n2} is termed an **L-estimator with smooth weight functions**. In general, we may set

$$L_n = L_{n1} + L_{n2}, \qquad (4.6.4)$$

where one of the two components may be degenerate. Further, in (4.6.1) and (4.6.2), the $X_{n:i}$ may also be replaced by $h(X_{n:i})$, for some suitable $h(\cdot)$, defined on $\mathbb{R}$. The form of L_{n2} or L_n, in general, may also arise in a nonparametric formulation, without attaching special importance to a specific p.d.f. f_0. To illustrate this point, let us consider two special L-estimators (of location):

i) **Trimmed mean:** For some non-negative integer k $(< n/2)$, let

$$T_{n,k} = (n - 2k)^{-1} \sum_{i=k+1}^{n-k} X_{n:i}. \qquad (4.6.5)$$

For $k = 0$, $T_{n,0} = \overline{X}_n$ (sample mean), whereas for $k = [(n-1)/2]$, $T_{n,k}$ is the sample median. Note that (4.6.5) conforms to (4.6.2) with $c_{ni} = 0$ for $i \leq k$ and $i \geq n - k + 1$ and $c_{ni} = (n - 2k)^{-1}$ for $k + 1 \leq i \leq n - k$. In fact, if we let $k \sim n\alpha$ for some α $(0 < \alpha < 1/2)$, we may identify easily that (4.6.3) holds, i.e., as $n \to \infty$,

$$J_n(u) \to J(u) = \begin{cases} (1-2\alpha)^{-1}, & \alpha \le u \le 1-\alpha, \\ 0, & u < \alpha, u > 1-\alpha. \end{cases} \qquad (4.6.6)$$

Also, for the case of $\overline{X}_n \ (= T_{n,0})$, $J_n(u) = J(u) = 1$, for all $u \in (0,1)$. Generally, $\overline{X}_n$ is not a very robust statistic, and, for large n, a suitable choice of (small) $\alpha \ [= (k/n)]$ may enhance the robustness of $T_{n,k}$ without any perceptible change in its asymptotic efficiency.

ii) **Winsorized mean:** For some $k(0 \le k < n/2)$, let

$$W_{n,k} = n^{-1}\left\{ k(X_{n:k} + X_{n:n-k+1}) + \sum_{i=k+1}^{n-k} X_{n:i} \right\}, \qquad (4.6.7)$$

where conventionally, for $k = 0$, $k(X_{n:k} + X_{n:n-k+1}) = 0$. Note that for $k = 0$, $W_{n,0} = \overline{X}_n$, whereas, for $k \ge 1$,

$$W_{n,k} = n^{-1}k(X_{n:k} + X_{n:n-k+1}) + (1 - 2k/n)T_{n,k}, \qquad (4.6.8)$$

so that (4.6.8) conforms to (4.6.4).

There are some other statistics, viz., the mid-range $\frac{1}{2}(X_{n:1} + X_{n:n})$ and range $(X_{n:n} - X_{n:1})$, which are of the form L_{n1}; but $X_{n:1}$ and $X_{n:n}$ are sample extreme values, not quantiles, and, hence, the study of their asymptotic properties may require the methodology considered in Section 4.4. In this section, we shall mainly confine our attention to L_{n2}.

Recall the definition of F_n in (4.1.1) and (4.1.13) and, as such, we have by (4.6.2) and (4.6.3)

$$L_{n2} = \int_R J_n(F_n(x))x\,dF_n(x). \qquad (4.6.9)$$

It is, therefore, quite intuitive to conceive of a parameter $\theta = \theta(F)$, a functional of F, of the form

$$\theta(F) = \int_R J(F(x))x\,dF(x), \qquad (4.6.10)$$

so that L_{n2} is a natural estimator of $\theta(F)$, which may also have some simple interpretation in a variety of models. For example, suppose that $F(x)$ is symmetric about θ [i.e., $F(\theta - y) + F(\theta + y) = 1$, for all $y \in \mathbb{R}$] and $J(u)$ is a symmetric function about $u = 1/2$ [i.e., $J(\frac{1}{2} - u) = J(\frac{1}{2} + u)$, for all $u \in (0, \frac{1}{2})$], and, further, $\int_0^1 J(u)du = 1$. Then, we have

$$\theta(F) - \theta = \int_R J(F(\theta + y))y\,dF(\theta + y) = 0, \qquad (4.6.11)$$

so that $\theta(F) = \theta$. This formulation leads to a large class of L-estimators of location of a symmetric d.f. F. Looking back at (4.6.9) and (4.6.10), is it quite natural to inquire whether L_{n2} converges (weakly/a.s.) to $\theta(F)$ as $n \to \infty$? The answer depends on both $J_n(\cdot)$ and $F(\cdot)$. To make this point clear, we may write

$$
\begin{aligned}
L_{n2} - \theta(F) &= \int_R x[J_n(F_n(x)) - J(F(x))]dF_n(x) \\
&\quad + \int_R xJ(F(x))d[F_n(x) - F(x)] \\
&= A_{n1} + A_{n2}, \quad \text{say.} \tag{4.6.12}
\end{aligned}
$$

We also note that the results of Chapter 2 are directly adoptable for A_{n2} $[= n^{-1}\sum_{i=1}^{n} X_i J(F(X_i)) - \mathrm{E}XJ(F(X))]$ whenever $XJ(F(X))$ is integrable with respect to F, i.e., $\theta(F)$ exists. But more elaborate analysis may generally be needed to show that A_{n1} has the desired convergence properties. This can be done by imposing additional regularity conditions on $F(\cdot)$ and/or $J_n(\cdot)$. In this context, we note that by the Glivenko-Cantelli Theorem (4.5.1) as n increases,

$$
\|F_n - F\| = \sup_x |F_n(x) - F(x)| \xrightarrow{\text{a.s.}} 0. \tag{4.6.13}
$$

As such, if F has a finite support, i.e., for some finite c $(0 < c < \infty)$, $P\{|X_i| \le c\} = 1$,

$$
\|J_n - J\| = \sup_{0 \le u \le 1} |J_n(u) - J(u)| \to 0 \quad \text{as} \quad n \to \infty
$$

and J is uniformly continuous, then $A_{n1} \xrightarrow{\text{a.s.}} 0$. On the other hand, if F does not have a compact support, the needed regularity conditions on $\{J_n(\cdot)\}$ are to be tailored according to the tail-behavior of F (i.e., appropriate moment conditions on F). Note that by the Hölder Inequality (1.4.27), for $r \ge 1$, $s \ge 1$, $r^{-1} + s^{-1} = 1$,

$$
|A_{n1}| \le \left\{ \int_R |x|^r dF_n(x) \right\}^{1/r} \left\{ \int_R |J_n\{F_n(x) - J(F(x))\}|^s dF_n(x) \right\}^{1/s}, \tag{4.6.14}
$$

where $\mathrm{E}_F|X|^r < \infty$ implies $\int_R |x|^r dF_n(x) \xrightarrow{\text{a.s.}} \mathrm{E}_F|X|^r$. Also, making use of the probability integral transformation $Y = F(X)$ and the reduced empirical d.f. $G_n(t) = n^{-1}\sum_{i=1}^{n} I_{\{Y_i \le t\}}, 0 \le t \le 1$, we have

$$
\int_R |J_n\{F_n(x)\} - J\{F(x)\}|^s dF_n(x) = \int_0^1 |J_n\{G_n(t)\} - J(t)|^s dG_n(t), \tag{4.6.15}
$$

and, hence, we may refer to properties of uniform order statistics [i.e., $U_{n:i} = F(X_{n:i}), 1 \le i \le n$] along with convergence of $J_n(u) - J(u)$ to show that (4.6.15) converges to 0 (weakly or a.s.). Basically, one then needs

$$n^{-1} \sum_{i=1}^{n-1} \left| J_n\left(\frac{i}{n}\right) - J\left(\frac{i}{n}\right) \right|^s \to 0, \qquad n^{-1} J_n(1) \to 0, \qquad (4.6.16)$$

and

$$\int_0^1 |J(t)|^s \, dt < \infty \quad \text{for some } s \text{ such that} \quad \frac{1}{r} + \frac{1}{s} = 1, \qquad (4.6.17)$$

where $E|X|^r < \infty$.

As in Chapter 3, one may also be interested in the asymptotic normality of $n^{1/2}\{L_{n2} - \theta(F)\}$, which may be used to provide a confidence interval for $\theta(F)$ or to test for a null hypothesis for $\theta(F)$, based on L_{n2} and a suitable estimate of its asymptotic variance. In this context, the decomposition in (4.6.12) may not be the most convenient means, as both $n^{1/2} A_{n1}$ and $n^{1/2} A_{n2}$ contribute to the asymptotic normality and they are generally not independent. We may introduce $\{\Psi_n(\cdot)\}$ and $\Psi(\cdot)$, such that $\Psi_n(u) \to \Psi(u)$ as $n \to \infty$, and

$$d\Psi_n\{F_n(x)\} = J_n\{F_n(x)\} dF_n(x), \qquad (4.6.18)$$

$$d\Psi\{F(x)\} = J\{F(x)\} dF(x). \qquad (4.6.19)$$

Then, by (4.6.9), (4.6.10) and integration by parts, we have

$$\begin{aligned}
\sqrt{n}\{L_{n2} - \theta(F)\} &= -\int_{-\infty}^{\infty} \sqrt{n}[\Psi_n\{F_n(x)\} - \Psi\{F(x)\}] dx \\
&= -\int_{-\infty}^{\infty} \sqrt{n}[\Psi_n\{F_n(x)\} - \Psi\{F_n(x)\}] dx \\
&\quad - \int_{-\infty}^{\infty} \sqrt{n}[\Psi\{F_n(x)\} - \Psi\{F(x)\}] dx.
\end{aligned}$$

$$(4.6.20)$$

In a majority of cases, $J_n(\cdot)$ and $J(\cdot)$ can be chosen such that on replacing F_n by $F_n^* = n(n+1)^{-1} F_n$, the first integral on the right-hand side of (4.6.20) can be made $o_p(1)$, whereas under a continuous differentiability condition on Ψ [granted by a smooth $J(\cdot)$], the second integral can be written as

$$-\int_{-\infty}^{\infty} \sqrt{n}\{F_n(x) - F(x)\} \Psi'\{F(x)\} dx$$

$$-\int_{-\infty}^{\infty} \left\{ \sqrt{n}[\Psi\{F_n(x)\} - \Psi\{F(x)\}] \right.$$

$$\left. -\Psi'\{F(x)\}\sqrt{n}[F_n(x) - F(x)] \right\} dx \qquad (4.6.21)$$

whence the second term can be made $o_p(1)$. In the first term, the weak convergence of $\sqrt{n}(F_n - F)$ can be incorporated (when Ψ' satisfies an integrability condition) to prove the desired asymptotic normality. This intuitive proof will be formalized in Chapter 8 in a comparatively more general setup. However, to be consistent with the level of presentation, at this stage, we shall not enter into these technicalities.

Looking at $\theta(F)$ in (4.6.10), we may remark that there are many applications, especially, in nonparametric statistics, where $\theta(F)$ may be expressed as

$$\theta(F) = \int \cdots \int g(x_1, \ldots, x_m) dF(x_1) \ldots dF(x_m), \qquad (4.6.22)$$

i.e., as **functional of** F of finite degree $m(\geq 1)$, where $g(\cdot)$ is a **kernel** (of known form) which may be taken symmetric in its m arguments. Here also, one may replace F by F_n, and the corresponding functional $\theta(F_n)$, termed the **von Mises' function**, is an estimator of $\theta(F)$. Such statistics, along with the related U-statistics, will be studied in Chapters 5 and 8.

In the context of robust estimation, the so called R- and M-estimators are defined as (implicit) functionals of F_n. To have some ideas about the related asymptotic theory, we need to formulate some general "differentiable statistical" functionals. Again, because of the technicalities involved, we shall only provide a broad outline of such functionals in Chapter 8. An expansion, similar to that in (4.6.21 but defined on a function space, provides the necessary tools.

The main thrust of this chapter has been on the interrelations of order statistics and empirical distributions, so that for their asymptotics, one can be studied via the other. As has been pointed out in Section 4.1, such relations may be used with advantage in problems of reliability and life-testing models. To illustrate this point further, let us consider the following simple reliability models.

Consider a **system** consisting of s units in **series**. Denoting the individual lifetimes of these units by $X_1, \ldots, X_s$ respectively, the lifetime of the system is $Y = \min\{X_1, \ldots, X_s\}$. Thus, if $\overline{F} \ (= 1 - F)$ stands for the **survival function** of X and $\overline{G} \ (= 1 - G)$ for that of Y, we have $\overline{G}(y) = \{\overline{F}(y)\}^s$, $y \in \mathbb{R}^+$. Suppose now that given a set $\{X_1, \ldots, X_n\}$ of n observations on the unit lifetimes, we want to estimate $\overline{G}$ as well as $\theta(G)$, the mean of the d.f. G, or some other parameters. Defining F_n as in (4.1.1), we have that $\overline{G}_n(y) = \{1 - F_n(y)\}^s$, $y \geq 0$, is a natural estimator of $\overline{G}$, although it is not unbiased. On the other hand,

$$\widehat{G}_n(y) = \binom{n}{s}^{-1} \sum_{1 \leq i_1 < \cdots < i_s \leq n} I_{\{\min(X_{i_1}, \ldots, X_{i_s}) \leq y\}}, \qquad y \in \mathbb{R}^+, \qquad (4.6.23)$$

estimates $G(y)$ unbiasedly and consistently since it is a U-statistic (for any given y). We may virtually repeat the proof of Theorem 4.5.1 (see Exercise 4.6.2) and verify that

$$\|\widehat{G}_n - G\| = \sup_y |\widehat{G}_n(y) - G(y)| \overset{\text{a.s.}}{\longrightarrow} 0. \qquad (4.6.24)$$

Similarly,

$$U_n = \binom{n}{s}^{-1} \sum_{1 \le i_1 < \cdots < i_s \le n} \min\{X_{i_1}, \ldots, X_{i_s}\} \qquad (4.6.25)$$

is also a U-statistic and, as such, is an unbiased estimator of

$$\theta(G) = \int_0^\infty x \, dG(x).$$

The asymptotic normality of U_n may be shown along the lines of Section 5.3 (see Exercise 4.6.3).

As before, we denote the ordered r.v.'s corresponding to $X_1, \ldots, X_n$ by $X_{n:1} < \cdots < X_{n:n}$ (ties neglected with probability 1 as F is assumed to be continuous a.e.). Then Exercise 4.6.3 is set to verify that

$$U_n = \binom{n}{s}^{-1} \sum_{i=1}^n \binom{n-i}{s-1} X_{n:i} \qquad (4.6.26)$$

so that, as in (4.6.2), U_n is an L-statistic. Also, verify that (Exercise 4.6.4) for U_n in (4.6.26), the score function $J(u)$, $0 < u < 1$, is given by

$$J(u) = s(1 - u)^{s-1}, \quad 0 \le u \le 1. \qquad (4.6.27)$$

Consider next s units in a system in **parallel**, so that the system's lifetime is given by $Y = \max\{X_1, \ldots, X_s\}$. Exercises 4.6.5 and 4.6.6 are posed to verify that $\widehat{G}_n$ and U_n can be defined analogously to the system in series case. Also, it may be seen that, here,

$$U_n = \binom{n}{s}^{-1} \sum_{i=1}^n \binom{i-1}{s-1} X_{n:i}. \qquad (4.6.28)$$

A third notable example relates to the strength of a bundle of filaments, which we have introduced in Exercise 2.4.1. In this case, the bundle strength is given by

$$\begin{aligned} Z_n &= \max_{1 \le i \le n} \left\{ \left(1 - \frac{i-1}{n}\right) X_{n:i} \right\} \\ &= \sup[x\{1 - F_n(x^-)\} : 0 \le x < \infty] \end{aligned}$$

which also depicts the intricate relationship between the order statistics and the empirical distribution functions. This can be used to provide a much simpler proof of the asymptotic convergence of Z_n to $\theta(F) = \sup[x\{1 - F(x)\}, x \in \mathbb{R}^+]$.

4.7 Concluding notes

We may start with the observation that for a continuous d.f. F, the joint distribution of $X_{n:1}, \ldots, X_{n:n}$ is given by

$$n! \, dF(X_{n:1}) \cdots dF(X_{n:n}),$$

so that the conditional distribution of $\mathbf{X}_n = (X_1, \ldots, X_n)^t$ given $\mathbf{t}_n = (X_{n:1}, \ldots, X_{n:n})^t$ is uniform over the $n!$ permutations

$$\{X_{n:i_1}, \ldots, X_{n:i_n}\}$$

with $(i_1, \ldots, i_n)$ being a permutation of $(1, \ldots, n)$. The same result holds for $\mathbf{t}_n$ defined in (4.2.5) for general F (with possible jump-points). This shows that $\mathbf{t}_n$ is, in general, a sufficient statistic for F (in a nonparametric setup), and in view of the one-to-one relationship between $\mathbf{t}_n$ and F_n, the same conclusion holds for F_n as well. Moreover, viewed from this perspective, U-statistics may also be characterized as conditional expectation of a kernel, given $\mathbf{t}_n$, and, hence, are optimal in this sense. However, in a parametric formulation, $\mathbf{t}_n$ is not necessarily minimal sufficient, and, hence, exact small sample optimality properties may not directly be derivable in a great number of cases. There are, of course, situations where one or more functions of $\mathbf{t}_n$ may be minimal sufficient. These are presented in the form of Exercises 4.7.1–4.7.4.

Although we have focused on both sample quantiles and extreme values, because of slower rates of convergence, the latter are less useful in practice, from an asymptotic point of view. An exception is the case where the end point is finite with a positive density; here, the rate of convergence is even better than that of quantiles. However, such extreme value distributions have good uses in reliability and systems analysis models, and our treatment could be useful in that case too. Exercises 4.7.5–4.7.8 are geared in this direction.

4.8 Exercises

Exercise 4.2.1: Use (4.2.16) to show that for $n = 2m + 1$ and $k = m + 1$, when F is symmetric about its median θ, then $Q_{n,k}(x)$ is also symmetric about θ.

Exercise 4.2.2: Define $q_{n,k}(\cdot)$ as in (4.2.20) and assume that $f(x) = F'(x)$ is symmetric about θ. Then show that

$$q_{n,k}(\theta + x) = q_{n,n-k+1}(\theta - x), \quad x \in \mathbb{R}.$$

Extend this symmetry result to the bivariate case, i.e. to the density of $(X_{n:k}, X_{n:n-k+1})$. Hence, or otherwise, show that for $n = 2m$ and $k = m$ when the median is defined as in (4.2.12), under the symmetry of F, it has a symmetric distribution too.

Exercise 4.2.3: Let θ be the median of F [i.e., $F(\theta) = 1/2$], and use (4.2.17) to verify that for every $n > 1$ and $0 \le r < s \le n + 1$ (where $X_{n:0} = -\infty$ and $X_{n:n+1} = +\infty$),

$$P\{X_{n:r} \le \theta \le X_{n:s}\} = 2^{-n} \sum_{i=r}^{s-1} \binom{n}{i}.$$

Choose $s = n - r + 1$, $r \le n/2$, and show that the above formula provides a distribution-free confidence interval for θ. What can be said about the confidence coefficient (i.e., coverage probability) when n is not large.

Exercise 4.2.4: Verify (4.2.29).

Exercise 4.3.1: Verify that the proof of Theorem 4.3.1 remains intact for all $k = k_n$ such that $n^{-1}k_n \to p$ as $n \to \infty$.

Exercise 4.3.2: Let $X_i, i \ge 1$ be i.i.d. r.v.'s with a continuous pdf:

$$f(x; \theta) = c(x - \theta)^2 I_{\{\theta-1 \le x \le \theta+1\}}, \quad c > 0.$$

Define the sample median $\tilde{X}_n$ by $X_{n:m}$ where $m = [n/2]$. Show that $\tilde{X}_n$ is a strongly consistent estimator of θ. Also, let $Z_n = n^{1/6}(\tilde{X}_n - \theta)$. Show that Z_n^3 has asymptotically a normal distribution with zero mean and a finite variance.

Exercise 4.3.3: Let $X_1, \ldots, X_n$ be i.i.d. r.v.'s with a pdf $f(x; \theta) = f(x-\theta)$, where $f(\cdot)$ is symmetric about 0. Let then $\hat{\xi}_{p,n} = [X_{n:[np]} + X_{n:n-[np]+1}]/2$, for $0 < p < 1$.

i) What is the asymptotic variance of $n^{1/2}(\hat{\xi}_{p,n} - \theta)$?

ii) Under what condition on $f(\cdot)$, does it have a minimum at $p = 1/2$?

iii) Verify ii) when $f(\cdot)$ is (a) normal and (b) double exponential.

Exercise 4.3.4: Suppose that $F(x)$ has the following (scale-perturbed) form:

$$F(x) = \begin{cases} \Phi((x - \theta)/a), & x > \theta, \quad a > 0 \\ \Phi((x - \theta)/b), & x < \theta, \quad b > 0, \quad b \ne a \end{cases}$$

where $\Phi(\cdot)$ is the standard normal d.f. Show that $F(\cdot)$ does not have a pdf at θ, although the rhs and lhs derivatives exist at 0. Verify that the median of F is still uniquely defined and the sample median converges a.s. to the population median, as $n \to \infty$. What can be said about the asymptotic law for $n^{1/2}(\widetilde{X}_n - \theta)$?

Exercise 4.4.1: Let $X_1, \ldots, X_n$ be i.i.d. r.v.'s with a pdf

$$f(x) = (1/2)\exp(-|x|), \quad x \in \mathbb{R}.$$

Define $M_n = (X_{n:1} + X_{n:n})/2$ and $Y_n = (X_{n:[np]} + X_{n:n-[np]+1})/2$ where $p \in (0,1)$. Obtain the asymptotic distribution of the normalized form of M_n and, hence, show that as $n \to \infty$, M_n does not converge to 0, in probability. Obtain the asymptotic distribution of the normalized form of Y_n, and show that $Y_n \xrightarrow{\text{P}} 0$, as $n \to \infty$, for every $p \in (0,1)$. Compare the asymptotic variance of Y_n for $p \in (0,1.2)$ with that of $p = 1/2$ (i.e., the median), and comment on their asymptotic relative efficiency.

Exercise 4.4.2: In the previous exercise, instead of the Laplace density, consider a logistic density function and conclude on the parallel results on M_n and Y_n.

Exercise 4.4.3: Consider a distribution F, symmetric about θ and having a finite support $[\theta - \delta, \theta + \delta]$, where $0 < \delta < \infty$. Suppose that F has a terminal contact of order m. Define the kth mid-range $M_{nk} = \frac{1}{2}(X_{n:k} + X_{n:n-k+1})$, for $k = 1, 2, \ldots$. Use (4.4.18)–(4.4.19) and Theorem 4.4.3 to derive the asymptotic distribution of the normalized form of M_{nk}. Compute the variance from this asymptotic distribution and denote this by $V_m(k)$, for $k > 1$. Show that, for $m = 1$, $V_1(k)$ is nondecreasing in k, whereas for $m \geq 2$, an opposite inequality may hold.

Exercise 4.5.1: Consider the Cramér-von Mises statistic

$$C_n = n \int_{-\infty}^{\infty} [F_n(x) - F(x)]^2 \, dF(x).$$

Derive the expression for the following: (i) EC_n, (ii) $\text{Var}(C_n)$, (iii) third and fourth central moments of C_n, and (iv) the usual Pearsonian measures of skewness and kurtosis of C_n. Hence, comment on the asymptotic distribution of C_n.

Exercise 4.5.2: Comment on the difficulties you may encounter if you want to study the Cramér-von Mises statistic in the bivariate (or multivariate) case where the d.f. $F(x)$ is defined on $\mathbb{R}^2$ (or $\mathbb{R}^p$, $\rho \geq 2$) and is not necessarily expressible as the product of its marginal distributions.

Exercise 4.5.3: Consider the Kolmogorov-Smirnov statistics for the one-sample and two-sample problems, and comment why you do not expect the asymptotic normality to be tenable in such cases. (We refer to Chapter 8 for more details on their asymptotic distributions.)

Exercise 4.6.1: Verify that under (4.6.16) and (4.6.17), (4.6.15) converges to 0 in an appropriate mode.

Exercise 4.6.3: Provide a formal proof of the asymptotic normality of U_n in (4.6.25) by using the projection technique in Section 3.4 (or the general results to be posed in Section 5.3).

Exercise 4.7.1: Let $X_{n:1}, \ldots, X_{n:n}$ be the order statistics of a sample of size n from a normal distribution with mean θ and variance σ^2. Show that the sum of these order statistics and the sum of squares of them are jointly minimal sufficient for the parameters.

Exercise 4.7.2: Consider a negative exponential distribution with pdf

$$f(x; \mu, \lambda) = \lambda^{-1} \exp((\mu - x)/\lambda) I_{\{x \geq \mu\}}.$$

Show that the minimal order statistics and the sum of order statistics constitute the minimal sufficient statistics.

Exercise 4.7.3: Consider the same model as in Exercise 4.7.2, but allow a Type II censoring, wherein the observable r.v.'s are $X_{n:k}$, $k = 1, \ldots, r$, for some $r \leq n$. Show that the minimal sufficiency characterization remains valid.

Exercise 4.7.4: Consider the uniform $[\theta - \delta, \theta + \delta]$ distribution. Show that $X_{n:1}$ and $X_{n:n}$ constitute the minimal sufficient statistics for (θ, δ).

Exercise 4.7.5: Consider the bundle strength of filaments model treated in Exercise 2.4.1. Show that $Z_n - \theta$ can be approximated by $x_0[F(x_0) - F_n(x_0)]$ where x_0 is defined (uniquely) by $\theta = x_0[1 - F(x_0)]$. Hence, or otherwise, use the results in Section 2.5 to establish the asymptotic normality of $n^{1/2}(Z_n - \theta)$.

Exercise 4.7.6: Consider a triangular density f on $(\theta - 1, \theta + 1)$, θ real, such that f is unimodal, $f(\theta) > 0$ and $f(\theta \pm 1) = 0$. Compare the asymptotic variance of the sample median and mid-range and derive suitable conditions under which one performs better than the other.

Exercise 4.7.7: Let $\{X_i; i \geq 1\}$ be a sequence of i.i.d. r.v.'s with a pdf $f(x)$, $x \in \mathbb{R}$. Define then $Y_n = X_n$ if X_n is $> X_1, \ldots, X_{n-1}$, and 0 otherwise, for $n \geq 1$. Thus, the Y_n are the record values in the sequence. What can you say about the asymptotic law of Y_n (suitably normalized)? Also, under what condition does a limit law exist for the normalized sum $Y_1 + \ldots + Y_n$?

Exercise 4.7.8: Construct an unbiased estimator of

$$P\{\min(X_1, \ldots, X_k) \leq x\}, \quad x \in \mathbb{R},$$

where the X_i, $i \leq n$, are i.i.d. r.v.'s with a d.f. F. Use U-statistics theory to find the asymptotic distribution of the estimator.

CHAPTER 5

Asymptotic Behavior of Estimators and Test Statistics

5.1 Introduction

Consider a family of probability spaces (Ω, A, P_θ) indexed by a parameter $\theta \in \Theta \subset \mathbb{R}$ and let $X_1, \ldots, X_n$ be a random sample from some specific (but unknown) distribution P_θ in this family. In this context, the objective of **parametric inference** is to propose and evaluate methods for choosing appropriate statistics $T_n = T_n(X_1, \ldots, X_n)$ to **estimate** θ (i.e., to guess the true value of θ) or to **test hypotheses** about θ (i.e., to decide whether $\theta \in \Theta_0 \subset \Theta$ or not); in the first case T_n is called an **estimator** of θ (and is usually denoted as $\widehat{\theta}_n$) and in the second case a **test statistic** .

Among the methods usually employed to obtain such estimators and test statistics, a few play a fundamental role, mainly because of their intuitive interpretation and computational ease. In many cases, however, their statistical properties in small samples are difficult to derive and, therefore, the study of their asymptotic behavior constitutes an appealing alternative. In this chapter we are concerned with the large sample properties of statistics obtained via Maximum Likelihood (ML), Least Squares (LS), Likelihood Ratio and other common methods in parametric inference. We first present the motivating concepts for estimation problems.

A sequence $\{\widehat{\theta}_n\}$ of estimators (of a parameter θ) is (weakly) **consistent** if $\widehat{\theta}_n - \theta \xrightarrow{\text{P}} 0$; alternatively, if $\widehat{\theta}_n - \theta \xrightarrow{\text{a.s.}} 0$, the sequence $\{\widehat{\theta}_n\}$ is said to be **strongly consistent**. This property of increasing accuracy of an estimator is certainly desirable and, in general, it may be shown to hold under rather mild regularity conditions; the proofs are essentially based on the techniques described in Chapter 2. A related, but stronger concept, that is frequently employed in consistency proofs is that of **asymptotic unbiasedness**: a sequence $\{\widehat{\theta}_n\}$ of estimators is asymptotically unbiased if $E\widehat{\theta}_n \to \theta$ as $n \to \infty$. For most practical applications, however, such properties of sequences of estimators are of limited value if not coupled with some form of convergence in distribution. For this reason, in this text, consistency is only addressed in connection with asymptotic normality, even though, in such cases, the proofs generally require more stringent assumptions; references involving consistency results under less restrictive conditions will be indicated. Even if we restrict ourselves to the class of asymptotically normal estimators, we still need some further criteria in order to choose the "best" candidate; among them lies the concept of **asymptotic relative efficiency**, which in the case of asymptotically normal estimators corresponds to the ratio of the respective asymptotic variances.

Although statistical tests and estimators were introduced primarily for specific parametric models, their nonparametric counterparts have also received adequate attention. The so-called U-statistics [viz, Hoeffding (1948)], introduced briefly in Section 2.4 [see (2.4.28)], are the precursors of nonparametric estimators, and they have important roles in parametric Large Sample Theory too. Many nonparametric test statistics are also based on such U-statistics. For this reason, we intend to cover the basic large sample properties of U-statistics along with other related ones.

In Section 5.2 we deal with MLE; in Section 5.3 we discuss the asymptotic behavior of estimators based on U-statistics; the large sample properties of estimators derived via other frequently used methods such as LS is considered in Section 5.4; Section 5.5 is devoted to the study of the asymptotic efficiency of estimators and other related concepts, and, finally, in Section 5.6 we deal with the asymptotic properties of likelihood ratio and other common test statistics.

5.2 Asymptotic behavior of maximum likelihood estimators

The method of Maximum Likelihood introduced by Fisher in 1922 is certainly one of the most commonly used techniques for parametric estimation. It generally leads to computationally tractable equations which yield estimates possessing good asymptotic statistical properties. Let $X_1, \ldots, X_n$ be i.i.d. random variables with density function $f(x; \theta)$, $\theta \in \Theta$. The **likelihood function** is defined as in (2.4.3) by

$$L_n(\theta) = \prod_{i=1}^{n} f(X_i; \theta), \quad \theta \in \Theta,$$

regarded as a function of θ. We say that $\widehat{\theta}_n$ is a MLE of θ if

$$L_n(\widehat{\theta}_n) = \sup_{\theta \in \Theta} L_n(\theta).$$

In many (regular) cases the MLE may be obtained via the maximization of the **log-likelihood** function which essentially reduces to solving the

estimating equation

$$U_n(\theta) = (\partial/\partial\theta)\log L_n(\theta) = 0.$$

The function $U_n(\theta) = U_n(\theta, X_1, \ldots, X_n)$ is known as the **estimating function** when considered as a function of θ and as the **score statistic** when considered as a function of $X_1, \ldots, X_n$; in (regular) cases where it has a single root it may actually replace the MLE for technical purposes. Let us now consider a few examples.

Example 5.2.1: Let $X_1, \ldots, X_n$ be i.i.d. r.v.'s such that $X_1 \sim N(\theta, 1)$. The likelihood function is

$$L_n(\theta) = (2\pi)^{-n/2} \exp\left\{-(1/2)\sum_{i=1}^{n}(X_i - \theta)^2\right\}$$

and $\sup_{\theta \in R} L_n(\theta)$ corresponds to $\min_{\theta \in R} \sum_{i=1}^{n}(X_i - \theta)^2$ which implies that the MLE of θ is $\widehat{\theta}_n = \overline{X}_n$. ∎

Example 5.2.2: Let $X_1, \ldots, X_n$ be i.i.d. r.v.'s following Bernoulli(θ) distributions. The likelihood function is

$$L_n(\theta) = \theta^{T_n}(1 - \theta)^{n - T_n}$$

where $T_n = \sum_{i=1}^{n} X_i$ and the estimating function $U_n(\theta) = (T_n - n\theta)/\theta(1 - \theta)$ is a nonlinear function of θ with a single root $\widehat{\theta}_n = T_n/n$. ∎

Example 5.2.3: Let $X_1, \ldots, X_n$ be i.i.d. r.v.'s with Exp(θ) distributions. The likelihood function is

$$L_n = \theta^{-n} \exp\{-T_n/\theta\}$$

where $T_n = \sum_{i=1}^{n} X_i$; here the estimating function $U_n(\theta) = (T_n - n\theta)/\theta^2$ is also a nonlinear function of θ with a single root $\widehat{\theta}_n = T_n/n$. ∎

Example 5.2.4: Let $X_1, \ldots, X_n$ be i.i.d. r.v.'s following Cauchy(θ) distributions, i.e., with density functions given by $f(x; \theta) = [\pi\{1 + (x - \theta)^2\}]^{-1}$. The likelihood function is

$$L_n(\theta) = \pi^{-n} \prod_{i=1}^{n}\{1 + (X_i - \theta)^2\}^{-1};$$

the estimating function

$$U_n(\theta) = 2\sum_{i=1}^{n}\left\{(X_i - \theta)/[1 + (X_i - \theta)^2]\right\}$$

behaves as a polynomial of degree $2n-1$ in θ, and, consequently, $L_n(\theta)$ may have several minima and maxima. In such a case, further analyses must be considered in order to determine the MLE $\widehat{\theta}_n$. ∎

Example 5.2.5: Let $X_1, \ldots, X_n$ be i.i.d. r.v.'s following the Unif$(0, \theta)$ distribution, i.e., with density function given by $f(x, \theta) = \theta^{-1} I_{[0,\theta]}(x)$. Here the likelihood function

$$L_n(\theta) = \theta^{-n} I_{[0,\theta]}(X_{n:1}, \ldots, X_{n:n}),$$

where $X_{n:1} < \cdots < X_{n:n}$ are the order statistics, is not differentiable, so that the estimating function does not exist. However, the likelihood function is clearly maximized at $\widehat{\theta}_n = X_{n:n}$, which is the MLE of θ. ∎

In cases where a sufficient statistic T_n (for θ) exists, we may write

$$L_n(\theta) = g(T_n \mid \theta)h(X_1, \ldots, X_n)$$

and it is clear that maximizing $L_n(\theta)$ is equivalent to maximizing $g(T_n \mid \theta)$ so that the MLE $\widehat{\theta}_n$ will be a function of T_n alone. Note that all the above examples with exception of Example 5.2.4 fall in this category.

Although in most of the examples considered so far, the MLE was obtained in explicit form, in general (viz., Example 5.2.4) we must rely on iterative methods to find a root of the likelihood equation. In cases where the loglikelihood is twice differentiable, the usual procedures for computing MLEs are based on the following Taylor expansion around some initial guess $\theta_n^{(0)}$:

$$
\begin{aligned}
0 &= \left. \frac{\partial}{\partial \theta} \log L_n(\theta) \right|_{\theta = \widehat{\theta}_n} \\
&= \left. \frac{\partial}{\partial \theta} \log L_n(\theta) \right|_{\theta = \theta_n^{(0)}} + (\widehat{\theta}_n - \theta_n^{(0)}) \left. \frac{\partial}{\partial \theta^2} \log L_n(\theta) \right|_{\theta = \theta_n^*}
\end{aligned}
$$

where θ_n^* lies between $\widehat{\theta}_n$ and $\theta_n^{(0)}$. Therefore, we have

$$\widehat{\theta}_n = \theta_n^{(0)} - \left\{ \left. \frac{\partial}{\partial \theta} \log L_n(\theta) \right|_{\theta = \theta_n^{(0)}} \right\} \left\{ \left. \frac{\partial^2}{\partial \theta^2} \log L_n(\theta) \right|_{\theta = \theta_n^*} \right\}^{-1}. \quad (5.2.1)$$

If we choose $\theta_n^{(0)}$ in a neighborhood of $\widehat{\theta}_n$ (i.e., if $\theta_n^{(0)}$ is based on some consistent estimator of θ) we may use (5.2.1) to consider a first step estimator:

$$\widehat{\theta}_n^{(1)} = \theta_n^{(0)} - \left\{ \left. \frac{\partial}{\partial \theta} \log L_n(\theta) \right|_{\theta = \theta_n^{(0)}} \right\} \left\{ \left. \frac{\partial^2}{\partial \theta^2} \log L_n(\theta) \right|_{\theta = \theta_n^{(0)}} \right\}^{-1}. \quad (5.2.2)$$

The well-known **Newton-Raphson method** consists of the repeated iteration of (5.2.2) with $\theta_n^{(0)}$ replaced by the value of $\widehat{\theta}_n^{(i)}$ obtained at the previous step. If, at each iteration, we replace $(\partial^2/\partial\theta^2)\log L_n(\theta)\big|_{\theta=\theta_n^{(i)}}$ by its expected value $\mathrm{E}\left\{(\partial^2/\partial\theta^2)\log L_n(\theta)\right\}\big|_{\theta=\theta_n^{(i)}}$, the procedure is known as the **Fisher-scoring method**.

The next theorem establishes conditions under which MLE are asymptotically normal. In this context, we assume that the d.f. F has a density function f with respect to the Lebesgue measure. The modifications for counting measures are fairly trivial: replace dx by $d\nu(x)$ in the integrals to follow.

Theorem 5.2.1: *Let* $X_1, \ldots, X_n$ *be i.i.d. r.v.'s with p.d.f.* $f(x;\theta)$, $\theta \in \Theta \subset \mathbb{R}$, *satisfying the following conditions:*

i) $(\partial/\partial\theta)f(x;\theta)$ *and* $(\partial^2/\partial\theta^2)f(x;\theta)$ *exist almost everywhere and are such that*

$$\left|\frac{\partial}{\partial\theta}f(x;\theta)\right| \le H_1(x) \quad and \quad \left|\frac{\partial^2}{\partial\theta^2}f(x;\theta)\right| \le H_2(x)$$

where $\int_{\mathbb{R}} H_j(x)dx < \infty$, $j = 1, 2$;

ii) $(\partial/\partial\theta)\log f(x;\theta)$ *and* $(\partial^2/\partial\theta^2)\log f(x;\theta)$ *exist almost everywhere and are such that*

(a) X_1 *has a finite* **Fisher information**, *i.e.*

$$\begin{aligned} 0 < I(\theta) &= \mathrm{E}\left\{(\partial/\partial\theta)\log f(X_1;\theta)\right\}^2 \\ &= \int_{\mathbb{R}} \left\{(\partial/\partial\theta)f(x;\theta)\right\}^2 \left\{f(x;\theta)\right\}^{-1} dx < \infty, \end{aligned}$$

(b) as $\delta \to 0$,

$$\mathrm{E}\{\sup_{\{h:|h|\le\delta\}} |(\partial^2/\partial\theta^2)\log f(X_1;\theta+h) - (\partial^2/\partial\theta^2)\log f(X_1;\theta)|\}$$
$$= \psi_\delta \to 0.$$

Then the MLE $\widehat{\theta}_n$ *of* θ *is such that* $\sqrt{n}(\widehat{\theta}_n - \theta) \xrightarrow{\mathcal{D}} N(0, I^{-1}(\theta))$.

Proof: First observe that

$$\begin{aligned} \mathrm{E}\left\{\frac{\partial}{\partial\theta}\log f(X_1;\theta)\right\} &= \int_{\mathbb{R}}\left\{\frac{\partial}{\partial\theta}\log f(x;\theta)\right\}f(x;\theta)dx \\ &= \int_{\mathbb{R}}\frac{\partial}{\partial\theta}f(x;\theta)dx \\ &= \frac{\partial}{\partial\theta}\int_{\mathbb{R}}f(x;\theta)dx = 0. \end{aligned} \qquad (5.2.3)$$

Note that the order of differentiation and integration in (5.2.3) may be interchanged since condition (i) allows a direct application of Lebesgue

Dominated Convergence Theorem (2.5.5). Next, observe that

$$
\begin{aligned}
E\left\{\frac{\partial^2}{\partial\theta^2}\log f(X_1;\theta)\right\} &= \int_R \frac{\partial}{\partial\theta}\left[\frac{\partial}{\partial\theta}f(x;\theta)\{f(x;\theta)\}^{-1}\right]f(x;\theta)dx \\
&= \int_R \left[\frac{\partial^2}{\partial\theta^2}f(x;\theta)\frac{1}{f(x;\theta)} - \left\{\frac{\partial}{\partial\theta}f(x;\theta)\frac{1}{f(x;\theta)}\right\}^2\right]f(x;\theta)dx \\
&= \int_R \frac{\partial^2}{\partial\theta^2}f(x;\theta)dx - I(\theta) \\
&= \int_R \frac{\partial}{\partial\theta}\left\{\frac{\partial}{\partial\theta}f(x;\theta)\right\}dx - I(\theta) \\
&= \frac{\partial}{\partial\theta}\int_R f(x;\theta)dx - I(\theta) \\
&= -I(\theta).
\end{aligned}
\tag{5.2.4}
$$

Recall that $U_n(\theta) = \sum_{i=1}^n (\partial/\partial\theta)\log f(X_i;\theta)$ and use (5.2.3) and condition (iia) in connection with the Central Limit Theorem 3.3.1 to see that

$$
n^{-1/2}U_n(\theta) \xrightarrow{\mathcal{D}} N(0, I(\theta)).
\tag{5.2.5}
$$

Also, let $V_n(\theta) = \sum_{i=1}^n (\partial^2/\partial\theta^2)\log f(X_i;\theta)$ and use (5.2.4) in connection with the Khintchine Weak Law of Large Numbers (Theorem 2.3.6) to conclude that

$$
n^{-1}V_n(\theta) \xrightarrow{P} -I(\theta).
\tag{5.2.6}
$$

Now, for $|u| \le K$, $0 < K < \infty$, write

$$
\begin{aligned}
\lambda_n(u) &= \log L_n(\theta + n^{-1/2}u) - \log L_n(\theta) \\
&= \sum_{i=1}^n \{\log f(X_i;\theta + n^{-1/2}u) - \log f(X_i;\theta)\}
\end{aligned}
$$

and consider a Taylor expansion of $\log f(X_i;\theta + n^{-1/2}u)$ around θ to see that

$$
\lambda_n(u) = u\frac{1}{\sqrt{n}}\sum_{i=1}^n \frac{\partial}{\partial\theta}\log f(X_i;\theta) + \frac{u^2}{2}\frac{1}{n}\sum_{i=1}^n \frac{\partial^2}{\partial\theta^2}\log f(X_i;\theta_n^*)
$$

where $\theta_n^* \in (\theta, \theta + n^{-1/2}u)$. Then, defining

$$
Z_n(u) = \frac{1}{n}\sum_{i=1}^n \left\{\frac{\partial^2}{\partial\theta^2}\log f(X_i;\theta_n^*) - \frac{\partial^2}{\partial\theta^2}\log f(X_i;\theta)\right\},
$$

we have

$$
\lambda_n(u) = u\frac{1}{\sqrt{n}}U_n + \frac{u^2}{2n}V_n + \frac{u^2}{2}Z_n(u).
\tag{5.2.7}
$$

Given $\delta > 0$, there exists $n_0 = n_0(\delta)$ such that $n^{-1/2}|u| \leq n^{-1/2}K < \delta$ for all $n \geq n_0$. Therefore, for sufficiently large n, we have

$$|Z_n(u)| \leq \frac{1}{n}\sum_{i=1}^{n} \sup_{\{h:|h|\leq|u|/\sqrt{n}\}} \left| \frac{\partial^2}{\partial\theta^2} \log f(X_i;\theta+h) - \frac{\partial^2}{\partial\theta^2} \log f(X_i;\theta) \right|$$

$$\leq \frac{1}{n}\sum_{i=1}^{n} \sup_{\{h:|h|\leq\delta\}} \left| \frac{\partial^2}{\partial\theta^2} \log f(X_i;\theta+h) - \frac{\partial^2}{\partial\theta^2} \log f(X_i;\theta) \right|.$$

$$(5.2.8)$$

By the Khintchine Strong Law of Large Numbers (Theorem 2.3.13) and condition (iib), it follows that, as $\delta \to 0$,

$$\frac{1}{n}\sum_{i=1}^{n} \sup_{\{h:|h|\leq\delta\}} \left| \frac{\partial^2}{\partial\theta^2} \log f(X_i;\theta+h) - \frac{\partial^2}{\partial\theta^2} \log f(X_i;\theta) \right| \xrightarrow{\text{a.s.}} \psi_\delta \to 0$$

and, therefore, from (5.2.8) we have that

$$\sup_{u:|u|\leq K} |Z_n(u)| \xrightarrow{\text{a.s.}} 0. \qquad (5.2.9)$$

Rewriting (5.2.7) as

$$\lambda_n(u) = u\frac{1}{\sqrt{n}}U_n - \frac{u^2}{2}I(\theta) + \frac{u^2}{2}\left\{\frac{1}{n}V_n(\theta) + I(\theta)\right\} + \frac{u^2}{2}Z_n(u)$$

and using (5.2.6) and (5.2.9), it follows that uniformly in $u \in [-K, K]$

$$\lambda_n(u) = u\frac{1}{\sqrt{n}}U_n - \frac{u^2}{2}I(\theta) + o_p(1). \qquad (5.2.10)$$

Disregarding the $o_p(1)$ term and maximizing $\lambda_n(u)$ with respect to u, we obtain

$$\hat{u} = \frac{U_n(\theta)}{\sqrt{n}I(\theta)} + o_p(1). \qquad (5.2.11)$$

From the definition of $\lambda_n(u)$ we conclude that $\hat{u}$ will also correspond closely to the maximum of $L_n(\theta + n^{-1/2}u)$ which is attained at the MLE $\hat{\theta}_n$ of θ. Therefore, we have

$$\hat{\theta}_n = \theta + n^{-1/2}\hat{u} + o_p(n^{-1/2}) = \theta + n^{-1}\left\{\frac{U_n(\theta)}{I(\theta)} + o_p(n^{1/2})\right\}$$

which implies that $\sqrt{n}(\hat{\theta}_n-\theta) = \{U_n(\theta)/\sqrt{n}I(\theta)\}+o_p(1)$. Using the Slutsky Theorem (3.4.2) we may conclude then $\sqrt{n}(\hat{\theta}_n - \theta) \xrightarrow{D} N(0, I^{-1}(\theta))$. ∎

This approach to prove the asymptotic normality of MLEs was developed by LeCam (1956), Hájek (1972) and Inagaki (1973). Their major contribution was to relax the assumption on the third derivative of $\log f(X_1;\theta)$

assumed in Cramér's (1946, Chap. 33) classical proof; it was essentially replaced by condition (iib) in the second derivative.

Note, also, that the consistency of MLEs follows directly from Theorems 5.2.1 and 3.2.5. Proofs of consistency under less restrictive conditions may be found in Cramér (1946, Chap. 33), Wald (1949) and Redner (1981), among others.

Example 5.2.6: Let $X_1, \ldots, X_n$ be i.i.d. r.v.'s following $N(\theta, 1)$ distributions; as we have seen in Example 5.2.1, the MLE of θ is $\overline{X}_n$. Since the corresponding density function is $f(x;\theta) = (2\pi)^{-1/2} \exp\{-(x-\theta)^2/2\}$, it follows that $(\partial/\partial\theta) \log f(x;\theta) = x - \theta$ and $(\partial^2/\partial\theta^2) \log f(x;\theta) = -1$ so that

$$ \mathrm{E}\left\{ \sup_{\{h:|h|\leq\delta\}} \left| \frac{\partial^2}{\partial\theta^2}\log f(X_1;\theta+h) - \frac{\partial^2}{\partial\theta^2}\log f(X_1;\theta) \right| \right\} = \psi_\delta = 0, $$

for every $\delta > 0$ and $I(\theta) = 1$; furthermore, $U_n(\theta) = \sum_{i=1}^{n}(X_i - \theta)$. Therefore, from Theorem 5.2.1 we have $\sqrt{n}(\overline{X}_n - \theta) \xrightarrow{D} N(0,1)$. ∎

Example 5.2.7: Let $X_1, \ldots, X_n$ be i.i.d. r.v.'s following a Cauchy(θ) distribution, i.e., with density functions given by $f(x;\theta) = [\pi\{1+(x-\theta)^2\}]^{-1}$, $x \in \mathbb{R}$, $\theta \in \mathbb{R}$. Thus,

$$ (\partial/\partial\theta) \log f(x;\theta) = 2(x-\theta)/\{1+(x-\theta)^2\} $$

and

$$ (\partial^2/\partial\theta^2) \log f(x;\theta) = -2\{1+(x-\theta)^2\}^{-1} + 4(x-\theta)^2\{1+(x-\theta)^2\}^{-2}. $$

Now, since $(\partial^2/\partial\theta^2) \log f(x;\theta)$ is a bounded and continuous function of x and θ, assumption (iib) of Theorem 5.2.1 is satisfied; also, $U_n(\theta) = 2\sum_{i=1}^{n}(X_i - \theta)/\{1+(X_i - \theta)^2\}$ and $I(\theta) = 2$. Therefore, we have, $\sqrt{n}(\hat{\theta}_n - \theta) \xrightarrow{D} N(0, 1/2)$. Note that in this case we have derived the asymptotic distribution of the MLE without having to compute it. ∎

Example 5.2.8: Consider the negative binomial law defined by (2.5.11). It is easy to show that $(\partial/\partial\pi) \log p(n \mid m, \pi)$ $[= m/\pi - (n-m)/(1-\pi)]$ and $(\partial^2/\partial\pi^2) \log p(n \mid m, \pi)$ satisfy all the regularity conditions of Theorem 5.2.1, and the MLE of π is $\hat{\pi} = m/n$, which is asymptotically normal. This conclusion can also be drawn from the Central Limit Theorem on $(n - m/\pi)/\sqrt{m}$ as is set to be verified in Exercise 5.1.2. Note that here we have a discrete distribution, so that the density $f(n;\pi)$ is with respect to a counting measure. ∎

In passing, we note that Theorem 5.2.1 may hold even if we do not start with i.i.d. random variables, provided we justify the use of some version of the Strong (Weak) Law of Large Numbers to show that $V_n(\theta) \to -I^*(\theta)$ almost surely or in probability [where $I^*(\theta)$ is the average information on θ] and some version of the Central Limit Theorem to show that $U_n(\theta)$ conveniently standardized converges in distribution to a normal random variable. We finally note that this theorem may be easily generalized to the multiparameter case:

Theorem 5.2.2: *Let* $X_1, \ldots, X_n$ *be i.i.d. r.v.'s with p.d.f.* $f(x; \boldsymbol{\theta})$, $x \in \mathbb{R}$, $\boldsymbol{\theta} \in \boldsymbol{\Theta} \subset \mathbb{R}^q$ *satisfying the following assumptions:*

  i) *for* $i, j = 1, \ldots, q$, $(\partial/\partial\theta_i)f(x, \boldsymbol{\theta})$ *and* $(\partial^2/\partial\theta_i\partial\theta_j)f(x; \boldsymbol{\theta})$ *exist almost everywhere and are such that*

$$|(\partial/\partial\theta_i)f(x; \boldsymbol{\theta})| \leq H_i(x)$$

  and

$$|(\partial^2/\partial\theta_i\partial\theta_j)f(x; \boldsymbol{\theta})| \leq G_{ij}(x)$$

  where $\int_{\mathbb{R}} H_i(x)dx < \infty$ *and* $\int_{\mathbb{R}} G_{ij}(x)dx < \infty$;

  ii) *for* $i, j = 1, \ldots, q$, $(\partial/\partial\theta_i)\log f(x; \boldsymbol{\theta})$ *and* $(\partial^2/\partial\theta_i\partial\theta_j)\log f(x; \boldsymbol{\theta})$ *exist almost everywhere and are such that:*

  (a) the **Fisher information matrix**,

$$\mathbf{I}(\boldsymbol{\theta}) = \mathrm{E}\left\{ \left[\frac{\partial}{\partial\boldsymbol{\theta}} \log f(X_1, \boldsymbol{\theta}) \right] \left[\frac{\partial}{\partial\boldsymbol{\theta}} \log f(X_1, \boldsymbol{\theta}) \right]^t \right\},$$

  is finite and positive definite;

  (b) as $\delta \to 0$, *we have*

$$\mathrm{E}_{\theta}\left\{ \sup_{\{h: \|h\| \leq \delta\}} \left\| \frac{\partial^2}{\partial\boldsymbol{\theta}\partial\boldsymbol{\theta}^t} \log f(X_1; \boldsymbol{\theta} + \mathbf{h}) - \frac{\partial^2}{\partial\boldsymbol{\theta}\partial\boldsymbol{\theta}^t} \log f(X_1; \boldsymbol{\theta}) \right\| \right\}$$
$$= \psi_\delta \to 0.$$

Then the MLE $\widehat{\boldsymbol{\theta}}$ *of* $\boldsymbol{\theta}$ *is such that* $\sqrt{n}(\widehat{\boldsymbol{\theta}} - \boldsymbol{\theta}) \xrightarrow{D} N(\mathbf{0}, [\mathbf{I}(\boldsymbol{\theta})]^{-1})$.

Proof: For $\|\mathbf{u}\| \leq K$, $0 < K < \infty$, let us define

$$\lambda_n(\mathbf{u}) = \sum_{i=1}^{n} \{\log f(X_i; \boldsymbol{\theta} + n^{-1/2}\mathbf{u}) - \log f(X_i, \boldsymbol{\theta})\}$$

and follow the same steps of Theorem 5.2.1 to show that

$$\lambda_n(\mathbf{u}) = n^{-1/2}[\mathbf{U}_n(\boldsymbol{\theta})]^t\mathbf{u} - \mathbf{u}^t\mathbf{I}(\boldsymbol{\theta})\mathbf{u}/2 + o_p(1)$$

where $\mathbf{U}_n(\boldsymbol{\theta}) = \sum_{i=1}^{n}(\partial/\partial\boldsymbol{\theta})\log f(X_1; \boldsymbol{\theta})$; maximizing $\lambda_n(\mathbf{u})$ with respect to $\mathbf{u}$, the result follows. ∎

Theorem 5.2.1. (or 5.2.2) provides only sufficient (but not necessary) regularity conditions relating to the asymptotic normality and efficiency of the MLE. There are notable cases where such regularity conditions may not all hold, but still the desired asymptotic results hold. Toward this, we consider the following.

Example 5.2.9 (CMRR procedure in Example 2.3.7): The MLE $\widehat{N}$ of N is given by (2.3.34) and the maximum of the likelihood function by (2.3.36). Since $N\ (=\theta)$ is integer valued and (2.3.35) relates to a discrete distribution (in r_2), conditions (i) and (ii) of Theorem 5.2.1 are not that appropriate. Nevertheless, in Exercise 3.3.5, we have noted that the asymptotic (as $N \to \infty$) normality of $N^{-1/2}(\widehat{N} - N)$ can be obtained by direct adaptation of the classical Central Limit Theorem (on r_2/n_2). ∎

In the concluding section, we consider some further remarks concerning the asymptotic behavior of the MLE in some nonregular cases.

5.3 Asymptotic properties of U-statistics and related estimators

The U-statistics have been introduced in Section 2.4 (via Example 2.4.7). We adopt the same notation as in (2.4.26)–(2.4.28). In contrast to the parametric case, here, a parameter $\theta\ [= \theta(F)]$ is formulated as a functional of the underlying d.f. F, so that as in (2.4.26) we write

$$\theta(F) = \int \cdots \int g(x_1, \ldots, x_m) dF(x_1) \cdots dF(x_m) \qquad (5.3.1)$$

where m is a positive integer [called the **degree** of $\theta(F)$] and $g(x_1, \ldots, x_m)$ is a symmetric function of its m arguments and is termed a **kernel** . We have already illustrated in Section 2.4 that the mean, variance, higher order (central) moments, etc., are all special cases of (5.3.1).Recall that by (5.3.1),

$$\theta(F) = \mathrm{E}_F g\{X_1, \ldots, X_m\}, \quad F \in \mathcal{F}, \qquad (5.3.2)$$

where $\mathcal{F}$ is a certain class of distribution functions, so that $\theta(F)$ is also termed an **estimable parameter** or a **regular functional** of F. As in (2.4.28), we define a U-statistic, U_n, by

$$U_n = \binom{n}{m}^{-1} \sum_{\{i \leq i_1 < \cdots < i_m \leq n\}} g(X_{i_1}, \ldots, X_{i_m}), \quad n \geq m. \qquad (5.3.3)$$

In (5.3.1) or (5.3.3), it is not necessary to assume that the d.f. F is defined on $\mathbb{R}$ (i.e., the X_i's are real-valued r.v.'s), and these definitions extend readily to the vector case (where F is defined on $\mathbb{R}^p$, for some $p \geq 1$). In Section 2.2 [viz., (2.2.10)], we have introduced the sample (empirical) d.f. F_n (see also Chapter 4), a natural estimator of the unknown F. Hence, it may be possible to estimate $\theta(F)$ by $\theta(F_n) = V_n$, say, which are usually termed V-statistics. Indeed, such functionals of the empirical d.f. [termed the von Mises (1947) functionals] are closely related to the U-statistic U_n. In this context, we note that

$$
\begin{aligned}
V_n &= \theta(F_n) = \int \cdots \int g(x_1, \ldots, x_m) dF_n(x_1) \cdots dF_n(x_m) \\
&= n^{-m} \sum_{i_1=1}^{n} \cdots \sum_{i_m=1}^{n} g(X_{i_1}, \ldots, X_{i_m}),
\end{aligned}
\tag{5.3.4}
$$

and in this setup, it is not necessary to assume that $n \geq m$. For $m = 1$, $U_n = V_n$, so that they are the same and estimate $\theta(F)$ unbiasedly. For $m \geq 2$, this equivalence is not generally true, and V_n may not be unbiased for $\theta(F)$. For example, for $m = 2$, by (5.3.3) and (5.3.4), we have

$$
\begin{aligned}
V_n &= n^{-2} \left\{ n(n-1)U_n + \sum_{i=1}^{n} g(X_i, X_i) \right\} \\
&= U_n + n^{-1} \left\{ \frac{1}{n} \sum_{i=1}^{n} [g(X_i, X_i) - U_n] \right\},
\end{aligned}
\tag{5.3.5}
$$

so that whenever $g(X_1, X_1)$ is not unbiased for $\theta(F)$, V_n fails to be so. However, this bias is generally $O(n^{-1})$, so that for large values of n, U_n and V_n share the same properties (see Exercise 5.3.1).

The U-statistics are unbiased estimators, symmetric in the sample observations, and they play a fundamental role in the theory of unbiased estimation. For any estimator $T_n = T(X_1, \ldots, X_n)$, not necessarily symmetric in $X_1, \ldots, X_n$, one can construct a symmetric version:

$$
U_n = (n!)^{-1} \sum_{\mathcal{P}_n} T(X_{i_1}, \ldots, X_{i_n}),
\tag{5.3.6}
$$

where $\mathcal{P}_n$ stands for the set of $n!$ permutations of the indices $\{i_1, \ldots, i_n\}$ over $(1, \ldots, n)$, so that $U_n = E\{T_n \mid \mathcal{P}_n\}$ and, therefore, U_n has a smaller (or at most equal) mean square error (or risk with a convex loss function) than T_n. Moreover, when the X_i are real-valued r.v.'s, an U-statistic U_n is a (symmetric) function of the order statistics, and, hence, under conditions permitting the **completeness** of the order statistics, U_n is the unique minimum variance unbiased estimator of $\theta(F)$ when F is allowed to belong to a large class $\mathcal{F}$.

Note that for $m \geq 2$, the summands in (5.3.3) are not all independent, so that the classical central limit theorems discussed in Chapter 3 may not be directly applicable for U-statistics. We have also seen in Section 2.4 that the U_n form a reverse martingale sequence, so that we may as well use the central limit theorems for reverse martingales to derive the corresponding asymptotic normality results. However, in this context, verification of the needed regularity conditions may require cumbrous manipulations. A simpler method, due to Hoeffding (1948), consists in projecting U_n into another statistic involving independent summands for which the classical central limit theorem works out well, and we shall provide an outline of this ingenious **(H-)projection method**. Let us denote by

$$g_c(x_1, \ldots, x_c) = \mathrm{E}_F g\{x_1, \ldots, x_c, X_{c+1}, \ldots, X_m\}, \qquad (5.3.7)$$

for $c = 0, 1, \ldots, m$. Note that $g_0 = \theta(F)$ and $g_m = g$. Then let

$$Y_n = \frac{m}{n} \sum_{i=1}^{n} [g_1(X_i) - \theta(F)]. \qquad (5.3.8)$$

Let us also denote by

$$\xi_c = \mathrm{E}_F g_c^2(X_1, \ldots, X_c) - \theta^2(F), \quad 0 \le c \le m; \quad \xi_0 = 0. \qquad (5.3.9)$$

Then, by the Lindeberg-Feller CLT (Theorem 3.3.3), as $n \to \infty$,

$$\sqrt{n} Y_n \xrightarrow{\mathcal{D}} N(0, m^2 \xi_1) \quad \text{whenever} \quad 0 < \xi_1 < \infty. \qquad (5.3.10)$$

Let us also note that for each $i \, (= 1, \ldots, n)$,

$$\mathrm{E}[U_n - \theta(F) \mid X_i]$$

$$= \binom{n}{m}^{-1} \sum_{\{1 \le i_1 < \cdots < i_m \le n\}} \mathrm{E}\left\{g(X_{i_1}, \ldots, X_{i_m}) - \theta(F) \mid X_i\right\}$$

$$= \binom{n}{m}^{-1} \left\{ \binom{n-1}{m-1} [g_1(X_i) - \theta(F)] + \binom{n-1}{m} [\theta(F) - \theta(F)] \right\}$$

$$= \frac{m}{n} [g_1(X_i) - \theta(F)]. \qquad (5.3.11)$$

Therefore,

$$\sum_{i=1}^{n} \mathrm{E}[U_n - \theta(F) \mid X_i] = Y_n \qquad (5.3.12)$$

which shows that Y_n is the projection of $U_n - \theta(F)$; also, (5.3.8) shows that Y_n has independent summands. Further, by (5.3.12),

$$\mathrm{E}\left\{[U_n - \theta(F)] Y_n\right\} = \frac{m}{n} \sum_{i=1}^{n} \mathrm{E}\left\{[U_n - \theta(F)][g_1(X_i) - \theta(F)]\right\}$$

$$= \frac{m}{n} \sum_{i=1}^{n} \mathrm{E}\left\{[g_1(X_i) - \theta(F)] \mathrm{E}([U_n - \theta(F)] \mid X_i)\right\}$$

$$= m^2 n^{-2} \sum_{i=1}^{n} \mathrm{E}\left\{[g_1(X_i) - \theta(F)]^2\right\}$$

$$= m^2 n^{-1} \mathrm{E}[g_1(X_1) - \theta(F)]^2$$

$$= m^2 n^{-1} \xi_1 = \mathrm{E}Y_n^2. \tag{5.3.13}$$

Thus,

$$n\mathrm{E}[U_n - \theta(F) - Y_n]^2 = n\mathrm{E}[U_n - \theta(F)]^2 - n\mathrm{E}Y_n^2, \tag{5.3.14}$$

where, by (5.3.3) and (5.3.7)–(5.3.9), we have

$$\mathrm{E}[U_n - \theta(F)]^2 = \binom{n}{m}^{-1} \sum_{c=1}^{m} \binom{m}{c}\binom{n-m}{m-c} \xi_c$$

$$= m^2 n^{-1} \xi_1 + O(n^{-2}), \tag{5.3.15}$$

so that, by (5.3.13)– (5.3.15), we obtain

$$n\mathrm{E}[U_n - \theta(F) - Y_n]^2 = O(n^{-1}), \tag{5.3.16}$$

and, hence, by the Chebyshev Inequality (2.2.31), it follows that

$$\sqrt{n}|U_n - \theta(F) - Y_n| \xrightarrow{\mathrm{P}} 0. \tag{5.3.17}$$

By (5.3.10), (5.3.17) and the Slutsky Theorem (3.4.2), we have

$$\sqrt{n}\left\{U_n - \theta(F)\right\} \xrightarrow{\mathcal{D}} N(0, m^2 \xi_1). \tag{5.3.18}$$

If, in addition [to $\mathrm{E}_F g^2(X_1, \ldots, X_m) < \infty$], we assume that

$$\mathrm{E}_F|g_F(X_{i_1}, \ldots, X_{i_m})| < \infty, \quad 1 \le i_1 \le i_2 \le \cdots \le i_m \le m, \tag{5.3.19}$$

by using the decomposition in (5.3.5) for $m = 2$ (or a similar one for $m \ge 2$), we obtain that

$$\mathrm{E}|U_n - V_n| = O(n^{-1}), \tag{5.3.20}$$

so that $\sqrt{n}|U_n - V_n| \xrightarrow{\mathrm{P}} 0$. Hence, under (5.3.19) and $\mathrm{E}_F g^2 < \infty$, we have

$$\sqrt{n}\left\{V_n - \theta(F)\right\} \xrightarrow{\mathcal{D}} N(0, m^2 \xi_1). \tag{5.3.21}$$

It may be remarked that (5.3.18), (5.3.20) and (5.3.21) reflect the asymptotic equivalence of U- and V-statistics. Therefore, for large sample sizes, they share the common properties (i.e., consistency, asymptotic normality, etc.). In many practical applications, it may be necessary to estimate ξ_1 from the sample [in a comparable way as in the estimation of $\sigma^2 = \mathrm{Var}(X)$ for the estimation of $\theta = \mathrm{E}X$]. Note that by (5.3.9),

$$\xi_1 = \mathrm{E}_F\{g(X_1, \ldots, X_m)g(X_m, \ldots, X_{2m-1})$$

$$-\mathrm{E}_F g(X_1, \ldots, X_m)g(X_{m+1}, \ldots, X_{2m})\} \tag{5.3.22}$$

and is, therefore, an estimable parameter of degree $2m$. However, the computation of the unbiased estimator of ξ_1 (i.e., the U-statistic corresponding to ξ_1), for $m \geq 2$, may be generally very cumbersome. This can be largely avoided by using a **jackknifing** method, which in this special case, reduces to the following one [considered earlier by Sen (1960)]. For each i $(= 1, \ldots, n)$, let

$$
U_{n(i)} = \binom{n-1}{m-1}^{-1} \sum_{S_{n,i}} g(X_i, X_{i_2}, \ldots, X_{i_m}),
\tag{5.3.23}
$$

where the summation $S_{n,i}$ extends over all $1 \leq i_2 < \cdots < i_m \leq n$ with $i_j \neq i$ for $j = 2, \ldots, m$. Then, note that by (5.3.3) and (5.3.23), $U_n = n^{-1} \sum_{i=1}^{n} U_{n(i)}$. Let then

$$
S_n^2 = \frac{1}{(n-1)} \sum_{i=1}^{n} \{U_{n(i)} - U_n\}^2.
\tag{5.3.24}
$$

It is easy to verify that S_n^2 is a (strongly) consistent estimator of ξ_1 (see Exercise 5.3.2). Thus, for large sample sizes, the standardized form

$$
m^{-1} S_n^{-1} n^{1/2} \{U_n - \theta(F)\}
\tag{5.3.25}
$$

can be used to test suitable hypotheses for $\theta(F)$ or to provide a confidence interval for $\theta(F)$. Exercise 5.3.3 is set to verify that for $m = 1$, S_n^2 in (5.3.24) reduces to the usual sample variance. By virtue of (5.3.20) and (5.3.21), in (5.3.25), U_n may as well be replaced by V_n. Intuitively speaking, in (5.3.4), writing $F_n = F + (F_n - F)$, we end up with 2^m terms which can be gathered into $(m+1)$ subsets, where the kth subset contains $\binom{m}{k}$ terms, for $k = 0, 1, \ldots, m$. This leads us to

$$
V_n - \theta(F) = \sum_{k=1}^{m} \binom{m}{k} V_n^{(k)},
\tag{5.3.26}
$$

where for $k = 1, \ldots, m$,

$$
V_n^{(k)} = \int \cdots \int g_k(x_1, \ldots, x_k) d\{F_n(x_1) - F(x_1)\} \cdots d\{F_n(x_k) - F(x_k)\}.
$$

A very similar decomposition holds for U_n:

$$
U_n - \theta(F) = \sum_{k=1}^{m} \binom{m}{k} U_n^{(k)}, \qquad \binom{m}{1} U_n^{(1)} = \binom{m}{1} V_n^{(1)} = Y_n; \tag{5.3.27}
$$

in the literature this is known as the **Hoeffding decomposition** of U-statistics. It may be remarked that $E\{U_n^{(k)}\} = 0$ and $E\{(U_n^{(k)})^2\} = O(n^{-k})$, $k \geq 1$, so that the successive terms in the decomposition in (5.3.27) are of stochastically smaller order of magnitude, and they are pairwise orthogonal too. Further, $\left\{ U_n, U_n^{(1)}, \ldots, U_n^{(m)}; n \geq m \right\}$ form a reverse martingale (vector) (see the proof in Section 2.4), so that asymptotic properties for such reverse martingales (such as a.s. convergence, asymptotic distribution, etc.) can be incorporated to study similar results for $\{U_n\}$ or $\{V_n\}$. Detailed study of these asymptotic properties is somewhat beyond the scope of the current treatise; we may, however, refer to Chapter 3 of Sen (1981) for some general accounts. The U-statistics have also been extended to multisample models where the parameter θ is viewed as a functional of several distribution functions. For example, in a two-sample model, with two d.f.'s F and G (say), we may write $\theta = \theta(F, G)$ as

$$\theta = \int \cdots \int g(x_1, \ldots, x_r; y_1, \ldots, y_s) dF(x_1) \cdots dF(x_r) dG(y_1) \cdots dG(y_s)$$
(5.3.28)

where $g(\cdot)$ is a kernel of degree (r, s) and $r \geq 1$, $s \geq 1$. For two samples $X_1, \ldots, X_{n_1}$, from F and $Y_1, \ldots, Y_{n_2}$ from G, we define the (generalized) U-statistic as

$$U_{n_1, n_2} = \binom{n_1}{r}^{-1} \binom{n_2}{s}^{-1} \sum g(X_{i_1}, \ldots, X_{i_r}; Y_{i_1}, \ldots, Y_{i_s}) \qquad (5.3.29)$$

where the summation extends over all $1 \leq i_1 < \cdots < i_r \leq n_1$ and $1 \leq j_1 < \cdots < j_s \leq n_2$. The Hoeffding projection and Hoeffding decomposition for U_{n_1, n_2} work out neatly, and the asymptotic theory of such generalized U-statistics has been developed on parallel lines; we again refer to Chapter 3 of Sen (1981) for some of these details.

In (5.3.1)–(5.3.2), we have conceived of a kernel $g(\cdot)$ of finite degree m (≥ 1) and defined $\theta(F)$ as an estimable parameter if it admits unbiased estimators for samples of sizes m or more. There are other situations where restriction to unbiased estimators may lead to a far less richer class. Also, there are some other problems, mostly arising in robust estimation theory, where $\theta(F)$ is well defined but it may not correspond to a kernel of finite degree. Nevertheless, $\theta(F)$ may be sufficiently "smooth" (in a functional sense), so that the sample or empirical d.f. F_n may be plugged in the functional $\theta(\cdot)$ to estimate $\theta(F)$ in a natural and convenient way. Specifically, one may have a simple Taylor's expansion of the form

$$
\begin{aligned}
T_n &= \theta(F_n) = \theta\left\{ F + (F_n - F) \right\} \\
&= \theta(F) + \theta'_F (F_n - F) + \mathrm{Rem}\left\{ F_n - F; \theta(\cdot) \right\}, \qquad (5.3.30)
\end{aligned}
$$

where θ'_F stands for the derivative (interpreted suitably) and $\mathrm{Rem}(\cdot)$ for the remainder term in this (first order) expansion. It is possible to write the (linear) functional θ'_F as

$$\theta'_F(F_n - F) \;=\; \int \theta^{(1)}(F;x)d\{F_n(x) - F(x)\}$$

$$= \; n^{-1}\sum_{i=1}^{n} \theta^{(1)}(F;X_i), \qquad (5.3.31)$$

where $\theta^{(1)}(F;x)$ is termed the **influence function** (at F) and it is so normalized that $\int \theta^{(1)}(F;x)dF(x) = 0$, for all $F \in \mathcal{F}$. The usual laws of large numbers, probability inequalities and central limit theorems, considered in Chapters 2 and 3, are all adoptable for (5.3.31) (as it involves independent summands), and hence, the crux of the problem is to show that the remainder term in (5.3.30) is negligible (up to the desired extent). This naturally depends on the behavior of $F_n - F$ as well as on the "differentiability" of $\theta(\cdot)$ at F. A precise formulation of such a "differentiability" condition is somewhat beyond the scope of our text. A popular formulation is based on the so-called **Hadamard** or **compact** differentiability of $\theta(\cdot)$ which yields that

$$\mathrm{Rem}\,\{F_n - F;\; \theta(\cdot)\} = o(\|F_n - F\|), \qquad (5.3.32)$$

so that using the results of Chapter 4 on the Kolmogorov norm $\|F_n - F\|$, we obtain that (5.3.32) is $o_p(n^{-1/2})$, and this suffices for the asymptotic normality of $n^{1/2}\{T_n - \theta(F)\}$. For deeper results, one may need to have a "second order" expansion of $\theta(F_n)$, which may, in turn, require more stringent conditions on $\theta(\cdot)$. Fernholz (1983) is a good source to explore these technicalities. We consider here a few simple illustrative examples.

Example 5.3.1 (Rank weighted mean) [Sen (1964)]: Let $X_{n:1} \le \cdots \le X_{n:n}$ be the order statistics in a sample of size n from a distribution F with mean θ. For every $k \ge 1$, define

$$T_{n,k} = \binom{n}{2k+1}^{-1} \sum_{i=1}^{n} \binom{i-1}{k}\binom{n-i}{k} X_{n:i}, \qquad n \ge 2k+1. \quad (5.3.33)$$

Note that $T_{n,0}\ (= \overline{X}_n)$ is the sample mean and $T_{n,[(n+1)/2]}$ is the sample median. On the ground of robustness, often $T_{n,1}$ or $T_{n,2}$ are preferred to $T_{n,0}$. For every $k\ (\ge 0)$, $T_{n,k}$ is a linear combination of order statistics, and the population counterpart of $T_{n,k}$ is

$$\theta_k(F) \;=\; \frac{\Gamma(2k+2)}{\{\Gamma(k+1)\}^2} \int_{-\infty}^{\infty} \{F(x)[1-F(x)]\}^k\, x\,dF(x)$$

$$= \; \frac{\Gamma(2k+2)}{\{\Gamma(k+1)\}^2} \int_{0}^{1} \{u(1-u)\}^k F^{-1}(u)du, \qquad (5.3.34)$$

where $F^{-1}(u) = \inf\{x : F(x) \ge u\}$, $0 \le u \le 1$. Thus, whenever F is symmetric about θ, we have $\theta_k = \theta$, for all $k \geq 0$. For $k = 0$, the results of Chapters 2 and 3 apply directly, and, hence, we shall consider only the case of $k \geq 1$. Note that for $T_{n,0} (= \overline{X}_n)$, one needs to assume that $\sigma^2 = V(X) < \infty$, so that the asymptotic normality of $n^{1/2}(T_{n,0} - \theta)$ holds. However, for $k \geq 1$, less stringent regularity conditions suffice. Note that whenever $E_F|X|^r < \infty$ for some $r > 0$, $|x|^r F(x)\{1 - F(x)\}$ is bounded for all $x \in \mathbb{R}$ and it converges to zero as $x \to \pm\infty$. Thus, for $k \geq 1$, $|x|F(x)[1 - F(x)]^k = \{|x|^{1/k} F(x)[1 - F(x)]\}^k$ is bounded (and smooth) whenever $E_F|X|^r < \infty$ for some $r \geq 1/k$. This explains why, for $k \geq 1$, $T_{n,k}$ is more robust than $\overline{X}_n$. Thus, whenever $F(u)$ is absolutely continuous and strictly monotone, we have $\{u(1-n)\}^k F^{-1}(u) = J(u)$, say, a bounded and continuous function on $[0, 1]$, provided $E_F|X|^{1/k} < \infty$. As such, we can proceed as in (5.3.30)–(5.3.31) and verify the details. Interestingly enough, $T_{n,k}$ may also be expressed as a U-statistic. Let

$$g(X_1, \ldots, X_{2k+1}) = \text{med}(X_1, \ldots, X_{2k+1}). \qquad (5.3.35)$$

Then, it is easy to verify (see Exercise 5.3.4) that

$$T_{n,k} = \binom{n}{2k+1}^{-1} \sum_{\{1 \leq i_1 < \cdots < i_{2k+1} \leq n\}} g(X_{i_1}, \ldots, X_{i_{2k+1}}). \qquad (5.3.36)$$

As such, the asymptotic properties of U-statistics are all transmittable to $T_{n,k}$, for any fixed $k (\geq 0)$. On the other hand, if k= k_n is made to depend on n, then the kernel in (5.3.35) depends on n, and, hence, we may need some additional regularity conditions. These may not be necessary if we use the functional approach in (5.3.30). ∎

Example 5.3.2 (Trimmed mean): For a positive k $(< n/2)$, we define as in (4.6.5) a trimmed mean by

$$T_{n,k}^* = (n - 2k)^{-1} \sum_{i=k+1}^{n-k} X_{n:i}; \qquad (5.3.37)$$

for $k = 0, T_{n,0}^*$ reduces to $\overline{X}_n$. Suppose now that we have $k \sim n\alpha$ for some $0 < \alpha < 1/2$ (usually α is small). Then if we let

$$J_\alpha(u) = \begin{cases} 0, & 0 \leq u < \alpha \\ (1 - 2\alpha)^{-1}, & \alpha \leq u \leq 1 - \alpha \\ 0, & 1 - \alpha < u \leq 1, \end{cases} \qquad (5.3.38)$$

the population counterpart of $T_{n,k}^*$ in (5.3.37) is

$$\theta_k^* = \theta_\alpha^* = \int_0^1 J_\alpha(u) F^{-1}(u) du, \qquad (5.3.39)$$

where $F^{-1}(u)$ is bounded for $\alpha \leq u \leq 1 - \alpha$ and $J(u)$ is a flat line on the same interval. Here also when F is symmetric, $\theta_\alpha = \theta$, for all $\alpha \in (0, 1/2)$. We may write

$$T^*_{n,k} = \int_0^1 J_{k/n} \{F_n(x)\} \, x \, dF_n(x),$$

and, hence, proceed as in (5.3.30)–(5.3.31) to verify the necessary regularity conditions for the asymptotic normality of $n^{1/2}(T^*_{n,k} - \theta)$.   ∎

Example 5.3.3 (Winsorized mean): As in (4.6.7), it is defined as

$$W_{n,k} = n^{-1} \left\{ k(X_{n:k} + X_{n:n-k+1}) + \sum_{i=k+1}^{n-k} X_{n:i} \right\}, \qquad (5.3.40)$$

where k is ≥ 0. For $k = 0$, $W_{n,0}$ reduces to the mean $\overline{X}_n$. Note that by (5.3.37) and (5.3.40),

$$W_{n,k} = kn^{-1}(X_{n:k} + X_{n:n-k+1}) + (1 - 2kn^{-1})T^*_{n,k}, \qquad (5.3.41)$$

so that the asymptotic representation for sample quantiles studied in Chapter 4 can be combined with that of $T^*_{n,k}$ to establish parallel results.   ∎

Toward this, we consider the following:

Example 5.3.4 [Sample quantile (median)]: For a given p, $0 < p < 1$, the sample p-quantile is defined as in (4.2.11) by

$$\widehat{\xi}_{p,n} = X_{n:k_n} \quad \text{where} \quad k_n \sim np. \qquad (5.3.42)$$

Typically, we take $k_n = \min\{k : k/n \geq p\}$, so that we have

$$\widehat{\xi}_{p,n} = \inf\{n : F_n(x) \geq p\} = F_n^{-1}(p), \qquad (5.3.43)$$

where F_n is the sample d.f. Since F_n is a step function (with jumps at each $X_{n:i}$), a direct Taylor's expansion of $F_n^{-1}(p)$ around $F^{-1}(p) = \xi_p$ may not be feasible, even if F is continuous at ξ_p. However, by virtue of the results to be presented in Chapter 8, we have, for $n \to \infty$,

$$\{\sqrt{n}[F_n(F^{-1}(t)) - t], \quad 0 \leq t \leq 1\} \xrightarrow{\mathcal{D}} \quad \text{a Brownian bridge}, \qquad (5.3.44)$$

so that invoking the tightness part of this weak convergence result, we obtain that for $\delta \, (> 0)$ sufficiently small, as $n \to \infty$,

$$\sup\{\sqrt{n}|F_n(F^{-1}(t)) - t - F_n(\xi_p) + p| : |t - p| \leq \delta\} \xrightarrow{P} 0. \qquad (5.3.45)$$

Further, by the consistency results studied in Chapter 4, $F(X_{n:k_n}) \xrightarrow{\text{a.s.}} p$. Thus, noting that

$$F_n(X_{n:k_n}) = n^{-1}k_n = p + o(n^{-1/2}),$$

$$F(\xi_p) = p, \; F_n(\xi_p) = n^{-1} \sum_{i=1}^{n} I_{\{X_i \leq \xi_p\}}$$

and assuming that in a neighborhood of ξ_p, F admits a finite and positive density $f(\cdot)$, we have from (5.3.45),

$$F(X_{n:k_n}) - F_n(X_{n:k_n}) = -\{F_n(\xi_p) - p\} + o_p\left(n^{-1/2}\right)$$

which is equivalent to

$$\begin{aligned}
(X_{n:k_n} - \xi_p) &= -\{f(\xi_p)\}^{-1}\{F_n(\xi_p) - p\} + o_p\left(n^{-1/2}\right) \\
&= -\frac{1}{nf(\xi_p)} \sum_{i=1}^{n}\{I_{\{X_i \leq \xi_p\}} - p\} + o_p\left(n^{-1/2}\right),
\end{aligned}$$

$$(5.3.46)$$

so that (5.3.30)–(5.3.31) hold with the influence function

$$IC(x; F) = -\{f(\xi_p)\}^{-1}\{I_{\{X \leq \xi_p\}} - p\}, \quad -\infty < x < \infty.$$

Exercise 5.3.5 is set to use (5.3.46) for $X_{n:k}$ and $X_{n:n-k+1}$ in (5.3.40) and to work out a similar representation for $W_{n,k}$, when $k \sim n\alpha$, for some $0 < \alpha \leq 1/2$. ∎

5.4 Asymptotic behavior of other classes of estimators

The method of Maximum Likelihood is, perhaps, the most frequently used method for parametric estimation when the functional form of the under-lying distribution is known. In this section we deal with the asymptotic properties of some alternative estimation procedures which are usually employed in situations where only limited knowledge about the parent distribution is available. First we consider the so-called Method of Moments estimators (MME).

Let $X_1, \ldots, X_n$ be i.i.d. r.v.'s with density function $f(x; \boldsymbol{\theta})$, $\boldsymbol{\theta} \in \boldsymbol{\Theta} \subset \mathbb{R}^q$, $q \geq 1$, $x \in \mathbb{R}$, and assume that $EX_1^k = \mu_k' = h_k(\boldsymbol{\theta}) < \infty$, $k = 1, \ldots, q$; also let $m_{kn}' = n^{-1}\sum_{i=1}^{n} X_i^k$, $k = 1, \ldots, q$. The MME $\tilde{\boldsymbol{\theta}}_n$ of $\boldsymbol{\theta}$ is a solution (in $\boldsymbol{\theta}$) to the equations $h_k(\boldsymbol{\theta}) = m_{kn}'$, $k = 1, \ldots, q$. The following theorem establishes conditions under which the asymptotic distribution of $\tilde{\boldsymbol{\theta}}_n$ may be obtained.

Theorem 5.4.1: *Under the setup described above, assume that $\mu_k' < \infty$, $k = 1, \ldots, 2q$. Also let $\mathbf{h}(\boldsymbol{\theta}) = [h_1(\boldsymbol{\theta}), \ldots, h_q(\boldsymbol{\theta})]^t$ and $\mathbf{H}(\boldsymbol{\theta}) = (\partial/\partial\boldsymbol{\theta})\mathbf{h}(\boldsymbol{\theta})$ be such that the rank of $\mathbf{H}(\boldsymbol{\theta}) = q$ and its elements, $H_{ij}(\boldsymbol{\theta}) = (\partial/\partial\theta_j)h_i(\boldsymbol{\theta})$, $i, j = 1, \ldots, q$, are continuous in $\boldsymbol{\theta}$. Then,*

$$\sqrt{n}(\tilde{\boldsymbol{\theta}}_n - \boldsymbol{\theta}) \xrightarrow{\mathcal{D}} N_q(\mathbf{0}, [\mathbf{H}(\boldsymbol{\theta})]^{-1}\boldsymbol{\Sigma}\{[\mathbf{H}(\boldsymbol{\theta})]^{-1}\}^t)$$

with $\boldsymbol{\Sigma}$ being a $q \times q$ matrix with (j,k)th element $\mu'_{j+k} - \mu'_j \mu'_k$.

Proof. Consider the Taylor expansion

$$\mathbf{h}(\boldsymbol{\theta} + n^{-1/2}\mathbf{u}) = \mathbf{h}(\boldsymbol{\theta}) \quad + \quad n^{-1/2}\mathbf{H}(\boldsymbol{\theta})\mathbf{u}$$
$$+ \quad n^{-1/2}\left\{\mathbf{H}(\boldsymbol{\theta}^*) - \mathbf{H}(\boldsymbol{\theta})\right\}\mathbf{u} \qquad (5.4.1)$$

where $\boldsymbol{\theta}^* = \boldsymbol{\theta} + n^{-1/2}\gamma\mathbf{u}$, $0 \le \gamma \le 1$. In view of the assumption on the continuity of $\mathbf{H}(\boldsymbol{\theta})$, (5.4.1) may be re-expressed as

$$\sqrt{n}\left\{\mathbf{h}(\boldsymbol{\theta} + n^{-1/2}\mathbf{u}) - \mathbf{h}(\boldsymbol{\theta})\right\} = \mathbf{H}(\boldsymbol{\theta})\mathbf{u} + o(1). \qquad (5.4.2)$$

Letting $\mathbf{u} = \sqrt{n}(\widetilde{\boldsymbol{\theta}}_n - \boldsymbol{\theta})$, we get $\mathbf{h}(\boldsymbol{\theta} + n^{-1/2}\mathbf{u}) = \mathbf{h}(\widetilde{\boldsymbol{\theta}}_n) = \mathbf{m}_n$. Thus, (5.4.2) may be written as

$$\sqrt{n}\left\{\mathbf{m}_n - \mathbf{h}(\boldsymbol{\theta})\right\} = \mathbf{H}(\boldsymbol{\theta})\sqrt{n}(\widetilde{\boldsymbol{\theta}}_n - \boldsymbol{\theta}) + o_p(1). \qquad (5.4.3)$$

Now let $\boldsymbol{\lambda} \in \mathbb{R}^q$, $\boldsymbol{\lambda} \ne \mathbf{0}$, be an arbitrary, but fixed vector, and set

$$\sqrt{n}\boldsymbol{\lambda}^t\left\{\mathbf{m}_n - \mathbf{h}(\boldsymbol{\theta})\right\}$$
$$= \sqrt{n}\left\{(\lambda_1 m'_{1n} + \cdots + \lambda_q m'_{qn}) - (\lambda_1\mu'_1 + cldots + \lambda_q\mu'_q)\right\}$$
$$= \frac{1}{\sqrt{n}}\sum_{i=1}^n\left\{\sum_{j=1}^q \lambda_j(X_i^j - EX_i^j)\right\} = \frac{1}{\sqrt{n}}\sum_{i=1}^n U_i$$

where $U_i = \sum_{i=1}^q \lambda_j(X_i^j - EX_i^j)$ is such that $EU_i = 0$ and

$$EU_i^2 = \sum_{j=1}^q\sum_{k=1}^q \lambda_j\lambda_k E(X_i^j - \mu'_j)(X_i^k - \mu'_k)$$
$$= \sum_{j=1}^q\sum_{k=1}^q \lambda_j\lambda_k\left\{\mu'_{j+k} - \mu'_j\mu'_k\right\} = \boldsymbol{\lambda}^t\boldsymbol{\Sigma}\boldsymbol{\lambda} < \infty.$$

Since the U_i's are i.i.d. r.v.'s, it follows from the Central Limit Theorem 3.3.1 that

$$n^{-1/2}\sum_{i=1}^n U_i = \sqrt{n}\boldsymbol{\lambda}^t\left\{\mathbf{m}_n - \mathbf{h}(\boldsymbol{\theta})\right\} \xrightarrow{\mathcal{D}} N(0, \boldsymbol{\lambda}^t\boldsymbol{\Sigma}\boldsymbol{\lambda});$$

using the Cramér-Wold Theorem (3.2.5) we may conclude that

$$\sqrt{n}\left\{\mathbf{m}_n - \mathbf{h}(\boldsymbol{\theta})\right\} \xrightarrow{\mathcal{D}} N_q(\mathbf{0}, \boldsymbol{\Sigma}).$$

Finally, from (5.4.3), we get

$$\sqrt{n}(\widetilde{\boldsymbol{\theta}}_n - \boldsymbol{\theta}) \xrightarrow{\mathcal{D}} N_q\left\{\mathbf{0}, [\mathbf{H}^{-1}(\boldsymbol{\theta})]\boldsymbol{\Sigma}[\mathbf{H}^{-1}(\boldsymbol{\theta})]^t\right\}.$$

∎

Example 5.4.1: If $X_1, \ldots, X_n$ are i.i.d. r.v.'s following a $N(\mu, \sigma^2)$ distribution, we have

$$\mathbf{m}_n = \frac{1}{n} \left(\sum_{i=1}^{n} X_i, \sum_{i=1}^{n} X_i^2 \right)^t, \quad \mathbf{h}(\boldsymbol{\theta}) = (\mu, \mu^2 + \sigma^2)^t$$

and

$$\mathbf{H}(\boldsymbol{\theta}) = \begin{pmatrix} 1 & 0 \\ 2\mu & 1 \end{pmatrix};$$

since $\mu_3' = EX_1^3 = \mu^3 + 3\mu\sigma^2$ and $\mu_4' = EX_1^4 = \mu^4 + 3\sigma^4 + 6\mu^2\sigma^2$ a direct application of Theorem 5.4.1 leads us to the conclusion that

$$n^{-1/2} \left\{ \left(\sum_{i=1}^{n} X_i, \sum_{i=1}^{n} X_i^2 - n^{-1}(\sum_{i=1}^{n} X_i)^2 \right) - (\mu, \sigma^2) \right\}^t \xrightarrow{D} N_2(\mathbf{0}, \boldsymbol{\Gamma})$$

with

$$\boldsymbol{\Gamma} = \begin{pmatrix} \sigma^2 & 0 \\ 0 & \sigma^4 \end{pmatrix}.$$

∎

Observe that the method described above may be used to obtain the asymptotic distribution of the statistic $\mathbf{T}_n = n^{-1} \sum_{i=1}^{n} \mathbf{g}(X_i)$ where $\mathbf{g}(x) = \{g_1(x), \ldots, g_q(x)\}^t$ is a vector-valued function, by making $\mathbf{h}(\boldsymbol{\theta}) = E\mathbf{T}_n$.

Example 5.4.2: Let $X_1, \ldots, X_n$ be i.i.d. r.v.'s following a $N(\mu, \sigma^2)$ distribution; also let $\mathbf{T}_n = n^{-1} \left\{ \sum_{i=1}^{n} X_i, (n/(n-1)) \sum_{i=1}^{n}(X_i - \overline{X}_n)^2 \right\}^t$ and $\mathbf{h}(\boldsymbol{\theta}) = E\mathbf{T}_n = (\mu, \sigma^2)^t$. Then $\mathbf{H}(\boldsymbol{\theta}) = \mathbf{I}_2$ and

$$\sqrt{n} \left\{ (\overline{X}_n, S_n^2) - (\mu, \sigma) \right\}^t \xrightarrow{D} N_2 \{\mathbf{0}, \boldsymbol{\Gamma}\}$$

with

$$\boldsymbol{\Gamma} = \begin{pmatrix} \sigma^2 & 0 \\ 0 & 2\sigma^4 \end{pmatrix}.$$

Many methods of estimation of a parameter θ may be formulated in terms of the minimization (with respect to θ) of expressions of the type $\sum_{i=1}^{n} \rho(X_i; \theta)$ where ρ is some arbitrary function. When the score function $\psi(x; \theta) = (\partial/\partial\theta)\rho(x; \theta)$ exists, such methods reduce to solving implicit equations of the type

$$\sum_{i=1}^{n} \psi(X_i; \theta) = 0. \tag{5.4.4}$$

This is generally the case with MLEs, for which $\rho(x; \theta) = -\log f(x; \theta)$ with a score function $\psi(x; \theta) = \{-(\partial/\partial\theta)f(x; \theta)\}/f(x; \theta)$, where $f(x; \theta)$ represents the underlying density function, or with LSE's, for which we have $\rho(x;\theta) = \{x - g(\theta)\}^2$ and $\psi(x;\theta) = -2\{x - g(\theta)\}\, g'(\theta)$ where $g(\theta)$ is some convenient function.

Estimators defined by implicit equations like (5.4.4) are generally called **M-estimators** (Maximum-likelihood type) and are of special interest for location problems, i.e., $\psi(x;\theta) = \psi(x-\theta)$, since they may be easily defined in a way to produce **robust** estimates of θ, i.e., estimates that have stable statistical properties under fairly general specifications for the underlying probability model. To clarify this issue we bring into perspective the so-called **influence function**, which loosely speaking, measures the effect of a single observation on the performance of an estimator. For M-estimators of location, the influence function is proportional to the **score function** ψ; thus, a single outlier (i.e., an observation which is distant from the "bulk" of the data) may severely affect M-estimators obtained via the use of unbounded ψ.

In Figure 5.4.1 we depict the score functions corresponding to $\rho(x) = -\log f(x)$ for:

i) a normal distribution, where, $\psi_1(x) = x$, $x \in \mathbb{R}$,

ii) a double-exponential distribution, where, $\psi_2(x) = \text{sign}x$, $x \in \mathbb{R}$,

iii) a logistic distribution, where $\psi_3(x) = (e^x - 1)/(e^x + 1)$, $x \in \mathbb{R}$.

The fact that ψ_1 is unbounded indicates that a single outlier may have a disastrous effect on the sample mean (the M-estimator based on ψ_1); on the other hand, M-estimators based on either ψ_2 or ψ_3 are not affected by outliers although they do not preserve the nice linear influence provided by ψ_1. With this picture in mind, Huber (1964) proposed the following score function, which incorporates the desirable features of ψ_1, ψ_2 and ψ_3.

$$\psi(t) = \begin{cases} t, & |t| \leq k \\ k\,\text{sign}(t), & |t| > k \end{cases} \qquad (5.4.5)$$

where k is a positive constant; this function corresponds to

$$\rho(t) = \begin{cases} \frac{1}{2}t^2, & |t| \leq k \\ k|t| - \frac{1}{2}k^2, & |t| > k. \end{cases}$$

More specifically, the Huber score function (5.4.5) assigns a limited influence for observations in the tails of the distribution (i.e., for $|x| > k$) like ψ_2 and linear for the central portion (i.e., for $|x| \leq k$), like ψ_1, while maintaining some similarity to ψ_3.

In general, equations like (5.4.4) may have many roots, and they have to be examined in order to verify which of them corresponds to the required minimum of $\sum_{i=1}^n \rho(X_i,\theta)$. In many location problems, however, the estimating function $M_n(\theta) = \sum_{i=1}^n \psi(X_i - \theta)$ is non-increasing in θ and even in the presence of multiple roots, the corresponding M-estimator may be easily defined, i.e., as any value $\widehat{\theta}_n \in [\widehat{\theta}_{n_1}, \widehat{\theta}_{n_2}]$ where $\widehat{\theta}_{n_1} =$

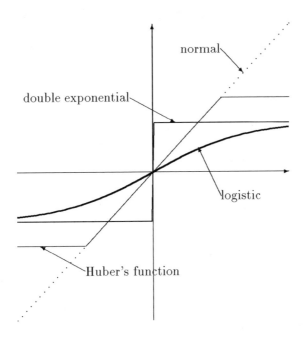

Figure 5.4.1. *Score functions for the normal, double exponential and logistic distributions and Huber's score function*

$\sup \{\theta : M_n(\theta) > 0\}$ and $\widehat{\theta}_{n_2} = \inf \{\theta : M_n(\theta) < 0\}$. For the sake of simplicity, we will restrict ourselves to cases where $M_n(\theta)$ has a single root $\widehat{\theta}_n$. The reader is referred to Huber (1981) or to Serfling (1980) for more general situations.

The following theorem illustrates a useful technique for proving the consistency and asymptotic normality of M-estimators.

Theorem 5.4.2 : *Let $X_1, \ldots, X_n$ be i.i.d. r.v.'s with d.f. F and $\psi(x - \theta)$ be nonincreasing in θ. Also let θ_0 be an isolated root of*

$$M(\theta) = \int \psi(x - \theta) dF(x) = 0$$

and assume that

i) in a neighborhood of θ, $M'(\theta_0) \neq 0$,

ii) $\int \psi^2(x - \theta) dF(x) < \infty$ and is continuous at θ_0.

Then, if $\widehat{\theta}_n$ is a solution to the estimating equation

$$M_n(\theta) = \sum_{i=1}^{n} \psi(X_i - \theta) = 0,$$

it follows that:

a) $\widehat{\theta}_n$ is consistent for θ_0;

b) $\sqrt{n}(\widehat{\theta}_n - \theta_0) \xrightarrow{D} N(0, \sigma^2(\theta_0)/[M'(\theta_0)]^2)$ with $\sigma^2(\theta_0) = \mathrm{Var}\psi(X_1 - \theta_0)$.

Proof: First note that since $\psi(x - \theta)$ is nonincreasing in θ, so are $M(\theta)$ and $M_n(\theta)$; thus, given $\epsilon > 0$, we have

$$
\begin{aligned}
P\left\{\widehat{\theta}_n - \theta_0 > \epsilon\right\} &\leq P\left\{M_n(\theta_0 + \epsilon) \geq 0\right\} \\
&= P\left\{\frac{1}{n}\sum_{i=1}^{n}\psi(X_i - \theta_0 - \epsilon) \geq 0\right\} \\
&= P\left\{\frac{1}{n}\sum_{i=1}^{n}[\psi(X_i - \theta_0 - \epsilon) - \mathrm{E}\psi(X_1 - \theta_0 - \epsilon)] \right. \\
&\qquad\qquad \left. \geq -\mathrm{E}\psi(X_1 - \theta_0 - \epsilon)\right\} \\
&= P\left\{\frac{1}{n}\sum_{i=1}^{n}Z_i \geq -\mathrm{E}\psi(X_1 - \theta_0 - \epsilon)\right\} \qquad (5.4.6)
\end{aligned}
$$

where $Z_i = \psi(X_i - \theta_0 - \epsilon) - \mathrm{E}\psi(X_1 - \theta_0 - \epsilon)$, $i = 1, \ldots, n$ are i.i.d. r.v.'s with zero means. Now consider a first order Taylor expansion of $M(\theta_0 + \epsilon) = \mathrm{E}\psi(X_1 - \theta_0 - \epsilon)$ around θ_0:

$$M(\theta_0 + \varepsilon) = M(\theta_0) + \varepsilon M'(\theta_0 + \varepsilon h)$$

where $0 \leq h \leq 1$. Since M is nonincreasing and $M(\theta_0) = 0$ we get $M(\theta_0 + \varepsilon) = \varepsilon\gamma(\varepsilon) < 0$. Therefore, given $\varepsilon > 0$, there exists $\delta = \delta(\varepsilon) > 0$ such that (5.4.6) may be expressed as

$$P\left\{\widehat{\theta}_n - \theta_0 > \varepsilon\right\} \leq P\left\{\frac{1}{n}\sum_{i=1}^{n}Z_i \geq \delta\right\}$$

which converges to zero as $n \to \infty$ by the Khintchine Weak Law of Large Numbers (Theorem 2.3.6). Repeating a similar argument for the term $P\{\widehat{\theta}_n - \theta_0 < -\varepsilon\}$ it is possible to show that $P\{|\widehat{\theta}_n - \theta_0| > \varepsilon\} \to 0$ as $n \to \infty$, completing the first part of the proof.

To proceed with the second part, note that for all $t \in \mathbb{R}$, we may write

$$
P\left\{\sqrt{n}(\widehat{\theta}_n - \theta_0) \leq t\right\} = P\left\{\widehat{\theta}_n \leq \theta_0 + \frac{t}{\sqrt{n}}\right\}
$$

$$
= P\left\{M_n\left(\theta_0 + \frac{t}{\sqrt{n}}\right) < 0\right\}
$$

$$
= P\left\{\frac{1}{\sqrt{n}}\sum_{i=1}^{n}\left[\frac{\psi(X_i - \theta_0 - t/\sqrt{n}) - \mathrm{E}\psi(X_1 - \theta_0 - t/\sqrt{n})}{\sqrt{\mathrm{Var}\psi(X_1 - \theta_0 - t/\sqrt{n})}}\right] \right.
$$

$$
\left. \leq -\frac{\sqrt{n}\mathrm{E}\psi(X_1 - \theta_0 - t/\sqrt{n})}{\sqrt{\mathrm{Var}\psi(X_1 - \theta_0 - t/\sqrt{n})}}\right\}
$$

$$
= P\left\{\frac{1}{\sqrt{n}}\sum_{i=1}^{n}Z_{ni} \leq -\frac{\sqrt{n}\mathrm{E}\psi(X_1 - \theta_0 - t/\sqrt{n})}{\sqrt{\mathrm{Var}\psi(X_1 - \theta_0 - t/\sqrt{n})}}\right\} \qquad (5.4.7)
$$

where

$$
Z_{ni} = \frac{1}{\sqrt{n}}\left\{\frac{\psi(X_i - \theta_0 - t/\sqrt{n}) - \mathrm{E}\psi(X_1 - \theta_0 - t/\sqrt{n})}{\sqrt{\mathrm{Var}\psi(X_1 - \theta_0 - t/\sqrt{n})}}\right\}, \quad i = 1, \ldots, n,
$$

are i.i.d. r.v.'s with $\mathrm{E}Z_{ni} = 0$ and $\mathrm{Var}Z_{ni} = 1$. Now use Taylor's expansion to see that, as $n \to \infty$,

$$
\sqrt{n}\mathrm{E}\psi(X_1 - \theta_0 - t/\sqrt{n}) = \sqrt{n}M(\theta_0 + t/\sqrt{n})
$$

$$
= \sqrt{n}\left\{M(\theta_0) + tM'(\theta_0)/\sqrt{n} + o(|t|/\sqrt{n})\right\}
$$

$$
= tM'(\theta_0) + \sqrt{n}o(|t|/\sqrt{n}) \longrightarrow tM'(\theta_0). \qquad (5.4.8)
$$

Also use conditions (i) and (ii) to see that, as $n \to \infty$,

$$
\mathrm{Var}\psi\left(X_1 - \theta_0 - \frac{t}{\sqrt{n}}\right) = \int \psi^2\left(x - \theta_0 - \frac{t}{\sqrt{n}}\right)dF(x)
$$

$$
-M^2\left(\theta_0 + \frac{t}{\sqrt{n}}\right) \to \sigma^2(\theta_0). \quad (5.4.9)
$$

From (5.4.8) and (5.4.9) it follows that, as $n \to \infty$,

$$
a_n = -\frac{\sqrt{n}\mathrm{E}\psi(X_1 - \theta_0 - t/\sqrt{n})}{\sqrt{\mathrm{Var}\psi(X_1 - \theta_0 - t/\sqrt{n})}} \longrightarrow -\frac{tM'(\theta_0)}{\sigma(\theta_0)}. \qquad (5.4.10)
$$

In view of (5.4.7) and (5.4.10), all we have to show in order to complete the proof is that as $n \to \infty$,

$$
P\left\{\frac{1}{\sqrt{n}}\sum_{i=1}^{n}Z_{ni} \leq -\frac{tM'(\theta_0)}{\sigma(\theta_0)}\right\} \longrightarrow \Phi\left\{-\frac{tM'(\theta_0)}{\sigma(\theta_0)}\right\}. \qquad (5.4.11)
$$

To see this, suppose that $\{Y_n\}$ is a sequence of random variables and that

$\{a_n\}$ is a sequence of real numbers such that $a_n \to a$ as $n \to \infty$, where a is a constant; then note that $Y_n - a_n = (Y_n - a) + (a_n - a)$ and, therefore, that if $Y_n - a \xrightarrow{D} N(-a, 1)$, the Slutsky Theorem (3.4.2) implies that $Y_n - a_n \xrightarrow{D} Y$. Thus, we have $P\{Y_n - a_n \leq 0\} = P\{Y_n \leq a_n\} \to \Phi(-a)$ as $n \to \infty$.

Now, to prove (5.4.11), we may relate to the Central Limit Theorem for Triangular Arrays 3.3.5 and it suffices to show that Lindeberg's condition holds, i.e., that

$$\mathrm{E}\left[Z_{n1}^2 I_{\{|Z_{n1}| > \sqrt{n}\varepsilon\}}\right] \to 0 \quad \text{as} \quad n \to \infty. \tag{5.4.12}$$

Let

$$B_n = \left\{|\psi(x - \theta_0 - t/\sqrt{n}) - M(\theta_0 + t/\sqrt{n})| > \sqrt{n}\varepsilon\sigma(\theta_0 + t/\sqrt{n})\right\}$$

where $\sigma^2(\theta_0 + t/\sqrt{n}) = \mathrm{Var}\psi(X_1 - \theta_0 - t/\sqrt{n})$ and note that

$$\mathrm{E}\left[Z_{n1}^2 I_{\{|Z_{n1}| > \sqrt{n}\varepsilon\}}\right] = \int_{B_n} \frac{\psi^2(x - \theta_0 - t/\sqrt{n})}{\sigma^2(\theta_0 + t/\sqrt{n})} dF(x)$$

$$-\frac{2M(\theta_0 + t/\sqrt{n})}{\sigma^2(\theta_0 + t/\sqrt{n})} \int_{B_n} \psi\left(x - \theta_0 - \frac{t}{\sqrt{n}}\right) dF(x)$$

$$+\frac{M^2(\theta_0 + t/\sqrt{n})}{\sigma^2(\theta_0 + t/\sqrt{n})} \int_{B_n} dF(x). \tag{5.4.13}$$

Since $M(\theta_0 + t/\sqrt{n})$ and $\sigma^2(\theta_0 + t/\sqrt{n})$ are continuous, it follows that the two last terms of the rhs of (5.4.13) converge to zero as $n \to \infty$. Thus, we have to show that as $n \to \infty$,

$$\int_{\{|\psi(x-\theta-t/\sqrt{n})| > \varepsilon\sqrt{n}\}} \psi^2\left(x - \theta_0 - \frac{t}{\sqrt{n}}\right) dF(x) \to 0. \tag{5.4.14}$$

In this direction, note that since ψ is non-increasing, given $\varepsilon > 0$, there exists $n_0 = n_0(\varepsilon)$ such that

$$\psi(x - \theta_0 + \varepsilon) \leq \psi\left(\frac{x - \theta_0 - t}{\sqrt{n}}\right) \leq \psi(x - \theta_0 - \varepsilon)$$

for all $n \geq n_0$ and for all $x \in \mathbb{R}$. Therefore, by assumption (ii) and the existence of the integral, (5.4.14) follows readily.   ∎

It is of interest to note that $M'(\theta) = (\partial/\partial\theta) \int \psi(x - \theta) dF(x)$ may be defined even if ψ is not differentiable. To see this, consider the change of variables $y = x - \theta$ and note that

$$(\partial/\partial\theta) \int \psi(x - \theta) f(x) dx = (\partial/\partial\theta) \int \psi(y) f(y + \theta) dy$$

$$= \int \psi(y)(\partial/\partial\theta)f(y+\theta)dy.$$

Thus, an integrability condition on $\{(\partial/\partial\theta)f(y+\theta)\}/f(y+\theta)$ should suffice. It is also noteworthy to mention that the asymptotic normality of M-estimators may be proved if the monotonicity condition of ψ is replaced by a smoothness assumption. The reader is referred to Huber (1981) for details.

Example 5.4.3: Let $X_1, \ldots, X_n$ be i.i.d. r.v.'s with distribution function F and let $\psi(x) = x$; thus, the solution to (5.4.4) defines the LSE $\hat\theta_n = \overline{X}_n$.

a) If F corresponds to the $N(\theta, \sigma^2)$ distribution, then, clearly, $M'(\theta) = -1$ and $\int \psi^2(x - \theta)dF(x) = \sigma^2$ so that conditions (i) and (ii) of Theorem 5.4.2 are satisfied and, consequently, $\sqrt{n}(\overline{X}_n - \theta) \xrightarrow{D} N(0, \sigma^2)$.

b) If F corresponds to a Pearson Type VII distribution with parameters $m = (\nu+1)/2$, $c = \sqrt{\nu}$ and θ, i.e., a Student t-distribution with $\nu \geq 3$ degrees of freedom, shifted by θ, we have

$$\begin{aligned}
M'(\theta) &= \frac{\partial}{\partial\theta} \int \psi(x-\theta)f(x)dx \\
&= \frac{\partial}{\partial\theta} \int (x-\theta)\frac{\Gamma\{(\nu+1)/2\}}{\sqrt{\pi\nu}\Gamma(\nu/2)} \left\{1 + \frac{(x-\theta)^2}{\nu}\right\}^{-(\nu+1)/2} dx \\
&= -1
\end{aligned}$$

and $\int \psi^2(x-\theta)f(x)dx = \nu/(\nu-2)$, so that by Theorem 5.4.1 we may conclude that $\sqrt{n}(\overline{X}_n - \theta) \xrightarrow{D} N(0, \nu/(\nu-2))$.

c) If F is as in (b), but $\nu = 2$, we may not apply Theorem 5.4.2, since condition (ii) does not hold; in such a case, a robust M-estimator for θ may be considered. ∎

Example 5.4.4: Let $X_1, \ldots, X_n$ be i.i.d. r.v.'s with density function f symmetric with respect to θ_0 and let ψ be the Huber function given by (5.4.5). Then we have

$$\begin{aligned}
M(\theta) &= \int_{\theta-k}^{\theta+k} (x-\theta)f(x)dx + k\int_{\theta+k}^{\infty} f(x)dx - k\int_{-\infty}^{\theta-k} f(x)dx \\
&= \int_{-k}^{k} xf(x+\theta)dx + k\left[P\{X \geq \theta+k\} - P\{X \leq \theta-k\}\right]
\end{aligned}$$

and, clearly, $M(\theta_0) = 0$ and

$$M'(\theta_0) = \int_{\theta_0-k}^{\theta_0+k} f(x)dx = P\{\theta_0 - k \leq X \leq \theta_0 + k\}.$$

Also, since ψ is bounded, we have

$$\sigma^2(\theta_0) = \int_{\theta_0-k}^{\theta_0+k} (x - \theta_0)^2 f(x)dx + 2k^2 P\{X \geq \theta_0 + k\} < \infty$$

irrespective of whether $\int_{-\infty}^{+\infty} x^2 f(x)dx$ is finite or not. Thus, the asymptotic normality of the M-estimator $\widehat{\theta}_n$ follows from Theorem 5.4.2.   ∎

The family of robust estimators (of location) consists of three important subfamilies: (i) L-estimators, (ii) M-estimators, and (iii) R-estimators. The L-estimators are linear functions of order statistics with (generally) smooth weight functions chosen skillfully to curb the influence of outliers. They also crop up with optimal linear estimation based on a subset of order statistics, as may be the case in censored samples. In Section 4.6, we have already provided an introduction to such L-estimators and have discussed the related large sample theory. Along with the current discussion of M-estimators, we include a very brief remark on R-estimators (of location/regression). R-estimators are based on appropriate rank statistics. For example, the classical Wilcoxon signed rank statistic can be used to derive a (globally) robust, translation-invariant, consistent and asymptotically normal estimator of location. This turns out to be the median of all the $\binom{n+1}{2}$ mid-ranges $(X_i + X_j)/2$, $1 \leq i \leq j \leq n$. We may as well use the sign statistic yielding the sample median as the appropriate estimator. To avoid repetition, we relegate the introduction to R-estimators to Section 7.3 where both the location and regression models are considered in the same vein.

5.5 Asymptotic efficiency of estimators

As we have seen, many types of estimators commonly employed in statistical analysis are asymptotically normally distributed around the true value of the parameter. Thus, in order to choose among competing estimators we need some further criterion. Consider, for example, sequences $\{\widehat{\theta}_n^{(1)}\}$ and $\{\widehat{\theta}_n^{(2)}\}$ of estimators such that $\sqrt{n}(\widehat{\theta}_n^{(1)} - \theta) \xrightarrow{\mathcal{D}} N(0, \sigma_1^2)$ and $\sqrt{n}(\widehat{\theta}_n^{(2)} - \theta) \xrightarrow{\mathcal{D}} N(0, \sigma_2^2)$. We may use the corresponding asymptotic variances σ_1^2 and σ_2^2 to define the **asymptotic relative efficiency** (ARE) of $\widehat{\theta}_n^{(2)}$ with respect to $\widehat{\theta}_n^{(1)}$ as

$$\text{ARE}(\widehat{\theta}_n^{(2)} \mid \widehat{\theta}_n^{(1)}) = \sigma_1^2/\sigma_2^2.$$

Obviously, if $\text{ARE}(\widehat{\theta}_n^{(2)} \mid \widehat{\theta}_n^{(1)}) < 1$, $\widehat{\theta}_n^{(1)}$ is more efficient than $\widehat{\theta}_n^{(2)}$.

Example 5.5.1: Let $X_1, \ldots, X_n$ be i.i.d. r.v.'s following a Pareto distribution, i.e., with density function

$$f(x;\theta) = \theta^{-1}x^{-(1+1/\theta)}, \quad x > 1, \quad 0 < \theta < 1/2.$$

Letting $\gamma = \theta^{-1}$, which implies $\gamma > 2$, we have

$$\mu_1' = EX_1 = \int_1^\infty x\gamma x^{-1-\gamma}dx = \frac{\gamma}{\gamma-1} = \frac{1}{1-\theta}.$$

Therefore, solving for θ and setting $\overline{X}_n = (1-\theta)^{-1}$, we obtain a MME of θ, namely, $\widetilde{\theta}_n = 1-1/\overline{X}_n$. Now let $h(\theta) = (1-\theta)^{-1}$ and $H(\theta) = (d/d\theta)h(\theta) = (1-\theta)^{-2}$. Also, observe that

$$\begin{aligned}
\mu_2' &= EX_1^2 = \int_1^\infty x^2\gamma x^{-1-\gamma}dx \\
&= \frac{\gamma}{2-\gamma}\int_1^\infty (2-\gamma)x^{1-\gamma}dx \\
&= \frac{\gamma}{\gamma-2} = \frac{1}{1-2\theta}.
\end{aligned}$$

Thus,

$$\operatorname{Var}X_1 = \sigma^2 = \frac{1}{1-2\theta} - \frac{1}{(1-\theta)^2} = \frac{\theta^2}{(1-2\theta)(1-\theta)^2}$$

and

$$\{H(\theta)\}^{-2}\sigma^2 = \frac{\theta^2(1-\theta)^2}{1-2\theta};$$

so from Theorem 5.4.1 it follows that

$$\sqrt{n}(\widetilde{\theta}_n - \theta) \xrightarrow{D} N\left(0, \theta^2(1-\theta)^2/(1-2\theta)\right).$$

Alternatively, the MLE of θ is a solution $\widehat{\theta}_n$ to

$$\frac{\partial}{\partial\theta}\log L_n(\theta) = -\frac{n}{\theta} + \frac{1}{\theta^2}\sum_{i=1}^n \log X_i = 0$$

which implies $\widehat{\theta}_n = n^{-1}\sum_{i=1}^n \log X_i$. Now observe that:

i)

$$\begin{aligned}
\left|\frac{\partial}{\partial\theta}f(x;\theta)\right| &= |\theta^{-2}x^{-(1+\theta^{-1})} + \theta^{-3}x^{-(1+\theta^{-1})}\log x| \\
&\leq \theta^{-2}x^{-(1+\theta^{-1})} + \theta^{-3}x^{-(1+\theta^{-1})} \\
&= H_1(x),
\end{aligned}$$

which is integrable;

ii)

$$
\begin{aligned}
\left|\frac{\partial^2}{\partial\theta^2}f(x;\theta)\right| &= \left|-2\theta^{-3}x^{-(1+\theta^{-1})}-2\theta^{-4}x^{-(1+\theta^{-1})}\log x\right.\\
&\qquad \left.+\theta^{-5}x^{-1(1+\theta^{-1})}\log^2 x\right|\\
&\leq 2\theta^{-3}x^{-(1+\theta^{-1})}\left\{1+\theta^{-1}\log x+\frac{1}{2}\theta^{-2}\log^2 x\right\}\\
&= H_2(x),
\end{aligned}
$$

which is integrable;

iii) $\mathrm{E}\log X_1=\theta$ and $\mathrm{E}(\log X_1)^2=2\theta^2$ which imply that

$$
\begin{aligned}
I(\theta) &= \mathrm{E}\left\{\frac{\partial}{\partial\theta}\log f(X_1;\theta)\right\}^2\\
&= \mathrm{E}\left\{\frac{1}{\theta^2}-\frac{2}{\theta^3}\log X_1+\frac{1}{\theta^4}\log^2 X_1\right\}\\
&= \frac{1}{\theta^2}<\infty;
\end{aligned}
$$

iv) $(\partial^2/\partial\theta^2)\log f(x;\theta)=(1/\theta^2)-(2/\theta^3)\log x$ which implies that

$$
\begin{aligned}
\psi_\delta &= \mathrm{E}\left\{\sup_{\{h:|h|\leq\delta\}}\left|\frac{1}{(\theta+h)^2}-\frac{1}{\theta^2}-\frac{2}{(\theta+h)^3}\log X_1+\frac{2}{\theta^3}\log X_1\right|\right\}\\
&\leq \sup_{\{h:|h|\leq\delta\}}\left|\frac{1}{(\theta+h)^2}-\frac{1}{\theta^2}\right|\\
&\qquad +\sup_{\{h:|h|\leq\delta\}}\left\{\left|\frac{1}{(\theta+h)^3}-\frac{1}{\theta^3}\right|2\mathrm{E}\log X_1\right\}\\
&= \left|\frac{1}{(\theta+\delta)^2}-\frac{1}{\theta^2}\right|+\left|\frac{1}{(\theta+\delta)^3}-\frac{1}{\theta^3}\right|2\theta\longrightarrow 0\quad\text{as}\quad \delta\to 0.
\end{aligned}
$$

Therefore, it follows from Theorem 5.2.1 that $\sqrt{n}(\widehat{\theta}_n-\theta)\overset{\mathcal{D}}{\longrightarrow}N(0,\theta^2)$. Consequently, we have

$$
\mathrm{ARE}(\widetilde{\theta}_n\mid\widehat{\theta}_n)=\theta^2\frac{(1-2\theta)}{\theta^2(1-\theta)^2}=\frac{1-2\theta}{(1-\theta)^2}<1,
$$

so that the MME is (asymptotically) less efficient than the MLE. ∎

Now let $X_1,\ldots,X_n$ be i.i.d. r.v.'s with density function $f(x;\theta)$, $\theta\in\Theta\subset\mathbb{R}$, such that $\mathrm{E}\{(\partial/\partial\theta)\log f(X_1;\theta)\}^2=I(\theta)<\infty$ and consider a sequence $\{\widehat{\theta}_n\}$ of unbiased estimators of θ such that $\mathrm{Var}\widehat{\theta}_n<\infty$. Then note that, by the Cauchy-Schwarz Inequality,

$$\mathrm{E}^2 \left\{ \sqrt{n}(\widehat{\theta}_n - \theta) \frac{1}{\sqrt{n}} \frac{\partial}{\partial \theta} \log L_n(\theta) \right\}$$

$$\leq \mathrm{E} \left\{ \sqrt{n}(\widehat{\theta}_n - \theta) \right\}^2 \mathrm{E} \left\{ \frac{1}{\sqrt{n}} \frac{\partial}{\partial \theta} \log L_n(\theta) \right\}^2. \qquad (5.5.1)$$

Expanding the left-hand side of (5.5.1) we obtain

$$\mathrm{E} \left\{ \sqrt{n}(\widehat{\theta}_n - \theta) \frac{1}{\sqrt{n}} \frac{\partial}{\partial \theta} \log L_n(\theta) \right\}$$

$$= \mathrm{E} \left\{ \widehat{\theta}_n \frac{\partial}{\partial \theta} \log L_n(\theta) \right\} - \theta \mathrm{E} \left\{ \frac{\partial}{\partial \theta} \log L_n(\theta) \right\}$$

$$= \int \widehat{\theta}_n(x) \frac{\partial}{\partial \theta} \log L_n(\theta; x) L_n(\theta; x) dx$$

$$= \int \widehat{\theta}_n(x) \frac{\partial}{\partial \theta} L_n(\theta; x) dx$$

$$= \frac{\partial}{\partial \theta} \int \widehat{\theta}_n(x) L_n(\theta; x) dx = \frac{\partial}{\partial \theta} \theta = 1. \qquad (5.5.2)$$

$$\mathrm{E} \left\{ \frac{1}{\sqrt{n}} \frac{\partial}{\partial \theta} \log L_n(\theta) \right\}^2 = I(\theta) < \infty,$$

it follows from (5.5.1) and (5.5.2) that:

$$\mathrm{E} \left\{ \sqrt{n}(\widehat{\theta}_n - \theta) \right\}^2 = n \mathrm{Var} \widehat{\theta}_n \geq \{I(\theta)\}^{-1} \qquad (5.5.3)$$

which is known as the **Information Inequality**; the number $\{I(\theta)\}^{-1}$ is usually referred to as the **Frechet-Cramér-Rao lower bound** (for the variances of all unbiased estimators of θ). This lower bound suggests the definition of the **asymptotic efficiency** of a sequence $\{\widehat{\theta}_n\}$ of estimators for which $\sqrt{n}(\widehat{\theta}_n - \theta) \xrightarrow{D} N(0, \sigma_\theta^2)$ as the ratio $\{I(\theta)\}^{-1}/\sigma_\theta^2$. Moreover, if $\sigma_\theta^2 = \{I(\theta)\}^{-1}$ the corresponding estimator T_n is said to be **asymptotically efficient**. Note that, in general, the asymptotic efficiency of an estimator as defined above may not coincide with the **Fisher asymptotic efficiency** given by $\{I(\theta)\}^{-1}/\lim_{n\to\infty} \mathrm{E}\{\sqrt{n}(\widehat{\theta}_n - \theta)^2\}$.

In view of Theorem 5.2.1, we may conclude that under quite general regularity conditions, MLEs are asymptotically efficient. More generally, if we recall that equality in (5.5.1) holds if and only if

$$\sqrt{n}(\widehat{\theta}_n - \theta) = K n^{-1/2}(\partial/\partial\theta) \log L_n(\theta),$$

with $K \neq 0$, it follows that all estimators for which

$$\sqrt{n}(\widehat{\theta}_n - \theta) = K n^{-1/2}(\partial/\partial\theta) \log L_n(\theta) + o_p(1)$$

are asymptotically efficient. Estimators in this class are also termed **Best Asymptotically Normal** (BAN).

Example 5.5.2: Let $X_1, \ldots, X_n$ be i.i.d. r.v.'s with a $N(\mu, \sigma^2)$ distribution. Since the assumptions of Theorem 5.2.1 are clearly satisfied in this case, it follows that both the (biased) MLE of σ^2, $\widehat{\sigma}_n^2 = n^{-1} \sum_{i=1}^n (X_i - \overline{X}_n)^2$, and the unbiased estimator S_n^2 are such that $\sqrt{n}(\widehat{\sigma}_n^2 - \sigma^2)$ or $\sqrt{n}(S_n^2 - \sigma^2)$ converge in distribution to a random variable following a $N(0, \mu_4 - \sigma^4)$ distribution (see Example 3.4.1), and consequently they are BAN estimators. In fact, note that $S_n^2 = \widehat{\sigma}_n^2 - \widehat{\sigma}_n^2/n = \widehat{\sigma}_n^2 + O_p(n^{-1})$. ∎

The above concepts may be readily extended to the multiparameter case; in this direction consider i.i.d. r.v.'s, $X_1, \ldots, X_n$ with density function $f(x; \boldsymbol{\theta})$ satisfying the regularity conditions of Theorem 5.2.2 and let $\{\widehat{\boldsymbol{\theta}}_n\}$ be a sequence of unbiased estimators of $\boldsymbol{\theta}$. Here, the Information Inequality (5.5.3) is replaced by the condition that:

$$\mathrm{E}\left\{ n(\widehat{\boldsymbol{\theta}}_n - \boldsymbol{\theta})(\widehat{\boldsymbol{\theta}}_n - \boldsymbol{\theta})^t \right\} - [\mathbf{I}(\boldsymbol{\theta})]^{-1} = n\mathbf{D}(\widehat{\boldsymbol{\theta}}_n) - [\mathbf{I}(\boldsymbol{\theta})]^{-1} \qquad (5.5.4)$$

be nonnegative definite. Then, if $\sqrt{n}(\widehat{\boldsymbol{\theta}}_n - \boldsymbol{\theta}) \xrightarrow{\mathcal{D}} N(\mathbf{0}, [\mathbf{I}(\boldsymbol{\theta})]^{-1})$, the estimator $\widehat{\boldsymbol{\theta}}_n$ is said to be asymptotically efficient. To prove (5.5.4), let $\boldsymbol{\lambda}$ and $\boldsymbol{\gamma}$ be arbitrary, but fixed vectors in $\mathbb{R}^q$, let

$$\mathbf{U}_n(\mathbf{X}; \boldsymbol{\theta}) = (\partial/\partial\boldsymbol{\theta}) \log L_n(\boldsymbol{\theta})$$

and use the Cauchy-Schwarz Inequality o see that

$$\mathrm{E}^2 \left\{ \sqrt{n}\boldsymbol{\lambda}^t (\widehat{\boldsymbol{\theta}}_n - \boldsymbol{\theta}) \frac{1}{\sqrt{n}} [\mathbf{U}_n(\mathbf{X}; \boldsymbol{\theta})]^t \boldsymbol{\gamma} \right\}$$

$$\leq \ \mathrm{E}\left\{ n\boldsymbol{\lambda}^t (\widehat{\boldsymbol{\theta}}_n - \boldsymbol{\theta})(\widehat{\boldsymbol{\theta}}_n - \boldsymbol{\theta})^t \boldsymbol{\lambda} \right\} \mathrm{E}\left\{ \frac{1}{n} \boldsymbol{\gamma}^t \mathbf{U}_n(\mathbf{X}; \boldsymbol{\theta})[\mathbf{U}_n(\mathbf{X}; \boldsymbol{\theta})]^t \boldsymbol{\gamma} \right\}.$$
$$(5.5.5)$$

The same argument employed to show (5.5.2) may be considered to prove that the left-hand side of (5.5.5) is equal to $\boldsymbol{\lambda}^t \boldsymbol{\gamma}$; consequently we may write

$$\mathrm{E}\left\{ n\boldsymbol{\lambda}^t (\boldsymbol{\theta}_n - \boldsymbol{\theta})(\boldsymbol{\theta}_n - \boldsymbol{\theta})^t \boldsymbol{\lambda} \right\} \geq \frac{\boldsymbol{\lambda}^t \boldsymbol{\gamma}}{\boldsymbol{\gamma}^t \mathbf{I}(\boldsymbol{\theta})\boldsymbol{\gamma}}. \qquad (5.5.6)$$

Maximizing the right-hand side of (5.5.6) with respect to $\boldsymbol{\gamma}$ we obtain

$$\boldsymbol{\lambda}^t \left\{ n\mathrm{E}(\boldsymbol{\theta}_n - \boldsymbol{\theta})(\boldsymbol{\theta}_n - \boldsymbol{\theta})^t \right\} \boldsymbol{\lambda} \geq \boldsymbol{\lambda}^t [\mathbf{I}(\boldsymbol{\theta})]^{-1} \boldsymbol{\lambda} \qquad (5.5.7)$$

and since $\boldsymbol{\lambda}$ is arbitrary, (5.5.4) follows. Furthermore, it follows that

$$\mathrm{E}\left\{ n(\boldsymbol{\theta}_n - \boldsymbol{\theta})(\boldsymbol{\theta}_n - \boldsymbol{\theta})^t \right\} = [\mathbf{I}(\boldsymbol{\theta})]^{-1}$$

if any only if
$$\sqrt{n}(\boldsymbol{\theta}_n - \boldsymbol{\theta}) = \mathbf{K}n^{-1/2}(\partial/\partial\boldsymbol{\theta})\log L_n(\boldsymbol{\theta})$$
where $\mathbf{K}$ is a nonsingular matrix; thus, all estimators for which
$$\sqrt{n}(\boldsymbol{\theta}_n - \boldsymbol{\theta}) = \mathbf{K}n^{-1/2}(\partial/\partial\boldsymbol{\theta})\log L_n(\boldsymbol{\theta}) + o_p(\mathbf{1})$$
are asymptotically efficient. This is the case with the MLE obtained in situations where the conditions of Theorem 5.2.2 are satisfied.

The asymptotic optimality (or efficiency) of estimators considered here has been tuned to the asymptotic mean squared error in the single parameter case and to the asymptotic dispersion matrix in the vector-parameter case. In the literature there are some alternative measures of efficiency of estimators. Among these a noteworthy case is the so-called **Pitman measure of closeness** (PMC). Let T_1 and T_2 be two rival estimators of a real-valued parameter θ. Then, T_1 is said to be closer to θ than T_2 in the Pitman (1937) sense if
$$P_\theta\left\{|T_1 - \theta| \le |T_2 - \theta|\right\} \ge \frac{1}{2}, \quad \text{for all} \quad \theta, \qquad (5.5.8)$$
with strict inequality holding for some θ. This definition is easily generalizable by replacing the simple Euclidean distance by a general loss function $L(a, b)$, and we may say that an estimator T_1 (possibly, vector-valued) is Pitman closer than a rival one T_2 [with respect to the loss function $L(\cdot)$] if
$$P_\theta\left\{L(T_1, \theta) \le L(T_2, \theta)\right\} \ge \frac{1}{2}, \quad \text{for all} \quad \theta, \qquad (5.5.9)$$
with strict inequality holding for some θ. In particular, in the vector case, usually, $L(\mathbf{a}, \mathbf{b})$ is taken as a quadratic loss, i.e., $L(\mathbf{a}, \mathbf{b}) = (\mathbf{a}-\mathbf{b})^t\mathbf{Q}(\mathbf{a}-\mathbf{b})$ for some given p.d. $\mathbf{Q}$. If (5.5.8) [or (5.5.9)] holds for all T_2 belonging to a class $\mathcal{C}$, then T_1 is said to be the Pitman closest estimator within the same class.

In an asymptotic shade, in (5.5.8) or (5.5.9), we may replace T_1 and T_2 by $\{T_{n1}\}$ and $\{T_{n2}\}$, respectively, and require that
$$\liminf_{n\to\infty}\left[P_\theta\left\{L(T_{n1}, \theta) \le L(T_{n2}, \theta)\right\}\right] \ge \frac{1}{2}, \quad \text{for all} \quad \theta. \qquad (5.5.10)$$
This relates to the asymptotic PMC.

Whereas a complete treatment of PMC or even the asymptotic PMC is somewhat beyond the scope of this book, we may refer to Keating, Mason and Sen (1993) for a very comprehensive treatise of the PMC. In particular, the asymptotic PMC has been treated in detail in their Chapter 6. It has been shown that within a general class of estimators which are asymptotic normal (AN), a BAN estimator in the sense of asymptotic variance being a minimum (as has been stressed earlier) is also the Pitman closest one in an asymptotic setup. Thus, whatever we have discussed about the MLE and other estimators in the light of their BAN characterization pertains to their asymptotic PMC characterization too. Actually, there is even a more general result: An asymptotic (partial) ordering of estimators in terms of their asymptotic mean square errors is isomorphic to an asymptotic PMC ordering. Thus, viewed from this stand, asymptotically, there is no need to prefer the particular definition of asymptotic efficiency/optimality, and it may even be interpreted in a more general setup.

We conclude this section by noting that even if the true underlying distribution function does not match the one utilized in deriving the MLE, it may still have an asymptotical normal distribution, although it may not be asymptotically efficient [see Singer and Sen (1986)]; this fact illustrates the lack of robustness of MLE and justifies the need of (distributionally) robust estimation methods.

5.6 Asymptotic behavior of some test statistics

Let $X_1, \ldots, X_n$ be i.i.d. r.v.'s with density function $f(x; \boldsymbol{\theta})$, $\boldsymbol{\theta} \in \boldsymbol{\Theta} \subset \mathbb{R}^q$. For testing $H_0 : \boldsymbol{\theta} = \boldsymbol{\theta}_0$ vs. $A : \boldsymbol{\theta} = \boldsymbol{\theta}_1$ (simple null hypothesis vs. simple alternative), the **most powerful** test is given by the **Neyman-Pearson criterion**: reject H_0 if $\lambda_n = \lambda_n(X_1, \ldots, X_n) = L_n(\boldsymbol{\theta}_1)/L_n(\boldsymbol{\theta}_0) \geq k_\alpha$ and accept H_0 otherwise, where $k_\alpha = k_\alpha(n, \alpha, \boldsymbol{\theta}_0, \boldsymbol{\theta}_1)$ is determined by $P\{\lambda_n \geq k_\alpha \mid H_0\} = P_{\boldsymbol{\theta}_0}\{\lambda_n \geq k_\alpha\} = \alpha$, $0 < \alpha < 1$, and α is the level of significance of the test.

Example 5.6.1: Let $X_1, \ldots, X_n$ be i.i.d. r.v.'s with a $N(\theta, 1)$ distribution and consider the problem of testing $H_0 : \theta = \theta_0$ vs. $A : \theta = \theta_1$. Here, we have $\log\{L_n(\theta_1)/L_n(\theta_0)\} = n(\theta_1 - \theta_0)\{\overline{X}_n - (\theta_0 + \theta_1)/2\}$, which implies that

$$
\begin{aligned}
P_{\theta_0}\{\lambda_n \geq k_\alpha\} &= P_{\theta_0}\left\{n\left[\overline{X}_n - \frac{\theta_0 + \theta_1}{2}\right] \geq \frac{\log k_\alpha}{\theta_1 - \theta_0}\right\} \\
&= P_{\theta_0}\left\{\sqrt{n}(\overline{X}_n - \theta_0) \geq \frac{\log k_\alpha}{\sqrt{n}(\theta_1 - \theta_0)} + \sqrt{n}\frac{(\theta_1 - \theta_0)}{2}\right\}
\end{aligned}
$$

and since $\sqrt{n}(\overline{X}_n - \theta_0) \sim N(0, 1)$, it follows that

$$
k_\alpha = \exp\left\{\sqrt{n}(\theta_1 - \theta_0)z_{1-\alpha} - n(\theta_1 - \theta_0)^2\right\},
$$

where $z_{1-\alpha}$ is the $100(1 - \alpha)$th percentile of the $N(0, 1)$ distribution.

The question is how to determine the constant k_α when the underlying distribution is non-normal. In this direction let

$$
Z_i = \log\{f(X_i; \theta_1)/f(X_i; \theta_0)\}
$$

and assume that $\mathrm{Var}_{\theta_0} Z_i = \sigma_0^2 < \infty$. Then

$$
\begin{aligned}
P_{\theta_0} \{\lambda_n \geq k_\alpha\} &= P_{\theta_0} \left\{ \sum_{i=1}^{n} \log \frac{f(X_i; \theta_1)}{f(X_i; \theta_0)} \geq \log k_\alpha \right\} \\
&= P_{\theta_0} \left\{ \sum_{i=1}^{n} Z_i \geq \log k_\alpha \right\} \\
&= P_{\theta_0} \left\{ \frac{1}{\sqrt{n}\sigma_0} \sum_{i=1}^{n} (Z_i - EZ_i) \geq \frac{\log k_\alpha}{\sqrt{n}\sigma_0} - \frac{\sqrt{n}}{\sigma_0} EZ_1 \right\}.
\end{aligned}
$$

By the Central Limit Theorem 3.3.1 it follows that

$$
n^{-1/2}\sigma_0^{-1} \sum_{i=1}^{n} (Z_i - EZ_i) \xrightarrow{D} N(0,1)
$$

and, then, for sufficiently large n we have

$$
\frac{\log k_\alpha}{\sqrt{n}\sigma_0} - \frac{\sqrt{n}}{\sigma_0} EZ_1 \simeq z_{1-\alpha}
$$

which implies that $k_\alpha \simeq \exp\{\sqrt{n}\sigma_0 z_{1-\alpha} + nEZ_1\}$. ∎

More generally, for testing $H_0 : \boldsymbol{\theta} = \boldsymbol{\theta}_0$ vs. $A : \boldsymbol{\theta} \neq \boldsymbol{\theta}_0$, most powerful or uniformly most powerful tests do not exist and we have to rely on other criteria for the choice of an appropriate test statistic. In particular, when the conditions of Theorem 5.2.1 are satisfied, three alternatives are more frequently employed:

i) **Wald's statistic:**

$$
Q_W = n(\widehat{\boldsymbol{\theta}}_n - \boldsymbol{\theta}_0)^t \mathbf{I}(\widehat{\boldsymbol{\theta}}_n)(\widehat{\boldsymbol{\theta}}_n - \boldsymbol{\theta}_0)
$$

where $\widehat{\boldsymbol{\theta}}_n$ is the MLE (or a BAN estimator) of $\boldsymbol{\theta}_0$ and

$$
\mathbf{I}(\widehat{\boldsymbol{\theta}}_n) = -\frac{1}{n} \sum_{i=1}^{n} \frac{\partial^2}{\partial\boldsymbol{\theta}\partial\boldsymbol{\theta}^t} \log f(X_i; \boldsymbol{\theta}) \bigg|_{\boldsymbol{\theta}=\widehat{\boldsymbol{\theta}}_n}.
$$

If $\boldsymbol{\theta}_0$ specifies the density completely, then we may use $\mathbf{I}(\boldsymbol{\theta}_0)$ instead of $\mathbf{I}(\widehat{\boldsymbol{\theta}}_n)$ in the above definition of Q_W.

ii) **Wilks' likelihood ratio statistic:**

$$
Q_L = -2\log \lambda_n = 2\left\{ \log L_n(\widehat{\boldsymbol{\theta}}_n) - \log L_n(\boldsymbol{\theta}_0) \right\}
$$

where $\lambda_n = L_n(\boldsymbol{\theta}_0)/\sup_{\boldsymbol{\theta}\in\boldsymbol{\Theta}} L_n(\boldsymbol{\theta})$;

iii) **Rao's efficient score statistic:**

$$
Q_R = n^{-1}[\mathbf{U}_n(\boldsymbol{\theta}_0)]^t [\mathbf{I}(\boldsymbol{\theta}_0)]^{-1} \mathbf{U}_n(\boldsymbol{\theta}_0)
$$

where $\mathbf{U}_n(\boldsymbol{\theta}_0) = \sum_{i=1}^{n}(\partial/\partial\boldsymbol{\theta})\log f(X_i;\boldsymbol{\theta})\big|_{\boldsymbol{\theta}=\boldsymbol{\theta}_0}$; note that this statistic does not require the computation of the MLE $\widehat{\boldsymbol{\theta}}_n$.

In the following theorem we derive the asymptotic distribution of the three statistics indicated above under the null hypothesis.

Theorem 5.6.1: *Let $X_1, \ldots, X_n$ be i.i.d. r.v.'s with p.d.f. $f(x;\boldsymbol{\theta})$, $\boldsymbol{\theta} \in \boldsymbol{\Theta} \subset \mathbb{R}^q$ satisfying the conditions of Theorem 5.2.1 and consider the problem of testing $H_0 : \boldsymbol{\theta} = \boldsymbol{\theta}_0$ vs. $A : \boldsymbol{\theta} \neq \boldsymbol{\theta}_0$. Then, each of the statistics Q_W, Q_L and Q_R indicated above has an asymptotic χ_q^2 distribution under H_0.*

Proof: From Theorem 5.2.1 we know that if $\boldsymbol{\theta} = \boldsymbol{\theta}_0$, the MLE of $\boldsymbol{\theta}$, say $\widehat{\boldsymbol{\theta}}_n$, is such that $\sqrt{n}(\widehat{\boldsymbol{\theta}}_n - \boldsymbol{\theta}_0) \xrightarrow{\mathcal{D}} N(\mathbf{0}, [\mathbf{I}(\boldsymbol{\theta})]^{-1})$ which implies that under H_0,

$$Q_W^* = n(\widehat{\boldsymbol{\theta}}_n - \boldsymbol{\theta}_0)^t \mathbf{I}(\boldsymbol{\theta}_0)(\widehat{\boldsymbol{\theta}}_n - \boldsymbol{\theta}_0) \xrightarrow{\mathcal{D}} \chi_q^2;$$

thus, putting $\mathbf{x} = \sqrt{n}(\widehat{\boldsymbol{\theta}}_n - \boldsymbol{\theta}_0)$ we have

$$\frac{Q_W}{Q_W^*} = \frac{\mathbf{x}^t \mathbf{I}(\widehat{\boldsymbol{\theta}}_n)\mathbf{x}}{\mathbf{x}^t \mathbf{I}(\boldsymbol{\theta}_0)\mathbf{x}}$$

and using Courant Theorem (1.4.2) it follows that

$$\text{ch}_q\left\{\mathbf{I}(\widehat{\boldsymbol{\theta}}_n)[\mathbf{I}(\boldsymbol{\theta}_0)]^{-1}\right\} \le \frac{Q_W}{Q_W^*} \le \text{ch}_1\left\{\mathbf{I}(\widehat{\boldsymbol{\theta}}_n)[\mathbf{I}(\boldsymbol{\theta}_0)]^{-1}\right\}.$$

Now, by the Khintchine Weak Law of Large Numbers (Theorem 2.3.6) we may conclude that $\mathbf{I}(\widehat{\boldsymbol{\theta}}_n)[\mathbf{I}(\boldsymbol{\theta}_0)]^{-1} \xrightarrow{P} \mathbf{I}_q$ and, consequently, we have

$$\text{ch}_1\{\mathbf{I}(\widehat{\boldsymbol{\theta}}_n)[\mathbf{I}(\boldsymbol{\theta}_0)]^{-1}\} \xrightarrow{P} 1$$

and

$$\text{ch}_q\{\mathbf{I}(\widehat{\boldsymbol{\theta}}_n)[\mathbf{I}(\boldsymbol{\theta}_0)]^{-1}\} \xrightarrow{P} 1.$$

Then it follows that $Q_W/Q_W^* \xrightarrow{P} 1$, and writing $Q_W = Q_W^* \times Q_W/Q_W^*$ we may conclude via the Slutsky Theorem (3.4.2) that $Q_W \xrightarrow{\mathcal{D}} \chi_q^2$ under H_0.

To obtain the asymptotic distribution of Q_L under H_0, first consider the following Taylor expansion:

$$\log L_n(\widehat{\boldsymbol{\theta}}_n) = \log L_n(\boldsymbol{\theta}_0) + \sqrt{n}(\widehat{\boldsymbol{\theta}}_n - \boldsymbol{\theta}_0)^t \left\{\frac{1}{\sqrt{n}}\frac{\partial}{\partial\boldsymbol{\theta}}\log L_n(\boldsymbol{\theta})\bigg|_{\boldsymbol{\theta}=\boldsymbol{\theta}_0}\right\}$$

$$+ \frac{n}{2}(\widehat{\boldsymbol{\theta}} - \boldsymbol{\theta}_0)^t \left\{\frac{1}{n}\frac{\partial^2}{\partial\boldsymbol{\theta}\partial\boldsymbol{\theta}^t}\log L_n(\boldsymbol{\theta})\bigg|_{\boldsymbol{\theta}=\boldsymbol{\theta}_n^*}\right\}(\widehat{\boldsymbol{\theta}}_n - \boldsymbol{\theta}_0)$$

$$(5.6.1)$$

where $\boldsymbol{\theta}_n^*$ belongs to the line segment joining $\boldsymbol{\theta}_0$ and $\widehat{\boldsymbol{\theta}}_n$. Then consider the following Taylor expansion of the first term within $\{\ \ \}$ on the right-hand side of (5.6.1):

$$
\begin{aligned}
\frac{1}{\sqrt{n}} \frac{\partial}{\partial \boldsymbol{\theta}} \log L_n(\boldsymbol{\theta}_0) &= \left. \frac{1}{\sqrt{n}} \frac{\partial}{\partial \boldsymbol{\theta}} \log L_n(\boldsymbol{\theta}) \right|_{\boldsymbol{\theta}=\widehat{\boldsymbol{\theta}}_n} \\
&+ \left\{ \left. -\frac{1}{n} \frac{\partial^2}{\partial \boldsymbol{\theta} \partial \boldsymbol{\theta}^t} \log L_n(\boldsymbol{\theta}) \right|_{\boldsymbol{\theta}=\boldsymbol{\theta}_n^{**}} \right\} \sqrt{n}(\widehat{\boldsymbol{\theta}}_n - \boldsymbol{\theta}_0) \\
&= \left\{ \left. -\frac{1}{n} \frac{\partial^2}{\partial \boldsymbol{\theta} \partial \boldsymbol{\theta}^t} \log L_n(\boldsymbol{\theta}) \right|_{\boldsymbol{\theta}=\boldsymbol{\theta}_n^{**}} \right\} \sqrt{n}(\widehat{\boldsymbol{\theta}}_n - \boldsymbol{\theta}_0)
\end{aligned}
$$

$$(5.6.2)$$

where $\boldsymbol{\theta}_n^{**}$ belongs to the line segment joining $\boldsymbol{\theta}_0$ and $\widehat{\boldsymbol{\theta}}_n$. Substituting (5.6.2) into (5.6.1) we get

$$
\begin{aligned}
\log L_n(\widehat{\boldsymbol{\theta}}_n) &= \log L_n(\widehat{\boldsymbol{\theta}}_0) \\
&- n(\widehat{\boldsymbol{\theta}}_n - \boldsymbol{\theta}_0)^t \left\{ \left. \frac{1}{n} \frac{\partial^2}{\partial \boldsymbol{\theta} \partial \boldsymbol{\theta}^t} \log L_n(\boldsymbol{\theta}) \right|_{\boldsymbol{\theta}=\boldsymbol{\theta}_n^{**}} \right\} (\widehat{\boldsymbol{\theta}}_n - \boldsymbol{\theta}_0) \\
&+ \frac{n}{2} (\widehat{\boldsymbol{\theta}}_n - \boldsymbol{\theta}_0)^t \left\{ \left. \frac{1}{n} \frac{\partial^2}{\partial \boldsymbol{\theta} \partial \boldsymbol{\theta}^t} \log L_n(\boldsymbol{\theta}) \right|_{\boldsymbol{\theta}=\boldsymbol{\theta}_n^*} \right\} (\widehat{\boldsymbol{\theta}}_n - \boldsymbol{\theta}_0).
\end{aligned}
$$

$$(5.6.3)$$

By the Khintchine Weak Law of Large Numbers (Theorem 2.3.6) under H_0, both terms within $\{\ \ \}$ in (5.6.3) converge in probability to $-\mathbf{I}(\boldsymbol{\theta}_0)$; therefore, we have

$$
\begin{aligned}
Q_L &= 2 \left\{ \log L_n(\widehat{\boldsymbol{\theta}}_n) - \log L_n(\boldsymbol{\theta}_0) \right\} \\
&= n(\widehat{\boldsymbol{\theta}}_n - \boldsymbol{\theta}_0)^t \mathbf{I}(\boldsymbol{\theta}_0)(\widehat{\boldsymbol{\theta}}_n - \boldsymbol{\theta}_0) + o_p(1)
\end{aligned}
$$

$$(5.6.4)$$

and by the Slutsky Theorem (3.4.2) it follows that $Q_L \xrightarrow{\mathcal{D}} \chi_q^2$, under H_0. Finally, note that, under H_0, the Central Limit Theorem 3.3.1 implies that $n^{-1/2} \mathbf{U}_n(\boldsymbol{\theta}_0) \xrightarrow{\mathcal{D}} N(\mathbf{0}, \mathbf{I}(\boldsymbol{\theta}_0))$; consequently, it follows that $Q_R \xrightarrow{\mathcal{D}} \chi_q^2$ under H_0. ∎

Let us now examine the distribution of the three test statistics above under some (fixed) alternative hypothesis of the form $A_{\boldsymbol{\Delta}} : \boldsymbol{\theta} = \boldsymbol{\theta}_0 + \boldsymbol{\Delta}$, where $\boldsymbol{\Delta}$ is a fixed vector in $\mathbb{R}^q$ such that $\boldsymbol{\theta} \in \boldsymbol{\Theta}$. Consider, for example, Wald's statistic Q_W and note that

$$
\sqrt{n}(\widehat{\boldsymbol{\theta}}_n - \boldsymbol{\theta}) = \sqrt{n}(\widehat{\boldsymbol{\theta}}_n - \boldsymbol{\theta}_0) - \sqrt{n}\boldsymbol{\Delta}
$$

which implies
$$\sqrt{n}(\widehat{\boldsymbol{\theta}}_n - \boldsymbol{\theta}_0) = \sqrt{n}(\widehat{\boldsymbol{\theta}}_n - \boldsymbol{\theta}) + \sqrt{n}\boldsymbol{\Delta}.$$
Now, since, under $A_{\boldsymbol{\Delta}}$,

$$\sqrt{n}(\widehat{\boldsymbol{\theta}}_n - \boldsymbol{\theta}) \xrightarrow{\mathcal{D}} N(\boldsymbol{\theta}, [\mathbf{I}(\boldsymbol{\theta} + \boldsymbol{\Delta})]^{-1})$$

and $\sqrt{n}\|\boldsymbol{\Delta}\| \to \infty$ as $n \to \infty$, it follows that $Q_W \to \infty$ as $n \to \infty$. Consequently, for all hypotheses $A_{\boldsymbol{\Delta}}$, $P_{\boldsymbol{\theta}}\{Q_W > K\} \to 1$ as $n \to \infty$, for all $K \in \mathbb{R}$. Therefore, in order to concentrate on meaningful comparisons, we restrict ourselves to **local Pitman-type alternatives** of the form

$$A_n : \boldsymbol{\theta}_n = \boldsymbol{\theta}_0 + n^{-\frac{1}{2}}\boldsymbol{\Delta}. \tag{5.6.5}$$

Theorem 5.6.2: *Assume that the conditions of Theorem 5.6.1 are satisfied. Then, under (5.6.5), each of the statistics Q_W, Q_L and Q_R defined above has an asymptotic $\chi_q^2(\boldsymbol{\Delta}^t\mathbf{I}(\boldsymbol{\theta}_0)\boldsymbol{\Delta})$ distribution.*

Proof: First note that $\sqrt{n}(\widehat{\boldsymbol{\theta}}_n - \boldsymbol{\theta}_0) = \sqrt{n}(\widehat{\boldsymbol{\theta}}_n - \boldsymbol{\theta}_n) + \boldsymbol{\Delta}$; now, from Theorem 5.2.1, it follows that under A_n,

$$\sqrt{n}(\widehat{\boldsymbol{\theta}}_n - \boldsymbol{\theta}_0) \xrightarrow{\mathcal{D}} N(\mathbf{0}, [\mathbf{I}(\boldsymbol{\theta}_0)]^{-1});$$

thus, using the multivariate version of the Slutsky Theorem (3.4.3) we have $\sqrt{n}(\widehat{\boldsymbol{\theta}}_n - \boldsymbol{\theta}_0) \xrightarrow{\mathcal{D}} N(\boldsymbol{\Delta}, [\mathbf{I}(\boldsymbol{\theta}_0)]^{-1})$ and

$$Q_W \xrightarrow{\mathcal{D}} \chi_q^2(\boldsymbol{\Delta}^t\mathbf{I}(\boldsymbol{\theta}_0)\boldsymbol{\Delta}). \tag{5.6.6}$$

Consider, then, the following Taylor expansion:

$$\frac{1}{\sqrt{n}}\sum_{i=1}^{n}\frac{\partial}{\partial\boldsymbol{\theta}}\log f(X_i;\boldsymbol{\theta})\Big|_{\boldsymbol{\theta}=\boldsymbol{\theta}_0+\boldsymbol{\Delta}/\sqrt{n}}$$

$$= \frac{1}{\sqrt{n}}\sum_{i=1}^{n}\frac{\partial}{\partial\boldsymbol{\theta}}\log f(X_i;\boldsymbol{\theta})\Big|_{\boldsymbol{\theta}=\boldsymbol{\theta}_0}$$

$$+ \left\{\frac{1}{n}\sum_{i=1}^{n}\frac{\partial}{\partial\boldsymbol{\theta}\partial\boldsymbol{\theta}^t}\log f(X_i;\boldsymbol{\theta})\Big|_{\boldsymbol{\theta}=\boldsymbol{\theta}_n^*}\right\}\boldsymbol{\Delta}$$

where $\boldsymbol{\theta}_n^*$ belongs to the line segment joining $\boldsymbol{\theta}_0$ and $\boldsymbol{\theta}_0 + (\boldsymbol{\Delta}/\sqrt{n})$. Using the Khintchine Weak Law of Large Numbers (Theorem 2.3.6), the Central Limit Theorem 3.3.1 and the Slutsky Theorem (3.4.2) we may conclude that, under A_n,

$$n^{-1/2}\sum_{i=1}^{n}(\partial/\partial\boldsymbol{\theta})\log f(X_i;\boldsymbol{\theta})|_{\boldsymbol{\theta}=\boldsymbol{\theta}_0} \xrightarrow{\mathcal{D}} N(\mathbf{I}(\boldsymbol{\theta}_0)\boldsymbol{\Delta}, \mathbf{I}(\boldsymbol{\theta}_0)),$$

which implies $Q_R \xrightarrow{\mathcal{D}} \chi_q^2(\boldsymbol{\Delta}^t \mathbf{I}(\boldsymbol{\theta}_0)\boldsymbol{\Delta})$.

To complete the proof, note that $\boldsymbol{\theta}_n^*$ and $\boldsymbol{\theta}_n^{**}$ in (5.6.3) are defined in such a way that (5.6.4) may be replaced by

$$Q_L = n(\widehat{\boldsymbol{\theta}}_n - \boldsymbol{\theta}_0)^t \mathbf{I}(\boldsymbol{\theta}_0)(\widehat{\boldsymbol{\theta}}_n - \boldsymbol{\theta}_0) + o_{p_{\boldsymbol{\theta}_n}}(1).$$

Therefore, using the Slutsky Theorem (3.4.2) and (5.6.6) we have

$$Q_L \xrightarrow{\mathcal{D}} \chi_q^2(\boldsymbol{\Delta}^t \mathbf{I}(\boldsymbol{\theta}_0)\boldsymbol{\Delta}).$$

∎

Next we consider tests of the (composite) hypothesis

$$H_0 : \mathbf{h}(\boldsymbol{\theta}) = \mathbf{0} \quad \text{vs.} \quad A : \mathbf{h}(\boldsymbol{\theta}) \neq \mathbf{0} \tag{5.6.7}$$

where $\mathbf{h} : I\!\!R^q \to I\!\!R^r$ is a vector-valued function such that the $(q \times r)$ matrix $\mathbf{H}(\boldsymbol{\theta}) = (\partial/\partial\boldsymbol{\theta})\mathbf{h}(\boldsymbol{\theta})$ exists and is continuous in $\boldsymbol{\theta}$ and $\text{rank}(\mathbf{H}(\boldsymbol{\theta})) = r$. In this direction, first note that (5.6.7) is equivalent to

$$H_0 : \boldsymbol{\theta} = \mathbf{g}(\boldsymbol{\beta}) \quad \text{vs.} \quad A : \boldsymbol{\theta} \neq \mathbf{g}(\boldsymbol{\beta}), \tag{5.6.8}$$

where $\mathbf{g} : I\!\!R^{q-r} \to I\!\!R^q$ is a vector-valued function such that the $(q \times q - r)$ matrix $\mathbf{G}(\boldsymbol{\beta}) = (\partial/\partial\boldsymbol{\beta})\mathbf{g}(\boldsymbol{\beta})$ exists and $\text{rank}[\mathbf{G}(\boldsymbol{\beta})] = q - r$ and $\boldsymbol{\beta} \in I\!\!R^{q-r}$. The expression (5.6.7) is known as the **constraint formulation** and (5.6.8) as the **freedom equation formulation** for the specified hypothesis. For example, let $q = 3$, $\boldsymbol{\theta} = (\theta_1, \theta_2, \theta_3)^t$ and $\mathbf{h}(\boldsymbol{\theta}) = \theta_1 - \theta_2$; then $\boldsymbol{\beta} = (\beta_1, \beta_2)^t$ and $\mathbf{g}(\boldsymbol{\beta}) = (\beta_1, \beta_1, \beta_2)^t$.

The three test statistics discussed above may also be considered in the present context. More specifically, letting $\widehat{\boldsymbol{\theta}}_n$ denote the (unrestricted) MLE of $\boldsymbol{\theta}$ and $\overline{\boldsymbol{\theta}}_n$ the restricted MLE of $\boldsymbol{\theta}$, i.e., subject to $\mathbf{h}(\overline{\boldsymbol{\theta}}_n) = \mathbf{0}$, we have:

i) **Wald's statistic:**

$$Q_W = n\mathbf{h}(\widehat{\boldsymbol{\theta}}_n)^t \left\{ [\mathbf{H}(\widehat{\boldsymbol{\theta}}_n)]^t [\mathbf{I}(\widehat{\boldsymbol{\theta}}_n)]^{-1} \mathbf{H}(\widehat{\boldsymbol{\theta}}_n) \right\}^{-1} \mathbf{h}(\widehat{\boldsymbol{\theta}}_n)$$

(actually, $\widehat{\boldsymbol{\theta}}_n$ may be replaced by any BAN estimator of $\boldsymbol{\theta}$);

ii) **Wilk's likelihood ratio statistic:**

$$Q_L = -2\log\lambda_n = 2\left\{ \log L_n(\widehat{\boldsymbol{\theta}}_n) - \log L_n(\overline{\boldsymbol{\theta}}_n) \right\}$$

where

$$\lambda_n = \sup_{\{\boldsymbol{\theta}\in\Theta:\mathbf{h}(\boldsymbol{\theta})=\mathbf{0}\}} L_n(\boldsymbol{\theta}) / \sup_{\boldsymbol{\theta}\in\Theta} L_n(\boldsymbol{\theta});$$

note that if we let $\widehat{\boldsymbol{\beta}}_n$ denote the MLE of $\boldsymbol{\beta}$, it follows that

$$\sup_{\boldsymbol{\beta}\in R^{q-r}} L_n(\boldsymbol{\beta}) = \sup_{\{\boldsymbol{\theta}\in\Theta:\mathbf{h}(\boldsymbol{\theta})=\mathbf{0}\}} L_n(\boldsymbol{\theta})$$

and the likelihood ratio statistic may be expressed as

$$Q_L = 2\left\{\log L_n(\widehat{\boldsymbol{\theta}}_n) - \log L_n(\widehat{\boldsymbol{\beta}}_n)\right\};$$

iii) **Rao's efficient score statistic:**

$$Q_R = n^{-1}[\mathbf{U}_n(\overline{\boldsymbol{\theta}}_n)]^t[\mathbf{I}(\overline{\boldsymbol{\theta}}_n)]^{-1}\mathbf{U}_n(\overline{\boldsymbol{\theta}}_n).$$

The analogue of Theorem 5.6.1 for composite hypotheses is the following:

Theorem 5.6.3: *Let $X_1,\ldots,X_n$ be i.i.d. r.v.'s with p.d.f. $f(x;\boldsymbol{\theta})$, $\boldsymbol{\theta} \in \boldsymbol{\Theta} \subset \mathbb{R}^q$, satisfying the conditions of Theorem 5.2.1 and consider the problem of testing (5.6.7). Then each of the statistics Q_W, Q_L and Q_R indicated above has an asymptotic χ_r^2 distribution under H_0.*

Proof: To show that $Q_W \xrightarrow{\mathcal{D}} \chi_r^2$ under H_0, use the same argument employed in Theorem 5.4.1 to see that

$$\sqrt{n}\left\{\mathbf{h}(\widehat{\boldsymbol{\theta}}_n) - \mathbf{h}(\boldsymbol{\theta})\right\} \xrightarrow{\mathcal{D}} N\left\{\mathbf{0}, [\mathbf{H}(\boldsymbol{\theta})]^t[\mathbf{I}(\boldsymbol{\theta})]^{-1}\mathbf{H}(\boldsymbol{\theta})\right\}$$

and then repeat the steps of the first part of Theorem 5.6.1.

To deal with Q_L, first recall that

$$n^{-1/2}\mathbf{U}_n(\boldsymbol{\theta}) \xrightarrow{\mathcal{D}} N(\mathbf{0},\mathbf{I}(\boldsymbol{\theta})). \tag{5.6.9}$$

Then consider the following Taylor expansion:

$$\begin{aligned}
0 &= \frac{1}{\sqrt{n}}\frac{\partial}{\partial\boldsymbol{\theta}}\log L_n(\boldsymbol{\theta})\Big|_{\boldsymbol{\theta}=\widehat{\boldsymbol{\theta}}_n} \\
&= \frac{1}{\sqrt{n}}\mathbf{U}_n(\boldsymbol{\theta}) + \left\{\frac{1}{n}\frac{\partial^2}{\partial\boldsymbol{\theta}\partial\boldsymbol{\theta}^t}\log L_n(\boldsymbol{\theta})\Big|_{\boldsymbol{\theta}=\boldsymbol{\theta}_n^*}\right\}\sqrt{n}(\widehat{\boldsymbol{\theta}}_n - \boldsymbol{\theta})
\end{aligned}$$

where $\boldsymbol{\theta}_n^*$ belongs to the line segment joining $\boldsymbol{\theta}$ and $\widehat{\boldsymbol{\theta}}_n$. Now, using the Khintchine Weak Law of Large Numbers (Theorem 2.3.6) we have

$$n^{-1}(\partial^2/\partial\boldsymbol{\theta}\partial\boldsymbol{\theta}^t)\log L_n(\boldsymbol{\theta})\big|_{\boldsymbol{\theta}=\boldsymbol{\theta}_n^*} \xrightarrow{\mathrm{P}} -\mathbf{I}(\boldsymbol{\theta})$$

and, therefore,

$$\sqrt{n}(\widehat{\boldsymbol{\theta}}_n - \boldsymbol{\theta}) = [\mathbf{I}(\boldsymbol{\theta})]^{-1}\frac{1}{\sqrt{n}}\mathbf{U}_n(\boldsymbol{\theta}) + o_p(\mathbf{1}). \tag{5.6.10}$$

Similarly, we may prove that

$$\sqrt{n}(\widehat{\boldsymbol{\beta}}_n - \boldsymbol{\beta}) = [\mathbf{I}^*(\boldsymbol{\beta})]^{-1}\mathbf{V}_n(\boldsymbol{\beta}) + o_p(\mathbf{1}). \tag{5.6.11}$$

where

$$\mathbf{V}_n(\boldsymbol{\beta}) = n^{-1/2}(\partial/\partial\boldsymbol{\beta})\log L_n(\boldsymbol{\beta})$$

and
$$\mathbf{I}^*(\boldsymbol{\beta}) = \mathrm{E}\left\{(\partial^2/\partial\boldsymbol{\beta}\partial\boldsymbol{\beta}^t)\log L_n(\boldsymbol{\beta})\right\}$$
is the corresponding Fisher information matrix. Then, using (5.6.10) we may write
$$
\begin{aligned}
Q_L^* &= 2\left\{\log L_n(\widehat{\boldsymbol{\theta}}_n) - \log L_n(\boldsymbol{\theta})\right\} \\
&= n(\widehat{\boldsymbol{\theta}}_n - \boldsymbol{\theta})^t \mathbf{I}(\boldsymbol{\theta})(\widehat{\boldsymbol{\theta}}_n - \boldsymbol{\theta}) \\
&= n^{-1}[\mathbf{U}_n(\boldsymbol{\theta})]^t [\mathbf{I}(\boldsymbol{\theta})]^{-1}\mathbf{U}_n(\boldsymbol{\theta}) + o_p(\mathbf{1}). \quad (5.6.12)
\end{aligned}
$$
Similarly, using (5.6.11) we obtain
$$
\begin{aligned}
Q_L^{**} &= 2\left\{\log L_n(\widehat{\boldsymbol{\beta}}_n) - \log L_n(\boldsymbol{\beta})\right\} \\
&= n(\widehat{\boldsymbol{\beta}}_n - \boldsymbol{\beta})^t [\mathbf{I}^*(\boldsymbol{\beta})](\widehat{\boldsymbol{\beta}}_n - \boldsymbol{\beta})[\mathbf{V}_n(\boldsymbol{\beta})]^t[\mathbf{I}^*(\boldsymbol{\beta})]^{-1}\mathbf{V}_n(\boldsymbol{\beta}) \\
&\quad + o_p(1). \quad (5.6.13)
\end{aligned}
$$
Now observe that
$$
\begin{aligned}
\mathbf{V}_n(\boldsymbol{\beta}) &= \frac{1}{\sqrt{n}}\frac{\partial}{\partial\boldsymbol{\beta}}\log L_n(\boldsymbol{\beta}) \\
&= \frac{\partial}{\partial\boldsymbol{\beta}}\mathbf{g}(\boldsymbol{\beta})\frac{1}{\sqrt{n}}\frac{\partial}{\partial\mathbf{g}(\boldsymbol{\beta})}\log L_n[\mathbf{g}(\boldsymbol{\beta})] \\
&= \frac{1}{\sqrt{n}}[\mathbf{G}(\boldsymbol{\beta})]^t\mathbf{U}_n(\boldsymbol{\beta}). \quad (5.6.14)
\end{aligned}
$$
Also, since
$$\mathbf{V}_n(\boldsymbol{\beta}) \xrightarrow{\mathcal{D}} N(\mathbf{0}, \mathbf{I}^*(\boldsymbol{\beta}))$$
and
$$n^{-1/2}[\mathbf{G}(\boldsymbol{\beta})]^t\mathbf{U}_n(\boldsymbol{\theta}) \xrightarrow{\mathcal{D}} N(\mathbf{0}, [\mathbf{G}(\boldsymbol{\beta})]^t[\mathbf{I}(\boldsymbol{\theta})]^{-1}\mathbf{G}(\boldsymbol{\beta}))$$
we may conclude that
$$\mathbf{I}^*(\boldsymbol{\beta}) = [\mathbf{G}(\boldsymbol{\beta})]^{-1}\mathbf{I}(\boldsymbol{\theta})([\mathbf{G}(\boldsymbol{\beta})]^t)^{-1}. \quad (5.6.15)$$
From the definition of Q_L and the equivalence between (5.6.7) and (5.6.8) we have $Q_L = Q_L^* - Q_L^{**}$; thus, using (5.6.12) – (5.6.14) it follows that
$$Q_L = n^{-1}[\mathbf{U}_n(\boldsymbol{\theta})]^t\left\{[\mathbf{I}(\boldsymbol{\theta})]^{-1} - \mathbf{G}(\boldsymbol{\beta})[\mathbf{I}^*(\boldsymbol{\beta})]^{-1}[\mathbf{G}(\boldsymbol{\beta})]^t\right\}\mathbf{U}_n(\boldsymbol{\theta}) + o_p(1).$$
$$(5.6.16)$$
Using (5.6.15) we have
$$
\begin{aligned}
&\left\{[\mathbf{I}(\boldsymbol{\theta})]^{-1} - \mathbf{G}(\boldsymbol{\beta})\mathbf{I}^*(\boldsymbol{\beta})\right\}\mathbf{I}(\boldsymbol{\theta})\left\{[\mathbf{I}(\boldsymbol{\theta})]^{-1} - \mathbf{G}(\boldsymbol{\beta})[\mathbf{I}^*(\boldsymbol{\beta})]^{-1}[\mathbf{G}(\boldsymbol{\beta})]^t\right\} \\
&= [\mathbf{I}(\boldsymbol{\theta})]^{-1}\mathbf{I}(\boldsymbol{\theta})[\mathbf{I}(\boldsymbol{\theta})]^{-1} - \mathbf{G}(\boldsymbol{\beta})[\mathbf{I}^*(\boldsymbol{\beta})]^{-1}[\mathbf{G}(\boldsymbol{\beta})]^t \\
&\quad - \mathbf{G}(\boldsymbol{\beta})[\mathbf{I}^*(\boldsymbol{\beta})]^{-1}[\mathbf{G}(\boldsymbol{\beta})]^t \\
&\quad + \mathbf{G}(\boldsymbol{\beta})[\mathbf{I}^*(\boldsymbol{\beta})]^{-1}[\mathbf{G}(\boldsymbol{\beta})]^t\mathbf{I}(\boldsymbol{\theta})\mathbf{G}(\boldsymbol{\beta})[\mathbf{I}^*(\boldsymbol{\beta})]^{-1}[\mathbf{G}(\boldsymbol{\beta})]^t \\
&= [\mathbf{I}(\boldsymbol{\theta})]^{-1} - \mathbf{G}(\boldsymbol{\beta})[\mathbf{I}(\boldsymbol{\beta})]^{-1}[\mathbf{G}(\boldsymbol{\beta})]^t. \quad (5.6.17)
\end{aligned}
$$

Furthermore,

$$
\begin{aligned}
\operatorname{tr}\left\{[\mathbf{I}(\boldsymbol{\theta})]^{-1}-\mathbf{G}(\boldsymbol{\beta})[\mathbf{I}^*(\boldsymbol{\beta})]^{-1}[\mathbf{G}(\boldsymbol{\beta})]^t\right\}\mathbf{I}(\boldsymbol{\theta}) & \\
= \operatorname{tr}\left\{\mathbf{I}_q-\mathbf{G}(\boldsymbol{\beta})[\mathbf{I}^*(\boldsymbol{\beta})]^{-1}[\mathbf{G}(\boldsymbol{\beta})]^t\mathbf{I}(\boldsymbol{\theta})\right\} & \\
= q-\operatorname{tr}\left\{[\mathbf{I}^*(\boldsymbol{\beta})]^{-1}[\mathbf{G}(\boldsymbol{\beta})]^t\mathbf{I}(\boldsymbol{\theta})\mathbf{G}(\boldsymbol{\beta})\right\} & \\
= q-\operatorname{tr}\left\{[\mathbf{I}^*(\boldsymbol{\beta})]^{-1}\mathbf{I}^*(\boldsymbol{\beta})\right\}=r. & \qquad (5.6.18)
\end{aligned}
$$

Then, using expressions (5.6.16)–(5.6.18), the Slutsky Theorem (3.4.2) and a well known result on the distribution of quadratic forms [see Searle (1971, Chap. 2), for example] it follows that, under H_0, $Q_L \xrightarrow{\mathcal{D}} \chi_r^2$.

To deal with Q_R, first recall that the (restricted) likelihood equations are

$$
\frac{\partial}{\partial\boldsymbol{\theta}}\log L_n(\boldsymbol{\theta})+\mathbf{H}(\boldsymbol{\theta})\boldsymbol{\lambda}=\mathbf{0},\qquad \mathbf{h}(\boldsymbol{\theta})=\mathbf{0}, \qquad (5.6.19)
$$

where $\boldsymbol{\lambda}\in\mathbb{R}^r$ is a vector of Lagrangian multipliers. Now, letting $\boldsymbol{\theta}_n=\boldsymbol{\theta}+\mathbf{u}/\sqrt{n}$ where $\|\mathbf{u}\|<K$, $0<K<\infty$, we may consider the following Taylor expansion:

$$
\begin{aligned}
\frac{1}{\sqrt{n}}\frac{\partial}{\partial\boldsymbol{\theta}}\log L_n(\boldsymbol{\theta})\bigg|_{\boldsymbol{\theta}=\boldsymbol{\theta}_n} & \\
=\frac{1}{\sqrt{n}}\mathbf{U}_n(\boldsymbol{\theta})+\left\{\frac{1}{n}\frac{\partial^2}{\partial\boldsymbol{\theta}\partial\boldsymbol{\theta}^t}\log L_n(\boldsymbol{\theta})\bigg|_{\boldsymbol{\theta}=\boldsymbol{\theta}_n^*}\right\}\sqrt{n}(\boldsymbol{\theta}_n-\boldsymbol{\theta}) &
\end{aligned}
$$

where $\boldsymbol{\theta}_n^*$ belongs to the line segment joining $\boldsymbol{\theta}$ and $\boldsymbol{\theta}_n$. Then, observing that

$$
n^{-1}(\partial^2/\partial\boldsymbol{\theta}\partial\boldsymbol{\theta}^t)\log L_n(\boldsymbol{\theta})\big|_{\boldsymbol{\theta}=\boldsymbol{\theta}_n^*}\xrightarrow{\mathrm{P}}-\mathbf{I}(\boldsymbol{\theta}),
$$

we may use the Khintchine Weak Law of Large Numbers (Theorem 2.3.6) to write

$$
\frac{1}{\sqrt{n}}\frac{\partial}{\partial\boldsymbol{\theta}}\log L_n(\boldsymbol{\theta})\bigg|_{\boldsymbol{\theta}=\boldsymbol{\theta}_n}=\frac{1}{\sqrt{n}}\mathbf{U}_n(\boldsymbol{\theta})-\mathbf{I}(\boldsymbol{\theta})\sqrt{n}(\boldsymbol{\theta}_n-\boldsymbol{\theta})+o_p(1).
$$

$$(5.6.20)$$

Also, since $\mathbf{H}(\boldsymbol{\theta})$ is continuous in $\boldsymbol{\theta}$, we have

$$
\mathbf{h}(\boldsymbol{\theta}_n)=[\mathbf{H}(\boldsymbol{\theta})]^t\sqrt{n}(\boldsymbol{\theta}_n-\boldsymbol{\theta})+o_p(1). \qquad (5.6.21)
$$

Since the (restricted) MLE $\overline{\boldsymbol{\theta}}_n$ must satisfy (5.6.19), and in view of (5.6.20) and (5.6.21), we may write

$$
\begin{aligned}
\frac{1}{\sqrt{n}}\mathbf{U}_n(\boldsymbol{\theta})-\mathbf{I}(\boldsymbol{\theta})\sqrt{n}(\overline{\boldsymbol{\theta}}_n-\boldsymbol{\theta})+\mathbf{H}(\boldsymbol{\theta})\frac{1}{\sqrt{n}}\overline{\boldsymbol{\lambda}}_n+o_p(1) &= \mathbf{0}, \\
[\mathbf{H}(\boldsymbol{\theta})]^t\sqrt{n}(\overline{\boldsymbol{\theta}}_n-\boldsymbol{\theta})+o_p(1) &= \mathbf{0}
\end{aligned}
$$

$$(5.6.22)$$

which may be re-expressed in matrix notation as

$$\begin{pmatrix} \mathbf{I}(\boldsymbol{\theta}) & -\mathbf{H}(\boldsymbol{\theta}) \\ -[\mathbf{H}(\boldsymbol{\theta})]^t & \mathbf{0} \end{pmatrix} \begin{pmatrix} \sqrt{n}(\overline{\boldsymbol{\theta}}_n - \boldsymbol{\theta}) \\ n^{-1/2}\overline{\boldsymbol{\lambda}}_n \end{pmatrix} = \begin{pmatrix} \mathbf{U}_n(\boldsymbol{\theta}) \\ \mathbf{0} \end{pmatrix} + o_p(\mathbf{1}).$$

Thus,

$$\begin{pmatrix} \sqrt{n}(\overline{\boldsymbol{\theta}}_n - \boldsymbol{\theta}) \\ n^{-1/2}\overline{\boldsymbol{\lambda}}_n \end{pmatrix} = \begin{pmatrix} \mathbf{P}(\boldsymbol{\theta}) & \mathbf{Q}(\boldsymbol{\theta}) \\ [\mathbf{Q}(\boldsymbol{\theta})]^t & \mathbf{R}(\boldsymbol{\theta}) \end{pmatrix} \begin{pmatrix} n^{-1/2}\mathbf{U}_n(\boldsymbol{\theta}) \\ \mathbf{0} \end{pmatrix} + o_p(\mathbf{1}),$$

$$(5.6.23)$$

where

$$\begin{pmatrix} \mathbf{P}(\boldsymbol{\theta}) & \mathbf{Q}(\boldsymbol{\theta}) \\ [\mathbf{Q}(\boldsymbol{\theta})]^t & \mathbf{R}(\boldsymbol{\theta}) \end{pmatrix} = \begin{pmatrix} \mathbf{I}(\boldsymbol{\theta}) & -\mathbf{H}(\boldsymbol{\theta}) \\ -[\mathbf{H}(\boldsymbol{\theta})]^t & \mathbf{0} \end{pmatrix}^{-1}$$

which implies

$$\begin{aligned} \mathbf{P}(\boldsymbol{\theta}) &= [\mathbf{I}(\boldsymbol{\theta})]^{-1} \left\{ \mathbf{I}_q - \mathbf{H}(\boldsymbol{\theta})([\mathbf{H}(\boldsymbol{\theta})]^t[\mathbf{I}(\boldsymbol{\theta})]^{-1}[\mathbf{H}(\boldsymbol{\theta})]^{-1}[\mathbf{H}(\boldsymbol{\theta})]^t[\mathbf{I}(\boldsymbol{\theta})]^{-1} \right\}, \\ \mathbf{Q}(\boldsymbol{\theta}) &= -[\mathbf{I}(\boldsymbol{\theta})]^{-1}\mathbf{H}(\boldsymbol{\theta})([\mathbf{H}(\boldsymbol{\theta})]^t[\mathbf{I}(\boldsymbol{\theta})]^{-1}\mathbf{H}(\boldsymbol{\theta}))^{-1}, \\ \mathbf{R}(\boldsymbol{\theta}) &= -([\mathbf{H}(\boldsymbol{\theta})]^t[\mathbf{I}(\boldsymbol{\theta})]^{-1}\mathbf{H}(\boldsymbol{\theta}))^{-1}. \end{aligned} \qquad (5.6.24)$$

Since by the Central Limit Theorem 3.3.1 and the Cramér-Wold Theorem (3.2.4), we have

$$\left[n^{-1/2}\mathbf{U}_n(\boldsymbol{\theta}), \mathbf{0} \right]^t \xrightarrow{\mathcal{D}} N \left\{ \mathbf{0}, \begin{pmatrix} \mathbf{I}(\boldsymbol{\theta}) & \mathbf{0} \\ \mathbf{0} & \mathbf{0} \end{pmatrix} \right\},$$

it follows from (5.6.23) and the Slutsky Theorem (3.4.2) that

$$\begin{pmatrix} \sqrt{n}(\overline{\boldsymbol{\theta}}_n - \boldsymbol{\theta}) \\ n^{-1/2}\overline{\boldsymbol{\lambda}}_n \end{pmatrix} \xrightarrow{\mathcal{D}} N(\mathbf{0}, \boldsymbol{\Sigma})$$

where

$$\boldsymbol{\Sigma} = ((\sigma_{ij})) = \begin{pmatrix} \mathbf{P}(\boldsymbol{\theta})\mathbf{I}(\boldsymbol{\theta})[\mathbf{P}(\boldsymbol{\theta})]^t & \mathbf{P}(\boldsymbol{\theta})\mathbf{I}(\boldsymbol{\theta})\mathbf{Q}(\boldsymbol{\theta}) \\ [\mathbf{Q}(\boldsymbol{\theta})]^t\mathbf{I}(\boldsymbol{\theta})[\mathbf{P}(\boldsymbol{\theta})]^t & [\mathbf{Q}(\boldsymbol{\theta})]^t\mathbf{I}(\boldsymbol{\theta})\mathbf{Q}(\boldsymbol{\theta}) \end{pmatrix}.$$

A few algebraic manipulations with the equations used to derive (5.6.24) will lead to

$$\mathbf{P}(\boldsymbol{\theta})\mathbf{I}(\boldsymbol{\theta})[\mathbf{P}(\boldsymbol{\theta})]^t = \mathbf{P}(\boldsymbol{\theta}),$$

$$\mathbf{P}(\boldsymbol{\theta})\mathbf{I}(\boldsymbol{\theta})\mathbf{Q}(\boldsymbol{\theta}) = \mathbf{0}$$

and

$$[\mathbf{Q}(\boldsymbol{\theta})]^t\mathbf{I}(\boldsymbol{\theta})\mathbf{Q}(\boldsymbol{\theta}) = -\mathbf{R}(\boldsymbol{\theta}).$$

Thus, we may write

$$\sigma_{11} = [\mathbf{I}(\boldsymbol{\theta})]^{-1} \left\{ \mathbf{I}_q - \mathbf{H}(\boldsymbol{\theta})([\mathbf{H}(\boldsymbol{\theta})]^t[\mathbf{I}(\boldsymbol{\theta})]^{-1}\mathbf{H}(\boldsymbol{\theta}))^{-1}[\mathbf{H}(\boldsymbol{\theta})]^t \right\} [\mathbf{I}(\boldsymbol{\theta})]^{-1},$$

$$\sigma_{22} = ([\mathbf{H}(\boldsymbol{\theta})]^t\mathbf{I}(\boldsymbol{\theta})\mathbf{H}(\boldsymbol{\theta}))^{-1},$$

and
$$\sigma_{12} = \sigma_{21} = \mathbf{0}.$$
Now, using (5.6.19), we have $\mathbf{U}_n(\overline{\boldsymbol{\theta}}_n) = -\mathbf{H}(\boldsymbol{\theta}_n)\overline{\boldsymbol{\lambda}}_n/\sqrt{n}$. Thus,
$$
\begin{aligned}
Q_R &= n^{-1}[\mathbf{U}_n(\overline{\boldsymbol{\theta}}_n)]^t[\mathbf{I}(\overline{\boldsymbol{\theta}}_n)]^{-1}\mathbf{U}_n(\overline{\boldsymbol{\theta}}_n) \\
&= n^{-1}\overline{\boldsymbol{\lambda}}_n^t[\mathbf{H}(\overline{\boldsymbol{\theta}}_n)]^t[\mathbf{I}(\overline{\boldsymbol{\theta}}_n)]^{-1}\mathbf{H}(\overline{\boldsymbol{\theta}}_n)\overline{\boldsymbol{\lambda}}_n \\
&= n^{-1}\overline{\boldsymbol{\lambda}}_n^t[\mathbf{R}(\overline{\boldsymbol{\theta}}_n)]^{-1}\overline{\boldsymbol{\lambda}}_n.
\end{aligned}
$$
From (5.6.24) we conclude that
$$Q_R^* = n^{-1}\overline{\boldsymbol{\lambda}}_n^t[\mathbf{R}(\boldsymbol{\theta})\overline{\boldsymbol{\lambda}}_n)]^{-1} \xrightarrow{D} \chi_r^2,$$
under H_0, and since $\mathbf{R}(\overline{\boldsymbol{\theta}}_n) \xrightarrow{P} \mathbf{R}(\boldsymbol{\theta})$, we may use an argument similar to that of Theorem 5.6.1 to conclude that
$$Q_R \xrightarrow{D} \chi_r^2 \quad \text{under} \quad H_0.$$

∎

The reader is referred to Silvey (1959) for details on the asymptotic distribution of Q_W, Q_L and Q_R under alternative hypotheses.

5.7 Concluding notes

It is virtually impossible to provide a complete coverage of the asymptotic theory of estimation and tests in a limited space and, especially, at the contemplated level of presentation. We have not been able to fathom into several important topics within this broad domain. Two most notable ones are the following:

i) statistical inference in nonregular models,

ii) asymptotically efficient (adaptive) procedures.

Nonregular models relate to the situation where the assumptions in Theorem 5.2.1 (5.2.2) may not hold. Example 5.2.5 is a notable case in this context. Note that the likelihood function $L_n(\theta)$ can be factorized here as $g_n(X_{n:n}, \theta)L_n^*(X_1, \ldots, X_n)$, where $g_n(x, \theta) = nx^{n-1}\theta^{-n}I_{[0,\theta]}(x)$ and L_n^* does not depend on θ. Thus, $X_{n:n}$ is a sufficient statistic, and the distribution theory of sample extreme values, studied in Section 4.4, may be used to conclude that

$$n(\theta - X_{n:n})/\theta \quad \text{has asymptotically the simple exponential law} \quad (5.7.1)$$

(which is not normal). This prescription works out well in a class of problems where the factorization of the likelihood function holds and, moreover,

$$g_n(x, \theta) = \frac{h_n(x)}{\rho_n(\theta)}, \tag{5.7.2}$$

for suitable $h_n(\cdot)$ and $\rho(\cdot)$. We pose a few problems to verify such a contention (Exercises 5.7.1 – 5.7.2).

We have tacitly assumed that the parameter space Θ is an open space and θ lies in the interior of Θ. This allows us to justify the quadratic approximation in (5.2.10) (or elsewhere) which plays a fundamental role in the studied asymptotics. There are problems (of good interest) where the parameter space may violate this condition. For example, consider the following simple (mixture) model where the density $f(x, \boldsymbol{\theta})$ can be written as

$$f(x; \theta_1, \theta_2, \theta_3) = \theta_3 \varphi(x - \theta_1) + (1 - \theta_3)\varphi(x - \theta_2),$$

with $\theta_3 \in [0, 1]$, $\theta_1 \in \mathbb{R}$, $\theta_2 \in \mathbb{R}$, and φ being the standard normal density. Contrary to the three-parameter model, one may have a single-parameter model when either of the following two situation arise:

i) $\theta_3 = 0$ or 1

ii) $\theta_1 = \theta_2$. Here $\boldsymbol{\Theta} = \mathbb{R} \times \mathbb{R} \times [0, 1]$. On the vertical diagonal **hyperplane**

$$\{\theta_1 = \theta_2 \in \mathbb{R}, 0 \le \theta_3 \le 1\}$$

as well as on the two **edges** $\theta_3 = 0$, $\theta_3 = 1$, we have a serious identifiability problem.

This, in turn, raises serious problems regarding the consistency of the MLE of $\boldsymbol{\theta}$ as well as in their possible asymptotic behavior. We may refer to Chernoff (1954) and Ghosh and Sen (1985) for some related studies. Another important area is hypothesis testing against restricted alternatives. The restricted model, such as orthant/ordered alternatives relating to the components of $\boldsymbol{\theta}$, generally do not satisfy the regularity conditions of Theorem 5.2.1 (or 5.2.2), so that the theory developed in Section 5.6 may not cover such problems too. There are some simplifications in some specialized models [viz., Barlow et al. (1972)], although a complete answer (especially in the non-null hypotheses case) is yet unknown.

The score function $(\partial/\partial\theta) \log f(x, \theta)$ plays the most fundamental role in the specification of asymptotically optimal estimators as well as test statistics. However, in a practical problem, rarely, is the form of $f(\cdot)$, and, hence, this score function, known. There has been, therefore, a spur of research activities in trying to estimate the score function itself from the data set at hand and then to base the estimators/test statistics on such estimated scores. This is known as an **adaptive procedure**. Generally a very large sample size is required to accomplish this dual job, and the mathematical manipulations are, indeed, very heavy and well beyond the contemplated level of this book. As such, we will not be tempted to include pertinent introductions to such a novel area of statistical research. However, Chapters 7 and 8 contain some discussions which might be of some help to the applications oriented readers.

5.8 Exercises

Exercise 5.2.1: For Example 5.2.8, show that $n = Y_1 + \cdots + Y_m$ where the Y_i are i.i.d. r.v.'s with mean π^{-1} and a finite variance. Hence, show that as $m \to \infty$, $(n - m/\pi)/\sqrt{m}$ is closely normally distributed. Use the transformation $T_m = m/n = g(n/m)$ to derive the asymptotic normality of $\sqrt{m}(T_m - \pi)$.

Exercise 5.3.1: Use the decomposition

$$V_n = n^{-m} n^{[m]} U_n + R_n, \quad n \geq m,$$

and show that under (5.3.19), $R_n = O(n^{-1})$ a.s. and $\mathrm{E}R_n^2 = O(n^{-2})$. Hence, or otherwise, show that the bias of V_n is $O(n^{-1})$.

Exercise 5.3.2: Use (5.3.23) and (5.3.24) to show that

$$S_n^2 = a_{n0} U_n^0 + a_{n1} U_n^1 + \cdots + a_{n2m} U_n^{2m},$$

where $a_{n0} = 1 + O(n^{-1})$, $a_{nk} = O(n^{-k})$, $k \geq 1$, and the U_n^j are all U-statistics with finite expectations and $\mathrm{E}U_n^0 = \xi_1$. Hence, or otherwise, use the reverse martingale property of U-statistics to verify that $S_n^2 \xrightarrow{\text{a.s.}} \xi_1$.

Exercise 5.3.3 (Continued): For $m = 1$, show that S_n^2 reduces to the sample variance of the $Y_i = g(X_i)$, $i \leq n$.

Exercise 5.3.4: Use (5.3.33) and (5.3.35) to verify (5.3.36).

Exercise 5.3.5: Work out the first order expansion for $W_{n,k}$ by using (5.3.46) along with (5.3.30)–(5.3.31) for $T_{n,k}^*$.

Exercise 5.7.1: For the exponential (location) model: $f(x, \theta) = \exp\{-(x - \theta)\}$, $x \geq \theta$, for a sample $X_1, \ldots, X_n$, show that the MLE of θ is $X_{n:1}$ and verify that (5.7.1) holds for $n(X_{n:1} - \theta)$.

Exercise 5.7.2: Consider a U-shaped distribution with density function

$$f(x, \theta) = c(\theta) \left\{ 1 - e^{-|x|} \right\} I_{\{|x| \leq \theta\}}.$$

Show that the MLE of θ is $\widehat{\theta}_n = (X_{n:n} - X_{n:1})/2$. Verify that (5.7.1) holds for $n(\widehat{\theta}_n - \theta)$, but with a Gamma distribution.

CHAPTER 6

Large Sample Theory for Categorical Data Models

6.1 Introduction

In general, categorical data models relate to **count data** corresponding to the classification of sampling units into **groups** or **categories** either on a qualitative or some quantitative basis. These categories may be defined by the essentially discrete nature of the phenomenon under study (see Example 1.2.2 dealing with the OAB blood classification model) or, often for practical reasons, by the grouping of the values of an essentially continuous underlying distribution (for example, shoe sizes: 5, $5\frac{1}{2}$, 6, $6\frac{1}{2}$, etc., corresponding to half-open intervals for the actual length of a foot). Even in the qualitative case there is often an implicit ordering in the categories resulting in **ordered categorical data** (viz., ratings: excellent, very good, good, fair and poor, for a research proposal under review). Except in some of the most simple cases, exact statistical analysis for categorical data models may not be available in a unified, simple form. Hence, asymptotic methods are important in this context. They not only provide a unified coverage of statistical methodology appropriate for large sample sizes but also suggest suitable modifications which may often be appropriate for moderate to small sample sizes. This chapter is devoted to the study of this related asymptotic theory.

Although there are a few competing probabilistic models for statistical analysis of categorical data sets, we shall find it convenient to concentrate on the so-called **product multinomial model** which encompasses a broad domain and plays a key role in the development of appropriate statistical analysis tools. Keeping in mind the OAB blood classification model (viz., Example 1.2.2), we may conceive of r ($\geq k$) categories (indexed as $1, \ldots, r$, respectively), so that an observation may belong to the jth category with a probability π_j, $j = 1, \ldots, r$, where the π_j are non-negative numbers adding up to 1. Thus, for a sample of n independent observations, we have the simple multinomial law for the count data $n_1, \ldots, n_r$ (the respective cell frequencies):

$$P\{n_1, \ldots, n_r\} = \left(n! \bigg/ \prod_{j=1}^{r} n_j! \right) \prod_{j=1}^{r} \pi_j^{n_j}, \qquad (6.1.1)$$

where $\sum_{j=1}^{r} n_j = n$ and $\sum_{j=1}^{r} \pi_j = 1$. Starting with this simple multinomial law, we may conceive of a more general situation corresponding to s (≥ 1) independent samples (of sizes $n_{1\cdot}, \ldots, n_{s\cdot}$, respectively) drawn (with replacement) from s populations whose elements may be classified into r_i (≥ 2) categories, $i = 1, \ldots, s$. Let n_{ij} denote the frequency (count) of sample units from the $n_{i\cdot}$ individuals in the ith sample classified into the jth category ($1 \leq j \leq r_i$) and let

$$\mathbf{n}_i = (n_{i1}, \ldots, n_{ir_i})^t$$

denote the corresponding vector of observed frequencies for the ith sample, $i = 1, \ldots, s$. Also, let

$$\boldsymbol{\pi}_i = (\pi_{i1}, \ldots, \pi_{ir_i})^t$$

denote the corresponding probabilities of classification, $i = 1, \ldots, s$, and let $\boldsymbol{\pi} = (\boldsymbol{\pi}_1^t, \ldots, \boldsymbol{\pi}_s^t)^t$ and $\mathbf{n} = (\mathbf{n}_1^t, \ldots, \mathbf{n}_s^t)^t$. Then the probability law for $\mathbf{n}$ is given by

$$P(\mathbf{n} \mid \boldsymbol{\pi}) = \prod_{i=1}^{s} \left\{ \left(n_i! \bigg/ \prod_{j=1}^{r_i} n_{ij}! \right) \prod_{j=1}^{r_i} \pi_{ij}^{n_{ij}} \right\}, \qquad (6.1.2)$$

where $\sum_{j=1}^{r_i} \pi_{ij} = \boldsymbol{\pi}_i^t \mathbf{1}_{r_i} = 1$, $i = 1, \ldots, s$. In the literature this is termed the product multinomial law. For $s = 1$, (6.1.2) reduces to the classical multinomial law in (6.1.1), and, hence, the genesis is easy to comprehend. In this setup, recall that $\boldsymbol{\pi}_i$ belongs to the r_i-dimensional simplex $\boldsymbol{\gamma}_i = \{\mathbf{x} \in E^{r_i} : \mathbf{x} \geq \mathbf{0}$ and $\mathbf{x}^t \mathbf{1}_{r_i} = 1\}$, $i = 1, \ldots, s$, so that $\boldsymbol{\pi} \in \boldsymbol{\Gamma} = \boldsymbol{\gamma}_1 \times \cdots \times \boldsymbol{\gamma}_s$. If all the r_i are equal to 2, we have a **product binomial model**.

In many problems of practical interest, within the general framework of (6.1.2), the π_{ij} can be expressed as functions of a vector of parameters

$$\boldsymbol{\theta} = (\theta_1, \ldots, \theta_q)^t$$

for some $q \leq \sum_{j=1}^{s} (r_j - 1)$. For example, in the blood groups model (treated in Example1.2.2) we have $s = 1$, $r_1 = r = 4$ and $\boldsymbol{\theta} = (\theta_1, \theta_2)^t$ where $0 \leq \theta_1 \leq 1$, $0 \leq \theta_2 \leq 1$ and $0 \leq \theta_1 + \theta_2 \leq 1$. In this setup we may write

$$\gamma = \left\{ (\pi_1, \pi_2, \pi_3, \pi_4) : \pi_j > 0, \ 1 \leq j \leq 4 \quad \text{and} \quad \sum_{j=1}^{4} \pi_j = 1 \right\}$$

and

$$\boldsymbol{\theta} = \left\{ (\theta_1, \theta_2, \theta_3) : \theta_i > 0, \ 1 \le i \le 3 \quad \text{and} \quad \sum_{i=1}^{3} \theta_i = 1 \right\},$$

and $\boldsymbol{\pi}$ has the domain $\boldsymbol{\gamma}$, whereas $\boldsymbol{\theta}$ has the domain $\boldsymbol{\Theta}$. Keeping this in mind, we may write, for the general model in (6.1.2),

$$\boldsymbol{\pi} = \boldsymbol{\pi}(\boldsymbol{\theta}), \quad \boldsymbol{\theta} \in \boldsymbol{\Theta}, \quad \boldsymbol{\pi} \in \boldsymbol{\Gamma}, \tag{6.1.3}$$

and treat the functional forms of the $\pi_{ij}(\boldsymbol{\theta})$ to be given and $\boldsymbol{\theta}$ as an unknown parameter (vector). Based on this formulation, the relevant questions may be expressed in the form of tests of goodness of fit [for the transformation model in (6.1.3)], or more generally of tests of hypotheses about $\boldsymbol{\theta}$ as well as of estimation of the elements of $\boldsymbol{\theta}$.

Section 6.2 is devoted to the study of goodness-of-fit tests for categorical data models. The classical Pearsonian goodness-of-fit test for a simple multinomial distribution is considered as a precursor for general goodness-of-fit tests for the product multinomial model in (6.1.2)–(6.1.3). The relevant parametric (asymptotic) theory (based on BAN estimators of $\boldsymbol{\theta}$) is presented in Section 6.3. The likelihood ratio and Wald tests for general hypotheses on the transformed model (6.1.3) and related large sample theory are presented in a unified, yet simple fashion. In the concluding section we briefly discuss the asymptotic properties of some other statistics commonly employed in the analysis of categorical data.

6.2 Nonparametric goodness-of-fit tests

To motivate the proposed large sample methodology, first, we consider the simple multinomial law in (6.1.1), so that we have

$$P\{\mathbf{n} \mid \boldsymbol{\pi}\} = \left(n! \Big/ \prod_{j=1}^{r} n_j! \right) \prod_{j=1}^{r} \pi_j^{n_j}, \quad n = \sum_{j=1}^{r} n_j, \quad \sum_{j=1}^{r} \pi_j = 1. \tag{6.2.1}$$

Note that, under (6.2.1),

$$\mathrm{E}(\mathbf{n}) = n\boldsymbol{\pi} \quad \text{and} \quad \mathrm{Var}\,(\mathbf{n}) = n\{\mathbf{D}_{\boldsymbol{\pi}} - \boldsymbol{\pi}\boldsymbol{\pi}^t\} \tag{6.2.2}$$

where $\boldsymbol{\pi} = (\pi_1, \ldots, \pi_r)^t$ and $\mathbf{D}_{\boldsymbol{\pi}} = \mathrm{Diag}\,(\pi_1, \ldots, \pi_r)$. Thus, $\mathbf{n} - n\boldsymbol{\pi}$ stands for the deviation of the observed frequencies from their model-based expectations. Hence, to assess the fit of (6.2.1), one may consider the following Pearsonian goodness-of-fit statistic (assuming that $\boldsymbol{\pi}$ is specified):

$$Q_P = Q_P(\boldsymbol{\pi}) = \sum_{j=1}^{r} (n_j - n\pi_j)^2 / (n\pi_j); \tag{6.2.3}$$

the case of $\boldsymbol{\pi}$ as in (6.1.3) with unknown $\boldsymbol{\theta}$ will be treated in the next section.

For a given $\boldsymbol{\pi}$, the exact distribution of Q_P can be enumerated by using (6.2.1) when n is not too large. However, this scheme not only depends on the specific $\boldsymbol{\pi}$ but also becomes prohibitively laborious as n becomes large. For this reason, we are interested in obtaining the asymptotic distribution of Q_P in a simpler form. In this context, we set

$$\mathbf{Z}_n = (Z_{n1}, \ldots, Z_{nr})^t$$

with $Z_{nj} = (n_j - n\pi_j)/\sqrt{n\pi_j}$, $1 \leq j \leq r$, so that $Q_P = \mathbf{Z}_n^t \mathbf{Z}_n = \|\mathbf{Z}_n\|^2$. We may also set $\boldsymbol{\pi}^{1/2} = (\pi_1^{1/2}, \ldots, \pi_r^{1/2})^t$ and write

$$\mathbf{Z}_n = n^{-1/2} \mathbf{D}_{\boldsymbol{\pi}^{1/2}}^{-1} (\mathbf{n} - n\boldsymbol{\pi}), \tag{6.2.4}$$

where $\mathbf{D}_{\boldsymbol{\pi}^{1/2}} = \mathrm{Diag}(\pi_1^{1/2}, \ldots, \pi_r^{1/2})$. Then, we have

$$\mathbf{Z}_n^t \boldsymbol{\pi}^{1/2} = n^{-1/2} (\mathbf{n} - n\boldsymbol{\pi})^t \mathbf{1}_r = n^{-1/2} \sum_{j=1}^r (n_j - n\pi_j) = 0$$

and

$$Q_P = \|\mathbf{Z}_n\|^2 = \mathbf{Z}_n^t \mathbf{Z}_n = \mathbf{Z}_n^t \{ \mathbf{I}_r - (\boldsymbol{\pi}^{1/2})(\boldsymbol{\pi}^{1/2})^t \} \mathbf{Z}_n,$$

i.e.,

$$Q_P = \mathbf{Z}_n^t \{ \mathbf{I}_r - (\boldsymbol{\pi}^{1/2})(\boldsymbol{\pi}^{1/2})^t \} \mathbf{Z}_n, \tag{6.2.5}$$

where

$$\left\{ \mathbf{I}_r - (\boldsymbol{\pi}^{1/2})(\boldsymbol{\pi}^{1/2})^t \right\} \left\{ \mathbf{I}_r - (\boldsymbol{\pi}^{1/2})(\boldsymbol{\pi}^{1/2})^t \right\} = \mathbf{I}_r - (\boldsymbol{\pi}^{1/2})(\boldsymbol{\pi}^{1/2})^t \tag{6.2.6}$$

by virtue of $(\boldsymbol{\pi}^{1/2})^t (\boldsymbol{\pi}^{1/2}) = \boldsymbol{\pi}^t \mathbf{1} = 1$. Let us define

$$\mathbf{X}_i = (X_{i1}, \ldots, X_{ir})^t, \quad i \geq 1,$$

by

$$\mathbf{X}_{ij} = \begin{cases} 1 & \text{if the } i\text{th sampling is classified in the } j\text{th cell} \\ 0 & \text{otherwise,} \end{cases} \tag{6.2.7}$$

$j = 1, \ldots, p$. This clearly implies that for all $i, j, l = 1, \ldots, r$:

i) $EX_{ij} = \pi_{ij}$,

ii) $EX_{ij} X_{il} = \begin{cases} \pi_j, & j = l \\ 0, & j \neq l, \end{cases}$

iii) $\mathrm{Cov}(X_{ij}, X_{il}) = \begin{cases} \pi_j(1 - \pi_j), & j = l \\ -\pi_j \pi_l, & j \neq l. \end{cases}$

Therefore, $\sum_{i=1}^{n} \mathbf{X}_i = \mathbf{n}$, $\mathbf{EX}_i = \boldsymbol{\pi}$ and $\text{Var} \, \mathbf{X}_i = \mathbf{D}_{\boldsymbol{\pi}} - \boldsymbol{\pi} \boldsymbol{\pi}^t$. Then write

$$\mathbf{Z}_n = n^{-1/2} \sum_{i=1}^{n} \mathbf{D}_{\boldsymbol{\pi}^{1/2}}^{-1} (\mathbf{X}_i - \boldsymbol{\pi}) = n^{-1/2} \sum_{i=1}^{n} \mathbf{Y}_i$$

where

$$\mathbf{Y}_i = \mathbf{D}_{\boldsymbol{\pi}^{1/2}}^{-1} (\mathbf{X}_i - \boldsymbol{\pi}),$$
$$\mathbf{EY}_i = 0,$$
$$\text{Var} \, \mathbf{Y}_i = \mathbf{D}_{\boldsymbol{\pi}^{1/2}}^{-1} \{ \mathbf{D}_{\boldsymbol{\pi}} - \boldsymbol{\pi} \boldsymbol{\pi}^t \} \mathbf{D}_{\boldsymbol{\pi}^{1/2}}^{-1} = \mathbf{I}_r - \boldsymbol{\pi}^{1/2} (\boldsymbol{\pi}^{1/2})^t$$

and use the (multivariate version of the) Central Limit Theorem 3.3.4 to see that

$$\mathbf{Z}_n \xrightarrow{\mathcal{D}} N_r(0, \mathbf{I}_r - \boldsymbol{\pi}^{1/2} (\boldsymbol{\pi}^{1/2})^t) \qquad (6.2.8)$$

where

$$
\begin{aligned}
\text{rank} \, \{ \mathbf{I}_r - \boldsymbol{\pi}^{1/2} (\boldsymbol{\pi}^{1/2})^t \} &= \text{tr} \, \{ \mathbf{I}_r - \boldsymbol{\pi}^{1/2} (\boldsymbol{\pi}^{1/2})^t \} \\
&= r - \text{tr} \, \{ \boldsymbol{\pi}^{1/2} (\boldsymbol{\pi}^{1/2})^t \} \\
&= r - \text{tr} \, \{ (\boldsymbol{\pi}^{1/2})^t \boldsymbol{\pi}^{1/2} \} \\
&= r - 1.
\end{aligned}
$$

Finally, using (6.2.5), (6.2.8) and Cochran Theorem (3.4.7) we may conclude that $Q_P \xrightarrow{\mathcal{D}} \chi^2_{(r-1)}$ when (6.2.1) fits.

Consider now the problem of testing

$$H : \boldsymbol{\pi} = \boldsymbol{\pi}_0 = (\pi_{01}, \dots, \pi_{0r})^t$$

against a fixed alternative

$$A : \boldsymbol{\pi} = \boldsymbol{\pi}_1 = (\pi_{11}, \dots, \pi_{1r})^t$$

using the statistic $Q_P(\boldsymbol{\pi}_0)$. We have seen earlier that when H is true, $Q_P(\boldsymbol{\pi}_0) \xrightarrow{\mathcal{D}} \chi^2_{(r-1)}$. Let us examine the power of the test.

When A is true, it follows by the Khintchine Weak Law of Large Numbers (Theorem 2.3.6) that $n^{-1} \sum_{i=1}^{n} \mathbf{X}_i \xrightarrow{P} \boldsymbol{\pi}_1$. Then, writing

$$Q_P(\boldsymbol{\pi}_0) = \sum_{j=1}^{r} (n_j/n - \pi_{0j})^2 / \pi_{0j}$$

we may conclude that

$$n^{-1} Q_P(\boldsymbol{\pi}_0) \xrightarrow{P} \sum_{j=1}^{n} (\pi_{1j} - \pi_{0j})^2 / \pi_{0j} = \Delta > 0. \qquad (6.2.9)$$

Thus, letting $\chi^2_{(r-1)}(\alpha)$ denote the $100(1 - \alpha)$th percentile of the $\chi^2_{(r-1)}$ distribution, we have

$$P\{ Q_P(\boldsymbol{\pi}_0) \geq \chi^2_{(r-1)}(\alpha) \mid A \} = P\{ n^{-1} Q_P(\boldsymbol{\pi}_0) \geq n^{-1} \chi^2_{(r-1)}(\alpha) \mid A \} \to 1$$

as $n \to \infty$, in view of (6.2.9) and the fact that $n^{-1}\chi^2_{(r-1)}(\alpha) \to 0$ as $n \to \infty$. Hence, for all fixed alternative hypotheses, the power of the test converges to 1, i.e., the test is consistent for any departure from the null hypothesis.

Now let us consider what happens to the power of the test when the distance between the alternative and the null hypotheses converges to zero as n increases, i.e., for Pitman alternatives of the form

$$A_n : \boldsymbol{\pi} = \boldsymbol{\pi}_0 + n^{-1/2}\boldsymbol{\gamma}$$

where $\boldsymbol{\gamma} = (\gamma_1, \ldots, \gamma_r)^t$ is a vector of constants such that $\mathbf{1}_r^t \boldsymbol{\gamma} = 0$. Note that

$$\mathrm{E}(\mathbf{Y}_i \mid A) = \mathbf{D}_{\boldsymbol{\pi}_0^{1/2}}^{-1} \mathrm{E}(\mathbf{X}_i - \boldsymbol{\pi}_0 \mid A) = n^{-1/2}\mathbf{D}_{\boldsymbol{\pi}_0^{1/2}}^{-1}\boldsymbol{\gamma}$$

and

$$
\begin{aligned}
\mathrm{Var}\,(\mathbf{Y}_i \mid A) &= \mathbf{D}_{\boldsymbol{\pi}_0^{1/2}}^{-1}\mathrm{Var}\,(\mathbf{X}_i - \boldsymbol{\pi}_0 \mid A)\mathbf{D}_{\boldsymbol{\pi}_0^{1/2}}^{-1} \\
&= \mathbf{D}_{\boldsymbol{\pi}_0^{1/2}}^{-1}\{\mathbf{D}_{\boldsymbol{\pi}_0+n^{-1/2}\boldsymbol{\gamma}} - (\boldsymbol{\pi}_0 + n^{-1/2}\boldsymbol{\gamma})(\boldsymbol{\pi}_0 + n^{-1/2}\boldsymbol{\gamma})^t\}\mathbf{D}_{\boldsymbol{\pi}_0^{1/2}}^{-1} \\
&= \mathbf{D}_{\boldsymbol{\pi}_0^{1/2}}^{-1}\{\mathbf{D}_{\boldsymbol{\pi}_0+n^{-1/2}\boldsymbol{\gamma}} - \boldsymbol{\pi}_0\boldsymbol{\pi}_0^t + \mathbf{1}_r O(n^{-1/2})\}\mathbf{D}_{\boldsymbol{\pi}_0^{1/2}}^{-1} \\
&= \mathbf{D}_{\boldsymbol{\pi}_0^{1/2}}^{-1}\{\mathbf{I}_r - \boldsymbol{\pi}_0\boldsymbol{\pi}_0^t + \mathbf{1}_r\mathbf{1}_r^t O(n^{-1/2})\}\mathbf{D}_{\boldsymbol{\pi}_0^{1/2}}^{-1} \\
&= \mathbf{I}_r - \boldsymbol{\pi}_0^{1/2}(\boldsymbol{\pi}_0^{1/2})^t + \mathbf{1}_r\mathbf{1}_r^t[O(n^{-1/2})].
\end{aligned}
$$

Therefore, we have

$$\mathrm{E}(\mathbf{Z}_n \mid A) = \mathbf{D}_{\boldsymbol{\pi}_0^{1/2}}^{-1}\boldsymbol{\gamma},$$

and

$$\mathrm{Var}\,(\mathbf{Z}_n \mid A) = \mathbf{I}_r - \boldsymbol{\pi}_0^{1/2}(\boldsymbol{\pi}_0^{1/2})^t + \mathbf{1}_r O(n^{-1/2})$$

and it follows from the (multivariate version of the) Central Limit Theorem 3.3.4 that, under $A_n : \boldsymbol{\pi} = \boldsymbol{\pi}_0 + n^{-1/2}\boldsymbol{\gamma}$,

$$\mathbf{Z}_n \xrightarrow{D} N_r\{\mathbf{D}_{\boldsymbol{\pi}_0^{1/2}}^{-1}\boldsymbol{\gamma},\ \mathbf{I}_r - \boldsymbol{\pi}_0^{1/2}(\boldsymbol{\pi}_0^{1/2})^t\}$$

which implies

$$Q_P = \mathbf{Z}_n^t(\mathbf{I}_r - \boldsymbol{\pi}_0^{1/2}(\boldsymbol{\pi}_0^{1/2})^t\}\mathbf{Z}_n \xrightarrow{D} \chi^2_{r-1}(\delta)$$

where the noncentrality parameter is given by

$$
\begin{aligned}
\delta &= \boldsymbol{\gamma}^t\mathbf{D}_{\boldsymbol{\pi}_0^{1/2}}^{-1}\{\mathbf{I}_r - \boldsymbol{\pi}_0^{1/2}(\boldsymbol{\pi}_0^{1/2})^t\}\mathbf{D}_{\boldsymbol{\pi}_0^{1/2}}^{-1}\boldsymbol{\gamma} \\
&= \boldsymbol{\gamma}^t\mathbf{D}_{\boldsymbol{\pi}_0^{1/2}}^{-1}\mathbf{D}_{\boldsymbol{\pi}_0^{1/2}}^{-1}\boldsymbol{\gamma} - 0 \\
&= \sum_{j=1}^{r}(\gamma_j^2/\pi_{0j}).
\end{aligned}
$$

The extension of the above results to the general case $(s > 1)$ is straightforward; using the same arguments as for the case $s = 1$, it is easy to show that under the hypothesis $H : \boldsymbol{\pi} = \boldsymbol{\pi}_0$, where $\boldsymbol{\pi}_0 = (\boldsymbol{\pi}_{01}^t, \ldots, \boldsymbol{\pi}_{0s}^t)^t$, $\boldsymbol{\pi}_{0i} = (\boldsymbol{\pi}_{0i1}, \ldots, \boldsymbol{\pi}_{0ir})^t$, $i = 1, \ldots, s$, we have

$$Q_P = Q_P(\boldsymbol{\pi}_0) = \sum_{i=1}^{s} \sum_{j=1}^{r} (n_{ij} - n\pi_{0ij})^2 / n\pi_{0ij} \xrightarrow{\mathcal{D}} \chi_{s(r-1)}^2$$

and that under $A_n : \boldsymbol{\pi} = \boldsymbol{\pi}_0 + n^{-1/2}\boldsymbol{\gamma}$, where $\boldsymbol{\gamma} = (\boldsymbol{\gamma}_1^t, \ldots, \boldsymbol{\gamma}_s^t)^t$, $\boldsymbol{\gamma}_i^t = (\gamma_{i1}, \ldots, \gamma_{ir})^t$, $\mathbf{1}_r^t \boldsymbol{\gamma}_i = 0$, $i = 1, \ldots, s$, we have

$$Q_P = Q_P(\boldsymbol{\pi}_0) \xrightarrow{\mathcal{D}} \chi_{s(r-1)}^2(\delta)$$

where $\delta = \sum_{i=1}^{s} \boldsymbol{\gamma}_i^t \mathbf{D}_{\boldsymbol{\pi}_0}^{-1/2} \boldsymbol{\gamma}_i = \sum_{i=1}^{s} \sum_{j=1}^{r} (\gamma_{ij}^2 / \pi_{0ij})$.

6.3 Estimation and goodness-of-fit tests: parametric case

In general, the "parametric" multinomial model corresponds to (6.1.2) with $\pi_{ij} = \pi_{ij}(\boldsymbol{\theta})$, $i = 1, \ldots, s$, $j = 1, \ldots, r$, where $\boldsymbol{\theta} = (\theta_1, \ldots, \theta_q)^t$ is a vector of underlying parameters. A typical example is the Hardy-Weinberg model for the ABO blood group classification system described in Example 1.2.2. There, model (6.1.2) holds with $s = 1$, $r = 4$, $\pi_1 = p_0$, $\pi_2 = p_A$, $\pi_3 = p_B$ and $\pi_4 = p_{AB}$; furthermore, we have $\pi_j = \pi_j(\boldsymbol{\theta})$ where $\boldsymbol{\theta} = (\theta_1, \theta_2, \theta_3)^t$ with $\theta_1 = q_0$, $\theta_2 = q_A$ and $\theta_3 = q_B$.

Example 6.3.1: Consider a two-way contingency table with rows and columns defined by random variables X (with a levels) and Y (with b levels), respectively. The question of interest is to verify whether X and Y are independent. A possible model to describe the corresponding frequency distribution based on n sampling units is (6.1.2) with $s = 1$, $r = ab$ and π_{ij} denoting the probability associated with the (i,j)th cell, $i = 1, \ldots, a$, $j = 1, \ldots, b$. If X and Y are independent, we may write $\pi_{ij} = \pi_{i\cdot}\pi_{\cdot j}$, $i = 1, \ldots, a$, $j = 1, \ldots, b$, where $\pi_{i\cdot}$ and $\pi_{\cdot j}$ represent the marginal probabilities corresponding to X and Y, respectively. Here $\pi_{ij} = \pi_{ij}(\boldsymbol{\theta})$ where $\boldsymbol{\theta} = (\theta_1, \ldots, \theta_{a+b})$ with $\theta_1 = \pi_{1\cdot}, \ldots, \theta_a = \pi_{a\cdot}, \theta_{a+1} = \pi_{\cdot 1}, \ldots, \theta_{a+b} = \pi_{\cdot b}$. ∎

We now consider the problem of estimating the parameter vector $\boldsymbol{\theta}$ and assessing goodness of fit of such "parametric" models via the three most common approaches, namely, the ML, Minimum chi-squared (MCS) and Modified Minimum chi-squared (MMCS) methods. For notational ease we restrict ourselves to the case $s = 1$, where the log-likelihood may be expressed as

$$\log L_n(\boldsymbol{\theta}) = \text{constant} + \sum_{j=1}^{r} n_j \log \pi_j(\boldsymbol{\theta}). \qquad (6.3.1)$$

We assume that $\pi_j(\boldsymbol{\theta})$, $j = 1, \ldots, r$, possess continuous derivatives up to the order 2.

The MLE of $\boldsymbol{\theta}$ is a solution $\widehat{\boldsymbol{\theta}}$ to

$$\sum_{j=1}^{r} \frac{n_j}{\pi_j(\boldsymbol{\theta})} \frac{\partial}{\partial \boldsymbol{\theta}} \pi_j(\boldsymbol{\theta}) = \mathbf{0} \quad \text{subject to} \quad \sum_{j=1}^{r} \pi_j(\boldsymbol{\theta}) = 1. \qquad (6.3.2)$$

Using the same technique employed in Theorem 5.2.1 we shall derive the asymptotic distribution of $\widehat{\boldsymbol{\theta}}$. First, let $\|\mathbf{u}\| \le K$, $0 < K < \infty$, and consider the Taylor expansion

$$
\begin{aligned}
\log L_n(\boldsymbol{\theta} + n^{-1/2}\mathbf{u}) &= \log L_n(\boldsymbol{\theta}) + \frac{1}{\sqrt{n}} \mathbf{u}^t \frac{\partial}{\partial \boldsymbol{\theta}} \log L_n(\boldsymbol{\theta}) \\
&\quad + \frac{1}{2n} \mathbf{u}^t \frac{\partial^2}{\partial \boldsymbol{\theta} \partial \boldsymbol{\theta}^t} \log L_n(\boldsymbol{\theta})\Big|_{\boldsymbol{\theta}=\boldsymbol{\theta}^*} \mathbf{u}
\end{aligned}
$$

where $\boldsymbol{\theta}^*$ belongs to the line segment joining $\boldsymbol{\theta}$ and $\boldsymbol{\theta} + n^{-1/2}\mathbf{u}$. Then we have

$$
\begin{aligned}
\lambda_n(\mathbf{u}) &= \log L_n(\boldsymbol{\theta} + n^{-1/2}\mathbf{u}) - \log L_n(\boldsymbol{\theta}) \\
&= \frac{1}{\sqrt{n}} \mathbf{u}^t \mathbf{U}_n + \frac{1}{2n} \mathbf{u}^t \mathbf{V}_n \mathbf{u} + \frac{1}{2n} \mathbf{u}^t \mathbf{W}_n \mathbf{u} \qquad (6.3.3)
\end{aligned}
$$

where

$$\mathbf{U}_n = \frac{\partial}{\partial \boldsymbol{\theta}} \log L_n(\boldsymbol{\theta}),$$

$$\mathbf{V}_n = \frac{\partial^2}{\partial \boldsymbol{\theta} \partial \boldsymbol{\theta}^t} \log L_n(\boldsymbol{\theta})$$

and

$$\mathbf{W}_n = \frac{\partial^2}{\partial \boldsymbol{\theta} \partial \boldsymbol{\theta}^t} \log L_n(\boldsymbol{\theta})\Big|_{\boldsymbol{\theta}=\boldsymbol{\theta}^*} - \frac{\partial^2}{\partial \boldsymbol{\theta} \partial \boldsymbol{\theta}^t} \log L_n(\boldsymbol{\theta}).$$

Now note that:

i) From the definition of the score statistic,

$$
\begin{aligned}
n^{-1/2}\mathbf{U}_n &= n^{-1/2} \sum_{j=1}^{r} \frac{n_j}{\pi_j(\boldsymbol{\theta})} \frac{\partial}{\partial \boldsymbol{\theta}} \pi_j(\boldsymbol{\theta}) \\
&= n^{-1/2} \sum_{j=1}^{n} \left\{ \sum_{i=1}^{n} \frac{X_{ij}}{\pi_j(\boldsymbol{\theta})} \frac{\partial}{\partial \boldsymbol{\theta}} \pi_j(\boldsymbol{\theta}) \right\} \\
&= \frac{1}{\sqrt{n}} \sum_{i=1}^{n} \mathbf{Y}_i
\end{aligned}
$$

where

$$\mathbf{Y}_i = \sum_{j=1}^{r} \frac{X_{ij}}{\pi_j(\boldsymbol{\theta})} \frac{\partial}{\partial(\boldsymbol{\theta})} \pi_j(\boldsymbol{\theta})$$

and X_{ij} is defined as in (6.2.7). Since

$$\mathbf{EY}_i = \sum_{j=1}^{r} \frac{\mathrm{E}X_{ij}}{\pi_j(\boldsymbol{\theta})} = \sum_{j=1}^{r} \frac{\partial}{\partial \boldsymbol{\theta}} \pi_j(\boldsymbol{\theta}) = \frac{\partial}{\partial \boldsymbol{\theta}} \sum_{j=1}^{r} \pi_j(\boldsymbol{\theta}) = 0$$

and

$$
\begin{aligned}
\mathbf{EY}_i \mathbf{Y}_i^t &= \sum_{j=1}^{r} \frac{\mathrm{E}X_{ij}^2}{\{\pi_j(\boldsymbol{\theta})\}^2} \frac{\partial}{\partial \boldsymbol{\theta}} \pi_j(\boldsymbol{\theta}) \frac{\partial}{\partial \boldsymbol{\theta}^t} \pi_j(\boldsymbol{\theta}) \\
&= \sum_{j=1}^{r} \frac{1}{\pi_j(\boldsymbol{\theta})} \frac{\partial}{\partial \boldsymbol{\theta}^t} \pi_j(\boldsymbol{\theta}) \frac{\partial}{\partial \boldsymbol{\theta}^t} \pi_j(\boldsymbol{\theta}) \\
&= \mathbf{I}(\boldsymbol{\theta}),
\end{aligned}
$$

it follows from the (multivariate version of the) Central Limit Theorem 3.3.4 that

$$n^{-1/2}\mathbf{U}_n \xrightarrow{\mathcal{D}} N(\mathbf{0}, \mathbf{I}(\boldsymbol{\theta})). \tag{6.3.4}$$

ii) Furthermore,

$$
\begin{aligned}
n^{-1}\mathbf{V}_n &= \frac{1}{n} \sum_{j=1}^{r} \frac{n_j}{\{\pi_j(\boldsymbol{\theta})\}^2} \frac{\partial}{\partial \boldsymbol{\theta}} \pi_j(\boldsymbol{\theta}) \frac{\partial}{\partial \boldsymbol{\theta}^t} \pi_j(\boldsymbol{\theta}) \\
&\quad + \frac{1}{n} \sum_{j=1}^{r} \frac{n_j}{\pi_j(\boldsymbol{\theta})} \frac{\partial^2}{\partial \boldsymbol{\theta} \partial \boldsymbol{\theta}^t} \pi_j(\boldsymbol{\theta}) \\
&= -\sum_{j=1}^{r} \frac{1}{n} \sum_{i=1}^{n} \frac{X_{ij}}{\{\pi_j(\boldsymbol{\theta})\}^2} \frac{\partial}{\partial \boldsymbol{\theta}} \pi_j(\boldsymbol{\theta}) \frac{\partial}{\partial \boldsymbol{\theta}^t} \pi_j(\boldsymbol{\theta}) \\
&\quad + \sum_{j=1}^{r} \frac{1}{n} \sum_{i=1}^{n} \frac{X_{ij}}{\pi_j(\boldsymbol{\theta})} \frac{\partial^2}{\partial \boldsymbol{\theta} \partial \boldsymbol{\theta}^t} \pi_j(\boldsymbol{\theta})
\end{aligned}
$$

where X_{ij} is defined as in (6.2.7). Then, using the Khintchine Weak Law of Large Numbers (Theorem 2.3.6), it follows that

$$n^{-1}\mathbf{V}_n \xrightarrow{\mathrm{P}} -\sum_{j=1}^{r} \frac{1}{\pi_j(\boldsymbol{\theta})} \frac{\partial}{\partial \boldsymbol{\theta}} \pi_j(\boldsymbol{\theta}) \frac{\partial}{\partial \boldsymbol{\theta}^t} \pi_j(\boldsymbol{\theta}) + \sum_{j=1}^{r} \frac{\partial^2}{\partial \boldsymbol{\theta} \partial \boldsymbol{\theta}^t} \pi_j(\boldsymbol{\theta}) = -\mathbf{I}(\boldsymbol{\theta}).$$

$$\tag{6.3.5}$$

iii) Since $(\partial^2/\partial \boldsymbol{\theta} \partial \boldsymbol{\theta}^t) \log L_n(\boldsymbol{\theta})$ is a continuous function of $\boldsymbol{\theta}$, we have

$$n^{-1}\mathbf{W} \to \mathbf{0} \quad \text{as} \quad n \to \infty. \tag{6.3.6}$$

From (6.3.3)–(6.3.6) we may conclude that

$$\lambda(\mathbf{u}) = n^{-1/2}\mathbf{u}^t\mathbf{U}_n - \frac{1}{2}\mathbf{u}^t\mathbf{I}(\boldsymbol{\theta})\mathbf{u} + o_p(1), \tag{6.3.7}$$

the maximum of which is attained at

$$\widehat{\mathbf{u}} = n^{-1/2}\mathbf{I}(\boldsymbol{\theta})\mathbf{U}_n + o_p(1).$$

This is also the maximum of $\log L_n(\boldsymbol{\theta})$, which corresponds to the MLE of $\boldsymbol{\theta}$. Thus,

$$\widehat{\boldsymbol{\theta}}_n = \boldsymbol{\theta} + n^{-1/2}\widehat{\mathbf{u}} = \boldsymbol{\theta} + n^{-1}[\mathbf{I}(\boldsymbol{\theta})]^{-1}\mathbf{U}_n + o_p(1)$$

which implies that

$$\sqrt{n}(\widehat{\boldsymbol{\theta}} - \boldsymbol{\theta}) = n^{-1/2}\mathbf{I}^{-1}(\boldsymbol{\theta})\mathbf{U}_n + o_p(1)$$

and by the Slutsky Theorem (3.4.2) we obtain

$$\sqrt{n}(\widehat{\boldsymbol{\theta}}_n - \boldsymbol{\theta}) \xrightarrow{\mathcal{D}} N(\mathbf{0}, [\mathbf{I}(\boldsymbol{\theta})]^{-1}). \tag{6.3.8}$$

Before we proceed with the next approach for the estimation of $\boldsymbol{\theta}$, it is convenient to note that, putting

$$n_j = n\pi_j(\boldsymbol{\theta}) + Z_{nj}(\boldsymbol{\theta})\sqrt{n\pi_j(\boldsymbol{\theta})}$$

where $Z_{nj}(\boldsymbol{\theta})$ is defined as in Section 6.2, the equations (6.3.2) may be re-expressed as

$$\sum_{j=1}^{r} Z_{nj}Z_{nj}(\boldsymbol{\theta})\frac{1}{\sqrt{\pi_j(\boldsymbol{\theta})}}\frac{\partial}{\partial\boldsymbol{\theta}}\pi_j(\boldsymbol{\theta}) = \mathbf{0}. \tag{6.3.9}$$

As we have mentioned before, the statistic

$$Q_P = Q_P(\boldsymbol{\theta}) = \sum_{j=1}^{r}\{n_j - n\pi_j(\boldsymbol{\theta})\}^2/n\pi_j(\boldsymbol{\theta})$$

may be used as a measure of the goodness of fit of the model $\pi_j = \pi_j(\boldsymbol{\theta})$, $j = 1, \ldots, r$. The value $\widetilde{\boldsymbol{\theta}}_n$ which minimizes $Q_P(\boldsymbol{\theta})$ is known as the MCS estimator of $\boldsymbol{\theta}$ and may be obtained as a solution to

$$\frac{\partial}{\partial\boldsymbol{\theta}}Q_P(\boldsymbol{\theta}) = -2\sum_{j=1}^{r}\frac{\{n_j - n\pi_j(\boldsymbol{\theta})\}}{\pi_j(\boldsymbol{\theta})}\frac{\partial}{\partial\boldsymbol{\theta}}\pi_j(\boldsymbol{\theta})$$

$$-\sum_{j=1}^{r}\frac{\{n_j - n\pi_j(\boldsymbol{\theta})\}^2}{n\{\pi_j(\boldsymbol{\theta})\}^2}\frac{\partial}{\partial\boldsymbol{\theta}}\pi_j(\boldsymbol{\theta}) = \mathbf{0}.$$

These equations may be expressed as

$$-2\sqrt{n}\left\{\sum_{j=1}^{r}Z_{nj}(\boldsymbol{\theta})\frac{1}{\sqrt{\pi_j(\boldsymbol{\theta})}}\frac{\partial}{\partial\boldsymbol{\theta}}\pi_j(\boldsymbol{\theta})\right.$$

$$\left.+\frac{1}{2\sqrt{n}}\sum_{j=1}^{r}Z_{nj}^2(\boldsymbol{\theta})\frac{1}{\pi_j(\boldsymbol{\theta})}\frac{\partial}{\partial\boldsymbol{\theta}}\pi_j(\boldsymbol{\theta})\right\}=0$$

which, in view of (6.2.8) and Theorem 3.2.5, reduces to

$$\sum_{j=1}^{r}Z_{nj}(\boldsymbol{\theta})\frac{1}{\sqrt{\pi_j(\boldsymbol{\theta})}}\frac{\partial}{\partial\boldsymbol{\theta}}\pi_j(\boldsymbol{\theta})+O_p(\mathbf{n}^{-1/2})=0. \qquad (6.3.10)$$

Note that the equations in (6.3.10) are equivalent to those in (6.3.9) and consequently to those in (6.3.2) up to the order $n^{-1/2}$. Now consider a Taylor expansion of (6.3.2) around the point $\tilde{\boldsymbol{\theta}}_n$:

$$\begin{aligned}O_p(\mathbf{1}) &= \sum_{j=1}^{r}\frac{n_j}{\pi_j(\boldsymbol{\theta})}\frac{\partial}{\partial\boldsymbol{\theta}}\pi_j(\boldsymbol{\theta})\Big|_{\boldsymbol{\theta}=\hat{\boldsymbol{\theta}}_n} - \sum_{j=1}^{r}\frac{n_j}{\pi_j(\boldsymbol{\theta})}\frac{\partial}{\partial\boldsymbol{\theta}}\pi_j(\boldsymbol{\theta})\Big|_{\boldsymbol{\theta}=\tilde{\boldsymbol{\theta}}_n}\\ &= \frac{1}{n}\left\{-\sum_{j=1}^{r}\frac{n_j}{\{\pi_j(\boldsymbol{\theta})\}^2}\frac{\partial}{\partial\boldsymbol{\theta}}\pi_j(\boldsymbol{\theta})\Big|_{\boldsymbol{\theta}=\boldsymbol{\theta}^*}\frac{\partial}{\partial\boldsymbol{\theta}^t}\pi_j(\boldsymbol{\theta})\Big|_{\boldsymbol{\theta}=\boldsymbol{\theta}^*}\right.\\ &\quad\left.+\sum_{j=1}^{r}\frac{n_j}{\pi_j(\boldsymbol{\theta})}\frac{\partial^2}{\partial\boldsymbol{\theta}\partial\boldsymbol{\theta}^t}\pi_j(\boldsymbol{\theta})\Big|_{\boldsymbol{\theta}=\boldsymbol{\theta}^*}\right\}n(\hat{\boldsymbol{\theta}}_n-\tilde{\boldsymbol{\theta}}_n)\end{aligned}$$

$$(6.3.11)$$

where $\boldsymbol{\theta}^*$ belongs to the line segment joining $\hat{\boldsymbol{\theta}}_n$ and $\tilde{\boldsymbol{\theta}}_n$. Since by (6.3.5) the term within $\{\ \}$ converges in probability to $-\mathbf{I}(\boldsymbol{\theta})$ which is finite, we may write

$$O_p(\mathbf{1}) = \{-\mathbf{I}(\boldsymbol{\theta})+o_p(\mathbf{1})\}n(\hat{\boldsymbol{\theta}}_n-\tilde{\boldsymbol{\theta}}_n).$$

This implies that $\tilde{\boldsymbol{\theta}}_n - \hat{\boldsymbol{\theta}}_n = O_p(\mathbf{n}^{-1})$ and we may conclude that

$$\sqrt{n}(\hat{\boldsymbol{\theta}}_n-\tilde{\boldsymbol{\theta}}_n) \xrightarrow{\mathrm{P}} \mathbf{0}. \qquad (6.3.12)$$

Along the lines of the previous case, we define the MMCS estimator of $\boldsymbol{\theta}$ as the value $\overline{\boldsymbol{\theta}}$ which minimizes

$$Q_N = Q_N(\boldsymbol{\theta}) = \sum_{j=1}^{r}\{n_j-n\pi_j(\boldsymbol{\theta})\}^2/n_j.$$

The corresponding estimating equations are

$$\frac{\partial}{\partial \boldsymbol{\theta}} Q_N(\boldsymbol{\theta}) = -2n \sum_{j=1}^{r} \frac{\{n_j - n\pi_j(\boldsymbol{\theta})\}}{n_j} \frac{\partial}{\partial \boldsymbol{\theta}} \pi_j(\boldsymbol{\theta})$$

or, equivalently,

$$\sum_{j=1}^{r} \frac{n\pi_j(\boldsymbol{\theta})}{n_j} Z_{nj}(\boldsymbol{\theta}) \frac{1}{\sqrt{\pi_j(\boldsymbol{\theta})}} \frac{\partial}{\partial \boldsymbol{\theta}} \pi_j(\boldsymbol{\theta}) = \mathbf{0}. \qquad (6.3.13)$$

Now, using the fact that for $x \neq 1$, $(1-x)^{-1} = 1 - x + O(x^2)$ in connection with (6.2.8) we have

$$
\begin{aligned}
\frac{n\pi_j(\boldsymbol{\theta})}{n_j} &= \frac{n\pi_j(\boldsymbol{\theta})}{n\pi_j(\boldsymbol{\theta}) + Z_{nj}(\boldsymbol{\theta})\sqrt{n\pi_j(\boldsymbol{\theta})}} \\
&= \left\{ 1 + \frac{Z_{nj}(\boldsymbol{\theta})}{\sqrt{n\pi_j(\boldsymbol{\theta})}} \right\}^{-1} \\
&= 1 - \frac{Z_{nj}(\boldsymbol{\theta})}{\sqrt{n\pi_j(\boldsymbol{\theta})}} + O_p\left\{ \frac{Z_{nj}^2(\boldsymbol{\theta})}{n\pi_j(\boldsymbol{\theta})} \right\} \\
&= 1 + O_p(n^{-1/2}). \qquad (6.3.14)
\end{aligned}
$$

Thus, from (6.3.13) and (6.3.14) it follows that the estimating equations corresponding to the MMCS estimator of $\boldsymbol{\theta}$ are equivalent to those in (6.3.2) up to the order $n^{-1/2}$. As in the case of the MCS estimator, it follows that

$$\sqrt{n}(\widetilde{\boldsymbol{\theta}} - \overline{\boldsymbol{\theta}}) \overset{\mathrm{P}}{\longrightarrow} \mathbf{0}. \qquad (6.3.15)$$

From (6.3.8) and (6.3.15) we may conclude that all three methods of estimation considered above produce BAN estimators. The choice among the three methods then relies on computational aspects and depends essentially on the intrinsic characteristics of the model $\pi_j = \pi_j(\boldsymbol{\theta})$, $j = 1, \ldots, r$. In general, this involves the solution of a set of nonlinear equations in $\boldsymbol{\theta}$. However, if the π_j's are linear functions of $\boldsymbol{\theta}$, i.e., $\pi_j = \mathbf{x}_j^t \boldsymbol{\theta}$, with $\mathbf{x}_j$ denoting a vector of constants, $j = 1, \ldots, r$, the MMCS estimator of $\boldsymbol{\theta}$ may be obtained as the solution to linear equations; the reader is referred to Koch et al. (1985) for details.

Also, note that the model $\pi_j = \pi_j(\boldsymbol{\theta})$, $j = 1, \ldots, r$, may be expressed in terms of the $r - q$ restrictions

$$H : \mathbf{F}(\boldsymbol{\pi}) = \mathbf{0} \qquad (6.3.16)$$

where $\mathbf{F}(\boldsymbol{\pi}) = [F_1(\boldsymbol{\pi}), \ldots, F_{r-q}(\boldsymbol{\pi})]^t$, by eliminating $\boldsymbol{\theta}$. In particular, if $\boldsymbol{\pi} = X\boldsymbol{\theta}$, (6.3.16) reduces to $H : \mathbf{W}\boldsymbol{\pi} = \mathbf{0}$ where $\mathbf{W}$ is a $(r - q \times r)$ matrix such that $\mathbf{W}X = \mathbf{0}$ (i.e., the rows of $\mathbf{W}$ are orthogonal to the columns of X).

If the model is nonlinear with $\mathbf{F}$ such that its elements possess continuous partial derivatives up to the second order in an open neighborhood of $\boldsymbol{\pi}$ and $\mathbf{H}(\boldsymbol{\pi}) = (\partial/\partial\boldsymbol{\pi})\mathbf{F}(\boldsymbol{\pi})$ is a $(r \times r - q)$ matrix of rank $r - q$, then the "linearized" version of (6.3.16) may be derived via a Taylor expansion of $\mathbf{F}(\boldsymbol{\pi})$ around $\widehat{\boldsymbol{\pi}} = n^{-1}\mathbf{n}$ the MLE of $\boldsymbol{\pi}$ under the unrestricted model and corresponds to

$$H_L : \mathbf{F}(\widehat{\boldsymbol{\pi}}) + \mathbf{H}^t(\widehat{\boldsymbol{\pi}})(\boldsymbol{\pi} - \widehat{\boldsymbol{\pi}}) = \mathbf{0}. \tag{6.3.17}$$

Neyman (1949) showed that BAN estimators for $\boldsymbol{\pi}$ may be obtained via the minimization of either $Q_P(\boldsymbol{\pi})$, $Q_N(\boldsymbol{\pi})$ or $Q_V(\boldsymbol{\pi}) = 2\sum_{j=1}^{r} n_j\{\log n_j - \log n\pi_j\}$ subject to (6.3.16) or (6.3.17). This result is particularly interesting since under the linearized restrictions, the MMCS estimates may be obtained as the solution to linear equations.

It is also worthwhile to mention that the bias associated with such BAN estimators is of the order $o(n^{-1/2})$. To reduce such bias as well as to improve the estimates of the associated asymptotic covariance matrix, one may rely on the classical jackknifing methods which will be briefly discussed in Chapter 8.

Goodness of fit for such models may be assessed via the statistics $\widehat{Q}_P$, $\widehat{Q}_N$ or $\widehat{Q}_V$, obtained by substituting any BAN estimator $\widehat{\boldsymbol{\pi}}$ for $\boldsymbol{\pi}$ in the expressions of Q_P, Q_N or Q_V respectively. We now show that these statistics follow asymptotic chi-squared distributions. In this direction first define

$$\mathbf{B} = \mathbf{B}(\boldsymbol{\theta}) = [\mathbf{b}_1, \ldots, \mathbf{b}_r]$$

$$= \begin{bmatrix} \dfrac{1}{\sqrt{\pi_1(\boldsymbol{\theta})}}(\partial/\partial\theta_1)\pi_1(\boldsymbol{\theta}) & \cdots & \dfrac{1}{\sqrt{\pi_r(\boldsymbol{\theta})}}(\partial/\partial\theta_1)\pi_r(\boldsymbol{\theta}) \\ \vdots & & \vdots \\ \dfrac{1}{\sqrt{\pi_1(\boldsymbol{\theta})}}(\partial/\partial\theta_q)\pi_1(\boldsymbol{\theta}) & \cdots & \dfrac{1}{\sqrt{\pi_r(\boldsymbol{\theta})}}(\partial/\partial\theta_q)\pi_r(\boldsymbol{\theta}) \end{bmatrix}$$

$$= \frac{\partial\boldsymbol{\pi}(\boldsymbol{\theta})}{\partial\boldsymbol{\theta}}\mathbf{D}_{\boldsymbol{\pi}^{1/2}(\boldsymbol{\theta})}^{-1}$$

and note that

i)

$$\mathbf{BB}^t = \frac{\partial}{\partial\boldsymbol{\theta}}\boldsymbol{\pi}(\boldsymbol{\theta})\mathbf{D}_{\boldsymbol{\pi}(\boldsymbol{\theta})}^{-1}\frac{\partial}{\partial\boldsymbol{\theta}^t}\boldsymbol{\pi}(\boldsymbol{\theta})$$

$$= \sum_{j=1}^{r}\frac{1}{\pi_j(\boldsymbol{\theta})}\frac{\partial}{\partial\boldsymbol{\theta}}\pi_j(\boldsymbol{\theta})\frac{\partial}{\partial\boldsymbol{\theta}^t}\pi_j(\boldsymbol{\theta}) = \mathbf{I}(\boldsymbol{\theta}),$$

ii)

$$
\begin{aligned}
n^{-1/2}\mathbf{U}_n &= n^{-1/2}\sum_{j=1}^{r}\frac{n_j}{\pi_j(\boldsymbol{\theta})}\frac{\partial}{\partial\boldsymbol{\theta}}\pi_j(\boldsymbol{\theta}) \\
&= \sum_{j=1}^{r} Z_{nj}(\boldsymbol{\theta})\frac{1}{\sqrt{\pi_j(\boldsymbol{\theta})}}\frac{\partial}{\partial\boldsymbol{\theta}}\pi_j(\boldsymbol{\theta}) = \mathbf{B}\mathbf{Z}_n,
\end{aligned}
$$

where $\mathbf{Z}_n$ is defined as in Section 6.2.

iii)

$$
\widehat{\pi}_j = \pi_j(\widehat{\boldsymbol{\theta}}_n) = \pi_j(\boldsymbol{\theta}) + (\widehat{\boldsymbol{\theta}}_n - \boldsymbol{\theta})^t\frac{\partial\pi_j(\boldsymbol{\theta})}{\partial\boldsymbol{\theta}}\Big|_{\boldsymbol{\theta}=\boldsymbol{\theta}^*},
$$

where $\boldsymbol{\theta}^*$ lies in the line segment joining $\boldsymbol{\theta}$ and $\widehat{\boldsymbol{\theta}}_n$.

Now, since $(\partial/\partial\boldsymbol{\theta})\pi_j(\boldsymbol{\theta})|_{\boldsymbol{\theta}=\boldsymbol{\theta}^*}$ is finite and using (6.3.8), we have

$$
\sqrt{n}\{\widehat{\pi}_j - \pi_j(\boldsymbol{\theta})\} = O_p(1) \Rightarrow \widehat{\pi}_j - \pi_j(\boldsymbol{\theta}) = O_p(n^{-1/2}). \tag{6.3.18}
$$

Then, for $j = 1, \ldots, r$, write

$$
\begin{aligned}
\frac{n_j - n\widehat{\pi}_j}{\sqrt{n\pi_j}} &= \frac{n_j - n\pi_j}{\sqrt{n\pi_j}} - \frac{\sqrt{n}(\widehat{\pi}_j - \pi_j)}{\sqrt{\pi_j}} \\
&= Z_{nj} - \sqrt{n}(\widehat{\boldsymbol{\theta}}_n - \boldsymbol{\theta})^t\mathbf{b}_j + o_p(1) \\
&= Z_{nj} - n^{-1/2}\mathbf{U}_n^t[\mathbf{I}(\boldsymbol{\theta})]^{-1}\mathbf{b}_j + o_p(1) \\
&= O_p(1). \tag{6.3.19}
\end{aligned}
$$

Using (6.3.18), (6.3.19) and employing an argument similar to that considered in (6.3.14) we have

$$
\begin{aligned}
\widehat{Q}_P &= \sum_{j=1}^{r}\frac{(n_j - n\widehat{\pi}_j)^2}{n\pi_j}\left\{1 + \frac{\widehat{\pi}_j - \pi_j}{\pi_j}\right\}^{-1} \\
&= \sum_{j=1}^{r}\frac{(n_j - n\widehat{\pi}_j)^2}{n\pi_j}\{1 + O_p(n^{-1/2})\} \\
&= \sum_{j=1}^{r}\frac{(n_j - n\widehat{\pi}_j)^2}{n\pi_j} + O_p(n^{-1/2}). \tag{6.3.20}
\end{aligned}
$$

Now, from (6.3.19) and (6.3.20) it follows that

$$
\begin{aligned}
\widehat{Q}_p &= \sum_{j=1}^{r}\{Z_{nj} - n^{-1/2}\mathbf{U}_n^t[\mathbf{I}(\boldsymbol{\theta})]^{-1}\mathbf{b}_j + o_p(1)\}^2 + O_p(n^{-1/2}) \\
&= \mathbf{Z}_n^t\mathbf{Z}_n + n^{-1}\mathbf{U}_t[\mathbf{I}(\boldsymbol{\theta})]^{-1}\mathbf{B}\mathbf{B}^t[\mathbf{I}(\boldsymbol{\theta})]^{-1}\mathbf{U}_n \\
&\quad - 2n^{-1/2}\mathbf{Z}_n^t\mathbf{B}^t[\mathbf{I}(\boldsymbol{\theta})]^{-1}\mathbf{U}_n + o_p(1)
\end{aligned}
$$

$$\begin{aligned}
&= \ \mathbf{Z}_n^t \mathbf{Z}_n + \mathbf{Z}_n^t \mathbf{B}^t [\mathbf{I}(\boldsymbol{\theta})]^{-1} \mathbf{B} \mathbf{Z}_n \\
&\quad -2\mathbf{Z}_n^t \mathbf{B}^t [\mathbf{I}(\boldsymbol{\theta})]^{-1} \mathbf{B} \mathbf{Z}_n + o_p(1) \\
&= \ \mathbf{Z}_n^t \{\mathbf{I}_r - \mathbf{B}^t [\mathbf{I}(\boldsymbol{\theta})]^{-1} \mathbf{B}\} \mathbf{Z}_n + o_p(1).
\end{aligned} \tag{6.3.21}$$

Also observe that

$$\{\mathbf{I}_r - \mathbf{B}^t [\mathbf{I}(\boldsymbol{\theta})]^{-1} \mathbf{B}\}\{\mathbf{I}_r - \boldsymbol{\pi}^{1/2}(\boldsymbol{\pi}^{1/2})^t\}\{\mathbf{I}_r - \mathbf{B}^t [\mathbf{I}(\boldsymbol{\theta})]^{-1} \mathbf{B}\} = \mathbf{I}_r - \mathbf{B}^t \mathbf{I}(\boldsymbol{\theta}) \mathbf{B}$$

so, from (6.2.6), the Slutsky Theorem (3.4.2) and Cochran Theorem (3.4.7), for example, we may conclude that if the model $\pi_j = \pi_j(\boldsymbol{\theta})$, $j = 1, \ldots, r$, holds, $\widehat{Q}_p$ follows asymptotically a chi-squared distribution with degrees of freedom given by

$$\begin{aligned}
\text{rank} \, [\{\mathbf{I}_r &- \mathbf{B}^t [\mathbf{I}(\boldsymbol{\theta})]^{-1} \mathbf{B}\}\{\mathbf{I}_r - \boldsymbol{\pi}^{1/2}(\boldsymbol{\pi}^{1/2})^t\}] \\
&= \text{tr} \, \{\mathbf{I}_r - \boldsymbol{\pi}^{1/2}(\boldsymbol{\pi}^{1/2})^t - \mathbf{B}^t [\mathbf{I}(\boldsymbol{\theta})]^{-1} \mathbf{B} \\
&\quad + \mathbf{B}^t [\mathbf{I}(\boldsymbol{\theta})]^{-1} \mathbf{B} \boldsymbol{\pi}^{1/2}(\boldsymbol{\pi}^{1/2})^t\} \\
&= \text{tr} \, \{\mathbf{I}_r\} - \text{tr} \, \{\boldsymbol{\pi}^{1/2}(\boldsymbol{\pi}^{1/2})^t\} - \text{tr} \, \{[\mathbf{I}(\boldsymbol{\theta})]^{-1} \mathbf{B} \mathbf{B}^t\} \\
&\quad + \text{tr} \, \{\mathbf{B}^t [\mathbf{I}(\boldsymbol{\theta})]^{-1} \mathbf{B} \boldsymbol{\pi}^{1/2}(\boldsymbol{\pi}^{1/2})^t\} \\
&= r - 1 - q.
\end{aligned}$$

The proof of similar results for $\widehat{Q}_N$ and $\widehat{Q}_V$ is left to the reader.

Goodness of fit of models defined by constraints of the form (6.3.16) may also be assessed by the method proposed by Wald (1943). In this direction let $\widehat{\boldsymbol{\pi}}$ denote the MLE of $\boldsymbol{\pi}$ under the unrestricted model and consider the following Taylor expansion:

$$\mathbf{F}(\widehat{\boldsymbol{\pi}}) = \mathbf{F}(\boldsymbol{\pi}) + [\mathbf{H}(\boldsymbol{\pi})]^t (\widehat{\boldsymbol{\pi}} - \boldsymbol{\pi}) + O_p(\mathbf{n}^{-1})$$

which implies

$$\sqrt{n}\{\mathbf{F}(\widehat{\boldsymbol{\pi}}) - \mathbf{F}(\boldsymbol{\pi})\} = [\mathbf{H}(\boldsymbol{\pi})]^t \sqrt{n}(\widehat{\boldsymbol{\pi}} - \boldsymbol{\pi}) + O_p(\mathbf{n}^{-1/2}). \tag{6.3.22}$$

Now let $\mathbf{X}_i$ be defined as in Section 6.2 and note that

$$\begin{aligned}
\sqrt{n}(\widehat{\boldsymbol{\pi}} - \boldsymbol{\pi}) &= \ n^{-1/2}(\mathbf{n} - n\boldsymbol{\pi}) \\
&= \ n^{-1/2} \sum_{i=1}^{n} (\mathbf{X}_i - \boldsymbol{\pi}) \xrightarrow{\mathcal{D}} N_r(\mathbf{0}, \mathbf{V}(\boldsymbol{\pi}))
\end{aligned} \tag{6.3.23}$$

where $\mathbf{V}(\boldsymbol{\pi}) = \mathbf{D}_{\boldsymbol{\pi}} - \boldsymbol{\pi}\boldsymbol{\pi}^t$ is a singular matrix of rank $r - 1$. From (6.3.22), (6.3.23) and the Slutsky Theorem (3.4.2) we get:

$$\sqrt{n}\{\mathbf{F}(\widehat{\boldsymbol{\pi}}) - \mathbf{F}(\boldsymbol{\pi})\} \xrightarrow{\mathcal{D}} N(\mathbf{0}, [\mathbf{H}(\boldsymbol{\pi})]^t \mathbf{V}(\boldsymbol{\pi}) \mathbf{H}(\boldsymbol{\pi}))$$

and, consequently,

$$Q_W(\boldsymbol{\pi}) = \ n[\mathbf{F}(\widehat{\boldsymbol{\pi}})]^t \{[\mathbf{H}(\boldsymbol{\pi})]^t \mathbf{V}(\boldsymbol{\pi}) \mathbf{H}(\boldsymbol{\pi})\}^{-1} \mathbf{F}(\widehat{\boldsymbol{\pi}}) \xrightarrow{\mathcal{D}} \chi^2_{r-1-q}(\delta) \tag{6.3.24}$$

where $\delta = [\mathbf{F}(\boldsymbol{\pi})]^t\{[\mathbf{H}(\boldsymbol{\pi})]^t\mathbf{V}(\boldsymbol{\pi})\mathbf{H}(\boldsymbol{\pi})\}^{-1}\mathbf{F}(\boldsymbol{\pi})$. Clearly, under $H : \mathbf{F}(\boldsymbol{\pi}) = 0$, $Q_W(\boldsymbol{\pi}) \xrightarrow{\mathcal{D}} \chi^2_{r-1-q}$. Since $\boldsymbol{\pi}$ is unknown, we may replace it with a BAN estimator in (6.3.24) and use a technique similar to that considered in Theorem 5.6.1 to show that

$$\widehat{Q}_W = n[\mathbf{F}(\widehat{\boldsymbol{\pi}})]^t\{[\mathbf{H}(\widehat{\boldsymbol{\pi}})]^t\mathbf{V}(\widehat{\boldsymbol{\pi}})\mathbf{H}(\widehat{\boldsymbol{\pi}})\}^{-1}\mathbf{F}(\widehat{\boldsymbol{\pi}}) \xrightarrow{\mathcal{D}} \chi^2_{r-1-q}(\delta).$$

Bhapkar (1966) showed that $\widehat{Q}_W$ is algebraically identical to $\widehat{Q}_N$ derived under the "linearized" constraints (6.3.17) and, thus, it shares the same asymptotic properties of the latter. The Wald statistic $\widehat{Q}_W$ is generated preferable for computational reasons. This interesting equivalence will be discussed under a broader perspective in Section 7.5.

6.4 Asymptotic theory for some other important statistics

First, let us consider a simple 2×2 contingency table relating to two rows (1 and 2) and two columns (1 and 2), so that the cell probability for the ith row, jth column is denoted by π_{ij}, $i, j = 1, 2$. In a sample of size n, the corresponding observed frequencies are n_{ij}, $i, j = 1, 2$. The **odds ratio** or the **cross-product ratio** is defined by

$$\psi = (\pi_{11}\pi_{22})/(\pi_{21}\pi_{12}), \tag{6.4.1}$$

and a departure of ψ from 1 indicates a non-null association between the r.v. defining rows and columns. Based on the sample data, a natural estimator of ψ is

$$\widehat{\psi}_n = (n_{11}n_{22}/(n_{21}n_{12}). \tag{6.4.2}$$

If we denote the sample proportions by

$$p_{ij} = n_{ij}/n, \quad i, j = 1, 2, \tag{6.4.3}$$

then $\widehat{\psi}_n$ is a multiplicative function of p_{ij} for which a multinomial probability law is associated. However, in this multiplicative form, the (asymptotic or exact) variability of $\widehat{\psi}_n$ about ψ depends on ψ and is of slowly convergent form. On the other hand, if we use the (multivariate version of the) Central Limit Theorem 3.3.4 and Theorem 3.4.5 then we may observe that for

$$\theta = \log \psi = \log \pi_{11} + \log \pi_{22} - \log \pi_{21} - \log \pi_{12} \tag{6.4.4}$$

the natural estimator is

$$\widehat{\theta}_n = \log \widehat{\psi}_n = \log p_{11} + \log p_{22} - \log p_{21} - \log p_{12}, \tag{6.4.5}$$

and, further, as n increases, the asymptotic mean square error for $\sqrt{n}(\widehat{\theta}_n - \theta)$ is given by

$$\sum_{i=1}^{2}\sum_{j=1}^{2} \pi_{ij}^{-1},$$ (6.4.6)

which can, as well, be estimated by

$$n\sum_{i=1}^{2}\sum_{j=1}^{2} n_{ij}^{-1}.$$ (6.4.7)

Thus,

$$\sqrt{n}(\widehat{\theta}_n - \theta)/\left(\frac{n}{n_{11}} + \frac{n}{n_{12}} + \frac{n}{n_{21}} + \frac{n}{n_{22}}\right)^{1/2} \xrightarrow{\mathcal{D}} N(0,1),$$ (6.4.8)

which may be used to draw a confidence interval for θ (or ψ) and to test a suitable hypothesis on θ (or ψ). Note the similarity with the Wald procedure for a specific choice of $F(\boldsymbol{\pi})$ termed the log-linear model. Instead of the logarithmic transformation in (6.4.4), another way to look into the association in a 2×2 table is to use Yule's measure

$$\varphi_Y = \frac{\psi - 1}{\psi + 1} = \frac{\pi_{11}\pi_{22} - \pi_{12}\pi_{21}}{\pi_{11}\pi_{22} + \pi_{12}\pi_{21}},$$ (6.4.9)

so that $-1 \leq \varphi_Y \leq 1$, and $\varphi_Y = 0$ relates to no association. Again Wald's method can be used to draw statistical conclusions on φ_Y through $\widehat{\varphi}_{Y_n} = (n_{11}n_{22} - n_{12}n_{21})/(n_{11}n_{22} + n_{21}n_{12})$.

For a general $I \times J$ contingency table with probabilities $\{\pi_{ij}\}$, Goodman and Kruskal's **concentration coefficient** (τ) is defined as

$$\tau = \left(\sum_{i=1}^{I}\sum_{j=1}^{J} \pi_{ij}^2/\pi_{i\cdot} - \sum_{j=1}^{J} \pi_{\cdot j}^2\right) \Big/ \left(1 - \sum_{j=1}^{J} \pi_{\cdot j}^2\right)$$ (6.4.10)

where $\pi_{i\cdot} = \sum_{i=1}^{I} \pi_{ij}$, and $\pi_{\cdot j} = \sum_{j=1}^{J} \pi_{ij}$, $1 \leq j \leq J$. A related **entropy-based** measure, called the **uncertainty coefficient**, is

$$\gamma = \left(\sum_{i=1}^{I}\sum_{j=1}^{J} \pi_{ij} \log(\pi_{ij}/\pi_{i\cdot}\pi_{\cdot j})\right) \Big/ \left(\sum_{j=1}^{J} \pi_{\cdot j} \log \pi_{\cdot j}\right).$$ (6.4.11)

The case of $\tau = \gamma = 0$ relates to independence. Large sample statistical inference for τ and γ can again be drawn by the methods discussed in the earlier sections. Particularly, the Wald method will be very adaptable in such a case of a nonlinear function reducible to a linear one by first order Taylor's expansion incorporated in this method.

Binary response variables constitute an important class of categorical models. Suppose that the response variable Y is binary (i.e., either $Y = 0$ or 1) and associated with Y there is a vector $\mathbf{x}$ of design (or regression) variables, such that $\pi(\mathbf{x}) = P\{Y = 1 \mid \mathbf{x}\} = 1 - P\{Y = 0 \mid \mathbf{x}\}$. In a variety of situations, it may be quite appropriate to let

$$\pi(\mathbf{x}) = (e^{\alpha + \mathbf{x}^t \boldsymbol{\beta}}) / (1 + e^{\alpha + \mathbf{x}^t \boldsymbol{\beta}}), \tag{6.4.12}$$

which resembles the classical **logistic function**, and, hence, the model is termed a **logit model** or a logistic regression model. Note that

$$\log\{\pi(\mathbf{x})/(1 - \pi(\mathbf{x}))\} = \alpha + \mathbf{x}^t \boldsymbol{\beta}, \tag{6.4.13}$$

so that we have essentially transformed the original model to a **log-linear model** (or a **generalized linear model**). The likelihood function for the n observations $Y_1, \ldots, Y_n$ is

$$L_n(\alpha, \beta) = \prod_{i=1}^{n} \left\{ e^{(\alpha + \mathbf{x}_i^t \boldsymbol{\beta}) Y_i} \left[1 + e^{\alpha + \mathbf{x}_i^t \boldsymbol{\beta}} \right]^{-1} \right\}, \tag{6.4.14}$$

so that the ML (or BAN) estimator of α, β and tests for suitable hypotheses on β can be worked out as in Chapter 5. On the other hand, as in quantal bioassay models, we may have the following situation. Suppose that there are k (≥ 2) design points $\mathbf{x}_1, \ldots, \mathbf{x}_k$ and that corresponding to the ith point, there are n_i binary responses Y_{ij}, $j = 1, \ldots, n_i$, with the model (6.4.12) for $\mathbf{x} = \mathbf{x}_i$, $i = 1, \ldots, k$. We let $p_i = n_i^{-1} \sum_{j=1}^{n_i} Y_{ij}$, $i \leq i \leq k$, and let

$$Z_i = \log\{p_i/(1 - p_i)\}, \quad i = 1, \ldots, k. \tag{6.4.15}$$

Then, one may consider the measure of dispersion

$$Q_n(\alpha, \beta) = \sum_{i=1}^{k} p_i (1 - p_i)[Z_i - \alpha - \mathbf{x}_i^t \boldsymbol{\beta}]^2 \tag{6.4.16}$$

and minimizing $Q_n(\alpha, \beta)$ with respect to α, β, one gets **minimum logit** estimates of α, β. Recall that the $n_i p_i$ have binomial distributions, and, hence, by (6.4.15), $\sqrt{n_i}(Z_i - \alpha - \mathbf{x}_i^t \boldsymbol{\beta})$, $i = 1, \ldots, k$, are asymptotically normally distributed with zero mean and variances $\pi(\mathbf{x}_i)[1 - \pi(\mathbf{x}_i)]$, $i = 1, \ldots, k$; note also that they are independent. Hence, (6.4.16) is a version of the weighted least-squares setup with the unknown weights $\pi(\mathbf{x}_i)[1 - \pi(\mathbf{x}_i)]$ replaced by their estimates $p_i(1 - p_i)$. As such, the logit estimators $\widehat{\alpha}, \widehat{\beta}$ are linear in the Z_i with coefficients depending on the $\mathbf{x}_i$ and p_i [which are approximable by the $\pi(\mathbf{x}_i)$], so that the results of Chapter 3 can be readily used to draw conclusions on the asymptotic multinormality of $\widehat{\alpha}, \widehat{\beta}$. A general discussion on the asymptotic properties of models similar to the ones considered here is given in Section 7.4.

We conclude this section with some remarks on the so called **Cochran-Mantel-Haenszel** test for comparing two groups on a binary response, adjusting for control variables. Suppose we have K set of 2×2 contingency tables with the cell probabilities $\{\pi_{ijk}, i, j = 1, 2\}$ and observed frequencies $\{n_{ijk}; i, j = 1, 2\}$, for $k = 1, \ldots, K$. Suppose that we want to test the null hypothesis

$$H_0 : \pi_{ij} = \pi_{i \cdot k} \pi_{\cdot j k}, \quad i, j = 1, 2, \quad \text{for every} \quad k = 1, \ldots, K. \quad (6.4.17)$$

Under H_0, given the marginals $n_{i \cdot k}$, $n_{\cdot j k}$, the expected value of n_{11k} is $m_{11k} = n_{1 \cdot k} n_{\cdot 1 k} / n_{\cdot \cdot k}$ and the conditional variance of n_{11k} is

$$\nu_k^2 = n_{1 \cdot k} n_{\cdot 1 k} n_{2 \cdot k} n_{\cdot 2 k} / n_{\cdot \cdot k}^2 (n_{\cdot \cdot k} - 1), \quad (6.4.18)$$

for $k = 1, \ldots, K$, and these n_{11k}, $k \geq 1$, are independent too. As such, it is quite intuitive to construct a test statistic

$$Q_T = \sum_{k=1}^{K} (n_{11k} - m_{11k})^2 / \nu_k^2 \quad (6.4.19)$$

which will have asymptotically (under H_0) a chi-squared distribution with K degrees of freedom. Although this test is consistent against any departure from independence, possibly in different models for different strata, in terms of power, it may not be the best (when K is particularly not very small). If, on the other hand, the pattern of dependence is concordant across the K strata, then one may use the test statistic

$$Q_{MH} = \left(\sum_{k=1}^{K} (n_{11k} - m_{11k}) \right)^2 \Bigg/ \sum_{k=1}^{K} \nu_k^2 \quad (6.4.20)$$

which will have asymptotically (under H_0) a chi-squared distribution with 1 degree of freedom. This result can be verified directly by showing that under H_0, $n^{-1/2}(n_{11k} - m_{1k})$ asymptotically has a multinormal distribution with null mean vector and dispersion matrix $\text{Diag}\{\nu_1^2, \ldots, \nu_k^2\}$, for $k = 1, \ldots, K$, where $n = \sum_{k=1}^{K} n_{\cdot \cdot k}$. Because of the reduction in the degrees of freedom (from K to 1) when the noncentrality of the $n_{11k} - m_{11k}$ are in the same direction, Q_{MH} is likely to have more power than Q_T. On the other hand, if the individual noncentralities are in possibly different directions, the sum $\sum_{k=1}^{n} (n_{11k} - m_{11k})$ may have a very small noncentrality resulting in very little power of Q_{MH} (compared to Q_T). In meta-analyses, the concordance picture is likely to hold, and, hence, Q_{MH} should be preferred to Q_T.

6.5 Concluding notes

Although product binomial models may be considered as special cases of product multinomial models, they require extra attention because of the specialized nature of some of the related analytical tools. In fact, they are more in tune with those employed in the study of the broader class of (univariate) generalized linear models; these, in turn, are usually analyzed with methods similar to (generalized) least squares, and, thus, we find it convenient to relegate a more detailed asymptotic theory to Chapter 7. In this regard, our brief discussion of some asymptotic results in Section 6.4 deserves some explanation. The major issue may be highlighted by observing that (6.4.16) is a slightly simplified version of a generalized linear model formulation in the sense that the variance function does not depend on the parameters of interest, leading to less complex estimating equations. Even though an asymptotic equivalence between the two approaches may be established, there is a subtle point which merits further discussion. In (6.4.16), usually k (≥ 1) is a fixed positive integer, whereas the sample sizes (n_i) are taken large to justify the asymptotics; in that sense, the p_i are close to the π_i, i.e., $|p_i - \pi_i| = O_p(n^{-1/2})$, whereas $(Z_i - \alpha - \mathbf{x}_i^t\boldsymbol{\beta})^2 = O_p(n^{-1})$. Hence, as in the modified minimum chi-squared method, ignoring the dependence of the $p_i(1 - p_i)$ on (α, β) does not cost much in terms of large sample properties relative to the generalized linear models approach, where such dependence is a part of the scheme. From the computational point of view, however, the picture favors the modified minimum chi-squared method. A similar discussion remains valid for a comparison between the generalized least-squares and the generalized estimating equation approaches and we shall pursue this issue in Section 7.5.

Generally, categorical data models permit easier verification of the regularity conditions of Theorems 5.2.1 and 5.2.2 which are set up for a broader class of situations. However, in the product multinomial case, if any subset of the cell probabilities approaches zero, the dimension of the parameter space is altered and thereby the whole asymptotic theory may be different. In such a case, the MLE may not be BAN and the likelihood ratio test may not follow an asymptotic chi-squared distribution. We shall present some special cases as exercises. In particular, we also include a good example of product Poisson models in the context of contingency tables as related to Example 6.3.1.

6.6 Exercises

Exercise 6.2.1 [Example 1.2.2]: Consider the following table relating to the OAB blood group model:

Blood Group	Observed Frequency	Probability
O	n_O	$p_O = q_O^2$
A	n_A	$p_A = q_A^2 + 2q_O q_A$
B	n_B	$p_B = q_B^2 + 2q_O q_B$
AB	n_{AB}	$p_{AB} = 2q_A q_B$
Total	n	1

The MLE of the gene frequencies $\mathbf{q} = (q_O, q_A, q_B)^t$, considered in (1.2.5), may not be expressible in a closed algebraic form.

i) Show that $q_O = p_O^{1/2}$, $q_A = 1 - (p_O + p_B)^{1/2}$ and $q_B = 1 - (p_O + p_A)^{1/2}$, and use this to propose some initial estimators:

$$\widehat{q}_O^{(0)} = (n_O/n)^{1/2},$$

$$\widehat{q}_A^{(0)} = 1 - (n_O + n_B)^{1/2} n^{-1/2},$$

$$\widehat{q}_B^{(0)} = 1 - (n_O + n_A)^{1/2} n^{-1/2}.$$

Hence, or otherwise, use the **method of scoring** to solve for the MLE (or one-step MLE) by an iteration method.

ii) Show that $\widehat{q}_O^{(0)} + \widehat{q}_A^{(0)} + \widehat{q}_B^{(0)}$ may not be exactly equal to 1. Suggest suitable modifications to satisfy this restriction, and use the method of scoring on this modified version.

iii) Use all the three methods, Wald statistic, likelihood ratio statistic and the Pearsonian goodness-of-fit test statistic, to test the null hypotheses that the Hardy-Wienberg equilibrium holds. Comment on the relative computational ease of these three methods.

Exercise 6.2.2 (Continued): A scientific worker wanted to study whether the gene frequencies ($\mathbf{q}$) varies from one country to another. For this testing for the homogeneity of the $\mathbf{q}$, he obtained the following data set:

	Sample				
Blood Group	1	2	$\cdots$	k	Total
O	n_{1O}	n_{2O}	$\cdots$	n_{kO}	n_O
A	n_{1A}	n_{2A}	$\cdots$	n_{kA}	n_A
B	n_{1B}	n_{2B}	$\cdots$	n_{kB}	n_B
AB	n_{1AB}	n_{2AB}	$\cdots$	n_{kAB}	n_{AB}
Total	n_1	n_2	$\cdots$	n_k	n

i) Treating this as a $4 \times k$ contingency table, construct a completely nonparametric test for homogeneity of the k samples.

ii) Use the Hardy-Weinberg equilibrium for each population, and in that parametric setup, use the Wald test statistic to test for the homogeneity of the gene frequencies.

iii) Use the likelihood ratio test under the setup in (ii).

iv) Comment on the associated degrees of freedom for all the three tests, and also compare them in terms of (a) consistency, (b) power properties and (c) computational ease.

Exercise 6.2.3 (Continued): The same scientific worker has the feeling that the gene frequencies may vary from one race to another (viz. Mongolian, Negroid, Caucasian) and, hence, from one geographical area to another depending on their relative components. He collected a data set from one of the Caribbean islands where Negroid and Caucasian mixtures are profound (with very little Mongolian presence). As such, he decided to work with the following mixture model:

$$p_X = \pi p_X^{(N)} + (1 - \pi)p_X^{(C)}, \quad \text{for} \quad X = \text{O, A, B, AB,}$$

where $0 \leq \pi \leq 1$, and for the $p^{(N)}$ (and $p^{(C)}$), the Hardy-Weinberg equilibrium holds with $\mathbf{q} = \mathbf{q}^{(N)}$ (and $\mathbf{q}^{(C)}$). In this setup, he encountered the following problems:

a) Under $H_0 : \mathbf{q}^{(N)} = \mathbf{q}^{(C)}$, there are only 2 unknown (linearly independent) parameters (π drops out); whereas under the alternative, $\mathbf{q}^{(N)} \neq \mathbf{q}^{(C)}$, there are five linearly independent ones (including π). Thus, he confidently prescribed a chi-squared test with $5 - 2$ ($= 3$) degrees of freedom. Do you support this prescription? If not, why?

b) He observed that there are only 3 (independent) cell probabilities, but
5 unknown parameters. So he concluded that his hypothesis was not
testable in a single sample model. This smart scientist, therefore, decided to choose two different islands (for which the π values are quite
different). From the 4×2 contingency table he had this way; he wanted
to estimate $\pi_1, \pi_2, q_O^{(j)}, q_A^{(j)}$, $j = N, C$. He had 6 linearly independent
cell probabilities and 6 unknown parameters, so that he was satisfied
with the model. Under $H_0 : \mathbf{q}^{(N)} = \mathbf{q}^{(C)}$, he had 2 parameters, whereas
he had 6 under the alternative. Hence, he concluded that the degrees of
freedom of his goodness-of-fit test would be equal to 4. Being so confi-
dent this time, he carried out a volume of simulation work to examine
the adequacy of the chi-squared approximation. Alas, the fit was very
poor! A smarter colleague suggested a fractional degrees of freedom
(2.3) for the chi-squared approximation – that showed some improve-
ment. However, he was puzzled why the degrees of freedom was not an
integer! Anyway, as there was no theoretical foundation, in frustration,
he gave up! Can you eliminate the impasse?

c) Verify that for this mixture model there is a basic **identifiability
issue**: if $\mathbf{q}^{(N)} = \mathbf{q}^{(C)}$ regardless of whether the π_j $(j \leq k)$ are on the
boundary (i.e., $\{0\}, \{1\}$) or not, the number of unknown parameters
is equal to 2, whereas this number jumps to $4 + k$ when H_0 does
not hold. Thus, the parameter point belongs to a boundary of the
parameter space under H_0. Examine the impact of this irregularity on
the asymptotics underlying the usual goodness-of-fit tests.

d) Consider the $4 \times k$ contingency table (for $k \geq 2$) and discuss how the
nonparametric test overcomes this problem?

e) Can you justify the Hardy-Weinberg equilibrium for the mixture model
from the random mating point of view?

Exercise 6.3.1: Consider the multinomial model in (6.1.1). Let $n_1, \ldots, n_r$
be r independent Poisson variables with parameters $m_1, \ldots, m_r$, respectively. Show that $n = n_1 + \cdots + n_r$ has the Poisson distribution with the
parameter $m = m_1 + \cdots + m_r$. Hence, or otherwise, show that the conditional probability law for $n_1, \ldots, n_r$, given n, is multinomial with the cell
probabilities $\pi_j = m_j/m, 1 \leq j \leq r$.

Exercise 6.3.2 (Continued): Using the Poisson to Normal convergence,
show that $(n_j - m_j)^2/m_j$ are independent and asymptotically distributed
as $\chi_1^2, 1 \leq j \leq r$. Hence, using the Cochran Theorem, verify that given n,
$\sum_{j=1}^{r}(n_j - m_j)^2/m_j$ has asymptotically chi-squared law with $r - 1$ degrees
of freedom.

Exercise 6.3.3: For the product multinomial model (6.1.2), show that treating the n_{ij} as independent Poisson variables with parameters m_{ij}, $1 \leq j \leq r_j$, $1 \leq i \leq s$, and conditioning on the n_i $(= \sum_{j=1}^{r_i} n_{ij})$, $1 \leq i \leq s$, one has the same multinomial law, where $\pi_{ij} = m_{ij}/m_i$, $1 \leq j \leq r_i$, $m_i = \sum_{j=1}^{r_i} m_{ij}$, $1 \leq i \leq s$.

Exercise 6.3.4: Let n_{ij} $(1 \leq i \leq r, 1 \leq j \leq c)$ be the observed cell frequencies of a $r \times c$ contingency table, and let $n_{i.} = \sum_{j=1}^{c} n_{ij}$ and $n_{.j} = \sum_{i=1}^{r} n_{ij}$ be the marginal totals and $n = \sum_{i=1}^{r} \sum_{j=1}^{c} n_{ij}$. Treat the n_{ij} as independent Poisson variables with parameters m_{ij}, and hence, or otherwise, show that conditional on the $n_{i.}$ and $n_{.j}$ being given, the n_{ij} have the same multinomial law as arising in the classical $r \times c$ table with marginals fixed. Use the Poisson to Normal convergence along with the Cochran Theorem to derive the asymptotic chi-squared distribution [with $(r - 1) \times (c - 1)$ degrees of freedom] for the classical contingency table test.

Exercise 6.3.5 (Continued): Consider the log-linear model

$$\log m_{ij} = \mu + \alpha_i + \beta_j + \gamma_{ij}, \quad 1 \leq i \leq r, \quad 1 \leq j \leq c,$$

where $\sum_{i=1}^{r} \alpha_i = 0$, $\sum_{j=1}^{c} \beta_j = 0$ and $\sum_{i=1}^{r} \gamma_{ij} = 0 = \sum_{j=1}^{c} \gamma_{ij}$, for all i, j. Show that the classical independence model relates to $\gamma_{ij} = 0$, for all i, j. Critically examine the computational aspects of the estimation of α's, β's and γ's by the ML, MCS and MMCS methods, and show that the test considered in Exercise 6.3.4 is appropriate in this setup too.

Exercise 6.4.1: The lifetime X of an electric lamp has the simple exponential survival function $\overline{F}(x) = e^{-x/\theta}$, $\theta > 0$, $x \geq 0$. A factory has n lamps which are lighted throughout the day and night, and every week the burnt out ones are replaced by new ones. At a particular inspection time, it was observed that out of n, n_1 were dead, and $n - n_1 = n_2$ were still alive. Find out a BAN estimator of θ based on this data set. Obtain the Fisher information on θ from this binary model as well as from the ideal situation when $X_1, \ldots, X_n$ would have been all observed. Compare the two and comment on the loss of information due to this grouping.

Exercise 6.4.2: Consider a 2×2 contingency table with the cell frequencies n_{ij} and probabilities π_{ij}, for $i, j = 1, 2$. Under the independence model, $\pi_{ij} = \pi_{i.}\pi_{.j}$, and its estimator is $n_{i.}n_{.j}/n_{..}^2 = \hat{\pi}_{ij}^*$, say.

i) Obtain the expressions for $E(\hat{\pi}_{ij}^*)$ and $\text{Var}(\hat{\pi}_{ij}^*)$ when the independence model may not hold.

ii) Hence, or otherwise, comment on the rationality of the test for independence based on the odds ratio in (6.4.1).

Exercise 6.4.3: Consider the genetical data

Type	AA	Aa	aA	aa	Total
Probability	$(1+\theta)/4$	$(1-\theta)/4$	$(1-\theta)/4$	$(1+\theta)/4$	1
Obs. frequency	n_1	n_2	n_3	n_4	n

Find the MLE of the linkage factor θ. Compute the mean and variance of the estimator, and find out the appropriate variance stabilizing transformation.

Exercise 6.4.4 (Continued): Use the results in the previous exercise to provide an asymptotic $100\gamma\%$ confidence interval for θ.

Exercise 6.4.5: Suppose that there are k (≥ 2) independent data sets each pertaining to the model in Exercise 6.4.3 (with respective θ-values as $\theta_1,\ldots,\theta_k$). Use the variance stabilizing transformation to test for the homogeneity of $\theta_1,\ldots,\theta_k$. Comment on the desirability of the transformation in this setup.

CHAPTER 7

Large Sample Theory for Regression Models

7.1 Introduction

The general denomination of **regression models** is used to identify statistical models for the relationship between one or more **explanatory** (independent) **variables** and one or more **response** (dependent) **variables**. Typical examples include the investigation of the influence of:

 i) the amount of fertilizer on the yield of a certain type of crop;

 ii) the type of treatment and age on the serum cholesterol levels of patients;

 iii) the driving habits and fuel type on the gas mileage of a certain make of automobile;

 iv) the type of polymer, extrusion rate and extrusion temperature on the tensile strength and number of defects/unit length of synthetic fibers.

Within this class, the so-called **linear models** play an important role for statistical applications; such models are easy to interpret, mathematically tractable and may be successfully employed for a variety of practical situations as in (i)–(iv) above. They include models usually considered in **Linear Regression Analysis, Analysis of Variance (ANOVA), Analysis of Covariance (ANCOVA)** and may easily be extended to include **Logistic Regression Analysis, Generalized Linear Models, Multivariate Regression, Multivariate Analysis of Variance (MANOVA)** or **Multivariate Analysis of Covariance (MANCOVA)**.

In matrix notation, the univariate linear model may be expressed as

$$\mathbf{Y}_n = \mathbf{X}_n \boldsymbol{\beta} + \boldsymbol{\varepsilon}_n \tag{7.1.1}$$

where $\mathbf{Y}_n = (Y_1, \ldots, Y_n)^t$ is an $(n \times 1)$ vector of observable response variables, $\mathbf{X}_n = ((x_{ij}^{(n)}))$ is an $(n \times q)$ matrix of known constants $x_{ij}^{(n)}$ (which for ease of notation we denote x_{ij}, omitting the superscript, represents the value of the jth explanatory variable for the ith sample unit), $\boldsymbol{\beta} = (\beta_1, \ldots, \beta_q)^t$ denotes the $(q \times 1)$ vector of unknown parameters, and

$\varepsilon_n = (\varepsilon_1, \ldots, \varepsilon_n)^t$ is an $(n \times 1)$ vector of unobservable random variables (errors) assumed to follow a (generally unknown) distribution with d.f. F such that $\mathrm{E}\varepsilon_n = \mathbf{0}$ and $\mathrm{Var}\,\varepsilon_n = \sigma^2\mathbf{I}_n$, $\sigma^2 < \infty$, i.e. the elements of ε_n are uncorrelated, have zero means and constant (but unknown) variance. This is known in the literature as the **Gauss-Markov setup**. It is usual to take $x_{i1} = 1$, $i = 1, \ldots, n$, and in the case of a single explanatory variable, (7.1.1) reduces to the **simple linear regression model**:

$$Y_i = \alpha + \beta x_i + \varepsilon_i, \quad i = 1, \ldots, n, \tag{7.1.2}$$

where α and β are the (unknown) intercept and slope, respectively, and x_i denotes the value of the explanatory variable for the ith sample unit, $i = 1, \ldots, n$.

Statistical analysis under (7.1.1) or (7.1.2) usually involve estimation of the parameter vector $\boldsymbol{\beta}$ $[\alpha, \beta$ in (7.1.2)$]$ and tests of hypotheses about its components. The well-known least-squares (LS) method for estimating $\boldsymbol{\beta}$ consists of minimizing the residual sum of squares

$$\varepsilon_n^t \varepsilon_n = (\mathbf{Y}_n - \mathbf{X}_n\boldsymbol{\beta})^t(\mathbf{Y}_n - \mathbf{X}_n\boldsymbol{\beta}) \tag{7.1.3}$$

which, for (7.1.2) reduces to minimizing

$$\sum_{i=1}^{n} \varepsilon_i^2 = \sum_{i=1}^{n}(Y_i - \alpha - \beta x_i)^2. \tag{7.1.4}$$

Since (7.1.3) is a quadratic function of $\boldsymbol{\beta}$ it is easy to show that its minimum is attained at the point $\widehat{\boldsymbol{\beta}}_n$, identifiable as a solution to the "normal" equations:

$$\mathbf{X}_n^t\mathbf{X}_n\widehat{\boldsymbol{\beta}}_n = \mathbf{X}_n^t\mathbf{Y}_n. \tag{7.1.5}$$

$\widehat{\boldsymbol{\beta}}_n$ is termed the **least-squares estimator** (LSE) of $\boldsymbol{\beta}$. Similarly we may show that the LSE of (α, β) in (7.1.2) is a solution to

$$\begin{aligned}\sum_{i=1}^{n} Y_i &= n\widehat{\alpha}_n + \widehat{\beta}_n \sum_{i=1}^{n} x_i, \\ \sum_{i=1}^{n} x_i Y_i &= \widehat{\alpha}_n \sum_{i=1}^{n} x_i + \widehat{\beta}_n \sum_{i=1}^{n} x_i^2.\end{aligned} \tag{7.1.6}$$

If $\mathrm{rank}\,(\mathbf{X}_n) = q$, (7.1.5) has a single solution, given by

$$\widehat{\boldsymbol{\beta}}_n = (\mathbf{X}_n^t\mathbf{X}_n)^{-1}\mathbf{X}_n^t\mathbf{Y}_n; \tag{7.1.7}$$

otherwise, there are infinite solutions which may be obtained by replacing $(\mathbf{X}_n^t\mathbf{X}_n)^{-1}$ with a generalized inverse $(\mathbf{X}_n^t\mathbf{X}_n)^-$ in (7.1.7). Similarly, if the

x_i's are not all equal in (7.1.2), the unique solution to (7.1.6) is given by

$$\widehat{\beta}_n = \sum_{i=1}^{n} Y_i(x_i - \overline{x}_n) / \sum_{i=1}^{n} (x_i - \overline{x}_n)^2,$$

$$\widehat{\alpha}_n = \overline{Y}_n - \widehat{\beta}_n \overline{x}_n. \tag{7.1.8}$$

Since, for $\mathbf{X}_n$ not of full rank, $\mathbf{X}_n^t \boldsymbol{\beta}$ can be written as $\mathbf{X}_n^{*t} \boldsymbol{\beta}^*$, where $\mathbf{X}_n^*$ is of full rank and $\dim(\boldsymbol{\beta}^*) \leq \dim(\boldsymbol{\beta})$, for all practical purposes, the **model specification matrix** $\mathbf{X}_n$ can be chosen of full rank q. We will restrict our attention to this case, unless otherwise stated.

Given the assumptions underlying (7.1.1), it follows immediately from (7.1.7) that $\mathrm{E}\widehat{\beta}_n = \boldsymbol{\beta}$ and $\mathrm{Var}\,\widehat{\beta}_n = \sigma^2(\mathbf{X}_n^t\mathbf{X}_n)^{-1}$. More specifically, for (7.1.2) we obtain

$$\widehat{\mathrm{Var}} \begin{pmatrix} \widehat{\alpha}_n \\ \widehat{\beta}_n \end{pmatrix} = \frac{\sigma^2}{\sum_{i=1}^{n}(x_i - \overline{x}_n)^2} \begin{bmatrix} \sum_{i=1}^{n} x_i^2/n & -\overline{x}_n \\ -\overline{x}_n & 1 \end{bmatrix}.$$

Moreover, under the same assumptions, the well-known **Gauss-Markov Theorem** states that for every fixed q-vector $\mathbf{c} \in \mathbb{R}^q$, $\mathbf{c}^t\widehat{\beta}_n$ is the **best linear unbiased estimator** (BLUE) of $\mathbf{c}^t\boldsymbol{\beta}$, in the sense that it has the smallest variance in the class of linear unbiased estimators of $\mathbf{c}^t\boldsymbol{\beta}$. Although this constitutes an important optimality property of the LSE, it is of little practical application unless we have some idea of the corresponding distribution. If we consider the additional assumption that the random errors in (7.1.1) [or (7.1.2)] are normally distributed, it follows that the LSE $\widehat{\beta}_n$ coincides with the MLE and that

$$\widehat{\beta}_n \sim N_q(\boldsymbol{\beta}, \sigma^2(\mathbf{X}_n^t\mathbf{X}_n)^{-1}).$$

This assumption, however, is too restrictive and rather difficult to verify in practice, indicating that some approximate results are desirable. This chapter is devoted to the study of the large sample properties of the LSE (and some other alternatives) of $\boldsymbol{\beta}$ in (7.1.1).

In Section 7.2 the asymptotic properties of the LSE are laid down. In Section 7.3 we motivate the use of alternative (robust) estimation methods for the linear model; essentially we expand the brief discussion presented in Section 5.4 to this more interesting case and outline the proofs of some asymptotic results. Section 7.4 deals with extensions of the results to Generalized Linear Models. The relation between Generalized Estimating Equations and Generalized Least Squares is outlined in Section 7.5 and, finally, in Section 7.6, asymptotic results for Nonparametric Regression Models are briefly described.

7.2 Generalized least-squares procedures

We begin this section with a result on the asymptotic distribution of the
LSE for the simple linear regression model. A detailed proof is presented
since similar steps may be successfully employed in other situations:

Theorem 7.2.1: *Consider the simple regression model (7.1.2) under the
Gauss-Markov setup with i.i.d. error terms ε_i, $i = 1, \ldots, n$, and let $\widehat{\alpha}_n$ and
$\widehat{\beta}_n$ be the LSE of α and β, respectively. Write*

$$t_n = \sum_{i=1}^{n} (x_i - \bar{x}_n)^2,$$

$$c_{ni} = (x_i - \bar{x}_n)/\sqrt{t_n}$$

and assume further that:

i) $\max_{1 \leq i \leq n} c_{ni}^2 \to 0$ *as* $n \to \infty$ *(Noether's condition);*

ii) $\lim_{n \to \infty} \bar{x}_n = \bar{x} < \infty;$

iii) $\lim_{n \to \infty} n^{-1} t_n = \lim_{n \to \infty} n^{-1} \sum_{i=1}^{n} (x_i - \bar{x}_n)^2 = t < \infty.$

Then

$$\sqrt{n} \begin{pmatrix} \widehat{\alpha}_n - \alpha \\ \widehat{\beta}_n - \beta \end{pmatrix} \xrightarrow{\mathcal{D}} N_2 \left(\mathbf{0}, \begin{bmatrix} 1 + \bar{x}^2/t & -\bar{x}/t \\ -\bar{x}/t & 1/t \end{bmatrix} \sigma^2 \right). \qquad (7.2.1)$$

Proof: From (7.1.8) and (7.1.2) we may write

$$\widehat{\beta}_n = t_n^{-1} \sum_{i=1}^{n} (x_i - \bar{x}_n)\{\alpha + \beta(x_i - \bar{x}_n) + \varepsilon_i\}$$

$$= \beta + t_n^{-1} \sum_{i=1}^{n} (x_i - \bar{x}_n)\varepsilon_i \qquad (7.2.2)$$

which implies that

$$\sqrt{t_n}(\widehat{\beta}_n - \beta) = \sum_{i=1}^{n} \frac{(x_i - \bar{x}_n)}{\sqrt{t_n}} \varepsilon_i = \sum_{i=1}^{n} c_{ni}\varepsilon_i. \qquad (7.2.3)$$

Observe that $\sum_{i=1}^{n} c_{ni} = 0$ and $\sum_{i=1}^{n} c_{ni}^2 = 1$; then, in view of (i), it follows
from the Hájek-Šidák Central Limit Theorem (3.3.6) that:

$$\sqrt{t_n}(\widehat{\beta}_n - \beta) \xrightarrow{\mathcal{D}} N(0, \sigma^2).$$

Using (iii) and the Slutsky Theorem (3.4.2) we obtain

$$\sqrt{n}(\widehat{\beta}_n - \beta) \xrightarrow{\mathcal{D}} N(0, \sigma^2/t).$$

Again, using (7.1.8) and (7.1.2) we may write

$$
\begin{aligned}
\widehat{\alpha}_n &= \overline{Y}_n - \widehat{\beta}_n \overline{x}_n \\
&= \alpha + \beta \overline{x}_n + \overline{\varepsilon}_n - \widehat{\beta}_n \overline{x}_n \\
&= \alpha - (\widehat{\beta}_n - \beta)\overline{x}_n + \overline{\varepsilon}_n
\end{aligned} \tag{7.2.4}
$$

which implies that

$$
\begin{aligned}
\sqrt{n}(\widehat{\alpha}_n - \alpha) &= \sqrt{n}\,\overline{\varepsilon}_n - \sqrt{n}(\widehat{\beta}_n - \beta)\overline{x}_n \\
&= \sqrt{n}\,\overline{\varepsilon}_n - \sqrt{n/t_n}\,\overline{x}_n \sqrt{t_n}(\widehat{\beta}_n - \beta)
\end{aligned}
$$

and which in view of (7.2.3) may be re-expressed as

$$
\sqrt{n}(\widehat{\alpha}_n - \alpha) = \frac{1}{\sqrt{n}}\sum_{i=1}^{n}\varepsilon_i - \sqrt{n/t_n}\,\overline{x}_n \sum_{i=1}^{n} c_{ni}\varepsilon_i = \sum_{i=1}^{n} d_{ni}\varepsilon_i \tag{7.2.5}
$$

where

$$
d_{ni} = n^{-1/2} - \sqrt{n/t_n}\,\overline{x}_n c_{ni}.
$$

Observing that $\sum_{i=1}^{n} d_{ni} = \sqrt{n}$ and $\sum_{i=1}^{n} d_{ni}^2 = 1 + n\,\overline{x}_n^2/t_n$ and applying the Hájek-Šidak Central Limit Theorem (3.3.6) we obtain

$$
\sqrt{n}\left(1 + \frac{n\overline{x}_n^2}{t_n}\right)^{-\frac{1}{2}}(\widehat{\alpha}_n - \alpha) \xrightarrow{\mathcal{D}} N(0, \sigma^2).
$$

Using (ii), (iii) and the Slutsky Theorem (3.4.2) it follows that

$$
\sqrt{n}(\widehat{\alpha}_n - \alpha) \xrightarrow{\mathcal{D}} N\left(0, \sigma^2\left(1 + \frac{\overline{x}^2}{t}\right)\right).
$$

Now let $\boldsymbol{\lambda} = (\lambda_1, \lambda_2)^t \in \mathbb{R}^2$, $\boldsymbol{\lambda} \neq \mathbf{0}$, be an arbitrary, but fixed vector and use (7.2.3) and (7.2.5) to see that

$$
\lambda_1 \sqrt{n}(\widehat{\alpha}_n - \alpha) + \lambda_2 \sqrt{n}(\widehat{\beta}_n - \beta) = \sum_{i=1}^{n} f_{ni}\varepsilon_i
$$

where

$$
f_{ni} = \lambda d_{ni} + \lambda_2 \sqrt{n/t_n}\, c_{ni} \quad \text{with} \quad \sum_{i=1}^{n} f_{ni} = \lambda_1 \sqrt{n}
$$

and

$$
\sum_{i=1}^{n} f_{ni}^2 = \lambda_1^2\left(1 + \overline{x}_n^2 n/t_n\right) + \lambda_2^2 n/t_n - 2\lambda_1\lambda_2 \overline{x}_n n/t_n.
$$

Letting

$$
\mathbf{V}_n = \begin{bmatrix} 1 + n\overline{x}_n^2/t_n & -n\overline{x}_n/t_n \\ -n\overline{x}_n/t_n & n/t_n \end{bmatrix}
$$

and applying the Hájek-Šidak Central Limit Theorem $(3.3.6)$ it follows that

$$\left\{(\lambda_1, \lambda_2)\mathbf{V}_n \begin{pmatrix} \lambda_1 \\ \lambda_2 \end{pmatrix}\right\}^{-\frac{1}{2}} \left\{\lambda_1 \sqrt{n}(\widehat{\alpha}_n - \alpha) + \lambda_2 \sqrt{n}(\widehat{\beta}_n - \beta)\right\} \xrightarrow{\mathcal{D}} N(0, \sigma^2)$$

and using the Slutsky Theorem $(3.4.2)$ we have, in view of (ii) and (iii),

$$\lambda_1 \sqrt{n}(\widehat{\alpha}_n - \alpha) + \lambda_2 \sqrt{n}(\widehat{\beta}_n - \beta) \xrightarrow{\mathcal{D}} N(0, \sigma^2 \boldsymbol{\lambda}^t \mathbf{V} \boldsymbol{\lambda})$$

where

$$\mathbf{V} = \begin{bmatrix} 1 + \overline{x}^2/t & -\overline{x}/t \\ -\overline{x}/t & 1/t \end{bmatrix}.$$

Since $\boldsymbol{\lambda}$ is arbitrary, the result follows directly from the Cramér-Wold Theorem $(3.2.4)$. ∎

It is possible to relax (iii) in Theorem 7.2.1 to a certain extent. For example, we have $t_n^{1/2}(\widehat{\beta}_n - \beta) \xrightarrow{\mathcal{D}} N(0, 1)$ whenever (i) holds in conjunction with

$$\text{iii'}) \; \liminf_{n \to \infty} n^{-1} t_n > 0.$$

On the other hand, if $n^{-1} t_n \to \infty$ as $n \to \infty$, $t_n^{1/2}(\widehat{\alpha}_n - \alpha)$ does not have a limiting law, and, hence, the bivariate result in $(7.2.1)$ may not hold.

As a direct consequence of the above theorem and a multivariate generalization of Theorem 3.2.5 it follows that $\sqrt{n}(\widehat{\alpha}_n - \alpha, \widehat{\beta}_n - \beta)^t = O_P(\mathbf{1})$, which implies that $\widehat{\alpha}_n$ and $\widehat{\beta}_n$ are weakly consistent. In fact, they are strongly consistent too. Also, for this consistency property, asymptotic normality is not needed, and, hence, some of the assumed regularity conditions for Theorem 7.2.1 may also be relaxed. To see this note that by $(7.2.3)$

$$\begin{aligned} t_n(\widehat{\beta}_n - \beta) &= \sum_{i=1}^{n}(x_i - \overline{x}_n)\varepsilon_i \\ &= \sum_{i=1}^{n}(x_i - \overline{x})\varepsilon_i + n(\overline{x} - \overline{x}_n)\overline{\varepsilon}_n. \end{aligned} \qquad (7.2.6)$$

Let $A_n = \sum_{i=1}^{n}(x_i - \overline{x})\varepsilon_i$, $B_n = n(\overline{x} - \overline{x}_n)\overline{\varepsilon}_n$ and observe that $\overline{\varepsilon}_n = n^{-1}\sum_{i=1}^{n}\varepsilon_i \xrightarrow{\text{a.s.}} 0$ by the Khintchine SLLN (Theorem 2.3.13). Then, using (ii) and (iii) as in Theorem 7.2.1 we have

$$\frac{B_n}{t_n} = \frac{n}{t_n}(\overline{x} - \overline{x}_n)\overline{\varepsilon}_n = O(1)o(1)o(1) \text{ a.s. } = o(1) \text{ a.s.} \qquad (7.2.7)$$

Also,

$$A_{n+1} = A_n + \varepsilon_{n+1}(x_{n+1} - \overline{x}), n \geq 1$$

so that for all $n \geq 1$,

$$E\{A_{n+1} \mid \varepsilon_1, \ldots, \varepsilon_n\} = A_n,$$

implying that $\{A_n\}$ is a martingale. Then, in view of (ii) and (iii) of Theorem 7.2.1 together with the fact that $E\varepsilon_i^2 = \sigma^2 < \infty$, it follows from the Kolmogorov SLLN for martingales (Theorem 2.4.2) that

$$A_n/t_n \xrightarrow{\text{a.s.}} 0. \tag{7.2.8}$$

Substituting (7.2.7) and (7.2.8) in (7.2.6) we may conclude that $\widehat{\beta}_n - \beta \xrightarrow{\text{a.s.}} 0$ and via (7.2.4) also that $\widehat{\alpha}_n - \alpha \xrightarrow{\text{a.s.}} 0$. In this context, we may even relax the second moment condition on the ε_i to a lower order moment condition, but the proof will be somewhat more complicated.

In order to make full use of Theorem 7.2.1 to draw statistical conclusions on (α, β), we need to provide a consistent estimator of the unknown variance σ^2. In this direction, an unbiased estimator of σ^2 is

$$S_n^2 = (n-2)^{-1} \sum_{i=1}^{n} (Y_i - \widehat{\alpha}_n - \widehat{\beta}_n x_i)^2. \tag{7.2.9}$$

Using (7.1.2), (7.2.2) and (7.2.4), we obtain, from (7.2.9), that

$$
\begin{aligned}
S_n^2 &= (n-2)^{-1} \sum_{i=1}^{n} \{\varepsilon_i - \bar{\varepsilon}_n - (\widehat{\beta}_n - \beta)(x_i - \bar{x}_n)\}^2 \\
&= (n-2)^{-1} \sum_{i=1}^{n} (\varepsilon_i - \bar{\varepsilon}_n)^2 - (\widehat{\beta}_n - \beta)^2 (n-2)^{-1} \sum_{i=1}^{n} (x_i - \bar{x}_n)^2 \\
&= \frac{(n-1)}{(n-2)} \frac{1}{(n-1)} \sum_{i=1}^{n} (\varepsilon_i - \bar{\varepsilon}_n)^2 - (\widehat{\beta}_n - \beta)^2 (n-2)^{-1} t_n.
\end{aligned}
\tag{7.2.10}
$$

Now, using (2.3.87) it follows that

$$(n-1)^{-1} \sum_{i=1}^{n} (\varepsilon_i - \bar{\varepsilon}_n)^2 \xrightarrow{\text{a.s.}} \sigma^2;$$

recalling (iii) of Theorem 7.2.1, we may write $n^{-1} t_n = O(1)$ and, finally, using the fact that $\widehat{\beta}_n - \beta \xrightarrow{\text{a.s.}} 0$ we may conclude from (7.2.10) that $S_n^2 \xrightarrow{\text{a.s.}} \sigma^2$. This, in turn, implies that $S_n^2 \xrightarrow{\text{P}} \sigma^2$.

Let us consider now an extension of Theorem 7.2.1 to the multiple regression model. In this context, note that (7.1.1) and (7.1.7) imply

$$\widehat{\beta}_n - \beta = (\mathbf{X}_n^t \mathbf{X}_n)^{-1} \mathbf{X}_n^t \varepsilon_n \tag{7.2.11}$$

so that $E\widehat{\beta}_n = \beta$ and $\text{Var}\,\widehat{\beta}_n = \sigma^2 (\mathbf{X}_n^t \mathbf{X}_n)^{-1}$. If we consider an arbitrary linear compound $Z_n = \widehat{\boldsymbol{\lambda}}^t (\widehat{\beta}_n - \beta)$, $\boldsymbol{\lambda} \in \mathbb{R}^q$, we get $Z_n = \mathbf{c}_n^t \varepsilon_n$ with $\mathbf{c}_n = \mathbf{X}_n (\mathbf{X}_n^t \mathbf{X}_n)^{-1} \boldsymbol{\lambda}$. Then, to obtain the asymptotic distribution of Z_n

(conveniently standardized) all we need is to verify that $\mathbf{c}_n$ satisfies the regularity condition of the Hájek-Šidak Central Limit Theorem (3.3.6). Denoting the kth row of $\mathbf{X}_n$ by $\mathbf{x}_{nk}^t$ and taking into account that the result must hold for every (fixed) but arbitrary $\boldsymbol{\lambda} \in {I\!\!R}^q$ such regularity condition may be reformulated by requiring that as $n \to \infty$,

$$\sup_{\boldsymbol{\lambda} \in {I\!\!R}^q} \left[\max_{1 \le i \le n} \left\{ \boldsymbol{\lambda}^t (\mathbf{X}_n^t \mathbf{X}_n)^{-1} \mathbf{x}_{ni} \mathbf{x}_{ni}^t (\mathbf{X}_n^t \mathbf{X}_n)^{-1} \boldsymbol{\lambda} / \boldsymbol{\lambda}^t (\mathbf{X}_n^t \mathbf{X}_n)^{-1} \boldsymbol{\lambda} \right\} \right] \to 0. \tag{7.2.12}$$

Now in view of Courant Theorem (1.4.2), we have

$$\sup_{\boldsymbol{\lambda} \in {I\!\!R}^q} \left\{ \boldsymbol{\lambda}^t (\mathbf{X}_n^t \mathbf{X}_n)^{-1} \mathbf{x}_{ni} \mathbf{x}_{ni}^t (\mathbf{X}_n^t \mathbf{X}_n)^{-1} \boldsymbol{\lambda} / \boldsymbol{\lambda}^t (\mathbf{X}_n^t \mathbf{X}_n)^{-1} \boldsymbol{\lambda} \right\}$$

$$= \mathrm{ch}_1 \left\{ (\mathbf{X}_n^t \mathbf{X}_n)^{-1} \mathbf{x}_{ni} \mathbf{x}_{ni}^t \right\}$$

$$= \mathbf{x}_{ni}^t (\mathbf{X}_n^t \mathbf{X}_n)^{-1} \mathbf{x}_{ni},$$

implying that (7.2.12) reduces to the generalized Noether condition

$$\max_{1 \le k \le n} \mathbf{x}_{nk}^t (\mathbf{X}_n^t \mathbf{X}_n)^{-1} \mathbf{x}_{nk} \to 0 \quad \text{as } n \to \infty. \tag{7.2.13}$$

Likewise, (ii) and (iii) in Theorem 7.2.1 extend to

$$\lim_{n \to \infty} n^{-1} (\mathbf{X}_n^t \mathbf{X}_n) = \mathbf{V}, \quad \text{finite and p.d.} \tag{7.2.14}$$

A direct application of Hájek-Šidak Theorem (3.3.6) to the Z_n's in conjunction with the Cramér-Wold Theorem (3.2.4) may then be used to prove the following:

Theorem 7.2.2: *Consider the model (7.1.1) and assume that the ε_i's are independent and identically distributed random variables with zero mean and finite positive variance σ^2. Then, under (7.2.13) and (7.2.14),*

$$\sqrt{n}(\widehat{\boldsymbol{\beta}}_n - \boldsymbol{\beta}) \xrightarrow{\mathcal{D}} N_q(\mathbf{0}, \sigma^2 \mathbf{V}^{-1}) \tag{7.2.15}$$

or, equivalently,

$$(\mathbf{X}_n^t \mathbf{X}_n)^{1/2} (\widehat{\boldsymbol{\beta}}_n - \boldsymbol{\beta}) \xrightarrow{\mathcal{D}} N_q(\mathbf{0}, \sigma^2 \mathbf{I}_q). \tag{7.2.16}$$

Loosely speaking, often (7.2.16) is written as

$$\widehat{\boldsymbol{\beta}}_n - \boldsymbol{\beta} \cong N(\mathbf{0}, \sigma^2 (\mathbf{X}_n^t \mathbf{X}_n)^{-1})$$

where $\cong$ denotes approximately distributed as (for large sample sizes).

Note that if the individual coordinates of $\widehat{\boldsymbol{\beta}}_n - \boldsymbol{\beta}$ converge to zero (in probability or almost surely), then $\|\widehat{\boldsymbol{\beta}}_n - \boldsymbol{\beta}\|$ does too. For such coordinate elements, the proof sketched for the simple regression model goes through neatly. Hence, omitting the details, we claim that $\widehat{\boldsymbol{\beta}}_n - \boldsymbol{\beta} \xrightarrow{\text{a.s.}} 0$ (and, hence in probability, also). Moreover, for the multiple regression model, an unbiased estimator of σ^2 is

$$
\begin{aligned}
S_n^2 &= (n-q)^{-1} \sum_{i=1}^{n} (Y_i - \mathbf{x}_{ni}^t \widehat{\boldsymbol{\beta}}_n)^2 \\
&= (n-q)^{-1} \left\{ \sum_{i=1}^{n} \varepsilon_i^2 - (\widehat{\boldsymbol{\beta}}_n - \boldsymbol{\beta})^t (\mathbf{X}_n^t \mathbf{X}_n)(\widehat{\boldsymbol{\beta}}_n - \boldsymbol{\beta}) \right\}
\end{aligned}
$$

$$(7.2.17)$$

so that using the Khintchine SLLN (Theorem 2.3.13) on $n^{-1} \sum_{i=1}^{n} \varepsilon_i^2$, using (7.2.14) on $n^{-1}(\mathbf{X}_n^t \mathbf{X}_n)$ and appealing to the fact that $\widehat{\boldsymbol{\beta}}_n - \boldsymbol{\beta} \xrightarrow{\text{a.s.}} 0$, we have that $S_n^2 \xrightarrow{\text{a.s.}} \sigma^2$.

Let us now examine the performance of the LSE $\widehat{\boldsymbol{\beta}}_n$ when the errors ε_i are independent but not necessarily identically distributed. For simplicity we go back to the simple regression model and let $\mathrm{E}\varepsilon_i^2 = \sigma_i^2$, $i \geq 1$, and $\mathbf{D}_n = \mathrm{diag}(\sigma_1^2, \ldots, \sigma_n^2)$. Then, since $\mathrm{E}\varepsilon_i = 0$, $i \geq 1$, it follows from (7.2.3) that $\mathrm{E}\widehat{\beta}_n = \beta$ and that

$$
\mathrm{Var}(\widehat{\beta}_n) = t_n^{-1} \sum_{i=1}^{n} (x_i - \overline{x}_n)^2 \sigma_i^2. \tag{7.2.18}
$$

The Noether condition in Theorem 7.2.1 can be modified to extend the asymptotic normality result to this heteroscedastic model, but there is a subtle point which we would like to make clear at this stage. If the ε_i's are normally distributed, then ε_i/σ_i are all standard normal variables, and, hence, they are identically distributed. On the other hand, if the distribution of ε_i, $i \geq 1$, does not have a specified form, the ε_i/σ_i's may not all have the same distribution [unless we assume that the distribution function F_i of ε_i, $i \geq 1$, is of the scale form $F_0(e/\sigma_i)$, where F_0 has a functional form independent of the scale factors]. Under such a scale family model,

$$
(x_i - \overline{x}_n)\varepsilon_i = \{(x_i - \overline{x}_n)\sigma_i\}\varepsilon_i/\sigma_i, \quad i \geq 1, \tag{7.2.19}
$$

so that the Noether condition, as well as (ii) and (iii) in Theorem 7.2.1 can be formulated in terms of the $(x_i - \overline{x}_n)\sigma_i$, $i \geq 1$, and the conclusion in (7.2.1) can be validated when the dispersion matrix is adjusted accordingly. However, these elements are then functions of the given x_i as well as of the unknown σ_i^2, $i \geq 1$, and, hence, to draw statistical conclusions from this asymptotic normality result, one needs to estimate the dispersion matrix from the observed sample. This may turn out to be much more involved. To illustrate this point, we consider the marginal law for $\widehat{\beta}_n - \beta$, so that we need to estimate $\operatorname{Var} \widehat{\beta}_n$ in (7.2.18). If the true ε_i, $i \geq 1$, were observable, we might have taken this estimator as

$$V_n^0 = t_n^{-1} \sum_{i=1}^{n} (x_i - \overline{x}_n)^2 \varepsilon_i^2. \tag{7.2.20}$$

Thus, if we assume that as $n \to \infty$,

$$\frac{\sum_{i=1}^{n} |(x_i - \overline{x}_n)|^{2+\delta} \operatorname{E} \left| \varepsilon_i^2 - \sigma_i^2 \right|^{1+\delta}}{\left(\sum_{i=1}^{n} (x_i - \overline{x}_n)^2 \sigma_i^2 \right)^{1+\delta}} \to 0 \tag{7.2.21}$$

for some $\delta > 0$, we may invoke the Markov LLN (Theorem 2.3.7) to conclude that

$$\frac{V_n^0}{\operatorname{Var} \widehat{\beta}_n} = \left\{ \sum_{i=1}^{n} (x_i - \overline{x}_n)^2 \varepsilon_i^2 \right\} \left\{ \sum_{i=1}^{n} (x_i - \overline{x}_n)^2 \sigma_i^2 \right\}^{-1} \xrightarrow{\text{P}} 1. \tag{7.2.22}$$

Note that (7.2.21) holds whenever the ε_i's have finite moments of order $r > 2$ and the x_i's are bounded, as in many practical situations.

Since the ε_i, $i \geq 1$, are unobservable, we may consider the residuals $\mathbf{e}_n = (e_{n1}, \ldots, e_{nn})^t$ given by

$$
\begin{aligned}
\mathbf{e}_n &= \mathbf{Y}_n - \widehat{\alpha}_n \mathbf{1}_n - \widehat{\beta}_n \mathbf{x}_n \\
&= \varepsilon_n + (\alpha - \widehat{\alpha}_n) \mathbf{1}_n + (\beta - \widehat{\beta}_n) \mathbf{x}_n \\
&= \varepsilon_n - \mathbf{A}_n \varepsilon_n = (\mathbf{I}_n - \mathbf{A}_n) \varepsilon_n
\end{aligned} \tag{7.2.23}
$$

where $\mathbf{A}_n$ is an $(n \times n)$ matrix. Thus we may define an estimator of $\operatorname{Var} \widehat{\beta}_n$ as

$$V_n = t_n^{-1} \sum_{i=1}^{n} (x_i - \overline{x}_n)^2 e_{ni}^2 = t_n^{-1} \mathbf{e}_n^t \mathbf{L}_n^t \mathbf{L}_n \mathbf{e}_n \tag{7.2.24}$$

where

$$\mathbf{L}_n = \operatorname{diag}\{x_1 - \overline{x}_n, \ldots, x_n - \overline{x}_n\}.$$

Also, letting $\boldsymbol{\sigma}_n = (\sigma_1^2, \ldots, \sigma_n^2)^t$ we have, along parallel lines,

$$\operatorname{Var} \widehat{\beta}_n = t_n^{-1} \boldsymbol{\sigma}_n^t \mathbf{L}_n^t \mathbf{L}_n \boldsymbol{\sigma}_n$$

and

$$V_n^0 = t_n^{-1} \varepsilon_n^t \mathbf{L}_n^t \mathbf{L}_n \varepsilon_n,$$

so that taking (7.2.23) and (7.2.24) into account we get

$$
\begin{aligned}
\frac{V_n - V_n^0}{\operatorname{Var} \widehat{\beta}_n} &= \frac{\varepsilon_n^t (\mathbf{I}_n - \mathbf{A}_n)^t \mathbf{L}_n^t \mathbf{L}_n (\mathbf{I}_n - \mathbf{A}_n) \varepsilon_n - \varepsilon_n^t \mathbf{L}_n^t \mathbf{L}_n \varepsilon_n}{\boldsymbol{\sigma}_n^t \mathbf{L}_n^t \mathbf{L}_n \boldsymbol{\sigma}_n} \\
&= \left\{ \frac{\varepsilon_n^t \mathbf{L}_n^t \mathbf{L}_n \varepsilon_n}{\boldsymbol{\sigma}_n^t \mathbf{L}_n^t \mathbf{L}_n \boldsymbol{\sigma}_n} \right\} \left\{ \frac{\varepsilon_n^t \mathbf{A}_n^t \mathbf{L}_n^t \mathbf{L}_n \mathbf{A}_n \varepsilon_n}{\varepsilon_n^t \mathbf{L}_n^t \mathbf{L}_n \varepsilon_n} - 2 \frac{\varepsilon_n^t \mathbf{A}_n^t \mathbf{L}_n^t \mathbf{L}_n \varepsilon_n}{\varepsilon_n^t \mathbf{L}_n^t \mathbf{L}_n \varepsilon_n} \right\}
\end{aligned}
$$

$$= \left\{ \frac{\varepsilon_n^t \mathbf{L}_n^t \mathbf{L}_n \varepsilon_n}{\sigma_n^t \mathbf{L}_n^t \mathbf{L}_n \sigma_n} \right\} \{a_n - b_n\} \qquad (7.2.25)$$

where

$$a_n = \frac{\varepsilon_n^t \mathbf{A}_n^t \mathbf{L}_n^t \mathbf{L}_n \mathbf{A}_n}{\varepsilon_n^t \mathbf{L}_n^t \mathbf{L}_n \varepsilon_n}$$

and

$$b_n = 2 \frac{\varepsilon_n^t \mathbf{A}_n^t \mathbf{L}_n^t \mathbf{L}_n \varepsilon_n}{\varepsilon_n^t \mathbf{L}_n^t \mathbf{L}_n \varepsilon_n}.$$

Since, by (7.2.22), the first term on the rhs of (7.2.25) converges in probability to 1, we may apply Courant Theorem (1.4.2) to a_n and b_n and conclude that sufficient conditions for $(V_n - V_n^0)/\mathrm{Var}\,\widehat{\beta}_n \xrightarrow{\mathrm{P}} 0$ and, consequently, for $V_n/\mathrm{Var}\,\widehat{\beta}_n \xrightarrow{\mathrm{P}} 1$ are

$$\mathrm{ch}_1\{\mathbf{A}_n^t \mathbf{L}_n^t \mathbf{L}_n \mathbf{A}_n (\mathbf{L}_n^t \mathbf{L}_n)^{-1}\} \to 0 \quad \text{and} \quad \mathrm{ch}_1\{\mathbf{A}_n + \mathbf{A}_n^t\} \to 0, \qquad (7.2.26)$$

as $n \to \infty$. The regularity conditions needed for this stochastic convergence result depend all on the set of (x_i, σ_i^2), $1 \le i \le n$, and can be verified for a given design $\{\mathbf{x}_n\}$ by imposing suitable variational inequality conditions on the σ_i as in the following:

Example 7.2.1 (Two-sample problem): Consider the simple regression model (7.1.2) with $n = 2m$, $x_i = 1$, $1 \le i \le m$, $x_i = -1$, $m + 1 \le i \le 2m$, and assume that $\mathrm{Var}\,\varepsilon_i = \sigma_i^2$, $1 \le i \le n$. Then $\bar{x}_n = 0$ and $t_n = 2m$. Also, using (7.2.3) and (7.2.4) we may write

$$\widehat{\beta}_n - \beta = t_n^{-1} \sum_{i=1}^{n} (x_i - \bar{x}_n)\varepsilon_i = \frac{1}{2m} \left(-\mathbf{1}_m^t, \mathbf{1}_m^t \right) \varepsilon_n$$

and

$$\widehat{\alpha}_n - \alpha = (\widehat{\beta}_n - \beta)\bar{x}_n + \bar{\varepsilon}_n = \frac{1}{2m} \left(\mathbf{1}_m^t, \mathbf{1}_m^t \right) \varepsilon_n$$

which, in connection with (7.2.23), yields

$$
\begin{aligned}
\mathbf{e}_n &= \varepsilon_n - \frac{1}{2m} \left(\mathbf{1}_m^t, \mathbf{1}_m^t \right) \varepsilon_n \begin{pmatrix} \mathbf{1}_m \\ \mathbf{1}_m \end{pmatrix} - \frac{1}{2m} \left(-\mathbf{1}_m^t, \mathbf{1}_m^t \right) \varepsilon_n \begin{pmatrix} -\mathbf{1}_m \\ \mathbf{1}_m \end{pmatrix} \\
&= \varepsilon_n - \frac{1}{2m} \begin{bmatrix} \mathbf{1}_m \mathbf{1}_m^t & \mathbf{1}_m \mathbf{1}_m^t \\ \mathbf{1}_m \mathbf{1}_m^t & \mathbf{1}_m \mathbf{1}_m^t \end{bmatrix} \varepsilon_n - \frac{1}{2m} \begin{bmatrix} \mathbf{1}_m \mathbf{1}_m^t & -\mathbf{1}_m \mathbf{1}_m^t \\ -\mathbf{1}_m \mathbf{1}_m^t & \mathbf{1}_m \mathbf{1}_m^t \end{bmatrix} \varepsilon_n \\
&= \varepsilon_n - \mathbf{A}_n \varepsilon_n
\end{aligned}
$$

where

$$\mathbf{A}_n = \frac{1}{m} \begin{bmatrix} \mathbf{1}_m \mathbf{1}_m^t & \mathbf{0} \\ \mathbf{0} & \mathbf{1}_m \mathbf{1}_m^t \end{bmatrix}.$$

Now, since $\mathbf{L}_n = \mathrm{diag}\{\mathbf{I}_m, -\mathbf{I}_m\}$, it follows that $\mathbf{L}_n^t \mathbf{L}_n = \mathbf{I}_{2m}$ so that $\mathrm{ch}_1\{\mathbf{A}_n^t \mathbf{L}_n^t \mathbf{L}_n \mathbf{A}_n (\mathbf{L}_n^t \mathbf{L}_n)^{-1}\} = m^{-2}$ and $\mathrm{ch}_1\{\mathbf{A}_n + \mathbf{A}_n^t\} = 2m^{-1}$, both converge to zero as $m \to \infty$, i.e., (7.2.26) holds. Now, if we assume further that $E|\varepsilon_i|^{2+\delta} < \infty$, the boundedness of $|x_i|$ implies (7.2.21) and, therefore, (7.2.24) is a strongly consistent estimate of Var $\widehat{\beta}_n$. ∎

The above discussion cast some light on the robustness of the estimator V_n with respect to a possible heteroscedasticity of the error components. A similar, but admittedly more complex analysis can be made for the general linear model (7.1.1) when the ε_i's have possibly different variances. For some of these details, we refer to Eicker (1967).

Let us go back to model (7.1.1) and consider another generalization wherein we assume that Var $\varepsilon_n = \sigma^2 \mathbf{G}_n$ with $\mathbf{G}_n$ being a known p.d. matrix. The **generalized least-squares estimator** (GLSE) of β is obtained by minimizing the quadratic form

$$(\mathbf{Y}_n - \mathbf{X}_n^t \beta)^t \mathbf{G}_n^{-1}(\mathbf{Y}_n - \mathbf{X}_n^t \beta) = \|\mathbf{Y}_n - \mathbf{X}_n^t \beta\|_{\mathbf{G}_n}^2 \qquad (7.2.27)$$

with respect to β and is given by:

$$\widetilde{\beta}_n = (\mathbf{X}_n^t \mathbf{G}_n^{-1} \mathbf{X}_n)^{-1} \mathbf{X}_n^t \mathbf{G}_n^{-1} \mathbf{Y}_n \qquad (7.2.28)$$

whenever $(\mathbf{X}_n^t \mathbf{G}_n^{-1} \mathbf{X}_n)$ is of full rank [otherwise, we may replace the inverse by a generalized inverse in (7.2.28)]. In the cases where $\mathbf{G}_n$ is a diagonal matrix, $\widetilde{\beta}_n$ in (7.2.28) is known as the **weighted least-squares estimators** (WLSE) of β. Before we proceed to consider the asymptotic properties of the GLSE/WLSE of β, we need some preliminary considerations.

If, in (7.1.1), ε_n has a n-variate multinormal distribution with null mean vector and dispersion matrix $\sigma^2 \mathbf{G}_n$, then $\widetilde{\beta}_n$ in (7.2.28), being linear in ε_n, has also a multinormal law with mean vector β and dispersion matrix $\sigma^2(\mathbf{X}_n^t \mathbf{G}_n^{-1} \mathbf{X}_n)^{-1}$. But this simple conclusion may not generally follow when ε_n does not have a multinormal law. We may notice that, by assumption, $\varepsilon_n^* = \mathbf{G}_n^{-1/2} \varepsilon_n$ has mean vector $\mathbf{0}$ and dispersion matrix $\sigma^2 \mathbf{I}_n$. Thus, the elements of ε_n^* are mutually uncorrelated. However, in general, uncorrelation may not imply independence, and homogeneity of variances may not imply identity of distributions. Thus, to adopt the GLSE/WLSE methods in a given context, we may need to spell out the model more precisely, so that the components of ε_n^* are independent and identically distributed. We do this by rewriting (7.1.1) as

$$\mathbf{Y}_n = \mathbf{X}_n \beta + \mathbf{G}_n^{1/2} \varepsilon_n^* \qquad (7.2.29)$$

where $\varepsilon_n^* = (\varepsilon_1^*, \ldots, \varepsilon_n^*)^t$ has independent and identically distributed components ε_i^*, such that $E\varepsilon_i^* = 0$ and $E(\varepsilon_i^*)^2 = \sigma^2$. For the WLSE, $G_n^{1/2}$ is a diagonal matrix, so that putting $\mathbf{G}_n^{1/2} = \mathbf{D}_n = \mathrm{diag}\,(\sigma_1, \ldots, \sigma_n)$ where the σ_i's are non-negative scalar constants, we may also write

$$\mathbf{G}_n^{1/2} \varepsilon_n^* = \mathbf{D}_n \varepsilon_n^* \qquad (7.2.30)$$

and conclude that $E(\varepsilon_i^*)^2 = 1$, $i \geq 1$. This is termed the heteroscedastic error model when the σ_i, $i \geq 1$, are not specified. In the current context, we assume that $\sigma_i^2 = \sigma^2 h_i$, $i = 1, \ldots, n$, where the h_i, $i \geq 1$, are given and σ^2 is unknown. Then let us write

$$\mathbf{Y}_n^* = \mathbf{G}_n^{-1/2}\mathbf{Y}_n \quad \text{and} \quad \mathbf{X}_n^* = \mathbf{G}_n^{-1/2}\mathbf{X}_n \qquad (7.2.31)$$

which, in conjunction with (7.2.29), leads to

$$\mathbf{Y}_n^* = \mathbf{X}_n^*\boldsymbol{\beta} + \boldsymbol{\varepsilon}_n^*. \qquad (7.2.32)$$

Now, from (7.2.28) and (7.2.31) we have

$$\tilde{\boldsymbol{\beta}}_n = (\mathbf{X}_n^{*t}\mathbf{X}_n^*)^{-1}\mathbf{X}_n^{*t}\mathbf{Y}_n^* \qquad (7.2.33)$$

which resembles (7.1.7) with $\mathbf{X}_n, \mathbf{Y}_n$ (and $\boldsymbol{\varepsilon}_n$) being replaced by $\mathbf{X}_n^*, \mathbf{Y}_n^*$ (and $\boldsymbol{\varepsilon}_n^*$), respectively. Thus, along the same lines as in (7.2.13) and (7.2.14), we assume that the following two conditions hold:

$$\max_{1 \leq i \leq n} \{\mathbf{x}_{ni}^{*t}(\mathbf{X}_n^{*t}\mathbf{X}_n^*)^{-1}\mathbf{x}_{ni}^*\} \to 0 \quad \text{as } n \to \infty \qquad (7.2.34)$$

and

$$\lim_{n \to \infty} n^{-1}(\mathbf{X}_n^{*t}\mathbf{X}_n^*) = \mathbf{V}^*, \quad \text{finite and p.d.} \qquad (7.2.35)$$

Then, Theorem 7.2.2 extends directly to the following:

Theorem 7.2.3: *For the model (7.2.29) with the ε_i^* being independent and identically distributed random variables with 0 mean and finite positive variance σ^2, it follows that under (7.2.34) and (7.2.35)*

$$\sqrt{n}(\tilde{\boldsymbol{\beta}}_n - \boldsymbol{\beta}) \xrightarrow{\mathcal{D}} N_q(\mathbf{0}, \sigma^2\mathbf{V}^{*-1}). \qquad (7.2.36)$$

The transformation model in (7.2.29) is a very convenient one for asymptotic theory, and in most of the practical problems (particularly for $\mathbf{G}_n$ diagonal), the assumed regularity conditions on the ε_n^* are easy to justify (without imposing normality on them). Also (7.2.34) and (7.2.35) are easy to verify for the specification matrix $\mathbf{X}_n$ and the matrix $\mathbf{G}_n$ usually considered in practice. In this context, a very common problem relates to the so-called **replicated model**, where each y_i is an average of a number of observations (say, m_i) which relate to a common $\mathbf{x}_i$, so that $\mathbf{G}_n = \mathrm{diag}\{m_1^{-1}, \ldots, m_n^{-1}\}$, with the m_i, $i \geq 1$, being known. This allows us to handle unbalanced designs in a general setup.

As may be the case with many practical problems, the observations y_i, $i \geq 1$, may be due to different investigators and/or made on different measuring instruments (whose variability may not be homogeneous) so that we may have a model similar to (7.2.29), but with $\mathbf{G}_n$ unknown. Thus, it may be more realistic to consider the following general model:

$$\mathbf{Y}_n = \mathbf{X}_n\boldsymbol{\beta} + \boldsymbol{\Sigma}_n^{1/2}\boldsymbol{\varepsilon}_n^* \qquad (7.2.37)$$

where the components of $\boldsymbol{\varepsilon}_n^*$ are independent and identically distributed with mean zero and unit variance, whereas $\boldsymbol{\Sigma}_n$ is a $(n \times n)$ unknown positive definite matrix. The fact that the dimension of $\boldsymbol{\Sigma}_n$ increases with n may cause some technical problems, which we illustrate with the following:

Example 7.2.2 (Neyman-Scott problem): Let $\mathbf{X}_n = \mathbf{1}_n$, and consider the model

$$\mathbf{Y}_n = \theta\mathbf{1}_n + \boldsymbol{\varepsilon}_n,$$

where $\theta \in \Theta \subset \mathbb{R}$ and $\varepsilon_1, \ldots, \varepsilon_n$ are independent random errors with zero means and unknown variances $\sigma_1^2, \ldots, \sigma_n^2$, respectively. Thus, here $\boldsymbol{\Sigma}_n = \text{diag}\{\sigma_1^2, \ldots, \sigma_n^2\}$. As in Neyman and Scott (1948), we assume that the ε_i, $i \geq 1$, are normally distributed, but, as we shall see, even this strong assumption does not help much in the study of the asymptotic properties of the estimator of the location parameter θ. If the σ_i^2, $i \geq 1$, were known, the WLSE of θ would have been

$$\widehat{\theta}_n^* = \sum_{i=1}^{n}\sigma_i^{-2}Y_i \Big/ \sum_{i=1}^{n}\sigma_i^{-2} \qquad (7.2.38)$$

$\widehat{\theta}_n^*$ is also the MLE of θ when the ε_i, $i \geq 1$, are normally distributed with known variances. When the σ_i, $i \geq 1$ are unknown, to obtain the MLE of θ (as well as of σ_i^2, $i \geq 1$), one needs to solve for $(n+1)$ simultaneous equations, which for the normal model reduce to

$$\sum_{i=1}^{n}(Y_i - \theta)/\sigma_i^2 = 0, \quad \sigma_i^{-1} = (Y_i - \theta)^2/\sigma_i^3, \quad 1 \leq i \leq n, \qquad (7.2.39)$$

and the solution $\widetilde{\theta}_n^*$, obtainable from

$$\sum_{i=1}^{n}(Y_i - \widetilde{\theta}_n^*)/(Y_i - \widetilde{\theta}_n^*)^2 = 0$$

or, equivalently, from

$$\sum_{i=1}^{n}|Y_i - \widetilde{\theta}_n^*|^{-1}\text{sign}\,(Y_i - \widetilde{\theta}_n^*) = 0, \qquad (7.2.40)$$

may not even be consistent; this is particularly due to the fact that the Y_i, $i \geq 1$, arbitrarily close (but not identical to) $\widetilde{\theta}_n^*$ may have unbounded influence on the estimating equation (7.2.40). Thus, it may be better to consider some alternative estimators $\widehat{\sigma}_1^2, \ldots, \widehat{\sigma}_n^2$ (of $\sigma_1^2, \ldots, \sigma_n^2$, respectively) and formulate an estimator of θ as

$$\check{\theta}_n^* = \sum_{i=1}^n \widehat{\sigma}_i^{-2} Y_i \bigg/ \sum_{i=1}^n \widehat{\sigma}_i^{-2}. \tag{7.2.41}$$

The question, however, remains open: is $\check{\theta}_n^*$ consistent for θ? If so, is it asymptotically normally distributed? Recall that by (7.2.38)

$$\begin{aligned}
E(\widehat{\theta}_n^* - \theta)^2 &= \sum_{i=1}^n \sigma_i^{-4} \sigma_i^2 \bigg/ \left(\sum_{i=1}^n \sigma_i^{-2}\right)^2 = \left(\sum_{i=1}^n \sigma_i^{-2}\right)^{-1} \\
&= n^{-1} \left(n^{-1} \sum_{i=1}^n \sigma_i^{-2}\right)^{-1} \leq n^{-1} \left(n^{-1} \sum_{i=1}^n \sigma_i^2\right)
\end{aligned} \tag{7.2.42}$$

where the last step is due to the AM/GM/HM Inequality (1.4.31) for non-negative numbers. Thus, whenever

$$\overline{\sigma}_n^2 = n^{-1} \sum_{i=1}^n \sigma_i^2 = o(n), \tag{7.2.43}$$

(7.2.42) converges to zero as $n \to \infty$ and, by Chebyshev Inequality (2.2.31), $\widehat{\theta}_n^* \xrightarrow{P} \theta$. In fact, (7.2.43) is only a sufficient condition; an even weaker condition is that the harmonic mean $\overline{\sigma}_{nH}^2 = (n^{-1} \sum_{i=1}^n \overline{\sigma}_i^2)^{-1}$ is $o(n)$. To see this, consider the particular case where $\sigma_i^2 \propto i$, $i \geq 1$; then $\overline{\sigma}_n^2 \propto (n+1)/2 = O(n)$, whereas $\overline{\sigma}_{nH}^2 \propto (n^{-1} \log n)^{-1} = n/\log n = o(n)$. On the other hand, in (7.2.41), the $\widehat{\sigma}_i^{-2}$, $i \geq 1$, are themselves random variables and, moreover, they may not be stochastically independent of the Y_i's. Thus, the computation of the mean square error of $\check{\theta}_n^*$ may be much more involved; it may not even be $o(1)$ under (7.2.43) or some similar condition. To illustrate this point, we assume further that the model corresponds to a replicated one, where $n = 2m$ and $\sigma_{2i}^2 = \sigma_{2i-1}^2 = \gamma_i$, $i = 1, \ldots, m$. Then $(Y_{2i} - Y_{2i-1})^2/2 = \widehat{\gamma}_i$, $i = 1, \ldots, m$, are unbiased estimators of the γ_i, $i = 1, \ldots, m$, and, moreover, $(Y_{2i} + Y_{2i-1})/2$ and $(Y_{2i} - Y_{2i-1})/2$ are mutually independent so that $\widehat{\gamma}_i$ is independent of $(Y_{2i} + Y_{2i-1})/2$, $i = 1, \ldots, m$. Thus, we may rewrite (7.2.41) as

$$\check{\theta}_n^* = \left\{\sum_{i=1}^n \widehat{\gamma}_i^{-1} (Y_{2i} + Y_{2i-1})/2\right\} \left\{\sum_{i=1}^m \widehat{\gamma}_i^{-1}\right\}^{-1}. \tag{7.2.44}$$

Using the aforesaid independence [between the $\widehat{\gamma}_i$ and $(Y_{2i} + Y_{2i-1})$], the conditional mean square error of $\breve{\theta}_n^*$ given $\widehat{\gamma}_1, \ldots, \widehat{\gamma}_m$ is equal to

$$\frac{1}{2} \left\{ \sum_{i=1}^m \widehat{\gamma}_i^{-2} \gamma_i \right\} \left\{ \sum_{i=1}^m \widehat{\gamma}_i^{-1} \right\}^{-2} \tag{7.2.45}$$

so that the unconditional mean square error is given by

$$\frac{1}{2} \mathrm{E} \left\{ \left(\sum_{i=1}^m \gamma_i / \widehat{\gamma}_i^2 \right) \left(\sum_{i=1}^m \widehat{\gamma}_i^{-1} \right)^{-2} \right\}. \tag{7.2.46}$$

Using again the AM/GM/HM Inequality (1.4.31), we may bound (7.2.45) from above by

$$\left\{ 1/2m^{-2} \sum_{i=1}^m \widehat{\gamma}_i^{-2} \gamma_i \right\} \left\{ m^{-1} \sum_{i=1}^m \widehat{\gamma}_i \right\}^2,$$

which is, in turn, bounded from above by

$$\left\{ \frac{1}{2m} \frac{1}{m} \sum_{i=1}^m \widehat{\gamma}_i^{-2} \gamma_i \right\} \left\{ \frac{1}{m} \sum_{i=1}^m \widehat{\gamma}_i^2 \right\}. \tag{7.2.47}$$

Now we could think of appealing to the Cauchy-Schwarz Inequality (1.4.26) to obtain an upper bound for (7.2.46), but even so, we may not have the required luck, as $\mathrm{E}\widehat{\gamma}_i^{-2} = +\infty$ for every $i \geq 1$, whereas $\mathrm{E}\widehat{\gamma}_i^2 = 2\gamma_i^2$. Thus inherent independence of the $\widehat{\gamma}_i$ and $(Y_{2i} + Y_{2i-1})$ as well as the normality of errors may not contribute much toward the use of Chebyshev Inequality (2.2.31) as a simple way to prove the consistency of $\breve{\theta}_n^*$. Also note that (7.2.45) is bounded from below by

$$\frac{1}{2} \left\{ \left(\sum_{i=1}^m \widehat{\gamma}_i^{-1} \gamma_i^{1/2} \right) \left(\sum_{i=1}^m \widehat{\gamma}_i^{-1} \right)^{-1} \right\}^2 \tag{7.2.48}$$

so that the fact that $\mathrm{E}\widehat{\gamma}_i^{-1} = +\infty$, $i \geq 1$, continues to pose difficulties in the computation of the mean squared error. This suggests that more structure on the σ_i^2, either in terms of a fixed pattern or possibly some Bayesian model, may be necessary to eliminate this drawback and to render some tractable solutions to the corresponding asymptotic study.   ∎

Let us return to the general linear model (7.2.37). If $\boldsymbol{\Sigma}_n$ were known, the GLSE of $\boldsymbol{\beta}$ would be

$$\widehat{\boldsymbol{\beta}}_n = (\mathbf{X}_n^t \boldsymbol{\Sigma}_n^{-1} \mathbf{X}_n)^{-1} \mathbf{X}_n^t \boldsymbol{\Sigma}_n^{-1} \mathbf{Y}_n. \tag{7.2.49}$$

Assuming that (7.2.34) and (7.2.35) with $\mathbf{G}_n$ replaced by $\boldsymbol{\Sigma}_n$ both hold, one may claim via Theorem 7.2.3 that

$$\sqrt{n}(\widehat{\boldsymbol{\beta}}_n - \boldsymbol{\beta}) \xrightarrow{\mathcal{D}} N_q(\mathbf{0}, \mathbf{V}^{-1}) \tag{7.2.50}$$

where $\mathbf{V} = \lim_{n \to \infty} n^{-1} \mathbf{V}_n$ and $\mathbf{V}_n = \mathbf{X}_n^t \boldsymbol{\Sigma}_n^{-1} \mathbf{X}_n$. Note that the consistency of $\widehat{\boldsymbol{\beta}}_n$ follows directly from (7.2.50). Since, in general, $\boldsymbol{\Sigma}_n$ is unknown, we may attempt to replace it by an estimate $\widehat{\boldsymbol{\Sigma}}_n$ in (7.2.49) obtaining the **two-step (Aitken) estimator**:

$$\widehat{\widehat{\boldsymbol{\beta}}}_n = (\mathbf{X}_n^t \widehat{\boldsymbol{\Sigma}}_n^{-1} \mathbf{X}_n)^{-1} \mathbf{X}_n^t \widehat{\boldsymbol{\Sigma}}_n^{-1} \mathbf{Y}_n. \tag{7.2.51}$$

Letting $\widehat{\mathbf{V}}_n = \mathbf{X}_n^t \widehat{\boldsymbol{\Sigma}}_n^{-1} \mathbf{X}_n$ it follows from (7.2.37) and (7.2.51) that

$$\widehat{\widehat{\boldsymbol{\beta}}}_n = \boldsymbol{\beta} + \widehat{\mathbf{V}}_n^{-1} \mathbf{X}_n^t \widehat{\boldsymbol{\Sigma}}_n^{-1} \boldsymbol{\Sigma}_n^{1/2} \boldsymbol{\varepsilon}_n^*, \tag{7.2.52}$$

whereas by (7.2.37) and (7.2.49) we have

$$\widehat{\boldsymbol{\beta}}_n = \boldsymbol{\beta} + \mathbf{V}_n^{-1} \mathbf{X}_n^t \boldsymbol{\Sigma}_n^{-1/2} \boldsymbol{\varepsilon}_n^*. \tag{7.2.53}$$

Therefore, from (7.2.52) and (7.2.53) we may write:

$$\begin{aligned}
\widehat{\widehat{\boldsymbol{\beta}}}_n - \widehat{\boldsymbol{\beta}}_n &= (\widehat{\mathbf{V}}_n^{-1} - \mathbf{V}_n^{-1}) \mathbf{X}_n^t \widehat{\boldsymbol{\Sigma}}_n^{-1} \boldsymbol{\Sigma}_n^{1/2} \boldsymbol{\varepsilon}_n^* \\
&\quad + \mathbf{V}_n^{-1} \mathbf{X}_n^t \left\{ \widehat{\boldsymbol{\Sigma}}_n^{-1} - \boldsymbol{\Sigma}_n^{-1} \right\} \boldsymbol{\Sigma}_n^{1/2} \boldsymbol{\varepsilon}_n^* \\
&= (\widehat{\mathbf{V}}_n^{-1} - \mathbf{V}_n^{-1}) \mathbf{X}_n^t \boldsymbol{\Sigma}_n^{-1/2} \boldsymbol{\varepsilon}_n^* \\
&\quad + (\widehat{\mathbf{V}}_n^{-1} - \mathbf{V}_n^{-1}) \mathbf{X}_n^t \left\{ \widehat{\boldsymbol{\Sigma}}_n^{-1} - \boldsymbol{\Sigma}_n^{-1} \right\} \boldsymbol{\Sigma}_n^{1/2} \boldsymbol{\varepsilon}_n^* \\
&\quad + \mathbf{V}_n^{-1} \mathbf{X}_n^t \left\{ \widehat{\boldsymbol{\Sigma}}_n^{-1} - \boldsymbol{\Sigma}_n^{-1} \right\} \boldsymbol{\Sigma}_n^{1/2} \boldsymbol{\varepsilon}_n^*.
\end{aligned} \tag{7.2.54}$$

The first term on the rhs of (7.2.54) may be expressed as

$$(\widehat{\mathbf{V}}_n^{-1} - \mathbf{V}_n^{-1}) \mathbf{V}_n (\widehat{\boldsymbol{\beta}}_n - \boldsymbol{\beta}) = (\widehat{\mathbf{V}}_n^{-1} \mathbf{V}_n - \mathbf{I}_n)(\widehat{\boldsymbol{\beta}}_n - \boldsymbol{\beta}). \tag{7.2.55}$$

Invoking (7.2.50) one may claim that $\widehat{\boldsymbol{\beta}}_n - \boldsymbol{\beta} \xrightarrow{\text{P}} \mathbf{0}$. Thus, (7.2.55) is $o_p(1)$ whenever $\widehat{\mathbf{V}}_n^{-1} \mathbf{V}_n - \mathbf{I}_n = O_p(1)$, i.e., whenever the characteristic roots of $\widehat{\mathbf{V}}_n^{-1} \mathbf{V}_n$ are bounded in probability. Similarly, the second term on the rhs of (7.2.54) may be re-expressed as

$$\begin{aligned}
(\widehat{\mathbf{V}}_n^{-1} \mathbf{V}_n - I_n) \mathbf{V}_n^{-1} \mathbf{X}_n^t & \left(\widehat{\boldsymbol{\Sigma}}_n^{-1} - \boldsymbol{\Sigma}_n^{-1} \right) \boldsymbol{\Sigma}_n^{1/2} \boldsymbol{\varepsilon}_n^* \\
&= (\widehat{\mathbf{V}}_n^{-1} \mathbf{V}_n - \mathbf{I}_n) \mathbf{V}_n^{-1} \mathbf{X}_n^t \boldsymbol{\Sigma}_n^{-1} \left(\boldsymbol{\Sigma}_n \widehat{\boldsymbol{\Sigma}}_n^{-1} - \mathbf{I}_n \right) \boldsymbol{\Sigma}_n^{1/2} \boldsymbol{\varepsilon}_n^*.
\end{aligned} \tag{7.2.56}$$

Thus, if all the characteristic roots of $\boldsymbol{\Sigma}_n \widehat{\boldsymbol{\Sigma}}_n^{-1} - \mathbf{I}_n$ are $o_p(1)$, whereas the characteristic roots of $\widehat{\mathbf{V}}_n^{-1} \mathbf{V}_n$ are $O_p(1)$, (7.2.56) converges stochastically to $\mathbf{0}$. A very similar treatment holds for the last term on the rhs of (7.2.54). Thus, we may conclude that $\widehat{\widehat{\boldsymbol{\beta}}}_n - \widehat{\boldsymbol{\beta}} \xrightarrow{P} \mathbf{0}$ whenever (7.2.34) and (7.2.35) hold in conjugation with

$$\mathrm{ch}_1(\mathbf{V}_n \widehat{\mathbf{V}}_n^{-1}) = O_p(1) \tag{7.2.57}$$

and

$$\mathrm{ch}_1\left(\boldsymbol{\Sigma}_n \widehat{\boldsymbol{\Sigma}}_n^{-1}\right) - 1 = o_p(1) = \mathrm{ch}_n\left(\boldsymbol{\Sigma}_n \widehat{\boldsymbol{\Sigma}}_n^{-1}\right) - 1. \tag{7.2.58}$$

On the other hand, if we replace (7.2.57) by

$$\mathrm{ch}_1(\mathbf{V}_n \widehat{\mathbf{V}}_n^{-1}) - 1 = o_p(1) = \mathrm{ch}_n(\mathbf{V}_n \widehat{\mathbf{V}}_n^{-1}) - 1, \tag{7.2.59}$$

we may conclude that

$$\sqrt{n}(\widehat{\widehat{\boldsymbol{\beta}}}_n - \widehat{\boldsymbol{\beta}}_n) \xrightarrow{P} \mathbf{0} \tag{7.2.60}$$

so that via (7.2.50) and (7.2.60) we obtain

$$\sqrt{n}(\widehat{\widehat{\boldsymbol{\beta}}}_n - \boldsymbol{\beta}) \xrightarrow{D} N_q(\mathbf{0}, \mathbf{V}^{-1}). \tag{7.2.61}$$

These regularity conditions framed for the consistency and asymptotic normality of $\widehat{\widehat{\boldsymbol{\beta}}}_n$ are sufficient ones and are easily verifiable in terms of the specification matrix $\mathbf{X}_n$, the covariance matrix $\boldsymbol{\Sigma}_n$ and its consistent estimator $\widehat{\boldsymbol{\Sigma}}_n$. Here we interpret the consistency of $\widehat{\boldsymbol{\Sigma}}_n$ (with respect to $\boldsymbol{\Sigma}_n$) in the sense of (7.2.58) or, more explicitly, as $\boldsymbol{\Sigma}_n \widehat{\boldsymbol{\Sigma}}_n^{-1} \xrightarrow{P} \mathbf{I}_n$.

In passing, we remark that both $\boldsymbol{\Sigma}_n$ and $\widehat{\boldsymbol{\Sigma}}_n$ are $(n \times n)$ matrices and, hence, in general, they may not satisfy this consistency property. However, in many problems of practical interest, both $\boldsymbol{\Sigma}_n$ and $\widehat{\boldsymbol{\Sigma}}_n$ are generated by a finite number of finite dimensional matrices for which it may be possible to verify the above conditions. This point is illustrated in the exercises.

We finalize this section by noting that a direct application of the Cramér-Wold Theorem (3.2.4) may be used to extend the above results to the standard **multivariate linear model**:

$$\mathcal{Y}_n = \mathbf{X}_n \mathbf{B} + \mathbf{E}_n \tag{7.2.62}$$

where $\mathcal{Y}_n = (\mathbf{Y}_1, \ldots, \mathbf{Y}_n)^t$ is a $(n \times p)$ matrix of observable response variables with rows $\mathbf{Y}_i^t = (Y_{i1}, \ldots, Y_{ip})$, $\mathbf{X}_n$ is defined as in (7.1.1), $\mathbf{B} = (\boldsymbol{\beta}_1, \ldots, \boldsymbol{\beta}_p)$ is a $(q \times p)$ matrix of unknown parameters [with the jth column, $j = 1, \ldots, p$, having the same interpretation as $\boldsymbol{\beta}$ in (7.1.1) with respect to the jth response variate] and $\mathbf{E}_n = (\boldsymbol{\varepsilon}_1, \ldots, \boldsymbol{\varepsilon}_n)^t$ is an $(n \times p)$ matrix of unobservable random errors, the rows of which are assumed independent and following a (generally) unknown distribution with d.f. F such that $\mathrm{E}\boldsymbol{\varepsilon}_i = \mathbf{0}$ and $\mathrm{Var}\,\boldsymbol{\varepsilon}_i = \boldsymbol{\Sigma}$, finite and p.d.

7.3 Robust estimators

The method of least squares and its ramifications are particularly appealing in the linear model setup because, generally, the corresponding estimators are computationally simple (i.e., the allied "normal equations" have explicit solutions) and are asymptotically normal under quite general regularity conditions (wherein the underlying error distribution function F belongs to a general class of which the normal law is an important member). From the general results presented in Section 7.2 we know that the LSE, $\widehat{\beta}_n$ of β in model (7.1.1) is such that $\operatorname{Var} \widehat{\beta}_n = \sigma^2 (\mathbf{X}_n^t \mathbf{X}_n)^{-1}$ where $\sigma^2 = \operatorname{Var}(\varepsilon)$, the variance of the error component ε, whereas $(\mathbf{X}_n^t \mathbf{X}_n)^{-1}$ depends solely on the specification matrix $\mathbf{X}_n$ and is independent of the d.f. F. This may well be a vintage point for looking at the picture from a wider perspective, when in (7.1.1) we do not necessarily assume that F is normal. In fact, all we have assumed is that $\mathrm{E}_F(\varepsilon) = \int x \, dF(x) = 0$ and that $\sigma_F^2 = \mathrm{E}(\varepsilon^2) = \int x^2 \, dF(x)$ is finite and positive. Also note that if we assume that F has a differentiable density function f vanishing at the two ends,

$$
\begin{aligned}
1 = \int f(x) \, dx &= \{x f(x)\} - \int x f'(x) \, dx \\
&= \int x \{-f'(x)/f(x)\} \, dF(x) \qquad (7.3.1)
\end{aligned}
$$

so that using the Cauchy-Schwarz Inequality (1.4.26), we have

$$
1 \le \left[\int x^2 \, dF(x) \right] \left[\int \{-f'(x)/f(x)\}^2 \, dF(x) \right] \quad \text{or} \quad \sigma_F^2 \ge 1/I(f) \quad (7.3.2)
$$

where $I(f) = \mathrm{E}_F[\{-f'(\varepsilon)/f(\varepsilon)\}^2]$ is the Fisher information. The equality in (7.3.2) holds when, for some $k \ne 0$,

$$
- f'(\varepsilon)/f(\varepsilon) = k\varepsilon, \quad \forall \varepsilon \quad \text{a.e.,} \qquad (7.3.3)
$$

i.e., when f is a normal density function. Thus, in the conventional linear model, where we assume that the errors are normally distributed, $\sigma_F^2 = 1/I(f)$ is a minimum, and, hence, the LSE is optimal. This rosy picture may disappear very fast if F departs from a normal form, even locally. To illustrate this point, we consider the following:

Example 7.3.1: Suppose that f corresponds to a double exponential distribution, i.e., $f(x) = \frac{1}{2} e^{-|x|}$, $-\infty < x < \infty$. In this case, $-f'(x)/f(x) = \operatorname{sign} x$, $x \in \mathbb{R}$, and $I(f) = 1$. On the other hand,

$$
\sigma_F^2 = \frac{1}{2} \int_{-\infty}^{+\infty} x^2 e^{-|x|} \, dx = \int_0^\infty x^2 e^{-x} \, dx = 2! = 2.
$$

Thus, σ_F^2 is twice as large as $1/I(f)$ and, therefore, the LSE $\widehat{\beta}_n$ of β in model (7.1.1) entails about 50% loss of efficiency in this case.   ∎

Example 7.3.2 (Error contamination model): Suppose that there exist an α $(0 < \alpha < 1)$ and a positive k (< 1) such that

$$f(\varepsilon) = (1 - \alpha)\varphi(\varepsilon) + \alpha k^{-1}\varphi(\varepsilon/k), \quad \varepsilon \in \mathbb{R},$$

where φ denotes the standard normal density function. According to this model, $100(1 - \alpha)\%$ of the observations come from a normal distribution (with zero mean and unit variance), whereas the remaining $100\alpha\%$ corresponds to a normal distribution with zero mean but a larger variance k^2; this accounts for the error contamination or presence of gross errors in a plausible manner. Here, we have [on using the fact that $\varphi'(x) = -x\varphi(x)$],

$$\frac{-f'(\varepsilon)}{f(\varepsilon)} = \frac{(1-\alpha)\varepsilon\varphi(\varepsilon) + \alpha k^{-2}\varepsilon\varphi(\varepsilon/k)}{(1-\alpha)\varphi(\varepsilon) + \alpha k^{-1}\varphi(\varepsilon/k)} = \varepsilon\left[\frac{1 - \alpha\{\varphi(\varepsilon) - k^{-2}\varphi(\varepsilon/k)\}}{1 - \alpha\{\varphi(\varepsilon) - k^{-1}\varphi(\varepsilon/k)\}}\right]$$

where the term within [ ] depends very much on ε (when $k \neq 1$), so that (7.3.3) does not hold, and, hence, $\sigma_F^2 > 1/I(f)$ for all $\alpha \in (0,1)$ when $k \neq 1$. The expression for $I(f)$ is not that simple to obtain.   ∎

Example 7.3.3: Suppose that F corresponds to the logistic distribution, i.e., $F(x) = (1 + e^{-x})^{-1}$, $-\infty < x < \infty$ so that $f(x) = e^{-x}(1 + e^{-x})^{-2} = F(x)[1 - F(x)]$, $x \in \mathbb{R}$. Thus, $-f'(x) = f(x)[2F(x) - 1]$, $x \in \mathbb{R}$, so that $I(f) = \int_0^1 (2u - 1)^2\, du = 1/3$. Since the density function f is symmetric about zero, it follows that

$$\sigma_F^2 = \int_{-\infty}^{+\infty} x^2 e^{-x}(1 + e^{-x})^2\, dx = \pi > 3 = 1/I(f).$$

∎

If the underlying error distribution in model (7.1.1) corresponds to one of the three in the examples discussed above, the LSE $\widehat{\beta}_n$ of β fails to achieve the lower bound for the variance, even asymptotically. This holds for almost all non-normal underlying error d.f. F even when they are quite close to a normal d.f., say Φ_σ in the sense that $\|F - \Phi_\sigma\|$ is small. Because of this discouraging factor, we say that the LSE is not **robust**.

We may appeal to the results in Chapter 5 and construct MLE/BAN estimators of β when the form of F is assumed to be given. For example, under the double-exponential distribution discussed in Example 7.3.1, the MLE of β is given by

$$\widehat{\beta}_n^* = \arg\min\left\{\sum_{i=1}^{n} |Y_i - \mathbf{x}_{ni}^t\beta| : \beta \in \mathbb{R}^q\right\}. \tag{7.3.4}$$

Although an explicit solution for (7.3.4) is not available (unlike the case of the LSE), we may note that $\sum_{i=1}^{n} |Y_i - \mathbf{x}_{ni}^t \boldsymbol{\beta}|$ is a convex function of $\boldsymbol{\beta}$ (for given $\mathbf{Y}_n$ and $\mathbf{X}_n$), so that, in principle, we may agree to abandon the role of the squared distance [i.e., $\sum_{i=1}^{n} |Y_i - \mathbf{x}_{ni}^t \boldsymbol{\beta}|^2$] in favor of the absolute distance [i.e., $\sum_{i=1}^{n} |Y_i - \mathbf{x}_{ni}^t \boldsymbol{\beta}|$]. The situation is somewhat more complicated for the other two examples, although it may be possible to use the distance function $\sum_{i=1}^{n} \{ -\log f(Y_i - \mathbf{x}_{ni}^t \boldsymbol{\beta}) \}$. This feature motivates us to consider an arbitrary non-negative function $\rho : \mathbb{R} \to \mathbb{R}^+$ and define

$$Q_n(\boldsymbol{\beta}) = \sum_{i=1}^{n} \rho(Y_i - \mathbf{x}_{ni}^t \boldsymbol{\beta}) \qquad (7.3.5)$$

so that minimization of $Q_n(\boldsymbol{\beta})$ with respect to $\boldsymbol{\beta}$ would lead to the desired estimator. The choice of ρ can, of course, be based on the assumed density function f; this ρ, however, may not be robust in the sense that it may generate inefficient estimators of $\boldsymbol{\beta}$ whenever the true underlying distribution does not exactly match the assumed one. The picture is very much similar to that of the LSE discussed above. In addition to this situation, there is the basic fact that there exists no single ρ which would lead to (at least asymptotically) optimal estimators for all F belonging to a certain class. An exception to this feature is the case where ρ is data dependent, i.e., chosen **adaptively** so as to yield asymptotically optimal estimators within a broad class. But this **adaptative estimation theory** requires some mathematical analysis which stands at a far higher level than that we wish to contemplate, and, hence, we shall not pursue this topic further.

From a practical point of view, the **robustness** picture can be judged from either a **local** or **global** perspective, depending on the level of confidence one would have on the assumed form of the density function f. If we refer to Example 7.3.2, choosing α close to zero implies that we have high confidence in f in the sense of it being close to a normal density, although we would like to have some protection against the presence of error contamination/gross errors/outliers. This would fit under the denomination of **local robustness**. On the other hand, we may suppose that the density f belongs to a class (viz., the class of all symmetric, unimodal, absolutely continuous pdf's with finite second moments), so that we desire to have an estimator of $\boldsymbol{\beta}$ which performs well for all underlying error distributions in this class. In this setup, we may have a larger class of estimators which are robust in a global sense, and within this class, we may want to choose specific ones which achieve high (asymptotic) efficiency for specific (important) members of the class of distributions under consideration.

Within the class of **M-estimators** (as defined in Section 5.4) one may easily construct robust estimators in a local sense, whereas rank based (or **R**)-estimators retain robustness to a higher degree of global perspective. We illustrate this for the parameter β in the simple regression model (7.1.2).

The LSE of β is given by

$$\widehat{\beta}_{LS} = \sum_{i=1}^{n}(x_i - \overline{x}_n)Y_i / \sum_{i=1}^{n}(x_i - \overline{x}_n)^2$$

and the corresponding ρ function is given by $\rho_{LS}(y) = y^2$, $y \in \mathbb{R}$. As has been mentioned before, $\widehat{\beta}_{LS}$ is not so robust against departures from the assumed normality of the error d.f. F. To clarify this picture, we first rewrite $\widehat{\beta}_{LS}$ as

$$\widehat{\beta}_{LS} = \sum_{1 \leq i < j \leq n}(x_i - x_j)(Y_i - Y_j) / \sum_{1 \leq i < j \leq n}(x_i - x_j)^2.$$

Note that by assumption, $x_1, \ldots, x_n$ are not all equal, but we are not ignoring the possibility of ties among them; thus, without loss of generality, we may set $x_1 \leq x_2 \leq \cdots \leq x_n$ with at least one strict inequality sign. Also, consider the set

$$S_n = \{(i,j) : 1 \leq i < j \leq n \quad \text{and} \quad x_i \neq x_j\} \qquad (7.3.6)$$

and let N be the cardinality of S_n. Then we may write

$$\begin{aligned}
\widehat{\beta}_{LS} &= \frac{\sum_{S_n}(x_j - x_i)(Y_j - Y_i)}{\sum_{S_n}(x_j - x_i)^2} \\
&= \frac{\sum_{S_n}(x_j - x_i)^2\{Y_j - Y_i)/(x_j - x_i)\}}{\sum_{S_n}(x_j - x_i)^2} \\
&= \sum_{S_n} w_{nij}Z_{ij}
\end{aligned}$$

where the summation $\sum_{S_n}$ extends over all possible pairs (i,j) belonging to the set S_n, $w_{nij} = (x_j - x_i)^2 / \sum_{S_n}(x_j - x_i)^2$ are non-negative weights adding up to 1 and the $Z_{ij} = (Y_j - Y_i)/(x_j - x_i)$, $(i,j) \in S_n$, are the so-called divided differences. Recall that by (7.1.2), for every $(i,j) \in S_n$,

$$Z_{ij} = \beta + (\varepsilon_j - \varepsilon_i)/(x_j - x_i) = \beta + \varepsilon_{ij}^*$$

where the $\varepsilon_{ij}^* = (\varepsilon_j - \varepsilon_i)/(x_j - x_i)$ have a d.f. symmetric about zero but with variances that may not all be the same. Thus, $\widehat{\beta}_{LS}$ is a weighted average of the basic estimators Z_{ij}, $(i,j) \in S_n$, with weights proportional to the distances $(x_j - x_i)^2$. This explains why outliers/gross errors have considerable influence on $\widehat{\beta}_{LS}$. To enhance robustness, we look at the set $\{Z_{ij}, (i,j) \in S_n\}$, order the Z_{ij}'s in ascending order and denote these ordered variables by $Z_{(1)} \leq Z_{(2)} \leq \cdots \leq Z_{(N)}$. Note that the Z_{ij}'s are not all mutually independent, so that the $Z_{(k)}$ do not relate to the conventional set of independent and identically distributed random variables). Now consider the alternative estimator

$$\widehat{\beta}_{TS} = \text{median}\{Z_{ij}, (i, j) \in S_n\}$$
$$= \begin{cases} Z_{(M+1)} & \text{if } N = 2M + 1 \\ \frac{1}{2}\left[Z_{(M)} + Z_{(M+1)}\right] & \text{if } N = 2M. \end{cases} \quad (7.3.7)$$

In the literature, $\widehat{\beta}_{TS}$ is known as the Theil-Sen [viz., Sen (1968a)] estimator of β. This is indeed a very robust estimator in a global sense and it is a member of the so-called class of R-estimators of β. To motivate $\widehat{\beta}_{TS}$, let us start with the usual **Kendall tau statistic**

$$T_n = \sum_{1 \le i < j \le n} \text{sign}\,(Y_j - Y_i)\,\text{sign}\,(x_j - x_i) = T_n(\mathbf{Y}_n, \mathbf{x}_n).$$

If we replace $\mathbf{Y}_n$ by $\mathbf{Y}_n - \mathbf{x}_n b$, $b \in \mathbb{R}$, then the corresponding T_n is denoted by $T_n(b)$ and may be written as

$$
\begin{aligned}
T_n(b) &= \sum_{1 \le i < j \le n} \text{sign}\,(Y_j - x_j b - Y_i + x_i b)\,\text{sign}\,(x_j - x_i) \\
&= \sum_{1 \le i < j \le n} \text{sign}\{Y_j - Y_i - b(x_j - x_i)\}\,\text{sign}\,(x_j - x_i) \\
&= \sum_{S_n} \text{sign}\left(\frac{Y_j - Y_i}{x_j - x_i} - b\right)\{\text{sign}\,(x_j - x_i)\}^2 \\
&= \sum_{S_n} \text{sign}\,(Z_{ij} - b),
\end{aligned}
$$

so that $T_n(b)$ is nonincreasing in $b \in \mathbb{R}$. Moreover, $T_n(\beta)$ has a distribution symmetric about zero. Thus, "equating $T_n(b)$ to zero" yields the estimator $\widehat{\beta}_{TS}$ (as the median of the Z_{ij}). Furthermore, for any $\varepsilon > 0$,

$$
\begin{aligned}
P_\beta\{\widehat{\beta}_{TS} > \beta + \varepsilon\} &\le P_\beta\{T_n(\beta + \varepsilon) \ge 0\} = P_0\{T_n(\varepsilon) \ge 0\}, \\
P_\beta\{\widehat{\beta}_{TS} < \beta - \varepsilon\} &\le P_\beta\{T_n(\beta - \varepsilon) \le 0\} = P_0\{T_n(-\varepsilon) \le 0\}.
\end{aligned}
\quad (7.3.8)
$$

Noting that $\binom{n}{2}^{-1} T_n(\pm\varepsilon)$ are U-statistics, we may use the results in Chapters 2, 3 and 5 to conclude that both of the above two terms converge to zero as $n \to \infty$, so that $\widehat{\beta}_{TS}$ is consistent. Similarly, we may approximate $P_\beta\{\sqrt{n}(\widehat{\beta}_{TS} - \beta) \le a\}$ by

$$
\begin{aligned}
&P_0\left\{T_n\left(-\frac{a}{\sqrt{n}}\right) \le 0\right\} \\
&= P_0\left\{T_n\left(-\frac{a}{\sqrt{n}}\right) - E_0 T_n\left(-\frac{a}{\sqrt{n}}\right) \le -E_0 T_n\left(-\frac{a}{\sqrt{n}}\right)\right\} \quad (7.3.9)
\end{aligned}
$$

and derive the asymptotic normality of $\sqrt{n}(\widehat{\beta}_{TS} - \beta)$. We pose most of these results in the form of exercises (see Exercises 7.3.3–7.3.4) and refer to Sen (1968a) for details.

In the same vein, we may consider the Wilcoxon signed rank statistic

$$W_n = W(Y_1, \ldots, Y_n) = \sum_{i=1}^{n} (\text{sign } Y_i) R_{ni}^+ \qquad (7.3.10)$$

where $R_{ni}^+ = R_{ni}^+(\mathbf{Y}_n)$ is the rank of $|Y_i|$ among $|Y_1|, \ldots, |Y_n|$, $i = 1, \ldots, n$. Now, using the identity

$$\sum_{i=1}^{n} I_{\{Y_i > 0\}} R_{ni}^+ = \sum_{1 \leq i \leq j \leq n} I_{\{Y_i + Y_j > 0\}}$$

we may conclude that

$$\sum_{1 \leq i \leq j \leq n} I_{\{Y_i + Y_j > 0\}} = \sum_{1 \leq i \leq j \leq n} I_{\{\frac{1}{2}(Y_i + Y_j) > 0\}}$$

$$= \frac{n(n+1)}{4} + \frac{1}{2} W_n. \qquad (7.3.11)$$

Then, letting $W_n(a) = W(Y_1 - a, \ldots, Y_n - a)$, $a \in \mathbb{R}$, it follows from (7.3.6) that $W_n(a)$ is nonincreasing in a (see Exercise 7.3.5). Also, taking $\beta = 0$ in model (7.1.2) it is clear that $W_n(\alpha)$ has a d.f. symmetric about zero. Thus, for the simple location model, we may equate $W_n(\alpha)$ to zero to obtain a robust estimator of the intercept α. If we write $M_{ij} = \frac{1}{2}(Y_i + Y_j)$, $1 \leq i \leq j \leq n$, and order these M_{ij}'s as

$$M_{(1)} \leq \cdots \leq M_{\left(\frac{n+1}{2}\right)},$$

then the estimator turns out to be the median $M_{(n^*)}$ where $n^* = \frac{1}{2}\binom{n+1}{2}$ if $\binom{n+1}{2}$ is odd and $\frac{1}{2}\{M_{n^*} + M_{(n^*+1)}\}$ if $\binom{n+1}{2}$ is even $(= 2n^*)$. Here also W_n is a linear function of two U-statistics (see Exercise 7.3.6), so that the consistency and asymptotic normality of $W_n(a)$ may be used to derive the parallel results for the estimator (see Exercise 7.3.7). When β is unknown, we consider the estimator $\widehat{\beta}_{TS}$ and write $\widehat{Y}_i = Y_i - x_i\beta_{TS}$, $i = 1, \ldots, n$; the Wilcoxon signed rank statistic based on the $\widehat{y}_i - a$ is denoted $\widehat{W}_n(a)$ and the derived estimator by $\widehat{\alpha}_n$.

Motivated by this simple case, we now consider a general class of rank statistics. Let

$$L_n = L_n(Y_i, \ldots, Y_n; \mathbf{x}_n) = \sum_{i=1}^{n} (x_i - \bar{x}_n) a_n(R_{ni})$$

where $a_n(1) \leq \cdots \leq a_n(n)$ are suitable scores and R_{ni} denotes the rank of

Y_i among $Y_1, \ldots, Y_n$, $1 \le i \le n$. Let then

$$L_n(b) = L_n(Y_1 - bx_1, \ldots, Y_n - bx_n, \mathbf{x}_n), \quad b \in \mathbb{R}.$$

Then it is easy to verify that $L_n(b)$ is nonincreasing in b and $L_n(\beta)$ has mean 0. Thus, we may set $\widehat{\beta}_{n1} = \sup\{b : L_n(b) > 0\}$, $\widehat{\beta}_{n2} = \inf\{b : L_n(b) < 0\}$ and then $\widehat{\beta}_{nR} = (\widehat{\beta}_{n1} + \widehat{\beta}_{n2})/2$, the **R-estimator** of β, is translation invariant and $\mathbf{x}_n$ regression equivariant. The consistency of $\widehat{\beta}_{nR}$ can be established (as in the case of $\widehat{\beta}_{TS}$) through that of $L_n(\beta \pm \varepsilon)$, $\varepsilon > 0$, and, similarly, the asymptotic normality of $\sqrt{n}(\widehat{\beta}_{nR} - \beta)$ can be studied via the asymptotic normality of $n^{-1/2}L_n(\beta + n^{-1/2}a)$. This calls for a study of the asymptotic properties of linear rank statistics, which is somewhat beyond the scope of this treatise. A fair amount of details is provided, however, in the related material of Section 8.6 [see (8.6.1)–(8.6.12)].

We may also extend the definition of L_n to the vector case by taking $\mathbf{L}_n = \sum_{i=1}^n (\mathbf{x}_i - \overline{\mathbf{x}}_n)a_n(R_{ni})$ and defining $\mathbf{L}_n(b)$ with $\mathbf{b} \in \mathbb{R}^q$. Here we may take

$$\widehat{\boldsymbol{\beta}}_{nR} = \arg\min\left\{\sum_{j=1}^q |L_{nj}(\mathbf{b})| : \mathbf{b} \in \mathbb{R}^q\right\}.$$

The asymptotic properties of such estimator requires a more sophisticated treatment and we may refer to Puri and Sen (1985, Chap. 6) for details.

Finally, we may consider a general signed rank statistic as

$$S_n = \sum_{i=1}^n \text{sign}(Y_i)a_n(R_{ni}^+) = S_n(\mathbf{Y}_n)$$

where R_{ni}^+ is the rank of $|Y_i|$ among $|Y_1|, \ldots, |Y_n|$, $1 \le i \le n$, and as in the case of W_n, we may work with $S_n(\mathbf{Y}_n - \mathbf{X}_n^t\widehat{\boldsymbol{\beta}}_{nR} - a\mathbf{1}_n)$ which, when equated to zero, yields an (aligned) R-estimator of α.

Up to this point, the main focus of our discussion in this section was on some asymptotic properties of regression parameters estimators which are robust in some global sense. In many practical situations, however, there is evidence in favor of some particular model and in such cases we would like to use this information to the extent possible, while providing some protection against mild departures from the proposed model, i.e., we want to deal with locally robust estimators. In this context, the M-methods introduced in Section 5.4 for the location model may be extended to the more general models under discussion here. We now proceed to examine some of their asymptotic properties.

An M-estimator of the parameter β in model (7.1.1) is defined as

$$\overline{\beta}_n = \arg\min \left\{ \sum_{i=1}^{n} \rho(Y_i - \mathbf{x}_{ni}^t \beta) : \beta \in \mathbb{R}^q \right\} \qquad (7.3.12)$$

where ρ is some arbitrary function chosen in accordance with the discussion in Section 5.4. When $\psi(z) = (\partial/\partial z)\rho(z)$ exists, (7.3.12) corresponds to a solution to the estimating equations

$$\mathbf{M}_n(\overline{\beta}_n) = \sum_{i=1}^{n} \psi(Y_i - \mathbf{x}_{ni}^t \overline{\beta}_n)\mathbf{x}_{ni} = \mathbf{0} \qquad (7.3.13)$$

which is generally obtained via some iterative procedure, like the Newton-Raphson method. We are interested in obtaining the asymptotic distribution of $\sqrt{n}(\overline{\beta}_n - \beta)$. Note that even if ψ is monotone, the elements of

$$\mathbf{M}_n(\mathbf{u}) = \sum_{i=1}^{n} \psi(Y_i - \mathbf{x}_{ni}^t \mathbf{u})\mathbf{x}_{ni}$$

are not monotone with respect to the elements of $\mathbf{u}$ and, therefore, we may not apply the same technique used to prove Theorem 5.4.2. Here we have to rely on an important linearity result due to Jurečková (1977) in order to establish the required asymptotic normality of the estimators. A complete rigorous treatment of this topic is beyond the scope of our text, but we present a general outline which may motivate the reader in understanding it. Consider the linear model (7.1.1) with the assumption that $\sigma^2 < \infty$ replaced by:

A1) The distribution function F of the error random variables is absolutely continuous with density function f such that $f'(\varepsilon)$ exists a.e.

A2) The distribution function F has a finite Fisher information with respect to location, i.e.,

$$I(\mu) = \int \{f'(\varepsilon)/f(\varepsilon)\}^2 f(\varepsilon)\, d\varepsilon < \infty.$$

Note that (A2) ensures that $\int |f'(x)|\, dx \leq I^{1/2}(\mu) < \infty$.

Assume further that:

B1) $\max_{1 \leq k \leq n} \mathbf{x}_{nk}^t (\mathbf{X}_n^t \mathbf{X}_n)^{-1} \mathbf{x}_{nk} \to 0$ as $n \to \infty$;

B2) $\lim_{n \to \infty} n^{-1}(\mathbf{X}_n^t \mathbf{X}_n) = \mathbf{V}$, finite and p.d.

Consider the M-estimator defined by (7.3.12) and assume that the score function ψ is a nonconstant function expressible in the form

$$\psi(x) = \sum_{\ell=1}^{s} \psi_\ell(x)$$

where $s \geq 1$ and each ψ_ℓ is monotone and is either an absolutely continuous function on any bounded interval in $\mathbb{R}$ with derivative ψ'_ℓ almost everywhere or is a step function. Also it satisfies

C1) $\int \psi(\varepsilon) f(\varepsilon) \, d\varepsilon = 0$,

C2) $\sigma_\psi^2 = \int \psi^2(\varepsilon) f(\varepsilon) \, d\varepsilon < \infty$,

C3) $\gamma = -\int \psi(\varepsilon) f'(\varepsilon) \, d\varepsilon < \infty$.

Note that whenever ψ is absolutely continuous, by partial integration, we have $\gamma = \int \psi'(\varepsilon) f(\varepsilon) \, d\varepsilon$.

Theorem 7.3.1: *For the linear model (7.1.1), under (A1), (A2), (B1), (B2), (C1), (C2) and (C3) it follows that*

$$\sqrt{n}(\overline{\beta}_n - \beta) \overset{\mathcal{D}}{\longrightarrow} N_q \left(0, \gamma^{-2} \sigma_\psi^2 \mathbf{V}^{-1} \right). \tag{7.3.14}$$

Outline of the proof: First we have to show that, given $K > 0$ and $\varepsilon > 0$, as $n \to \infty$

$$P \left\{ \sup_{\sqrt{n}\|\overline{\beta}_n - \beta\| \leq K} n^{-1/2} \|\mathbf{M}_n(\overline{\beta}_n) - \mathbf{M}_n(\beta) + n\gamma \mathbf{V}(\overline{\beta}_n - \beta)\| > \varepsilon \right\} \to 0. \tag{7.3.15}$$

The proof of this linearity result for the standard multivariate linear model is given in Singer and Sen (1985) and is based on a similar result proved by Jurečková (1977) under more restrictive conditions. Next we have to show that given $\eta > 0$ and $K > 0$,

$$P \left\{ \min_{\sqrt{n}\|\mathbf{u} - \beta\| \leq K} \frac{1}{\sqrt{n}} \|\mathbf{M}_n(\mathbf{u})\| > \eta \right\} \to 0 \quad \text{as } n \to \infty \tag{7.3.16}$$

or, in other words, that for every $K > 0$, $\frac{1}{\sqrt{n}} \mathbf{M}_n(\mathbf{u}) \overset{\mathrm{P}}{\longrightarrow} \mathbf{0}$ uniformly in the compact set $\sqrt{n}\|\mathbf{u} - \beta\| \leq K$, which is equivalent to showing that $\sqrt{n}\|\overline{\beta}_n - \beta\|$ is bounded in probability. The reader is referred to Jurečková (1977) for a proof of (7.3.16).

Now, in view of (7.3.15) and (7.3.16), it follows that $\sqrt{n} \mathbf{V} \gamma (\overline{\beta}_n - \beta)$ has the same asymptotic distribution as $n^{-1/2} \mathbf{M}_n(\beta) = n^{-1/2} \sum_{i=1}^n \mathbf{Z}_{ni}$, where $\mathbf{Z}_{ni} = \psi(Y_i - \mathbf{x}_{ni}^t \beta) \mathbf{x}_{ni}$, $i = 1, \ldots, n$, are independent r.v.'s such that $\mathbf{E}\mathbf{Z}_{ni} = \mathbf{0}$ and $\mathrm{Var}\, \mathbf{Z}_{ni} = \sigma_\psi^2 \mathbf{x}_{ni} \mathbf{x}_{ni}^t$. Using assumptions (B1) and (B2) in connection with Cramér-Wold Theorem (3.2.4) and Hájek-Šidák Theorem (3.3.6) we may conclude that $n^{-1/2} \mathbf{M}_n(\beta) \overset{\mathcal{D}}{\longrightarrow} N_q(\mathbf{0}, \mathbf{V}\sigma_\psi^2)$, and (7.3.14) follows. ∎

Since M-estimators are not scale invariant, the above result is of little practical utility; however, it is easily extended to cover situations where (7.3.12) is replaced by

$$\mathbf{M}_n(\overline{\beta}_n) = \sum_{i=1}^{n} \psi \left(\frac{Y_i - \mathbf{x}_{ni}^t \overline{\beta}_n}{\widehat{\sigma}_n} \right) \mathbf{x}_{ni} = \mathbf{0} \qquad (7.3.17)$$

where $\widehat{\sigma}_n$ is a $\sqrt{n}$ consistent estimate of the scale parameter σ associated to the distribution of the error r.v.'s provided that F has a finite Fisher information with respect to scale, i.e,

$$I(\sigma) = -1 + \int \varepsilon^2 \left\{ \frac{f'(\varepsilon)}{f(\varepsilon)} \right\}^2 f(\varepsilon) \, d\varepsilon < \infty.$$

The reader is referred to Singer and Sen (1985) for a proof as well as for extensions of the result to multivariate linear models.

7.4 Generalized linear models

The terminology **generalized linear models**, due to Nelder and Wedder-burn (1972), has become a household word in the statistical community. A systematic study, initiated by them, showed how linearity in common statistical models could be exploited to extend classical linear models (as treated in earlier sections) to more general statistical models, cropping up in diverse applications, with a view to unifying the apparently diverse statistical inference techniques commonly employed in such situations. Generalized linear models include, as special cases, linear regression and analysis of variance models, log-linear models for categorical data, product multinomial response models, logit and probit models in biological assays (quantal responses) as well as some simple statistical models arising in survival analysis. An excellent treatise of this subject matter is due McCullagh and Nelder (1989), where a variety of useful numerical illustrations has also been considered along with the relevant methodology. Since most of the inferential results based on such models are valid for large samples, there is a need to look into the allied asymptotic theory so that applications can be governed by a sound methodological basis. At a level consistent with the rest of this book, we plan to provide here the desired asymptotics, which would be a good supplement to the general methodology available so far and reported in McCullagh and Nelder (1989) among other places.

We start by introducing some notation and assumptions underlying the definition of generalized linear models. Consider a vector of observations $\mathbf{y}_N = (y_1, \ldots, y_N)^t$ corresponding to N independent random variables Y_i, $i = 1, \ldots, N$, where Y_i has a distribution in the **exponential family**, i.e., with density function

$$f(y_i, \theta_i, \phi) = c(y_i, \phi) \exp[\{y_i\theta_i - b(\theta_i)\}/a(\phi)] \qquad (7.4.1)$$

for $i = 1, \ldots, N$, so that the corresponding joint density is

$$\prod_{i=1}^{N} c(y_i, \phi) \exp[\{y_i\theta_i - b(\theta_i)\}/a(\phi)]$$

$$= c_N(\mathbf{y}_N, \phi) \exp\left[\sum_{i=1}^{N}\{y_i\theta_i - b(\theta_i)\}/a(\phi)\right]. \qquad (7.4.2)$$

Here θ_i, $i = 1, \ldots, N$, are the parameters of interest, ϕ (> 0) is a scale (nuisance) parameter and $a(\cdot), b(\cdot)$ and $c(\cdot)$ are functions of known form. Note that by (7.4.1), for each i,

$$\int c(y, \phi) \exp\{y\theta_i/a(\phi)\}d\nu(y) = \exp\{b(\theta_i)/a(\phi)\} \qquad (7.4.3)$$

so that differentiating both sides with respect to θ, we have

$$\int c(y, \phi) \left\{\frac{y}{a(\phi)}\right\} \exp\left\{\frac{y\theta_i}{a(\phi)}\right\} d\nu(y)$$

$$= \frac{b'(\theta_i)}{a(\phi)} \exp\left\{\frac{b(\theta_i)}{a(\phi)}\right\}$$

or, equivalently

$$\mathrm{E}Y_i/a(\phi) = b'(\theta_i)/a(\phi), \quad i = 1, \ldots, N, \qquad (7.4.4)$$

where $b'(\theta) = (\partial/\partial\theta)b(\theta)$. Differentiating one more time with respect to θ_i we have

$$\int c(y, \phi) \left\{\frac{y}{a(\phi)}\right\}^2 \exp\left\{\frac{y\theta_i}{a(\phi)}\right\} d\nu(y)$$

$$= \left[\frac{b''(\theta_i)}{a(\phi)} + \left\{\frac{b'(\theta_i)}{a(\phi)}\right\}^2\right] \exp\left\{\frac{b(\theta_i)}{a(\phi)}\right\}$$

which yields $\mathrm{E}Y_i^2 = b''(\theta_i)a(\phi) + (\mathrm{E}Y_i)^2$, so that

$$\mathrm{E}Y_i = \mu_i = \mu_i(\theta_i) = b'(\theta_i), \quad \mathrm{Var}\,Y_i = b''(\theta_i)a(\phi). \qquad (7.4.5)$$

We may also write

$$\mathrm{Var}\,Y_i = a(\phi)\frac{\partial}{\partial\theta_i}\mu_i(\theta_i) = a(\phi)v_i\{\mu_i(\theta_i)\} \qquad (7.4.6)$$

where $v_i\{\mu_i(\theta_i)\}$ is known as the **variance function** and it depends solely on $\mu_i(\theta_i)$, $1 \le i \le N$. In the particular case of a normal density, $\mu_i = \theta_i$

so that $b'(\theta_i) \equiv \theta_i$ and, hence, $b''(\theta_i) = 1$, i.e., $v_i\{\mu_i(\theta_i)\} = 1$ implying
$\operatorname{Var} Y_i = a(\phi)$. This explains why $a(\phi)$ is referred to as the **scale factor**. Further, since the form of $b(\theta_i)$ is assumed to be the same for every
i (they differ only in the θ_i), $v_i(y) = v(y)$ for all i, so that we have a
common variance function $v\{\mu(\theta_i)\}$, $i = 1, \ldots, N$. The $\mu_i(\theta_i)$ play a basic
role in this formulation; in the usual case of linear models, $\mu_i(\theta_i) = \theta_i$,
$i = 1, \ldots, N$, are expressible linearly in terms of a finite parameter vector
$\boldsymbol{\beta} = (\beta_1, \ldots, \beta_q)^t$, say, and of a given $(N \times q)$ specification matrix $\mathbf{X}_N$, so
that $\boldsymbol{\theta}_N = (\theta_1, \ldots, \theta_N)^t = \mathbf{X}_N\boldsymbol{\beta}$. Moreover, by (7.4.5), $\mu_i(\theta_i) = b'(\theta_i) = \mu(\theta_i)$. This suggests that it may be possible to conceive of a transformation

$$g\{\mu(\theta_i)\} = g(\mu_i), \quad i = 1, \ldots, N, \qquad (7.4.7)$$

where $g(t)$ is a monotone and differentiable function of t, such that

$$\mathbf{G}_N = \{g(\mu_1), \ldots, g(\mu_N)\}^t = \mathbf{X}_N\boldsymbol{\beta} \qquad (7.4.8)$$

with $\mathbf{X}_N = (\mathbf{x}_{N1}, \ldots, \mathbf{x}_{NN})^t$ denoting a known $(N \times q)$ matrix of known
constants and $\boldsymbol{\beta} = (\beta_1, \ldots, \beta_q)^t$ is a q-vector of unknown parameters. Thus,

$$g\{\mu(\theta_i)\} = \mathbf{x}_{Ni}^t\boldsymbol{\beta}, \quad i = 1, \ldots, N, \qquad (7.4.9)$$

and, hence, $g(\cdot)$ provides the link between the mean μ_i and the linear
predictor parameter $\mathbf{x}_{Ni}^t\boldsymbol{\beta}$, $i = 1, \ldots, N$. That is why $g(\cdot)$ is termed a **link
function** [viz., Nelder and Wedderburn (1972)]. In view of (7.4.9), we have

$$\theta_i = (g \circ \mu)^{-1}(\mathbf{x}_{Ni}^t\boldsymbol{\beta}), \quad i = 1, \ldots, N, \qquad (7.4.10)$$

so that when $g = \mu^{-1}$, $g \circ \mu$ is the identity function, and, hence, $\theta_i = \mathbf{x}_{Ni}^t\boldsymbol{\beta}$,
in such a case, $g(\cdot)$ is termed a **canonical link function**.

Example 7.4.1: Let Z_i, $i = 1, 2$, be independent r.v.'s with the $\operatorname{Bin}(n, \pi_i)$
distribution and take $Y_i = Z_i/n$, $i = 1, 2$. Then Y_i, $i = 1, 2$, has probability
function

$$f(y; n, \pi_i) = \binom{n}{ny} \pi_i^{ny}(1 - \pi_i)^{n-ny}, \quad y = 0, 1/n, 2/n, \ldots, 1.$$

This may be re-expressed as

$$\begin{aligned}
f(y; n, \pi_i) &= \binom{n}{ny} \exp\{ny \log \pi_i + (n - ny)\log(1 - \pi_i)\} \\
&= \binom{n}{ny} \exp\left[\left\{y \log \frac{\pi_i}{1 - \pi_i} - \log(1 - \pi_i)\right\}/n^{-1}\right] \\
&= c(y, \phi) \exp[\{y\theta_i - b(\theta_i)\}/a(\phi)]
\end{aligned}$$

where $\theta_i = \log\{\pi_i/(1 - \pi_i)\}$, $b(\theta_i) = \log(1 + e^{\theta_i})$, $\phi = n^{-1}$, $c(y, \phi) = \binom{n}{ny}$
and $a(\phi) = n^{-1}$. Two generalized linear models for this setup are given by

i) $\theta_i = \log\{\pi_i/(1 - \pi_i)\} = \alpha + \beta x_i$, $i = 1, 2$, with $x_1 = 1$, $x_2 = -1$ and α and β denoting two unknown scalars, if we choose the canonical link;

ii) $\pi_i = e^{\theta_i}/(1 + e^{\theta_i}) = \alpha + \beta x_i$, $i = 1, 2$, with x_1, x_2, α and β as above, if we consider the link function $g(z) = z$.

Either model would serve the purpose of verifying the homogeneity of the two distributions, which is implied by $\beta = 0$. ∎

Example 7.4.2: Let Y_i, $i = 1, \ldots, N$, be independent r.v.'s with the Poisson (λ_i) distribution, i.e., with probability function

$$f(y; \lambda_i) = e^{-\lambda_i} \lambda_i^y / y!, \quad y = 0, 1, 2 \ldots$$

which may be rewritten as

$$f(y; \lambda_i) = (y!)^{-1} \exp\{y \log \lambda_i - \lambda_i\} = c(y, \phi) \exp\{y\theta_i - b(\theta_i)\}$$

where $\theta_i = \log \lambda_i$, $b(\theta_i) = \exp \theta_i$, $\phi = 1$, $c(y, \phi) = (y!)^{-1}$ and $a(\phi) = 1$. Assume further that to each Y_i we associate a q-vector $\mathbf{x}_i$ corresponding to the observed values of q explanatory variables. Two generalized linear models for the association between the response and explanatory variables may be expressed as

i) $\theta_i = \log \lambda_i = \mathbf{x}_i^t \boldsymbol{\beta}$, $i = 1, \ldots, N$, where $\boldsymbol{\beta}$ is a q-vector of unknown constants, if we choose the canonical link;

ii) $\lambda_i = \exp \theta_i = \mathbf{x}_i^t \boldsymbol{\beta}$, $i = 1, \ldots, N$, with $\boldsymbol{\beta}$ as above, if we consider the link function $g(z) = z$.

The model in (i) is a log-linear model for Poisson counts whereas that in (ii) is a linear model similar in spirit to the usual (normal theory) regression models. ∎

Example 7.4.3: Let Y_i, $i = 1, \ldots, N$, be independent r.v.'s following Bernoulli(π_i) distributions, which may be expressed as

$$f(y; \pi_i) = c(y, \phi) \exp\{y\theta_i - b(\theta_i)\}$$

where $\theta_i = \log\{\pi_i/(1 - \pi_i)\}$, $b(\theta_i) = \log(1 + e^{\theta_i})$, $\phi = 1$, $c(y, \phi) = 1$ and $a(\phi) = 1$. Also, let $\mathbf{x}_i$ be as in Example 7.4.2. A generalized linear model commonly used to describe the relation between the Y_i's and $\mathbf{x}_i$'s is the logistic regression model

$$\theta_i = \log \frac{\pi_i}{1 - \pi_i} = \mathbf{x}_i^t \boldsymbol{\beta}, \quad i = 1, \ldots, N, \quad (7.4.11)$$

which may also be expressed as

$$\pi_i = \pi(\mathbf{x}_i) = \{1 + \exp(-\mathbf{x}_i^t \boldsymbol{\beta})\}^{-1}, \quad i = 1, \ldots, N.$$

∎

Although we have defined generalized linear models in terms of the two-parameter exponential family of distributions (7.4.1), it is possible to consider extensions to the multiparameter case. Typical examples are the log-linear models for product multinomial data discussed in Chapter 6 and Cox's proportional hazards models for survival data as discussed in Cox and Oakes (1984), for example.

In what follows, we define $h(\cdot) = (g \circ \mu)^{-1}(\cdot)$, so that (7.4.10) reduces to

$$\theta_i = h(\mathbf{x}_{Ni}^t \boldsymbol{\beta}), \quad i = 1, \dots, N, \tag{7.4.12}$$

where h is monotone and differentiable. In view of (7.4.9) and (7.4.12), given the explanatory vectors $\mathbf{x}_{Ni}$, $i = 1, \dots, N$, on the same spirit as in classical linear models, here, also, our main interest centers around the parameter vector $\boldsymbol{\beta}$. Looking back at (7.4.1) and (7.4.2), we may note that the nuisance parameter ϕ does not affect the estimating equations leading to the MLE of $\boldsymbol{\beta}$ and that it influences the Fisher information matrix, say, $\mathbf{I}(\boldsymbol{\beta})$ by a multiplicative factor, $\{a(\phi)\}^{-2}$ which may be estimated consistently; thus, for the sake of simplicity and without loss of generality, we take $a(\phi) \equiv 1$, so that the log-likelihood function (in terms of $\boldsymbol{\beta}$) corresponding to (7.4.2) and (7.4.12) is given by

$$\log L_N(\boldsymbol{\beta}) = \sum_{i=1}^{N} \left[Y_i h(\mathbf{x}_{Ni}^t \boldsymbol{\beta}) - b\left\{ h(\mathbf{x}_{Ni}^t \boldsymbol{\beta}) \right\} \right] - \text{constant} \tag{7.4.13}$$

where the constant term does not depend on $\boldsymbol{\beta}$. Also let

$$\mu_i(\boldsymbol{\beta}) = \mu\{h(\mathbf{x}_{Ni}^t \boldsymbol{\beta})\} = b'\{h(\mathbf{x}_{Ni}^t \boldsymbol{\beta})\} \tag{7.4.14}$$

and

$$v_i(\boldsymbol{\beta}) = v_i\{h(\mathbf{x}_{Ni}^t \boldsymbol{\beta})\} = b''\{h(\mathbf{x}_{Ni}^t \boldsymbol{\beta})\} \tag{7.4.15}$$

and note that, on letting $y = h(x)$, i.e. $x = h^{-1}(y) = (g \circ \mu)(y)$,

$$
\begin{aligned}
(d/dx)h(x) &= (d/dx)(g \circ \mu)^{-1}(x) = \{(d/dy)(g \circ \mu)(y)\}^{-1} \\
&= [g'\{\mu(y)\}\mu'(y)\}]^{-1} = [g'\{\mu[h(x)]\}\mu'(y)]^{-1} \\
&= [g'\{\mu[h(x)]\}b''\{h(x)\}]^{-1} \tag{7.4.16}
\end{aligned}
$$

so that

$$
\begin{aligned}
(\partial/\partial\boldsymbol{\beta})h(\mathbf{x}_{Ni}^t \boldsymbol{\beta}) &= [g'\{b'[h(\mathbf{x}_{Ni}^t \boldsymbol{\beta})]\}b''\{h(\mathbf{x}_{Ni}^t \boldsymbol{\beta})\}]^{-1}\mathbf{x}_{Ni} \\
&= [g'\{\mu_i(\boldsymbol{\beta})\}v_i(\boldsymbol{\beta})]^{-1}\mathbf{x}_{Ni} \tag{7.4.17}
\end{aligned}
$$

and

$$
\begin{aligned}
(\partial/\partial\boldsymbol{\beta})b\{h(\mathbf{x}_{Ni}^t \boldsymbol{\beta})\} &= b'\{h(\mathbf{x}_{Ni}^t \boldsymbol{\beta})\}(\partial/\partial\boldsymbol{\beta})h(\mathbf{x}_{Ni}^t \boldsymbol{\beta}) \\
&= \mu_i(\boldsymbol{\beta})[g'\{\mu_i(\boldsymbol{\beta})\}]v_i(\boldsymbol{\beta})]^{-1}\mathbf{x}_{Ni}. \tag{7.4.18}
\end{aligned}
$$

Therefore, from (7.4.13)–(7.4.18), the score statistic is given by:

$$
\mathbf{U}_N(\boldsymbol{\beta}) = (\partial/\partial\boldsymbol{\beta})\log L_N(\boldsymbol{\beta})
$$

$$
= \sum_{i=1}^{N}[Y_i\{g'[\mu_i(\boldsymbol{\beta})]v_i(\boldsymbol{\beta})\}^{-1}\mathbf{x}_{Ni} - \mu_i(\boldsymbol{\beta})[g'\{\mu_i(\boldsymbol{\beta})v_i(\boldsymbol{\beta})]^{-1}\mathbf{x}_{Ni}]
$$

$$
= \sum_{i=1}^{N}\frac{\{Y_i - \mu_i(\boldsymbol{\beta})\}}{[g'\{\mu_i(\boldsymbol{\beta})\}v_i(\boldsymbol{\beta})]}\mathbf{x}_{Ni}. \tag{7.4.19}
$$

Note that if g is a canonical link function, $h(x) = x$, so that, in (7.4.16), $(d/dx)h(x) = 1$, in (7.4.17), $(\partial/\partial\boldsymbol{\beta})h(\mathbf{x}_{Ni}^t\boldsymbol{\beta}) = \mathbf{x}_{Ni}$ and also in (7.4.18), $(\partial/\partial\boldsymbol{\beta})b\{h(\mathbf{x}_{Ni}^t\boldsymbol{\beta})\} = b'(\mathbf{x}_{Ni}^t\boldsymbol{\beta})\mathbf{x}_{Ni}$ implying that the denominators of the individual terms on the rhs of (7.4.19) are all equal to 1. Thus, the **estimating equations** (for the MLE $\widehat{\boldsymbol{\beta}}_N$) reduce to

$$
\sum_{i=1}^{N}\{Y_i - b'(\mathbf{x}_{Ni}^t\boldsymbol{\beta})\}\mathbf{x}_{Ni} = \mathbf{0}. \tag{7.4.20}
$$

As a check, the reader is invited to verify that for the classical linear model in (7.1.1) with the ε_i normally distributed, the link function is canonical and, further, that $b'(y) = y$, implying that (7.4.20) leads to the MLE (or LSE)

$$
\widehat{\boldsymbol{\beta}}_N = \left(\sum_{i=1}^{N}\mathbf{x}_{Ni}\mathbf{x}_{Ni}^t\right)^{-1}\left(\sum_{i=1}^{N}\mathbf{x}_{Ni}Y_i\right).
$$

In the general case of $h(\cdot)$ being not necessarily an identity function, by virtue of the assumed monotonicity ($\nearrow$) and differentiability of $g(\cdot)$, $g'\{\mu_i(\boldsymbol{\beta})\}v_i(\boldsymbol{\beta})$ is a non-negative function of the unknown $\boldsymbol{\beta}$ and the given $\mathbf{x}_{Ni}$, so that (7.4.19) may be interpreted as an extension of the corresponding "normal" equations in the weighted least-squares method [see (7.1.5)], where the weights, besides being non-negative, may, in general, depend on the unknown parameter $\boldsymbol{\beta}$. For this reason, we term

$$
\sum_{i=1}^{N}[g'\{\mu_i(\widehat{\boldsymbol{\beta}}_N)\}v_i(\widehat{\boldsymbol{\beta}}_N)]^{-1}\{Y_i - \mu_i(\widehat{\boldsymbol{\beta}}_N)\}\mathbf{x}_{Ni} = \mathbf{0} \tag{7.4.21}
$$

a **generalized estimating equation** (GEE). If we let

$$
\mathbf{r}_N(\boldsymbol{\beta}) = \{Y_1 - \mu_1(\boldsymbol{\beta}), \ldots, Y_N - \mu_N(\boldsymbol{\beta})\}^t \tag{7.4.22}
$$

and

$$
D_N(\boldsymbol{\beta}) = \mathrm{diag}\,[g'\{\mu_1(\boldsymbol{\beta})\}v_1(\boldsymbol{\beta}), \ldots, g'\{\mu_N(\boldsymbol{\beta})\}v_N(\boldsymbol{\beta})],
$$

then the GEE in (7.4.21) may also be written as

$$
\mathbf{X}_N^t\mathbf{D}_N^{-1}(\boldsymbol{\beta})\mathbf{r}_N(\boldsymbol{\beta})\big|_{\boldsymbol{\beta}=\widehat{\boldsymbol{\beta}}_N} = \mathbf{0}. \tag{7.4.23}
$$

Drawing a parallel between the general asymptotic results on the MLE
in Section 5.2 and the GLSE in Section 7.2, we study here the desired
large sample properties of $\hat{\boldsymbol{\beta}}_N$, the MLE for the generalized linear model in
(7.4.8). In this context, we like to clarify the role of N in the asymptotics to
follow. Compared to the linear model (7.1.1), usually here also we assume
that in the asymptotic case, $N \to \infty$. However, there may be a situation
where for each i, the Y_i may be a statistic based on a subsample of size n_i,
$i \geq 1$ (see Example 7.4.1). In such a case, we might consider a second type
of asymptotics where it is not crucial to have N large, provided the n_i's
are themselves large. For simplicity of presentation we take $N = 1$, and
recalling (7.4.1), we consider a sequence $\{f_{(n)}\}$ of densities of the form

$$
\begin{aligned}
f_{(n)}(y; \theta, \phi) &= f(y; \theta_{(n)}, \phi_{(n)}) \\
&= c(y, \phi_{(n)}) \exp[\{y\theta_{(n)} - b_{(n)}(\theta_{(n)})\}/a_{(n)}(\phi_{(n)})]
\end{aligned}
\tag{7.4.24}
$$

for $n \geq 1$ and observe that in many situations, as $n \to \infty$

$$
\frac{Y - b'_{(n)}(\theta_{(n)})}{\{b''_{(n)}(\theta_{(n)})a_{(n)}(\phi_{(n)})\}^{1/2}} \xrightarrow{\mathcal{D}} N(0,1).
\tag{7.4.25}
$$

This is the case with each Y_i in Example 7.4.1, where $b'_{(n)}(\theta_{(n)}) = \pi_i$,
$b''_{(n)}(\theta_{(n)}) = \pi_i(1 - \pi_i)$ and $a_{(n)}(\phi_{(n)}) = n^{-1}$. In Example 7.4.2 we may have
a similar picture for each Y_i, if we observe that $b'_{(n)}(\theta_{(n)}) = b''_{(n)}(\theta_{(n)}) = \lambda_i$,
$a_{(n)}(\phi_{(n)}) = 1$ and interpret $n \to \infty$ as $\lambda_i = \lambda_i(n) \to \infty$ as in the well-
known approximation of the binomial distribution by the Poisson distribu-
tion. Since this second type of asymptotics may be examined directly by
the methods described in Chapter 3, we will not elaborate on them further
and intend to concentrate on the regular case, where $N \to \infty$.

Note that the "weight matrix" $\mathbf{D}_N(\boldsymbol{\beta})$ in (7.4.23) may depend on the
parameter vector $\boldsymbol{\beta}$, as opposed to the GLSE case where $\mathbf{G}_n$ was independent of $\boldsymbol{\beta}$. This introduces additional complications, and, hence, further
regularity conditions may be needed to formulate the general asymptotics.
It follows from (7.4.19) that whenever $g'(\cdot)$ and $v(\cdot) \equiv b'(\cdot)$ are both differentiable,

$$
-\frac{\partial^2}{\partial\boldsymbol{\beta}\partial\boldsymbol{\beta}^t} \log L_N(\boldsymbol{\beta}) = \sum_{i=1}^{N} \frac{b''\{h(\mathbf{x}_{Ni}^t\boldsymbol{\beta})\}}{[g'\{\mu_i(\boldsymbol{\beta})\}v_i(\boldsymbol{\beta})]^2}\mathbf{x}_{Ni}\mathbf{x}_{Ni}^t
$$

$$
+ \sum_{i=1}^{N} \frac{\{Y_i - \mu_i(\boldsymbol{\beta})\}}{[g'\{\mu_i(\boldsymbol{\beta})\}v_i(\boldsymbol{\beta})]^2}
$$

$$
\times \left[v_i^2(\boldsymbol{\beta})g''\{\mu_i(\boldsymbol{\beta})\} + \frac{g'\{\mu_i(\boldsymbol{\beta})\}b'''\{h(\mathbf{x}_{Ni}^t\boldsymbol{\beta})\}}{g'\{\mu_i(\boldsymbol{\beta})\}v_i(\boldsymbol{\beta})} \right]\mathbf{x}_{Ni}\mathbf{x}_{Ni}^t
$$

$$
= \sum_{i=1}^{N} [g'\{\mu_i(\boldsymbol{\beta})\}]^{-2} \{v_i(\boldsymbol{\beta})\}^{-1} \mathbf{x}_{Ni} \mathbf{x}_{Ni}^{t}
$$

$$
+ \sum_{i=1}^{N} \{Y_i - \mu_i(\boldsymbol{\beta})\}
$$

$$
\times \left[\frac{g''\{\mu_i(\boldsymbol{\beta})\}}{(g'\{\mu_i(\boldsymbol{\beta})\})^2} + \frac{b'''\{h(\mathbf{x}_{Ni}^{t}\boldsymbol{\beta})\}}{(g'\{\mu_i(\boldsymbol{\beta})\})^2 \{v_i(\boldsymbol{\beta})\}^3} \right] \mathbf{x}_{Ni} \mathbf{x}_{Ni}^{t}. \quad (7.4.26)
$$

Since $EY_i = \mu_i(\boldsymbol{\beta})$ for all $i \geq 1$, we obtain from (7.4.26) that

$$
\mathbf{I}_N(\boldsymbol{\beta}) = \mathrm{E}_{\boldsymbol{\beta}} \left\{ -\frac{\partial^2}{\partial \boldsymbol{\beta} \partial \boldsymbol{\beta}^t} \log L_N(\boldsymbol{\beta}) \right\}
$$

$$
= \sum_{i=1}^{N} \left\{ [g'\{\mu_i(\boldsymbol{\beta})\}]^{-2} v_i(\boldsymbol{\beta})^{-1} \right\} \mathbf{x}_{Ni} \mathbf{x}_{Ni}^{t}. \quad (7.4.27)
$$

Thus we may re-express (7.4.26) as

$$
-\frac{\partial^2}{\partial \boldsymbol{\beta} \partial \boldsymbol{\beta}^t} \log L_N(\boldsymbol{\beta}) = \mathbf{I}_N(\boldsymbol{\beta}) + \mathbf{R}_N(\boldsymbol{\beta}) \quad (7.4.28)
$$

where

$$
\mathbf{R}_N(\boldsymbol{\beta}) = \sum_{i=1}^{N} \{Y_i - \mu_i(\boldsymbol{\beta})\} \left[\frac{g''\{\mu_i(\boldsymbol{\beta})\}}{(g'\{\mu_i(\boldsymbol{\beta})\})^2} + \frac{b'''\{h(\mathbf{x}_{Ni}^{t}\boldsymbol{\beta})\}}{(g'\{\mu_i(\boldsymbol{\beta})\})^2 \{v_i(\boldsymbol{\beta})\}^3} \right] \mathbf{x}_{Ni} \mathbf{x}_{Ni}^{t}.
$$

$$(7.4.29)$$

Note that for canonical link functions, $\mathbf{R}_N(\boldsymbol{\beta}) = \mathbf{0}$.

Now, in order to obtain the asymptotic distribution of the estimator of the parameter vector $\boldsymbol{\beta}$ under the setup outlined above, we first discuss the required assumptions, comparing them to those employed previously in similar contexts. First, along the same lines considered in the case of the GLSE in Section 7.2 we require that, when divided by N, the lhs of (7.4.28) satisfies at least a weak consistency property. In this direction we must first assume that

$$
\lim_{N \to \infty} \frac{1}{N} \mathbf{I}_N(\boldsymbol{\beta}) = \mathbf{I}(\boldsymbol{\beta}), \quad \text{finite and p.d.} \quad (7.4.30)
$$

Also, from (7.4.29) we see that $\mathbf{R}_N(\boldsymbol{\beta})$ is a sum of independent centered random variables with finite variances $v_i\{\mu_i(\boldsymbol{\beta})\}$ and nonstochastic matrix coefficients,

$$
\mathbf{G}_i = \left[\frac{g''\{\mu_i(\boldsymbol{\beta})\}}{(g'\{\mu_i(\boldsymbol{\beta})\})^2} + \frac{b'''\{h(\mathbf{x}_{Ni}^{t}(\boldsymbol{\beta}))\}}{(g'\{\mu_i(\boldsymbol{\beta})\})^2 \{v_i(\boldsymbol{\beta})\}^3} \right] \mathbf{x}_{Ni} \mathbf{x}_{Ni}^{t}.
$$

Thus, if we consider the assumption

$$\lim_{N \to \infty} N^{-2} \sum_{i=1}^{N} v_i\{\mu_i(\boldsymbol{\beta})\} \mathrm{tr}(\mathbf{G}_i \mathbf{G}_i^t) = 0, \qquad (7.4.31)$$

a direct application of Chebyshev Inequality (2.2.31) to (7.4.29) will enable us to show that $N^{-1} \mathbf{R}_N(\boldsymbol{\beta}) \xrightarrow{\mathrm{P}} \mathbf{0}$ and, thus, that the required consistency holds when the model is true.

Recalling (7.4.27), observe that $\mathbf{I}_N(\boldsymbol{\beta})$ depends on $g(\cdot)$ and $v(\cdot)$ as well as on the $\mathbf{x}_{Ni}$; thus, it is natural to expect some further assumptions related to the link function and to the variance function in addition to those on the explanatory vectors in order to establish the regularity conditions under which the asymptotic properties of the estimators will follow. In this direction we recall that in Theorem 5.2.1, dealing with the large sample properties of MLEs, we have used a compactness condition on the second derivative of the log-likelihood function, which we extend to the generalized linear model setup under consideration. This condition essentially rests on the uniform continuity of $[g'\{\mu_i(\boldsymbol{\beta}^*)\}]^{-2}$, $\{v_i(\boldsymbol{\beta}^*)\}^{-3}$, $g''\{\mu_i(\boldsymbol{\beta}^*)\}$ and $b'''\{h(\mathbf{x}_{Ni}^t \boldsymbol{\beta}^*)\}$ in an infinitesimal neighborhood of the true $\boldsymbol{\beta}$, i.e., in the set

$$B(\delta) = \{\boldsymbol{\beta}^* \in I\!\!R^q : \|\boldsymbol{\beta}^* - \boldsymbol{\beta}\| < \delta\}, \quad \delta \downarrow 0. \qquad (7.4.32)$$

Essentially, if we let

$$w_{1i}(\boldsymbol{\beta}) = [g'\{\mu_i(\boldsymbol{\beta})\}]^{-2}\{v_i(\boldsymbol{\beta})\}^{-1}$$

and

$$\begin{aligned} w_{2i}(\boldsymbol{\beta}) = \mu_i(\boldsymbol{\beta}) \Big\{ & g''\{\mu_i(\boldsymbol{\beta})\}[g'\{\mu_i(\boldsymbol{\beta})\}]^{-2} \\ & + b'''\{h(\mathbf{x}_{Ni}^t \boldsymbol{\beta})\}[g'\{\mu_i(\boldsymbol{\beta})\}]\{v_i(\boldsymbol{\beta})\}^{-3} \Big\}, \end{aligned}$$

the required compactness condition may be expressed as

i) for $k = 1, 2$, as $\delta \downarrow 0$,

$$\sup_{\boldsymbol{\beta}^* \in B(\delta)} N^{-1} \sum_{i=1}^{N} \|\{w_{ki}(\boldsymbol{\beta}^*) - w_{ki}(\boldsymbol{\beta})\}\mathbf{x}_{Ni}\mathbf{x}_{Ni}^t\| \to 0, \qquad (7.4.33)$$

ii) as $\delta \downarrow 0$,

$$\mathrm{E}_{\boldsymbol{\beta}}\Big\{ \sup_{\boldsymbol{\beta}^* \in B(\delta)} N^{-1} \sum_{i=1}^{N} |Y_i| \|\{w_{2i}(\boldsymbol{\beta}^*) - w_{2i}(\boldsymbol{\beta})\}\mathbf{x}_{Ni}\mathbf{x}_{Ni}^t\| \to 0. \quad (7.4.34)$$

Then we may consider the following theorem, the proof of which is based on the same technique as that used to demonstrate Theorem 5.2.1.

Theorem 7.4.1: *Consider the generalized linear model (7.4.8) and let $\widehat{\boldsymbol{\beta}}_N$ denote the MLE of $\boldsymbol{\beta}$ under the likelihood corresponding to (7.4.13). Then if (7.4.30)–(7.4.34) hold, it follows that*

$$\sqrt{N}(\widehat{\boldsymbol{\beta}}_N - \boldsymbol{\beta}) \xrightarrow{\mathcal{D}} N_q(\mathbf{0}, \mathbf{I}^{-1}(\boldsymbol{\beta})). \tag{7.4.35}$$

Proof: First let $\|\mathbf{u}\| < K$, $0 < K < \infty$ and consider a Taylor expansion of $\log L_N(\boldsymbol{\beta} + N^{-1/2}\mathbf{u})$ around $\log L_N(\boldsymbol{\beta})$ to define

$$
\begin{aligned}
\lambda_N(\mathbf{u}) &= \log L_N(\boldsymbol{\beta} + N^{-1/2}\mathbf{u}) - \log L_N(\boldsymbol{\beta}) \\
&= \frac{1}{\sqrt{N}}\mathbf{u}^t \mathbf{U}_N(\boldsymbol{\beta}) + \frac{1}{N}\mathbf{u}^t \frac{\partial^2}{\partial\boldsymbol{\beta}\partial\boldsymbol{\beta}^t}\log L_N(\boldsymbol{\beta})\Big|_{\boldsymbol{\beta}^*} \mathbf{u}
\end{aligned}
\tag{7.4.36}
$$

where $\boldsymbol{\beta}^*$ belongs to the line segment joining $\boldsymbol{\beta}$ and $\boldsymbol{\beta}^*$. Then let

$$
\begin{aligned}
Z_n(\mathbf{u}) &= \frac{1}{N}\Bigg\{\mathbf{u}^t \frac{\partial^2}{\partial\boldsymbol{\beta}\partial\boldsymbol{\beta}^t}\log L_N(\boldsymbol{\beta})\Big|_{\boldsymbol{\beta}^*}\mathbf{u} - \mathbf{u}^t\frac{\partial^2}{\partial\boldsymbol{\beta}\partial\boldsymbol{\beta}^t}\log L_N(\boldsymbol{\beta})\Big|_{\boldsymbol{\beta}}\mathbf{u} \\
&\quad + \mathbf{u}^t \frac{\partial^2}{\partial\boldsymbol{\beta}\partial\boldsymbol{\beta}^t}\log L_N(\boldsymbol{\beta})\Big|_{\boldsymbol{\beta}}\mathbf{u} + \mathbf{u}^t\mathbf{I}_N(\boldsymbol{\beta})\mathbf{u}\Bigg\}
\end{aligned}
$$

and note that (7.4.36) may be re-expressed as

$$\lambda_N(\mathbf{u}) = \frac{1}{\sqrt{N}}\mathbf{u}^t\mathbf{U}_N(\boldsymbol{\beta}) - \frac{1}{2N}\mathbf{u}^t\mathbf{I}_N(\boldsymbol{\beta})\mathbf{u} + Z_N(\mathbf{u}). \tag{7.4.37}$$

Now observe that

$$
\begin{aligned}
\sup_{\|\mathbf{u}\|<K} &\|Z_N(\mathbf{u})\| \\
\leq \frac{1}{2}&\sup_{\boldsymbol{\beta}^* \in B(K/\sqrt{n})}\left\|\frac{1}{N}\frac{\partial^2}{\partial\boldsymbol{\beta}\partial\boldsymbol{\beta}^t}\log L_N(\boldsymbol{\beta})\Big|_{\boldsymbol{\beta}^*} - \frac{1}{N}\frac{\partial^2}{\partial\boldsymbol{\beta}\partial\boldsymbol{\beta}^t}\log L_N(\boldsymbol{\beta})\Big|_{\boldsymbol{\beta}}\right\| \\
&+ \frac{K^2}{2}\left\|\frac{1}{N}\frac{\partial^2}{\partial\boldsymbol{\beta}\partial\boldsymbol{\beta}^t}\log L_N(\boldsymbol{\beta}) + \frac{1}{N}\mathbf{I}_N(\boldsymbol{\beta})\right\|.
\end{aligned}
\tag{7.4.38}
$$

Using assumptions (7.4.30) and (7.4.31) in connection with (7.4.28) we may conclude that the last term on the rhs of (7.4.38) converges to zero as $N \to \infty$. On the other hand, assumptions (7.4.33) and (7.4.34) may be employed to show that the first term on the rhs of (7.4.38) also converges to zero as $N \to \infty$. Thus, it follows that $Z_N(\mathbf{u}) = o_p(1)$ uniformly in $\mathbf{u}$, $\|\mathbf{u}\| < K$, so that (7.4.37) may be written as

$$\lambda_n(\mathbf{u}) = \frac{1}{\sqrt{N}}\mathbf{u}^t\mathbf{U}_N(\boldsymbol{\beta}) - \frac{1}{2N}\mathbf{u}^t\mathbf{I}_N(\boldsymbol{\beta})\mathbf{u} + o_p(1)$$

so that an argument similar to that used in Theorem 5.2.1 completes the proof. In this regard, note that, by (7.4.19), the score statistic $\mathbf{U}_N(\boldsymbol{\beta})$ involves independent centered summands, and, hence, we may directly appeal to the Lindeberg-Feller Central Limit Theorem (3.3.3) to establish its asymptotic normality. [If the $\mathbf{x}_{Ni}$ are dependent on N as well, we may need to use the Central Limit Theorem 3.3.5 for Triangular Schemes.] In this context, it may be easier to verify the Liapunov condition as in Theorem 3.3.2. Also note that in a generalized linear model setup, the $\mathbf{x}_{Ni}$ need not be all the same, and, hence, the $\mu_i(\boldsymbol{\beta})$ may be different. Thus, the Y_i may not be identically distributed, so that the Hájek-Šidák Theorem (3.3.6) may not be appropriate here. ∎

Suppose we assume that $E[Y_i] = \mu_i$ and $\text{Var}(Y_i) = \sigma^2 V_i(\mu_i)$ where σ^2 is a positive scalar, possibly unknown, and $V_i(.)$ is a completely known variance function, but make no specific assumptions about the functional form of the density. Then there is no (finite dimensional) likelihood function and so we can't obtain maximum-likelihood estimates in the usual way. However the equation

$$\sum_{i=1}^N \frac{d\mu_i}{d\boldsymbol{\beta}} V_i(\mu_i)^{-1}(Y_i - \mu_i) = 0$$

can be solved to obtain estimates of $\boldsymbol{\beta}$. Under this setup the above equation is referred to as the **quasi-score** or **quasi-likelihood** estimating equation (Wedderburn, 1974).

In the above setup the response $\mathbf{Y}_i$ can be an $n_i \times 1$ vector and the variance function $V_i(.)$ would be a mapping into an $n_i \times n_i$ matrix.

Note that in generalized linear models the roots of the score equation may may not be unique or not exist. See Wedderburn (1976), Pratt (1981), Silvapulle (1981) and Silvapulle and Burridge (1986). The conditions needed for existence and uniqueness are essentially similar to those for the MLE.

So for large-sample properties the roots are assumed to exist, at least with high probability, in addition to further regularity conditions. See Fahrmeir and Kaufmann (1985).

The large sample consistency and asymptotic normality of the quasi-likelihood estimators are obtained with assumptions about the quasi-score rather than the likelihood function. See Godambe and Heyde (1987).

It is important to point out that in the above section the variance function involves no unknown parameters. The "generalized estimating equations (GEE)" extension of Liang and Zeger (1986) allows for $V_i(.)$ to include a finite dimensional vector $\boldsymbol{\alpha}$ of unknown parameters to be estimated. Specifically, the variance matrix is decomposed into the correlation matrix and a diagonal matrix of variances. The diagonal matrix is a completely known function of μ_i. The correlation matrix is parametrized as a function of $\boldsymbol{\alpha}$. Usually specific patterns of correlation such as exchangeable and autoregressive ones are assumed. Liang and Zeger (1986) showed consistency and asymptotic normality of the GEE estimators. Their arguments require a $\sqrt{N}$ consistent estimator of $\boldsymbol{\alpha}$ which may be obtained by using simple moment estimators and this is in line with the one-step procedure mentioned earlier.

Example 7.4.4: Consider the setup described in Example 7.4.2, where for the sake of simplicity we assume that there is only one explanatory variable, i.e., $q = 1$ (to avoid a degenerate Poisson distribution, we take $x_i \neq 0$). For the canonical link function $g(z) = \log z$, we get $g'(z) = z^{-1}$ and $v_i(\beta) = e^{\beta x_i}$, so that $w_{1i}(\beta) = (e^{-\beta x_i})^{-2}(e^{\beta x_i})^{-1} = e^{\beta x_i}$ and the assumptions required for Theorem 7.4.1 reduce to

i)

$$\lim_{N \to \infty} \frac{1}{N} \sum_{i=1}^{N} e^{\beta x_i} x_i^2 = I(\beta) < \infty \qquad (7.4.39)$$

ii) and for some $K: 0 < K < \infty$, as $N \to \infty$

$$\sup_{|h| \leq K/\sqrt{N}} N^{-1} \sum_{i=1}^{N} |\{e^{(\beta+h)x_i} - e^{\beta x_i}\}x_i^2|$$

$$= N^{-1} \sum_{i=1}^{N} \{e^{(\beta+K/\sqrt{N})x_i} - e^{\beta x_i}\}x_i^2 \to 0. \qquad (7.4.40)$$

If we suppose that the range of the explanatory variable is bounded, then clearly both (7.4.39) and (7.4.40) hold, and it follows that the MLE $\hat{\beta}_N$ of β is such that

$$\sqrt{N}(\hat{\beta}_N - \beta) \xrightarrow{D} N(0, I^{-1}(\beta)).$$

Take, for example, the two-sample location problem, where $x_i = 1$ if the ith sample unit is obtained from the first population and $x_i = -1$, otherwise, $i = 1, \ldots, N$. Let $N_1 = \sum_{i=1}^{N} I_{\{x_i=1\}}$ and assume that as $N \to \infty$, $N_1/N \to \pi$, $0 < \pi < 1$. Then

$$I(\beta) = \lim_{N \to \infty} \left\{ \frac{N_1}{N} e^{\beta} + \frac{(N - N_1)}{N} e^{-\beta} \right\} = \pi e^{\beta} + (1 - \pi)e^{-\beta} < \infty$$

and

$$N^{-1} \sum_{i=1}^{N} \{e^{(\beta+K/\sqrt{N})x_i} - e^{\beta x_i}\}x_i^2$$

$$= \frac{N_1}{N} \{e^{\beta+K/\sqrt{N}} - e^{\beta}\} + \frac{N - N_1}{N} \left\{ e^{-(\beta+K/\sqrt{N})} - e^{-\beta} \right\}$$

which clearly converges to zero as $N \to \infty$, implying that

$$\sqrt{N}(\hat{\beta}_N - \beta) \xrightarrow{D} N(0, [\pi e^{\beta} + (1-\pi)e^{-\beta}]^{-1}).$$

The extension of such result to the multisample location problem is immediate (see Exercise 7.4.3).

If we consider the noncanonical link function $g(z) = z$, a convenient model is given by $\mu_i = e^{\theta_i} = \alpha + \beta x_i$. In such a case we have $g'(z) = 1$, $g''(z) = 0$, $v_i(\beta) = \alpha + \beta x_i$, $h(x) = \log x$ and $b'''(x) = e^x$, implying that $w_{1i}(\beta) = (\alpha + \beta x_i)^{-1}$; also,

$$\mathbf{x}_i \mathbf{x}_i^t = \begin{pmatrix} 1 & x_i \\ x_i & x_i^2 \end{pmatrix},$$

so that defining

$$A(N) = \sup_{\mathbf{h} \in H} \frac{1}{N} \sum_{i=1}^{N} \left\| \left\{ \frac{1}{(\alpha + h_1) + (\beta + h_2)x_i} - \frac{1}{\alpha + \beta x_i} \right\} (\mathbf{x}_i \mathbf{x}_i^t) \right\|$$

where $H = \{\mathbf{h} \in \mathbb{R}^2 : \|\mathbf{h}\| \le K/\sqrt{N}\}$ and

$$B(N) = \mathrm{E}_{\beta} \sup_{\mathbf{h} \in H} \frac{1}{N} \sum_{i=1}^{N} |Y_i| \left\| \left\{ \frac{1}{(\alpha + h_1) + (\beta + h_2)x_i} - \frac{1}{\alpha + \beta x_i} \right\} (\mathbf{x}_i \mathbf{x}_i^t) \right\|$$

the assumptions required for Theorem 7.4.1 reduce to

$$\lim_{N \to \infty} \frac{1}{N} \sum_{i=1}^{N} \frac{1}{\alpha + \beta x_i} \begin{pmatrix} 1 & x_i \\ x_i & x_i^2 \end{pmatrix} = \mathbf{I}(\beta) \quad \text{is finite and p.d.,} \quad (7.4.41)$$

$$\lim_{N \to \infty} \frac{1}{N^2} \sum_{i=1}^{N} \frac{1 + x_i^2}{\alpha + \beta x_i} = 0, \quad (7.4.42)$$

$$A(N) \to 0 \quad \text{as } N \to \infty, \quad (7.4.43)$$

and

$$B(n) \to 0 \quad \text{as } N \to \infty. \quad (7.4.44)$$

Again let us return to the two-sample location problem where, for convenience, we let $x_i = 0$ if the ith sample unit is obtained from the first population and $x_i = 1$, otherwise, $i = 1, \ldots, N$. Let N_1 and π have the same interpretation as before. Then we have

i)

$$\begin{aligned}
\mathbf{I}(\beta) &= \lim_{N \to \infty} \frac{N_1}{N} \frac{1}{\alpha} \begin{pmatrix} 1 & 0 \\ 0 & 0 \end{pmatrix} + \lim_{n \to \infty} \frac{N - N_1}{N} \frac{1}{\alpha + \beta} \begin{pmatrix} 1 & 1 \\ 1 & 1 \end{pmatrix} \\
&= \begin{bmatrix} \pi/\alpha + (1-\pi)/(\alpha + \beta) & (1-\pi)/(\alpha + \beta) \\ (1-\pi)/(\alpha + \beta) & (1-\pi)/(\alpha + \beta) \end{bmatrix}
\end{aligned}$$

is finite and p.d.,

ii)

$$
\lim_{N \to \infty} \frac{1}{N^2} \sum_{i=1}^{N} \frac{1 + x_i^2}{\alpha + \beta x_i}
$$

$$
= \lim_{N \to \infty} \left[\frac{N_1}{N} \left\{ \frac{1}{N\alpha} \right\} + \frac{N - N_1}{N} \left\{ \frac{2}{N(\alpha + \beta)} \right\} \right] = 0,
$$

iii)

$$
A(N) \leq \frac{N_1}{N} \left| \frac{1}{\alpha + K/\sqrt{N}} - \frac{1}{\alpha} \right|
$$

$$
+ \frac{N - N_1}{N} \left| \frac{1}{\alpha + \beta + 2K/\sqrt{N}} - \frac{1}{\alpha + \beta} \right| \to 0 \quad \text{as} \quad N \to \infty,
$$

iv)

$$
B(N) \leq \frac{N_1}{N} \alpha \left| \frac{1}{\alpha + K/\sqrt{N}} - \frac{1}{\alpha} \right|
$$

$$
+ \frac{N - N_1}{N} \left| \frac{1}{\alpha + \beta + 2K/\sqrt{N}} - \frac{1}{\alpha + \beta} \right| \to 0 \quad \text{as} \quad N \to \infty.
$$

Since (7.4.41)–(7.4.44) hold, the asymptotic normality of $\sqrt{N}(\hat{\beta}_N - \beta)$ follows from Theorem 7.4.1. It is important to note that the linear regression model imposes a restriction on the parameter space, induced by the fact that $\alpha + \beta x_i$ must be positive. Thus, the assumption of Theorem 7.4.1 remain valid when β lies in the interior of the (restricted) parameter space. If we allow β to lie on the boundary of this parameter space, the computations in (7.4.17), (7.4.18) and (7.4.26) are to be modified according to the boundary restraints on β, and the resulting equations may be highly complex. In fact, for such boundary points, we have a different solution which may be regarded as analogous to that of the classical restricted MLE case. This is outside the scope of our current treatise, but we like to refer to Barlow et al. (1972) for a very useful account relating to simple models.

As a final remark, we point that even though the generalized linear models approach permits a great flexibility, the associated asymptotic results rely heavily on the underlying exponential family distribution as well as on the choice of the link function. Therefore, we may question the robustness of such procedures with respect to departures from the highly structured models on which they are based, and look for alternative approaches. This topic will be briefly discussed in the next section.

7.5 Generalized least-squares versus generalized estimating equations

As we have seen in Section 7.4, the MLE of β under the generalized linear model $\mathbf{G}_N(\boldsymbol{\mu}_n) = \mathbf{X}_N\boldsymbol{\beta}$ is a solution to the GEE (7.4.21). A convenient way to solve each equations is via the Fisher scoring algorithm briefly introduced in Section 5.2. In particular, for the case under investigation here, the $(\nu+1)^{\text{th}}$ step of the algorithm corresponds to

$$\boldsymbol{\beta}^{(\nu+1)} = \boldsymbol{\beta}^{(\nu)} + \mathbf{I}_N^{-1}(\boldsymbol{\beta}^{(\nu)})\mathbf{U}_N(\boldsymbol{\beta}^{(\nu)}) \tag{7.5.1}$$

where $\mathbf{I}_N(\boldsymbol{\beta}^{(\nu)})$ and $\mathbf{U}_N(\boldsymbol{\beta}^{(\nu)})$ are, respectively, the Fisher information matrix (7.4.27) and the score statistic (7.4.9) evaluated at the estimate $\boldsymbol{\beta}^{(\nu)}$ obtained at the νth step. From (7.5.1) we may write

$$\mathbf{I}_N(\boldsymbol{\beta}^{(\nu)})\boldsymbol{\beta}^{(\nu+1)} = \mathbf{I}_N(\boldsymbol{\beta}^{(\nu)})\boldsymbol{\beta}^{(\nu)} + \mathbf{U}_N(\boldsymbol{\beta}^{(\nu)}). \tag{7.5.2}$$

Now, letting $\omega_i(\boldsymbol{\beta}^{(\nu)}) = [g'\{\mu_i(\boldsymbol{\beta}^{(\nu)})\}]^{-2}\{v_i(\boldsymbol{\beta}^{(\nu)})\}^{-1}$, $i = 1, \ldots N$, we have

$$\mathbf{I}_N(\boldsymbol{\beta}^{(\nu)})\boldsymbol{\beta}^{(\nu)} + \mathbf{U}_N(\boldsymbol{\beta}^{(\nu)}) = \left\{\sum_{i=1}^N \omega_i(\boldsymbol{\beta}^{(\nu)})\mathbf{x}_{Ni}\mathbf{x}_{Ni}^t\right\}\boldsymbol{\beta}^{(\nu)}$$

$$+ \sum_{i=1}^N \omega_i(\boldsymbol{\beta}^{(\nu)})\mathbf{x}_{Ni}\{Y_i - \mu(\boldsymbol{\beta}^{(\nu)})\}g'\{\mu_i(\boldsymbol{\beta}^{(\nu)})\}$$

which, upon considering the first order Taylor expansion,

$$\mathbf{G}_N(\mathbf{Y}_N) \cong \mathbf{G}_N\{\boldsymbol{\mu}_N(\boldsymbol{\beta}^{(\nu)})\} + \dot{G}_N\{\boldsymbol{\mu}_N(\boldsymbol{\beta}^{(\nu)})\}\{\mathbf{Y}_N - \boldsymbol{\mu}_N(\boldsymbol{\beta}^{(\nu)})\} \tag{7.5.3}$$

where

$$\mathbf{Y}_N = (Y_1, \ldots, Y_N)^t$$

and

$$\dot{G}_N\{\boldsymbol{\mu}_N(\boldsymbol{\beta}^{(\nu)})\} = \text{diag}\,[g'\{\mu_1(\boldsymbol{\beta}^{(\nu)})\}, \ldots, g'\{\mu_N(\boldsymbol{\beta}^{(\nu)})\}]$$

may be re-expressed as

$$\mathbf{I}_N(\boldsymbol{\beta}^{(\nu)})\boldsymbol{\beta}^{(\nu)} + \mathbf{U}_N(\boldsymbol{\beta}^{(\nu)})$$

$$= \sum_{i=1}^N \omega_i(\boldsymbol{\beta}^{(\nu)})\mathbf{x}_{Ni}[\mathbf{x}_{Ni}\boldsymbol{\beta} - \{Y_i - \mu_i(\boldsymbol{\beta}^{(\nu)})\}g'\{\mu_i(\boldsymbol{\beta}^{(\nu)})\}]$$

$$= \mathbf{X}_N^t \mathbf{W}_N(\boldsymbol{\beta}^{(\nu)})\mathbf{G}_N(\mathbf{Y}_N) \tag{7.5.4}$$

where $\mathbf{W}_N\{\boldsymbol{\mu}_N(\boldsymbol{\beta}^{(\nu)})\} = \text{diag}\,\{\omega_1(\boldsymbol{\beta}^{(\nu)}), \ldots, \omega_N(\boldsymbol{\beta}^{(\nu)})\}$. Thus, (7.5.2) reduces to the "normal" equations:

$$\mathbf{X}_N^t \mathbf{W}_N\{\boldsymbol{\mu}_N(\boldsymbol{\beta}^{(\nu)})\}\mathbf{X}_N\boldsymbol{\beta}^{(\nu+1)} = \mathbf{X}_N^t \mathbf{W}_N\{\boldsymbol{\mu}_N(\boldsymbol{\beta}^{(\nu)})\}\mathbf{G}_N(\mathbf{Y}_N)$$

which lead to

$$\boldsymbol{\beta}^{(\nu+1)} = [\mathbf{X}_N^t \mathbf{W}_N \{\boldsymbol{\mu}_N(\boldsymbol{\beta}^{(\nu)})\} \mathbf{X}_N]^{-1} \mathbf{X}_N^t \mathbf{W}_N \{\boldsymbol{\mu}_N(\boldsymbol{\beta}^{(\nu)})\} \mathbf{G}_N(\mathbf{Y}_N).$$
$$(7.5.5)$$

Since (7.5.5) has the same form of the GLSE (7.2.51), with weights that are recomputed at each step, the algorithm is known as **iterative reweighted least squares**. The MLE $\widehat{\boldsymbol{\beta}}_N$ is obtained by letting $\nu \to \infty$.

Suppose that consistent [but not necessarily $O_p(n^{-1/2})$] estimators $\widehat{\mu}_i$ of μ_i such that $\text{Var}(\widehat{\mu}_i) = v_i(\mu_i)$, $i = 1, \ldots, N$, are available. Then, writing

$$\widehat{\mathbf{G}}_N = \{g(\widehat{\mu}_1), \ldots, g(\widehat{\mu}_N)\}^t,$$

$$\mathbf{V}_N(\widehat{\boldsymbol{\mu}}_N) = \text{diag}\{v_1(\mu_1), \ldots, v_N(\mu_N)\}$$

and using a first order Taylor expansion similar to (7.5.3) we may consider the model

$$\widehat{\mathbf{G}}_N = \mathbf{X}_N \boldsymbol{\beta} + \varepsilon_N \qquad (7.5.6)$$

where $\varepsilon_N = \dot{\mathbf{G}}_N(\boldsymbol{\mu}_N)(\widehat{\boldsymbol{\mu}}_N - \boldsymbol{\mu}_N) + o_p(1)$ is such that $\text{E}\varepsilon_N = \mathbf{0} + o(1)$ and $\text{Var}\,\varepsilon_N = \mathbf{W}_N^{-1}(\boldsymbol{\mu}_N) + o(1)$ [here we remind the reader to recall that $\mathbf{W}_N^{-1}(\boldsymbol{\mu}_N) = \dot{\mathbf{G}}_N(\boldsymbol{\mu}_N)\mathbf{V}_N(\boldsymbol{\mu}_N)\dot{\mathbf{G}}_N^t(\boldsymbol{\mu})$]. Neglecting the $o(1)$ terms, we may relate (7.5.6) to the model (7.2.37) with $\boldsymbol{\Sigma}_N = \mathbf{W}_N^{-1}(\boldsymbol{\mu}_N)$ and, therefore, consider the GLSE

$$\widehat{\widehat{\boldsymbol{\beta}}}_N = \{\mathbf{X}_N^t \mathbf{W}_N(\widehat{\boldsymbol{\mu}}_N) \mathbf{X}_N\}^{-1} \mathbf{X}_N^t \mathbf{W}_N(\widehat{\boldsymbol{\mu}}_N) \widehat{\mathbf{G}}_N \qquad (7.5.7)$$

as an alternative to the MLE $\widehat{\boldsymbol{\beta}}_N$. Then, in view of (7.2.57) and (7.2.58) we may conclude that $\sqrt{N}(\widehat{\widehat{\boldsymbol{\beta}}}_N - \boldsymbol{\beta})$ has the same asymptotic distribution as $\sqrt{(}\widehat{\boldsymbol{\beta}}_N - \boldsymbol{\beta})$, namely, $N(\mathbf{0}, \mathbf{I}^{-1}(\boldsymbol{\beta}))$. But, (7.2.57)–(7.2.58) may be needed to justify the last result.

This approach is particularly appealing when $\mathbf{Y}_N$ itself is a consistent estimate of $\boldsymbol{\mu}_N$. In such cases, (7.5.7) constitutes the first step of the Fisher scoring algorithm corresponding to (7.5.5). Some further relations between the two approaches will be discussed in the following examples.

Example 7.5.1: Let Y_i, $i = 1, \ldots, s$, be independent r.v.'s with the Poisson (λ_i) distribution and consider the generalized linear model $\lambda_i = N_i h(\mathbf{x}_i^t \boldsymbol{\beta})$, $i = 1, \ldots, s$, where h is a convenient function, $\mathbf{x}_i$ is a q-vector of explanatory variables and N_i is a measure of exposure (i.e., geographic area, volume, etc.) In this context, the parameters $\nu_i = \lambda_i/N_i$, $i = 1, \ldots, s$, may be interpreted as rate parameters at which the events are counted in Y_i per unit of exposure N_i. Under this Poisson regression model, the log-likelihood may be expressed as

$$\log L_N(\boldsymbol{\beta}) = \sum_{i=1}^{s} \{Y_i \log \lambda_i(\boldsymbol{\beta}) - \lambda_i(\boldsymbol{\beta})\} + \text{ constant} \qquad (7.5.8)$$

so that, here, the MLE $\widehat{\beta}_N$ may be obtained as a solution to the GEE (7.4.21) which reduces to

$$\sum_{i=1}^{s} \left\{ \frac{Y_i - \lambda_i(\boldsymbol{\beta})}{\lambda_i(\boldsymbol{\beta})} \right\} \frac{\partial \lambda_i}{\partial \boldsymbol{\beta}}(\boldsymbol{\beta}) = \mathbf{0}. \tag{7.5.9}$$

On the other hand, we may rewrite (7.5.8) as

$$\log L_N(\boldsymbol{\beta}) = \sum_{i=1}^{s} \left\{ Y_i \log \left[1 + \frac{\lambda_i(\boldsymbol{\beta}) - Y_i}{Y_i} \right] - \lambda_i(\boldsymbol{\beta}) \right\} + \text{constant}$$

and consider the expansion

$$\log(1 + x) = x - \frac{1}{2}x^2 + \frac{1}{3}x^3 - \cdots, \quad |x| < 1,$$

to write

$$\log L_N(\boldsymbol{\beta}) = -\sum_{i=1}^{s} \frac{\{Y_i - \lambda_i(\boldsymbol{\beta})\}^2}{Y_i} - \sum_{i=1}^{s} \frac{\{Y_i - \lambda_i(\boldsymbol{\beta})\}^3}{Y_i^2} - \cdots + \text{constant}.$$

Taking $N = \max(N_1, \ldots, N_s)$ and using (7.4.25), we may see that

$$\sum_{i=1}^{s} \frac{\{Y_i - \lambda_i(\boldsymbol{\beta})\}^3}{Y_i^2} = \sum_{i=1}^{s} \frac{\{Y_i - N_i h(\mathbf{x}_i^t \boldsymbol{\beta})\}^3}{Y_i^2} = O_P(N^{-3/2}).$$

Thus, by neglecting the $O_P(N^{-3/2})$ terms, maximization of (7.5.8) will be equivalent to the minimization of $\sum_{i=1}^{s} \{Y_i - \lambda_i(\boldsymbol{\beta})\}^2 / Y_i$, which, in turn, corresponds to solving

$$\sum_{i=1}^{s} \left\{ \frac{Y_i - \lambda_i(\boldsymbol{\beta})}{Y_i} \right\} \frac{\partial \lambda_i(\boldsymbol{\beta})}{\partial \boldsymbol{\beta}} = \mathbf{0}. \tag{7.5.10}$$

If we consider the linear model $h(\mathbf{x}_i^t \boldsymbol{\beta}) = \mathbf{x}_i^t \boldsymbol{\beta}$, (7.5.10) reduces to

$$\sum_{i=1}^{s} N_i \mathbf{x}_i = \sum_{i=1}^{s} \frac{N_i^2}{Y_i} \mathbf{x}_i^t \widehat{\widehat{\boldsymbol{\beta}}}_N \mathbf{x}_i,$$

a solution of which is given by the GLSE

$$\widehat{\widehat{\boldsymbol{\beta}}}_N = (\mathbf{X}^t \mathbf{D_N} \mathbf{D_Y}^{-1} \mathbf{D_N} \mathbf{X})^{-1} \mathbf{X}^t \mathbf{D_N} \mathbf{1}_s$$

where $\mathbf{D_N} = \text{diag}\{N_1, \ldots, N_s\}$ and $\mathbf{D_Y} = \text{diag}\{Y_1, \ldots, Y_s\}$. For nonlinear models, such as $h(\mathbf{x}_i^t \boldsymbol{\beta}) = \exp(\mathbf{x}_i^t \boldsymbol{\beta})$, (7.5.10) are also nonlinear, but may be approximated by the linear equations which define the corresponding GLSE. The reader is referred to Koch et al. (1985) for details. ■

Example 7.5.2: Consider the situation described in Example 7.4.1 in a more general setup where we allow for s (≥ 2) groups and for a q-vector of covariates $\mathbf{x}_i$ associated with each Z_i. Writing $Y_i = \sum_{i=1}^{n} Y_{ij}$ where $Y_{ij} = n^{-1}$ with probability π_i and $Y_{ij} = 0$ with probability $(1 - \pi)$, the GEE (7.4.21) may be expressed as

$$\sum_{i=1}^{s}\sum_{j=1}^{n}\left\{\frac{Y_{ij} - \pi_i(\boldsymbol{\beta})}{\omega_i(\boldsymbol{\beta})}\right\}\mathbf{x}_{N_i} = \mathbf{0} \qquad (7.5.11)$$

where $\omega_i(\boldsymbol{\beta}) = g'\{\pi_i(\boldsymbol{\beta})\}\pi_i(\boldsymbol{\beta})[1 - \pi_i(\boldsymbol{\beta})]$, $i = 1, \ldots, s$. Whenever a consistent preliminary estimate $\widetilde{\boldsymbol{\beta}}$ of $\boldsymbol{\beta}$ is available, we may consider the GLSE, which is a solution to

$$\sum_{i=1}^{s}\sum_{j=1}^{n}\left\{\frac{Y_{ij} - \pi_i(\boldsymbol{\beta})}{\omega_i(\widetilde{\boldsymbol{\beta}})}\right\}\mathbf{x}_{N_i} = \mathbf{0}. \qquad (7.5.12)$$

Now, since

$$\sum_{i=1}^{s}\sum_{j=1}^{n}\left\{\frac{Y_{ij} - \pi_i(\boldsymbol{\beta})}{\omega_i(\widetilde{\boldsymbol{\beta}})}\right\}\mathbf{x}_{N_i} = \sum_{i=1}^{s}\sum_{j=1}^{n}\left\{\frac{Y_{ij} - \pi_i(\boldsymbol{\beta})}{\omega_i(\boldsymbol{\beta})}\right\}\frac{\omega_i(\boldsymbol{\beta})}{\omega_i(\widetilde{\boldsymbol{\beta}})}\mathbf{x}_{N_i}$$

$$= \sum_{i=1}^{s}\sum_{j=1}^{n}\left\{\frac{Y_{ij} - \pi_i(\boldsymbol{\beta})}{\omega_i(\boldsymbol{\beta})}\right\}\mathbf{x}_{N_i}$$

$$+ \sum_{i=1}^{s}\frac{\omega_i(\boldsymbol{\beta}) - \omega_i(\widetilde{\boldsymbol{\beta}})}{\omega_i(\widetilde{\boldsymbol{\beta}})}\sum_{j=1}^{n}\left\{\frac{Y_{ij} - \pi_i(\boldsymbol{\beta})}{\omega_i(\boldsymbol{\beta})}\right\}\mathbf{x}_{N_i},$$

it follows that (7.5.11) and (7.5.12) will be asymptotically equivalent whenever

$$\max_{1\leq i \leq s}\left|\frac{\omega_i(\boldsymbol{\beta}) - \omega_i(\widetilde{\boldsymbol{\beta}})}{\omega_i(\widetilde{\boldsymbol{\beta}})}\right| \xrightarrow{\text{P}} 0.$$

Note that this is true even if $n = 1$, although in such a case it may be difficult to obtain the preliminary estimate $\widetilde{\boldsymbol{\beta}}$. ∎

7.6 Nonparametric regression

It seems clear from the previous sections that in parametric statistical modeling (be it in a classical normal setup or via the generalized linear model approach), the form of the underlying density of the error component plays a vital role in establishing the asymptotic properties of the conventional statistical inference procedures. To a certain extent, the M-methods discussed in Sections 5.4 and 7.3 can accommodate mild violations of the proposed model, but in some practical problems it may be quite unreasonable to make even the less restrictive assumptions required by these procedures. As a simple illustration, consider a set of n independent and identically distributed random vectors (X_i, Y_i), $i = 1, \ldots, n$, and suppose that our primary interest centers on

$$\mu(x) = \mathrm{E}\{Y \mid X = x\}, \quad x \in \mathbb{R}, \tag{7.6.1}$$

where $\mu(x)$ is referred to as the regression function of Y on X. In the spirit of Section 7.2, we may set

$$\mu(x) = \alpha + \beta x, \quad x \in \mathbb{R}, \tag{7.6.2}$$

where α and β are unknown parameters, so that (7.6.1) is completely specified as a function of (α, β). Such a model is, of course, justified when linearity of regression and homoscedasticity, i.e., $\mathrm{Var}(Y \mid x) = \sigma_*^2$ for all x hold. This amounts to adopting the bivariate normal model as the underlying distribution. A departure from the assumed normality may not only distort (7.6.2) but may also affect considerably the homogeneity of the conditional variance. To enhance more generality, we may as well replace the linear model (7.6.2) by a generalized linear model, as described in Section 7.4; this allows us to cover a much broader scope of applications, not only by permitting a nonlinear link between the mean response μ and the explanatory variable X but also by relaxing the homoscedasticity assumption. Even though, such models are still not robust, in the sense that small departures from the assumed form of the conditional density of Y given $X = x$, $f(y \mid x)$, may cause considerable damage to the efficiency of the related statistical inference procedures. Alternative models that are not as dependent on a specific form for the underlying random error distribution have deserved a fair amount of research in the recent past. They can be classified into the following two categories:

i) **Semiparametric models**, where one assumes that the conditional density $f(y \mid x)$ has a parametric structure, yet allowing a nonparametric formulation for the underlying error distribution. More specifically, such models may be expressed as

$$f(y \mid x) = f_0\{y - \beta(x)\}, \quad x \in \mathbb{R}, \quad y \in \mathbb{R}, \tag{7.6.3}$$

where $\beta(x)$ is a function of x of known form, involving a finite number of parameters and f_0 has an arbitrary form. A typical example is the Cox (1972) proportional hazards model frequently employed in the analysis of survival data. Here the hazard at time t for a sample unit with covariate vector $\mathbf{x}$ is given by $h(t, \mathbf{x}) = \lambda(t) \exp(\mathbf{x}^t \boldsymbol{\beta})$, where $\lambda(\cdot)$ is an arbitrary function defined only at the points where the events of interest occur and which does not have any relevance with respect to the estimation of the parameter vector $\boldsymbol{\beta}$. These models may work out better for nonstochastic explanatory variables, as in the case of R-estimators, or the Cox proportional hazards model mentioned earlier.

ii) **Nonparametric regression models**, where no parametric structures are imposed on $f(y \mid x)$. Thus, we may express

$$\mu(x) = \int y f(y \mid x)\, dy, \quad x \in \mathbb{R}, \tag{7.6.4}$$

and then try to draw statistical conclusions about $\mu(x)$ through the
estimation of the density $f(y \mid x)$ in a complete nonparametric fashion.
In this formulation we may even generalize further, by replacing the
conditional mean $\mu(x)$ by comparable functionals of $f(y \mid x)$ [or the
corresponding conditional distribution function $F(y \mid x)$]. In this setup
we allow the explanatory variables to be stochastic as well.

For the semiparametric linear model in the simplest case, where $\beta(x) = \alpha + \beta(x)$ in (7.6.3), the R-estimators briefly introduced in Section 7.3 may
work out quite well; see, for example, Sen (1968b). More general method-
ology, developed during the past twelve years only, involves some novel
counting processes approaches which need a much higher mathematical
level than that intended here. Hence, we shall not go into these details.

In the nonparametric regression problem, the crucial step is the estima-
tion of the density $f(y \mid x)$, $x \in \mathbb{R}$. Note that even in the case of marginals,
the empirical d.f. of the Y_i, say, $F_n(y)$, $y \in \mathbb{R}$, is a step function, so that
the corresponding empirical density function $f_n(y)$ assumes the value zero
at all y for which $y \neq Y_i$ for some $i = 1, \ldots, n$. On the other hand, at each
of these Y_i, $f_n(y) = (d/dy)F_n(y)$ is not defined. The situation is more complicated for the conditional density $f(y \mid x)$. Because of this fundamental
problem, the first requirement of a statistical approach to this problem is to
formulate suitable **smooth** estimators for the density function. In this direction we describe some of such **smoothing techniques** for the problem
of estimating the density function of a random variable at a given point.
Let $Y_1, \ldots, Y_n$ be i.i.d. r.v.'s with density $f(y)$, $y \in \mathbb{R}$, and suppose that
we are interested in estimating $f(a)$, where $a \in \mathbb{R}$ is a continuity point
of f. Let $F_n(y) = n^{-1} \sum_{i=1}^{n} I_{\{Y_i \leq y\}}$ and for a small h (> 0), consider the
estimator

$$\widehat{f}_{n,h}(a) = \frac{1}{2h}\{F_n(a+h) - F_n(a-h)\}. \tag{7.6.5}$$

Recall that

$$(2nh)\widehat{f}_{n,h}(a) = \sum_{i=1}^{n} I_{\{a-h \leq Y_i \leq a+h\}} \sim \mathrm{Bin}(n, \pi_h(a)) \tag{7.6.6}$$

where

$$\pi_n(a) = F(a+h) - F(a-h) = 2hf(a) + o(h). \tag{7.6.7}$$

We may improve the order of the remainder term from $o(h)$ to $o(h^2)$ by
assuming further that $f'(x)$ exists and is finite at $x = a$. Using the Central

Limit Theorem 3.3.1 for the binomial law, whenever

$$n\pi_h(a)\{1 - \pi_h(a)\} \to \infty \text{ with } n \to \infty,$$

i.e., $nh \to \infty$, we obtain

$$n^{-1/2}[\pi_h(a)\{1 - \pi_h(a)\}]^{-1/2}\{2nh\widehat{f}_{n,h}(a) - n\pi_h(a)\} \xrightarrow{\mathcal{D}} N(0,1) \quad (7.6.8)$$

which can be rewritten as

$$\frac{2nh\{\widehat{f}_{n,h}(a) - f(a) - o(1)\}}{\{2nhf(a) + o(nh)\}^{1/2}} \xrightarrow{\mathcal{D}} N(0,1). \quad (7.6.9)$$

This, in turn, is equivalent to saying that

$$(2nh)^{1/2}\{\widehat{f}_{n,h}(a)/f^{1/2}(a) - f^{1/2}(a) - o(1)\}$$

converges in law to a standard normal variate as $n \to \infty$. On the other hand, in this setup $(nh)^{1/2}o(1)$ is not generally $o(1)$, so that the bias term in (7.6.7) may have a significant role. The picture is slightly better when it is possible to assume that $f'(x)$ exists and is continuous at $x = a$, so that $o(h)$ can be replaced by $o(h^2)$. In such a case, to control the bias term we need to choose h $(= h_n)$, such that, as $n \to \infty$, $(nh_n)^{1/2}h_n \nrightarrow \infty$, i.e., $nh_n^3 \nrightarrow \infty$, i.e. $h_n = O(n^{-1/3})$. Further improvements depend on additional smoothness assumptions on the density function. If we assume that $f''(x)$ exists and is finite in a neighborhood of a, then, in (7.6.7), we may replace $o(h)$ by $O(h^3)$, so that we may even take h $(= h_n)$ such that as $n \to \infty$, $(nh_n)^{1/2}h_n \nrightarrow \infty$, i.e., $nh_n^5 = O(1)$ or $h_n = O(n^{-1/5})$, a rate that is generally recommended in practice. In this case, $nh_n \sim n^{4/5}$, so that, on writing

$$\pi_{h_n}(a) = 2h_n f(a) + \frac{1}{3}h_n^3 f''(a) + o(h_n^3) \quad (7.6.10)$$

we have

$$\sqrt{2}n^{2/5}\left\{\widehat{f}_{n,h_n}(a) - f(a) - \frac{1}{6}h_n^2 f''(a)\right\} - o(n^{2/5}h_n^{2/5}) \xrightarrow{\mathcal{D}} N(0, f(a))$$

$$(7.6.11)$$

so that, for $h_n \sim n^{-1/5}$,

$$n^{2/5}\{\widehat{f}_{n,h_n}(a) - f(a)\} \xrightarrow{\mathcal{D}} N(f''(a)/6, f(a)/2). \quad (7.6.12)$$

Although for any practical purposes, both parameters of the limiting distribution must be estimated, the crucial problem here relates to the estimation of the bias term $f''(a)/6$; this may be accomplished either by estimating $f''(a)$ or by choosing $h_n = o(n^{1/5})$, so that $n^{2/5}h_n^2 \to 0$ as $n \to \infty$. In either case, the rate of convergence (i.e., $n^{2/5}$ or less) is slower than the conventional rate of $n^{1/2}$.

A more popular method, known as the **kernel method**, can be described as follows. Choose some known density $K(x)$ possessing some smoothness properties (usually one takes a unimodal symmetric density density such as the normal) and let $\{h_n\}$ be a sequence of non-negative real numbers such that $h_n \downarrow 0$ as $n \to \infty$. Then the required estimator is

$$\hat{f}_{n,K}(a) = \int_R \frac{1}{h_n} K\left(\frac{a-y}{h_n}\right) dF_n(y)$$

$$= \frac{1}{nh_n} \sum_{i=1}^n K\left(\frac{a-Y_i}{h_n}\right)$$

$$= \frac{1}{n} \sum_{i=1}^n Z_{ni}(Y_i; a, h_n) \qquad (7.6.13)$$

where

$$Z_{ni}(Y_i; a, h_n) = \frac{1}{h_n} K\left(\frac{a-Y_i}{h_n}\right), i = 1, \ldots, n,$$

are i.i.d. r.v.'s (in a triangular scheme). Thus, we may use the Central Limit Theorem for triangular arrays (3.3.5), carry out a similar analysis as in the case of (7.6.5) and, thus, obtain the asymptotic normality of the standardized form of $\hat{f}_{n,K}(a)$. The resulting picture is very much comparable to (7.6.12).

We now return to the original nonparametric regression model (7.6.4). Here the situation is somewhat more complex, since to estimate the conditional density $f(a \mid x)$ for given a and x, we need to choose appropriate neighborhoods of both points. Before commenting on the use of the nonparametric methods outlined above, we introduce a third one, known as the **k nearest neighborhood (k-NN) method**, which is specially appealing in this bivariate context. Corresponding to a given x, we consider the distances $|X_i - x|$, $i = 1, \ldots, n$, and denote by $Z_1 \leq \cdots \leq Z_n$ their ordered values. Choose $k (= k_n)$ such that $k_n \sim n^\lambda$ for some $\lambda \in (1/2, 4/5]$. Also, let $O_1, \ldots, O_{k_n}$ be the subscripts of the X_i's corresponding to the smallest k_n Z_i's, so that $Z_j = |X_{O_j} - x|$, $j = 1, \ldots, k_n$. To estimate $\mu(x)$ in (7.6.4) we take the empirical distribution function $F_{k_n}^*(\cdot)$ of the Y_{O_j}, $j = 1, \ldots, k_n$, and consider the estimator

$$\hat{\mu}_n(x) = \int_R y \, dF_{k_n}^*(y) = \frac{1}{k_n} \sum_{j=1}^{k_n} Y_{O_j}. \qquad (7.6.14)$$

As we have noted earlier, under this approach, it is not necessary to choose $\mu(x)$ as the regression function. Based on robustness considerations, it may as well be appropriate to choose a conditional quantile function, say the median, i.e., $\mu(x) = F_{Y|x}^{-1}(1/2) = \inf\{y : F(y \mid x) \geq 1/2\}$ so that we may select the corresponding sample quantile of Y_{O_j}, $j = 1, \ldots, n$, as our estimator. Recall that the Y_{O_j}, $j = 1, \ldots, k_n$, given the Z_j, $j = 1, \ldots, k_n$, are conditionally independent, but not necessarily identically distributed r.v.'s, so that the results of Chapters 3 and 4 can be used to study the asymptotic normality of $k_n^{1/2}\{\hat{\mu}_n(x) - \mu(x)\}$. Here also, the rate of convergence is $k_n^{1/2}(\sim n^{2/5})$ and not $n^{1/2}$. Moreover, we may note that $F_{k_n}^*(y)$, even in a conditional setup, does not estimate $F_{Y|x}(y)$ unbiasedly; the bias is of the order $n^{-2/5}$ and, as such, for $k_n \sim n^{4/5}$, $k_n^{1/2}$ is of the same order as $n^{2/5}$. This makes it necessary to incorporate a small bias correction, or to choose $k_n \sim o(n^{4/5})$. Again, from a robustness point of view, we may argue that $\hat{\mu}_n(x)$ in (7.6.14) may not be preferred to $\tilde{\mu}_n(x) = \inf\{y : F_{k_n}^*(y) \geq 1/2\}$. We may also remark that instead of the $k - NN$ method, we could have used the kernel method with $h_n \sim n^{-1/5}$, and both estimators would have the same asymptotical behavior. For some details on the related asymptotics, we refer to Gangopadhyay and Sen (1990, 1992), where a list of other pertinent references is also given.

Finally, we note that the above discussion was centered on the case of stochastic explanatory variables; in the nonstochastic case, the appropriate treatment reduces to grouping the observations according to the values of the explanatory variables and applying the density estimation techniques presented above to each of these groups.

7.7 Concluding notes

The asymptotic theory for linear models may be focused in an unified manner to cover the cases of weighted least squares and generalized least squares as well as of some robust procedures. Along with the general asymptotic results developed in Section 5.2, this methodology provides the foundations for the large sample theory of generalized linear models. In this setup, the generalized (weighted) least-squares method plays a central role, especially for categorical data, given their popularization by Grizzle, Starmer and Koch (1969). These authors applied generalized least-squares methods to many of the models usually considered by the generalized linear models approach. Since, in both cases, we must rely on their asymptotic properties for inferential purposes, the analysis of the relationship between the two methodologies requires some attention. In this context, it is proper for us to clarify a subtle point of difference between the two approaches when viewed from an asymptotic perspective. In the generalized least-squares theory for categorical data models, we have an asymptotic situation comparable to that of (7.4.25) where the sample size N is usually fixed (and not large) but the observations themselves correspond to large subsamples. On the other hand, the general thrust on the asymptotics of generalized linear models is mainly on N being large. We hope that the examples and exercises set in this context will contribute to the clarification of this issue.

Although nonlinear statistical models have deserved a good deal of attention in the past few years, the developments so far have not yet crossed the fence of pure academics. In order to adopt such methodology in applied research, one may need sample sizes so large that they do not lie within limits of practical consideration. The generalized linear models or the robust procedures discussed in this chapter are generally applicable for moderately large sample sizes and have good efficiency properties, too. The nonparametric regression methods examined in Section 7.6 require somewhat larger sample sizes, but still are far better than completely arbitrary nonlinear models. **Spline methodology** is an important omission on our part, but, again, its inclusion would require a much higher level of mathematical sophistication than that for which we aimed. The reader seeking more information on this topic is referred to the excellent expository monograph by Thompson and Tapia (1990). We are also somewhat skeptical about the genuine prospects of the so-called semiparametric models in applied research. As with parametric models, robustness may be a main issue, in the sense that even a small departure from an assumed semiparametric model may lead to a serious bias in the large sample context.

7.8 Exercises

Exercise 7.2.1: Under the general linear model (7.1.1) show that $\widehat{\beta}_n - \beta \xrightarrow{\text{a.s.}} \mathbf{0}$.

Exercise 7.2.2: Consider the general linear model (7.1.1) and S_n^2 defined by (7.2.17). Show that $S_n^2 \xrightarrow{\text{a.s.}} \sigma^2$.

Exercise 7.2.3: Show that (7.2.21) implies (7.2.22).

Exercise 7.2.4: Consider the simple regression model (7.1.2) and assume further that $\overline{x}_n \to 0$ as $n \to \infty$ and that $\sigma_i^2 \propto |x_i - \overline{x}_n|^a$ for some $a \geq 0$, $1 \leq i \leq n$. Verify (7.2.21) under appropriate conditions on the x_i, $1 \leq i \leq n$. (For $a = 0$, 1 or 2, we may end up with the LSE, ratio estimator and regression estimator formulations.)

Exercise 7.2.5: Consider the estimator $\check{\theta}_n^*$ in (7.2.44). Let $\boldsymbol{\Sigma}_n = \mathbf{I}_2 \otimes \text{diag}\{\sigma_1^2, \ldots, \sigma_m^2\}$ and $\widehat{\boldsymbol{\Sigma}}_n = \mathbf{I}_2 \otimes \text{diag}\{\widehat{\gamma}_1, \ldots, \widehat{\gamma}_m\}$; show that (7.2.58) does not hold.

Exercise 7.2.6 (Continued): Under a setup similar to that of Example 7.2.2, consider a model where for some k (≥ 1), $Y_{(m-1)k+1}, \ldots, Y_{mk}$ are i.i.d. r.v.'s with a $N(\theta, \sigma_m^2)$ distribution for $m = 1, \ldots, n^*$ and $n = mn^*$. Verify that (7.2.58) and (7.2.59) hold whenever k is large and n^* is fixed.

Exercise 7.3.1: For the location model, i.e., $\mathbf{x}_{ni} = 1$, $i = 1, \ldots, n$, show that $\widehat{\beta}_n^*$ in (7.3.4) reduces to the sample median.

Exercise 7.3.2: For the Theil-Sen estimator $\widehat{\beta}_{TS}$, show that $n - 1 \leq$ cardinality(S_n) $\leq \binom{n}{2}$. When are the lower and upper bounds attained?

Exercise 7.3.3: Show that the rhs of both inequalities in (7.3.8) converge to zero as $n \to \infty$.

Exercise 7.3.4: Use (7.3.9) to show that the Theil-Sen estimator $\widehat{\beta}_{TS}$ (conveniently standardized) is asymptotically normal.

Exercise 7.3.5: Use (7.3.11) to show that the statistic $W_n(a)$ is nonincreasing in a.

Exercise 7.3.6: Show that W_n defined in (7.3.10) is such that

$$W_n = \frac{n-1}{n+1}U_n + \frac{2}{n+1}U_n^*$$

where

$$U_n = \binom{n}{2}^{-1} \sum_{1 \leq i < j \leq n} \text{sign}\,(Y_i + Y_j)$$

and

$$U_n^* = n^{-1} \sum_{i=1}^{n} \text{sign}\,(Y_i).$$

Exercise 7.3.7 (Continued): Use the result of Exercise 7.3.6 to verify the asymptotic normality of W_n.

Exercise 7.3.8 (Continued): Use the result of Exercise 7.3.7 to derive the asymptotic normality of the estimator of the location model parameter α, obtained by equating $W_n(\alpha)$ to zero.

Exercise 7.4.1: For normally distributed response variables, show that (7.4.20) reduce to the classical normal equations.

Exercise 7.4.2: For the logistic regression model (7.4.11) obtain the GEE as given by (7.4.21) and compare it with the minimum logit equations obtained from (6.4.16).

Exercise 7.4.3: Consider an extension of Example 7.4.2 to the case where the number of populations is greater than 2. Obtain the asymptotic distribution of the corresponding parameter vector.

Exercise 7.5.1: Let Z_i, $i = 1, \ldots, s$, be independent r.v.'s following the Bin (n, π_i) distribution and assume that associated to each Z_i there is a q-vector of explanatory r.v.'s $\mathbf{x}_i$. Consider the logistic regression model $\log \pi_i/(1 - \pi_i) = \mathbf{x}_i^t \boldsymbol{\beta}$, $i = 1, \ldots, s$. Obtain the expressions for the iterative reweighted least-squares procedure and for the generalized least-squares estimator of $\boldsymbol{\beta}$, as well as for the variances of the corresponding asymptotic distributions.

Exercise 7.6.1: For (7.6.3), show that if x is a discrete r.v having finitely many mass points, $\beta(x)$ can be treated rather arbitrarily, but the estimators may have slower rate of convergence.

Exercise 7.6.2: Consider the Cox (1972) model with a single explanatory variable x assuming only two values, 0 and 1. Show that $h(t, x)$ relates equivalently to the Lehmann (1953) model $\overline{G} \equiv \overline{F}^\lambda$.

Exercise 7.6.3: In (7.6.13), let $K(\cdot)$ be the standard normal density. Derive an expression for the asymptotic bias of $\widehat{f}_{n,K}(a)$.

CHAPTER 8

Invariance Principles in Large Sample Theory

8.1 Introduction

The principal tools in modern probability theory play a significant role in the development of large sample methods for statistical inference. In the previous chapters, attempts have been made to depict this interplay of these two main themes, albeit at an intermediate level. During this discourse, it has been observed from time to time (viz., the last two sections of Chapter 4) that there are certain important cases which may require a more delicate analysis wherein **weak convergence in metric spaces** (or the so-called **weak invariance principles**) may generally provide the key to a unified approach. Much of these developments on invariance principles have taken place only during the past three decades, and a complete and rigorous treatment of this vital area is admittedly beyond the scope of our intended intermediate level of presentation. Nevertheless, we felt that without an introduction to this novel and important area, our treatise would remain somewhat incomplete. For this reason, we present in this chapter some of these basic results, although in a somewhat diluted form to match the level of presentation of this book and to eliminate some of the usual abstractions associated with them. The probability inequalities, laws of large numbers and central limit theorems have been incorporated in diverse modes to exploit fully the methodological aspects of the large sample theory of statistics. The probability inequalities led to the so-called **maximal inequalities** (some of which are discussed in Chapter 2) which play a basic role in the generalization of the concept of **weak convergence** in the real or finite dimensional problem arising typically in statistical inferences for a broad class of **stochastic processes**. Two important members of this class are the so-called **partial sum processes** and **empirical distributional processes**, and these will be considered in Sections 8.3 and 8.4, respectively; the basic idea of weak invariance principles is outlined in Section 8.2. Function(al)s of distribution functions were introduced in Chapters 4 and 5; in Section 8.5 their weak convergence results are presented in an alternative (unified) form based on the results of Sections 8.2 and 8.4. Section 8.6 deals with robust (viz., R- and M-) estimators of location/regression parameters, which were introduced in Sections 5.3, 5.4 and 7.3. The results of Section 8.5 play an important role in this context. The concluding section deals with some **strong invariance principles** and presents an outline of the so-called **embedding of Wiener processes**. These results have important applications in the asymptotic theory of sequential analysis. References to more advanced texts and monographs dealing with these topics are made in the respective contexts.

8.2 Weak invariance principles

A **stochastic process** $X = \{X_t, \ t \in I\}$ is a collection of r.v.'s indexed by the parameter t belonging to an index set I; t is usually termed the **time parameter** and depending on t being discrete or continuous, X is termed a discrete or continuous time parameter stochastic process. As an example, consider first the simple **random walk** model where $I = \{k \geq 0\}$, and

$$X_k = Y_0 + \cdots + Y_k, \quad k = 0, 1, 2, \ldots \qquad (8.2.1)$$

with $Y_0 = 0$ with probability one, and for $k \geq 1$,

$$Y_k = \begin{cases} +1 & \text{with probability} 1/2 \\ -1 & \text{with probability} 1/2; \end{cases} \qquad (8.2.2)$$

all these Y's are also assumed to be mutually independent. Then $X = \{X_t, t = 0, 1, 2, \ldots\}$ is a discrete time parameter stochastic process. As a second example consider the empirical d.f. F_n, defined by (4.1.1), and define the reduced empirical d.f. $\{G_n(t), \ t \in [0, 1]\}$ as in (4.5.1). Define then $Z_n = \{Z_n(t), \ t \in [0, 1]\}$ by letting

$$Z_n(t) = n^{1/2}[G_n(t) - t], \quad 0 \leq t \leq 1. \qquad (8.2.3)$$

Then, for every $n \geq 1$, Z_n is a continuous time parameter stochastic process. In both the examples, at any particular time point t, the r.v. X_t [or $Z_n(t)$] has a discrete distribution. It is possible to introduce some other stochastic processes which have pointwise continuous distributions, and we shall discuss some of these in what follows. In the second example, we have actually a sequence $\{Z_n, n \geq 1\}$ of stochastic processes, where for any fixed $t \in [0, 1]$, as we have observed in Chapter 4, $Z_n(t)$ has asymptotically a normal distribution with zero mean and variance $t(1 - t)$. We have also observed that for different time points, say, s and t (both belonging to $[0, 1]$), $Z_n(s)$ and $Z_n(t)$ are not independent, even asymptotically. On the other hand, the probability law (say, P_n) for Z_n is generated by the n-fold uniform distribution of the $Y_i = F(X_i)$, $i = 1, \ldots, n$, on $[0, 1]^n$, and, hence, it should be possible to define $\{P_n\}$ in a convenient manner and to study its convergence properties, if it has any. Thus, it seems desirable to have some process Z with a probability law, say P, and to show that in a meaningful manner, as $n \to \infty$,

$$\{P_n\} \quad \text{converges weakly to} \quad P. \tag{8.2.4}$$

Recall that for any positive integer m and time points $t_1, \ldots, t_m$ all belonging to I, the vector $\{Z_n(t_1), \ldots, Z_n(t_m)\}$ constitutes a subset of Z_n, and, hence, there exists a projection

$$P_n(t_1, \ldots, t_m) \tag{8.2.5}$$

which pertains to the probability law for $\{Z_n(t_1), \ldots, Z_n(t_m)\}$. In a similar manner, for $\{Z(t_1), \ldots, Z(t_m)\}$ we have the projection $P(t_1, \ldots, t_m)$ for P. Thus, it is quite intuitive to argue that if (8.2.4) holds, then, as $n \to \infty$,

$$\{P_n(t_1, \ldots, t_m)\} \quad \text{converges weakly to} \quad P(t_1, \ldots, t_m). \tag{8.2.6}$$

Moreover, this should be true for all possible choices of $(t_1, \ldots, t_m)$ as well as m. Incidentally, (8.2.5) represents a finite dimensional probability law (for a finite m), and, hence, the weak convergence results of Chapter 3 should be directly applicable. In the literature, this property is termed the **convergence of finite dimensional distributions (f.d.d.)** of $\{Z_n\}$ to Z. This is a necessary condition for (8.2.4) to hold, but is not generally a sufficient one. This technical difficulty may again be explained with the help of the reduced empirical d.f. process in (8.2.3). Let $Y_{n:1} < \cdots < Y_{n:n}$ be the ordered r.v's corresponding to $Y_i = F(X_i)$, $1 \leq i \leq n$. Then, $Z_n = \{Z_n(t), 0 \leq t \leq 1\}$ is completely specified by the n-dimensional vector $\{Z_n(Y_{n:i}), \ldots, Z_n(Y_{n:n})\}$. There are two features to note in this context. First, the $Y_{n:i}$ are themselves r.v's and not fixed. Second, as n increases, so does the cardinality of this set, so that for any fixed m and $t_1, \ldots, t_m$, we may not be able to approximate this n-dimensional joint distribution law by an m-dimensional one when $n \to \infty$. The set of points $\{Y_{n:1}, \ldots, Y_{n:n}\}$, however, becomes **dense** in $[0, 1]$ in a meaningful way, and, hence, if m is chosen large enough and $t_{i+1} - t_i$ small enough, uniformly in i ($\leq m$), $t_0 = 0$ and $t_{m+1} = 1$, then the convergence of f.d.d.'s may actually ensure (8.2.4) provided $\{Z_n\}$ is sufficiently approximable by a continuous function. This, in turn, depends on how Z_n is actually defined and what type of **compactness** conditions it satisfies. Motivated by these remarks, first, let us try to define suitable $\{Z_n\}$, and then introduce suitable compactness or tightness conditions which along with (8.2.6) will ensure (8.2.4). In this context, we may as well note that, as in (8.2.3), the process Z_n may not have continuous paths. In fact, in (8.2.3), $Z_n(t)$ has jump discontinuities at each $Y_{n:i}$, $1 \leq i \leq n$. Thus, the sample paths of Z_n are continuous from the right, have left hand-limits, but admit jumps of magnitude $n^{-1/2}$ at each of the n time points $Y_{n:1} < \cdots < Y_{n:n}$. For this reason, we need to define suitable **spaces** of functions on I, which can be used to verify (8.2.4) under suitable regularity conditions.

Let $C[0, 1]$ be the space of all continuous functions $f = \{f(t), \ t \in [0, 1]\}$ on $[0, 1]$. We may define the **modulus of continuity** $w_f(\delta)$ as

$$w_f(\delta) = \sup\{|f(t) - f(s)| : \ 0 \le s < t \le s + \delta \le 1\}, \quad \delta > 0, \qquad (8.2.7)$$

so that $f \in C[0, 1]$ iff

$$w_f(\delta) \to 0 \quad \text{as} \quad \delta \downarrow 0. \qquad (8.2.8)$$

Since $[0, 1]$ is a compact interval and $f \in C[0, 1]$ is uniformly continuous [by (8.2.8)], it follows that

$$\|f\| = \sup\{|f(t)| : \ 0 \le t \le 1\} < \infty. \qquad (8.2.9)$$

Perhaps, it would be convenient to introduce at this stage the **uniform metric** $\rho(\cdot)$ as

$$\rho(f_1, f_2) = \|f_1 - f_2\| = \sup\{|f_1(t) - f_2(t)| : \ 0 \le t \le 1\} \qquad (8.2.10)$$

[the topology specified by $\rho(\cdot)$ is termed the **uniform topology**].

In our context, $f = \{f(t), \ 0 \le t \le 1\}$ is generally a stochastic function. If P stands for the probability law (measure) (on $C[0, 1]$) associated with f, we say that P is **tight** or **relatively compact**, if for every $\varepsilon > 0$, there exists a compact K_ε (depending on ε), such that

$$P\{f \in K_\varepsilon\} > 1 - \varepsilon. \qquad (8.2.11)$$

In view of (8.2.8), (8.2.9) and (8.2.11), we may formalize the tightness concept in the following:

Definition 8.2.1: *A sequence $\{X_n = [X_n(t), \ t \in [0, 1]]\}$ of stochastic processes with probability measures $\{P_n\}$ (on $C[0, 1]$) satisfies the tightness axiom iff*

i) for every $\varepsilon > 0$, there exists an $M_\varepsilon < \infty$, such that

$$P_n\{\|X_n\| > M_\varepsilon\} \le \varepsilon, \quad \forall n \ge n_o, \qquad (8.2.12)$$

ii) for every $\varepsilon > 0$,

$$\lim_{\delta \downarrow 0}(\limsup_{n \to \infty} P_n\{w_{X_n}(\delta) > \varepsilon\}) = 0. \qquad (8.2.13)$$

Let us now go back to (8.2.3), and in the same fashion, we consider a sequence $\{Z_n = [Z_n(t), \ t \in [0, 1]]\}$ of stochastic processes (on $C[0, 1]$). Let $\{P_n\}$ be a sequence of probability laws (measures) on $C[0, 1]$ (induced by $\{Z_n\}$). Similarly, let Z be another random element of $C[0, 1]$ with a probability law (measure) P. [As in Chapter 3, $\{P_n\}$ and P are not necessarily defined on the same measure space.] Then, we have the following:

Theorem 8.2.1: *If $\{P_n\}$ is tight and (8.2.6) holds whenever $t_1, \ldots, t_m$ all belong to $[0,1]$ $(m \geq 1)$, then (8.2.4) holds (i.e., $\{P_n\} \Rightarrow P$).*

We may refer to Billingsley (1968) for an excellent treatise of this weak convergence result (along with others). With this, we may formalize the definition of weak convergence in a general metric space S as follows.

Definition 8.2.2: *A sequence $\{P_n, n \geq 1\}$ of probability measures (on S) is said to converge weakly to a measure P (denoted by $P_n \Rightarrow P$) if*

$$\int g \, dP_n \to \int g \, dP \quad as \quad n \to \infty, \tag{8.2.14}$$

for all $g(\cdot)$ belonging to the space $C(S)$, the class of all bounded, continuous, real functions on S.

This definition implies that for every continuous functional $h(Z_n)$ of Z_n,

$$P_n \Rightarrow P \quad \Longrightarrow \quad h(Z_n) \xrightarrow{\mathcal{D}} h(Z). \tag{8.2.15}$$

This last result is of immense statistical importance, and we will term it the **weak invariance principle** for $\{P_n\}$ (or $\{Z_n\}$). This definition is not confined only to the $C[0,1]$ space, and more general metric spaces can readily be accommodated by formalizing the concept of tightness in the respective context and then rephrasing Theorem 8.2.1 accordingly. This was the main theme in Parthasarathy (1967) and Billingsley (1968), the two most notable texts in this novel area, and we like to draw the attention of readers to this supplementary material.

The space $C[0,1]$ can be trivially enlarged to the space $C[0,T]$ for any finite T; even the space $C[0,\infty)$ can be included in this setup by a simple modification of the uniform metric $\rho(\cdot)$ in (8.2.10); see, for example, Sen (1981, p. 23). Let us consider now another important metric space, namely, $D[0,1]$, which includes $C[0,1]$ as a subspace and is rich enough to cover most of the important statistical applications we have in mind.

Definition 8.2.3: *$D[0,1]$ is the space of all $f = \{f(t),\ t \in [0,1]\}$ such that $f(t)$ is right-continuous and has a left-hand limit i.e., $f(t) = \lim_{\delta \downarrow 0} f(t+\delta)$ exists for every $0 \leq t < 1$, $f(t-0) = \lim_{\delta \downarrow 0} f(t-\delta)$ exists for every $0 < t \leq 1$, but $f(t)$ and $f(t-0)$ may not agree everywhere. Thus, f has only discontinuities of the first kind. A function $f \in D[0,1]$ has a jump at $t \in [0,1]$ if $|f(t) - f(t-0)| > 0$.*

Recall that for the $C[0,1]$ space, $f(t) = f(t-0)$ for every $t \in [0,1]$, so that $C[0,1]$ is a subspace of $D[0,1]$. If is easy to verify that $\{Z_n\}$ in (8.2.3) belongs to $D[0,1]$ but not to $C[0,1]$. In view of possible jumps (and jump-points) in the $D[0,1]$ space, we need to modify the definitions in (8.2.7) and (8.2.10). Compared to (8.2.10), we introduce the **Skorokhod metric**

$$d_S(f_1, f_2) = \inf_{\varepsilon > 0} \{\|\lambda(t) - t\| < \varepsilon, \ \|f_1(t) - f_2(\lambda(t))\| < \varepsilon\}, \qquad (8.2.16)$$

where $\lambda = \{\lambda(t), \ t \in [0,1]\} \in \Lambda$, the class of all strictly increasing, continuous mappings $\lambda \colon [0,1] \to [0,1]$. (This metric generates the **Skorokhod (J_1-) topology** on $D[0,1]$.) For a function $f \in D[0,1]$, the definition of tightness in (8.2.11) is valid, but, in view of the Skorokhod metric in (8.2.16) [replacing the uniform metric in (8.2.10)], in the formulation of a compact K_ε, we may need to replace the modulus of continuity $\omega_f(\delta)$ in (8.2.7) by the following:

$$\omega'_f(\delta) = \sup\{\min[|f(u) - f(s)|, |f(t) - f(u)|] : 0 \le s \le u \le t \le s + \delta \le 1\}, \qquad (8.2.17)$$

which takes into account the possible jump discontinuities of $f(\in D[0,1])$. It is easy to verify that for every $0 < \delta < 1/2$ and $f \in D[0,1]$,

$$\omega'_f(\delta) \le \omega_f(2\delta), \qquad (8.2.18)$$

so that the definition of compactness of $\{X_n\}$ ($\in C[0,1]$) in (8.2.12)–(8.2.13) goes through for $D[0,1]$-valued functions as well; we may replace, $\omega_{X_n}(\delta)$ by $\omega'_{X_n}(\delta)$ in (8.2.13). In practice, often this may not be that crucial. As such, Theorem 8.2.1 and (8.2.14) hold for the $D[0,1]$ space too, so that we may use (8.2.15) with advantage for plausible statistical applications. The inequality in (8.2.18) and Theorem 8.2.1 ensure that if $\{Z_n\}$ does not belong to the $C[0,1]$ space (belongs to $D[0,1]$), we may still use (8.2.6), (8.2.12) and (8.2.13) to ensure the weak convergence of $\{P_n\}$ to P. The crux of the problem is, therefore, to verify (8.2.12)–(8.2.13) along with (8.2.6).

As in the case with $C[0,1]$, the space $D[0,1]$ can also be readily extended to $D[0,T]$, for an arbitrary T $(< \infty)$. Even $D[0,\infty)$ can be covered with a modification of the Skorokhod metric in (8.2.16).

8.3   Weak convergence of partial sum processes

Let us consider first the following simple model which includes the random walk model in (8.2.1)–(8.2.2) as a particular case. Let $\{X_i, \ i \ge 1\}$ be a sequence of i.i.d. r.v.'s with finite first and second moments, and without any loss of generality, we let $\mu = EX = 0$ and $\sigma^2 = EX^2 = 1$. Also, conventionally, we let $X_0 = 0$. Let then

$$S_k = \sum_{j \le k} X_j, \quad k = 0, 1, 2, \ldots . \qquad (8.3.1)$$

Define $\lfloor a \rfloor = \max\{k \colon k \le a\}$ and for each $n(\ge 1)$,

$$W_n(t) = n^{-1/2} S_{\lfloor nt \rfloor}, \quad 0 \le t \le 1. \qquad (8.3.2)$$

Note that

$$W_n(k/n) = n^{-1/2}S_k, \quad k = 0, 1, 2, \ldots, n, \tag{8.3.3}$$

and

$$W_n(t) = W_n(k/n), \quad \frac{k}{n} \le t < \frac{k+1}{n}, \quad k \ge 0. \tag{8.3.4}$$

Thus, for each n, $W_n = \{W_n(t), \ t \in [0,1]\}$ belongs to the $D[0,1]$ space (with the jump points k/n, $1 \le k \le n$). By virtue of (8.3.1) and (8.3.3), we note that

$$EW_n(k/n) = 0, \quad EW_n^2(k/n) = k/n \tag{8.3.5}$$

and that, for $q \ge k$,

$$
\begin{aligned}
E[W_n(k/n)W_n(q/n)] &= n^{-1}E(S_k S_q) \\
&= n^{-1}E[S_k\{S_k + (S_q - S_k)\}] \\
&= n^{-1}[ES_k^2 + 0] \\
&= k/n. \tag{8.3.6}
\end{aligned}
$$

Therefore, by (8.3.4), (8.3.5) and (8.3.6), we have

$$EW_n(t) = 0 \quad \text{and} \quad E[W_n(s)W_n(t)] = n^{-1}\lfloor n(s \wedge t)\rfloor, \tag{8.3.7}$$

So, we obtain that for every $s, t \in [0, 1]$

$$\lim_{n \to \infty} E[W_n(t)W_n(s)] = \lim_{n \to \infty} \text{Cov}[W_n(t), W_n(s)] = s \wedge t \tag{8.3.8}$$

Now, for every (fixed) t, $0 < t \le 1$, $W_n(t) = n^{-1/2}S_{\lfloor nt \rfloor}$ involves a sum of $\lfloor nt \rfloor$ i.i.d. r.v.'s where, as $n \to \infty$, $\lfloor nt \rfloor \to \infty$ and $n^{-1}\lfloor nt \rfloor \to t$. Hence, by the Central Limit Theorem 3.3.1, as $n \to \infty$,

$$W_n(t) \xrightarrow{\mathcal{D}} N(0, t), \quad t \in (0, 1]. \tag{8.3.9}$$

Let us next consider two points t_1, t_2, $0 < t_1 < t_2 \le 1$, and take $k_{n_1} = \lfloor nt_1 \rfloor$ and $k_{n_2} = \lfloor nt_2 \rfloor$. Then, note that

$$W_n(t_2) - W_n(t_1) = n^{-1/2}\sum_{i=k_{n_1}+1}^{k_{n_2}} X_i \tag{8.3.10}$$

is independent of $W_n(t_1)$, and further, by (8.3.6), that as $n \to \infty$

$$E\{[W_n(t_2) - W_n(t_1)]^2\} = (k_{n_2} - k_{n_1})/n \to (t_2 - t_1). \tag{8.3.11}$$

In general, for any (fixed) $m \ (\ge 1)$ and $0 < t_1 < \cdots < t_m \le 1$, by the same decomposition as in (8.3.10), it follows that

$$W_n(t_j) - W_n(t_{j-1}), \quad 1 \le j \le m, \tag{8.3.12}$$

with $t_0 = 1$, are mutually stochastically independent, and as in (8.3.11), for each j $(= 1, \ldots, m)$, as $n \to \infty$

$$E\{[W_n(t_j) - W_n(t_{j-1})]^2\} = n^{-1}\{\lfloor nt_j \rfloor - \lfloor nt_{j-1} \rfloor\} \to (t_j - t_{j-1}). \quad (8.3.13)$$

As such, we may consider the m-vector

$$[W_n(t_1), \ldots, W_n(t_m)]^t \quad (8.3.14)$$

and writing $U_{nj} = W_n(t_j) - W_n(t_{j-1})$, $1 \le j \le m$, we rewrite (8.3.14) as

$$(U_{n1}, U_{n1} + U_{n2}, \ldots, U_{n1} + \cdots + U_{nm})^t, \quad (8.3.15)$$

where the U_{nj} are mutually independent and asymptotically normal with 0 mean and variance $t_j - t_{j-1}$, for $j = 1, \ldots, m$. Thus, (8.3.14) has asymptotically a multinormal law with null mean vector and dispersion matrix $((t_j \wedge t_{j'}))$. Let us now consider a Gaussian random function $W = \{W(t), t \ge 0\}$ having the following properties:

i) The process has independent and homogeneous increments.

ii) For every $s < t$, $W(t) - W(s)$ is Gaussian with 0 mean and variance $t - s$.

Then, W is termed a standard **Brownian motion** or **Wiener process** on $\mathbb{R}^+ = [0, \infty)$. Note that by virtue of (i) and (ii), we have

$$\begin{aligned} E\{W(t)W(s)\} &= E\{W^2(s) + W(s)[W(t) - W(s)]\} \\ &= EW^2(s) = s, \quad s \le t. \end{aligned}$$

Also, by virtue of (i) and (ii), for every m (≥ 1) and $t_1 < \cdots < t_m$, the vector $[W(t_1), \ldots, W(t_m)]^t$ has the m-variate normal distribution with null mean vector and dispersion matrix $((t_j \wedge t_{j'}))$. Also, since for every $t > s$, $W(t) - W(s)$ is $N(0, t - s)$, we can argue that the sample paths of W are continuous, with probability one. By an abuse of notation, we denote by $W = \{W(t), t \in [0, 1]\}$, the standard Brownian motion process on $[0, 1]$. Then W belongs to the $C[0, 1]$ space, with probability one. Moreover, for every positive integer k and $t > s$, by virtue of (ii)

$$E|W(t) - W(s)|^{2k} = \frac{(2k)!}{2^k k!}(t - s)^k, \quad (8.3.16)$$

so that by using (8.3.16) with $k = 2$, in a version of the Billingsley (1968) Inequality (with continuous time parameter), it follows that the Gaussian probability measure (on $C[0, 1]$) induced by W is tight.

By the discussion following (8.3.15) and the above characterization of W, it follows that for every (fixed) m (≥ 1) and $0 < t_1 < \cdots < t_m \le 1$, as $n \to \infty$,

$$[W_n(t_1), \ldots, W_n(t_m)] \xrightarrow{\mathcal{D}} [W(t_1), \ldots, W(t_m)] \quad (8.3.17)$$

and $W_n(0) = W(0) = 0$, with probability one. Thus, (8.2.6) holds. There-fore, to verify Theorem 8.2.1, in this case, we need to verify the two condi-tions in (8.2.12) and (8.2.13) for $\{W_n\}$. Recall that

$$\|W_n\| = n^{-1/2} \max_{1 \leq k \leq n} |S_k|, \tag{8.3.18}$$

where the S_k, $k > 0$, form a zero mean martingale sequence. Hence, by the Kolmogorov Maximal Inequality (Theorem 2.4.1), for every K (> 0)

$$P\{\|W_n\| > K\} \leq K^{-2} n^{-1} E(S_n^2) = K^{-2} < \varepsilon, \tag{8.3.19}$$

by choosing K adequately large. Thus, (8.2.12) holds. Next, we may note that, if for every $0 < \delta \leq \delta_0$, $n \geq n_0$,

$$P\{ \sup_{0 \leq u \leq \delta} |W_n(s + u) - W_n(s)| > \varepsilon\} < \eta \delta \tag{8.3.20}$$

for every $s \in [0, 1)$, where $\varepsilon > 0$ and $\eta > 0$ are arbitrary, then (8.2.13) holds for $\{W_n\}$. Thus, if the X_i in (8.3.1) have a finite fourth order moment, we may again use the Kolmogorov Maximal Inequality for the (submartingale) $[W_n(s + u) - W_n(s)]^2$, $0 \leq u \leq \delta$, and verify that (8.3.20) holds. So that, by Theorem 8.2.1, we may conclude that as $n \to \infty$

$$W_n \xrightarrow{D} W, \quad \text{on} \quad D[0, 1]; \tag{8.3.21}$$

or, equivalently, writing $P_n = P(W_n)$, $n \geq n_0$,

$$P_n \Rightarrow P = P(W). \tag{8.3.22}$$

The fourth moment condition, as assumed above, is a sufficient but not necessary condition; (8.3.20) holds even under the finiteness of the sec-ond moment. To verify this, we may need the following inequality [due to Brown (1971b)], which is presented without proof.

Theorem 8.3.1. Let $\{X_n, \ n \geq 1\}$ be a submartingale. Then, for every $t > 0$,

$$P\{ \max_{1 \leq k \leq n} |X_k| > 2t\} \leq t^{-1} E[|X_n| I_{\{|X_n| \geq t\}}]. \tag{8.3.23}$$

For any s (≥ 0), let $a = \lfloor ns \rfloor$ and $a + b = \lfloor n(s + \delta) \rfloor$. Then

$$\sup\{|W_n(s + u) - W_n(s)| : 0 \leq u \leq \delta\}$$
$$= n^{-1/2} \max_{1 \leq k \leq b} |X_{a+1} + \cdots + X_{a+k}|$$
$$\overset{D}{=} n^{-1/2} \max_{1 \leq k \leq b} |S_k|, \tag{8.3.24}$$

where $\overset{D}{=}$ stands for the equality of distributions, and where the $|S_k|$, $k \geq 0$, form a submartingale sequence. Thus, by (8.3.23) and (8.3.24), we have for every $\varepsilon > 0$,

$$P\{\sup_{0 \le u \le \delta} |W_n(s+u) - W_n(s)| > \varepsilon\}$$

$$= P\{\max_{1 \le k \le b} |S_k| > \varepsilon\sqrt{n}\}$$

$$\le \frac{2}{\varepsilon\sqrt{n}} E[|S_b| I_{\{|S_b| \ge \varepsilon\sqrt{n}/2\}}]$$

$$\le \frac{2}{\varepsilon\sqrt{n}} \sqrt{E(S_b^2)} [P\{|S_b| \ge \varepsilon\sqrt{n}/2\}]^{1/2}. \qquad (8.3.25)$$

Note that $n^{-1}E(S_b^2) \le \delta$, whereas for any $\delta > 0$, as $n \to \infty$,

$$n^{-1/2}S_b \sim N(0, \delta), \qquad (8.3.26)$$

so that as $n \to \infty$ we have

$$P\left\{|S_b| \ge \frac{1}{2}\varepsilon\sqrt{n}\right\} = P\left\{n^{-1/2}\frac{|S_b|}{\sqrt{\delta}} \ge \frac{\varepsilon}{2\sqrt{\delta}}\right\}$$

$$\cong 2\left[1 - \Phi\left(\frac{\varepsilon}{2\sqrt{\delta}}\right)\right], \qquad (8.3.27)$$

where $\Phi(x)$ stands for the d.f. of a $N(0,1)$ variable. Recall that as $x \to \infty$, $1 - \Phi(x) \cong (2\pi)^{-1/2} x^{-1} e^{-1/2x^2}$, so that for every (fixed) $\varepsilon > 0$, choosing δ (> 0) sufficiently small (so that $\varepsilon/2\sqrt{\delta}$ is large), we may take the right-hand side of (8.3.25), for large n, as

$$\frac{4\delta^{3/4}\varepsilon^{-3/2}}{(2\pi)^{1/4}} e^{-\varepsilon^2/16\delta} < \eta\delta, \quad \delta \le \delta_0, \qquad (8.3.28)$$

as $4\delta^{-1/4}\varepsilon^{-3/2}e^{-\varepsilon^2/16\delta} \to 0$ as $\delta \downarrow 0$. Therefore, (8.3.20) holds under the classical second moment condition (needed for the Central Limit Theorem), and, hence, (8.3.21) holds under the same condition as in Theorem 3.3.1.

The weak convergence result in (8.3.21) is not confined only to the simplest case treated in (8.3.1)–(8.3.4). This holds under the full generality of the Lindeberg-Feller Central Limit Theorem (3.3.3). The formulation is quite simple and is presented below.

Let $\{X_i, i \ge 1\}$ be a sequence of independent r.v.'s with $\mu_i = EX_i$ and $\sigma_i^2 = \text{Var}(X_i)$ $(< \infty)$, $i \ge 1$. Let $X_0 = \mu_0 = 0$, and for every $k \ge 1$, let

$$S_k^0 = \sum_{j \le k}(X_j - \mu_j), \, s_k^2 = \sum_{j \le k}\sigma_j^2. \qquad (8.3.29)$$

For every n (≥ 1), define $W_n = \{W_n(t), t \in [0,1]\}$ by letting

$$W_n(t) = s_n^{-1}S_{k_n(t)}^0, \quad 0 \le t \le 1, \qquad (8.3.30)$$

$$k_n(t) = \max\{k : s_k^2 \le ts_n^2\}, \quad 0 \le t \le 1. \qquad (8.3.31)$$

Then W_n belongs to the $D[0,1]$ space. Assume that the classical Lindeberg condition holds, i.e., for every positive ε, as $n \to \infty$,

$$\frac{1}{s_n^2} \sum_{i=1}^{n} E[(X_i - \mu_i)^2 I_{\{|X_i - \mu_i| > \varepsilon s_n\}}] \to 0. \tag{8.3.32}$$

The computations leading to (8.3.8) and (8.3.13) are routine in this setup too, whereas the Lindeberg-Feller Central Limit Theorem (3.3.3) ensures (8.3.9) as well as its extension to the f.d.d.'s in (8.3.14). The tightness part of $\{W_n\}$ follows from the martingale property of the S_k^0 and, hence, we arrive at the following:

Theorem 8.3.2: *Under the same hypotheses of the Lindeberg-Feller Theorem (3.3.3), for $\{W_n\}$ defined by (8.3.30)–(8.3.31), the weak convergence result in (8.3.21) holds.*

Because of the implications in (8.2.15) [i.e., for arbitrary continuous $f(\cdot)$], Theorem 8.3.2 may also be termed a functional central limit theorem. Let us illustrate the scope of such a functional central limit theorem by some examples.

Example 8.3.1 (CLT for random sample size): As in (8.3.1) consider a sequence $\{X_i,\ i \geq 1\}$ of i.i.d. r.v.'s whose mean μ and variance σ^2 $(< \infty)$ are unknown. Based on a sample of size n (i.e., $X_1, \ldots, X_n$), conventional estimators of μ and σ^2 are respectively

$$\overline{X}_n = n^{-1} \sum_{i=1}^{n} X_i \quad \text{and} \quad S_n^2 = \frac{1}{n-1} \sum_{i=1}^{n} (X_i - \overline{X}_n)^2. \tag{8.3.33}$$

In a variety of situations, the sample size n is not fixed a priori, but is based on a **stopping rule**, i.e., $n \, (= N)$ is a positive integer value r.v., such that for every $n \geq 1$, the event $[N = n]$ depends on $(X_1, \ldots, X_n)$. For example, suppose that one wants to find a confidence interval for μ (based on $\overline{X}_n, S_n^2$) such that (a) the coverage probability is equal to some fixed $1-\alpha$ and (b) the width of the interval is bounded from above by some (fixed) $2d, d > 0$. For this problem, no fixed-sample size solution exists [Dantzig (1940)] and one may have to consider two- (or multi-) stage procedures [Stein (1945)] or some sequential procedures [Chow and Robbins (1965)]. In the Stein (1945) two-stage procedure, an initial sample of size n_0 (≥ 2) is incorporated in the definition of the stopping number N as

$$N_d = \max\{n_0, \lfloor t_{\alpha/2}^2 d^{-2} S_{n_0}^2 \rfloor + 1\}, \tag{8.3.34}$$

where $t_{\alpha/2}$ is the upper $50\alpha\%$ percentile point of a Student t-distribution with $n_0 - 1$ degrees of freedom, whereas in the Chow-Robbins (1965) sequential case, we have

$$N_d = \min\{k : k \geq n_0 \quad \text{and} \quad S_k^2 \leq d^2 k / a_k^2\}, \qquad (8.3.35)$$

where n_0 (≥ 2) is the starting sample size and $a_k \to \tau_{\alpha/2}$, the upper $50\alpha\%$ percentile point of the standard normal d.f. (as $k \to \infty$); it may as well be taken equal to $\tau_{\alpha/2}$.

If σ^2 were known and d (> 0) is small, by virtue of the Central Limit Theorem 3.3.1, a prefixed sample size (n_d) would have led to the desired solution, where

$$n_d = d^{-2} \sigma^2 \tau_{\alpha/2}^2. \qquad (8.3.36)$$

By (2.3.87), $S_n^2 \xrightarrow{\text{a.s.}} \sigma^2$, whenever $\sigma^2 < \infty$. Hence, comparing (8.3.35) and (8.3.36), we note that

$$N_d / n_d \xrightarrow{\text{a.s.}} 1 \quad \text{as} \quad d \downarrow 0. \qquad (8.3.37)$$

On the other hand, $\{N_d\}$ is stochastic, whereas $\{n_d\}$ is not. Moreover, as $d \downarrow 0$, by Theorem 3.3.1,

$$n_d^{1/2} (\overline{X}_{n_d} - \mu)/\sigma \sim N(0, 1).$$

Thus, it may be of interest to see whether by virtue of (8.3.37), n_d can be replaced by N_d, when $d \to 0$. The answer is in the affirmative. Note that

$$N_d^{1/2} \frac{\overline{X}_{N_d} - \mu}{S_{N_d}} = \left(\frac{\sigma}{S_{N_d}}\right) N_d^{-1/2} \left(\frac{S_{N_d}^0}{\sigma}\right)$$

$$= \left(\frac{\sigma}{S_{N_d}}\right) \left(\frac{n_d}{N_d}\right)^{1/2} W_{n_d}(N_d/n_d), \qquad (8.3.38)$$

where the S_k^0 and $W_n(k/n)$ are defined as in (8.3.29) and (8.3.30). Recall that by the tightness property of the probability measure $\{P_{n_d}\}$ induced by $\{W_{n_d}\}$, for every $\varepsilon > 0$ and $\eta > 0$, there exist a $\delta_0 < 1$ and a sample size n_0, such that for $n \geq n_0$

$$P\{|W_n(t) - W_n(1)| > \varepsilon \text{ for some } t : |t - 1| < \delta\} < \eta. \qquad (8.3.39)$$

Since $N_d / n_d \xrightarrow{\text{P}} 1$ and $\sigma / S_{N_d} \xrightarrow{\text{P}} 1$, as $d \downarrow 0$, by (8.3.38) and (8.3.39), we have for $d \downarrow 0$,

$$N_d^{1/2} (\overline{X}_{N_d} - \mu)/S_{N_d} \overset{\text{P}}{\sim} W_{n_d}(1) \sim N(0, 1). \qquad (8.3.40)$$

A similar result holds for (8.3.34) whenever n_0 [$= n_0(d)$] increases with $d \downarrow 0$. Let us now formalize the asymptotic normality result in (8.3.40) in a general form. Define $\{\overline{X}_n, n \geq 1\}$ as in (8.3.33) and let N_n be a positive integer-valued r.v. indexed by n, a nonstochastic positive integer such that

$$N_n / n \xrightarrow{\text{P}} C \ (> 0). \qquad (8.3.41)$$

Then, whenever the Central Limit Theorem holds for $\{X_n\}$, it holds for $\{\overline{X}_{N_n}\}$ as well. The key to this solution is given by (8.3.39) which is a by-product of the tightness part of the weak convergence of $\{W_n\}$ to W (which holds under no extra conditions on the Central Limit Theorem). In this way, we do not need to verify the Anscombe (1952) condition

$$P\{\max_{m:|m-n|<\delta n} \sqrt{n}|\overline{X}_m - \overline{X}_n| > \varepsilon\} < \eta \qquad (8.3.42)$$

for $n \geq n_0(\varepsilon, \eta)$ and $\delta \leq \delta_0$. In fact, the Anscombe condition is itself a by-product of the tightness part of $\{W_n\}$. Actually, there are even stronger results relating to the weak convergence of $\{W_{N_n}\}$ to W, for which we may refer to Billingsley (1968). ∎

Example 8.3.2 [Repeated significance test (RST)]: Consider the same model as in Example 8.3.1, and suppose that one may want to test for

$$H_0 : \mu = \mu_0 = 0 \quad \text{vs.} \quad H_1 : \mu \neq 0. \qquad (8.3.43)$$

In order that the test has a good power (when μ is not too far away from the null value), one generally requires that the sample size n be large. In clinical trials/medical studies and other areas, often, the sample units are not all available at the same time point, and, hence, their collection may demand a **waiting time**. In this context, **interim analyses** are sometimes made on the accumulating data set whereby either periodically over time a test is made for H_0 vs. H_1 in (8.3.43) or after each observation is available, the test statistic is updated. We may refer to Armitage (1975) for an excellent statistical account of RST in medical studies. In either way, the target sample size n provides an upper bound for the actual sample size, and we have a sequence $\{n_1, \ldots, n_k = n\}$ of increasing sample sizes, such that, at each n_j, the test statistic is used to draw statistical conclusions; the number k may be fixed in advance or may even be equal to n (i.e., $n_j = j$, $1 \leq j \leq n$). Also, if periodically over time is the prescription, then the n_j may also be stochastic. For the model (8.3.43), if σ is assumed to be given, based on a sample of size m, an appropriate test statistic is

$$T_m^* = m^{1/2}\overline{X}_m/\sigma. \qquad (8.3.44)$$

Recall that $\{T_m, \ m \leq n\}$ contains a set of test statistics which are not independent. Even if they were independent, making multiple tests, each one at a level of significance, say, α $(0 < \alpha < 1)$, can easily make the actual overall level of significance quite large (compared to α). Thus, the basic issue is how to control the overall significance level in this RST scheme. Such a scheme may also be regarded as a version of a Truncated Sequential Probability Ratio Test (TSPRT) with truncation at $N = n$. However, there are generally some subtle differences in the boundaries relating to the

RST and TSPRT schemes. Nevertheless, we may motivate the large sample theory in a unified manner as follows.

Define W_n as in (8.3.30)–(8.3.31) with $\mu_j = 0$, $\sigma_j = \sigma$, for all $j \geq 1$. Plot $W_n(t)$ against $t \in [0, 1]$ as indicated in Figure 8.3.1. Then we may describe the RST in terms of W_n as follows. First, consider the RST in the form of a TSPRT, resulting in horizontal boundaries. For a two-sided alternative [as in (8.3.43)], we take the test statistic as

$$Z_n^{(1)} = \max_{0 \leq k \leq n} |W_n(k/n)| = \sup_{0 \leq t \leq 1} |W_n(t)|, \qquad (8.3.45)$$

so that the stopping time T_1 is given by

$$T_1 = [\min\{k \geq 1 : |W_n(k/n)| \geq C_{n\alpha}^{(1)}\}] \wedge n \qquad (8.3.46)$$

where $C_{n\alpha}$ specifies the boundary, such that

$$P\{Z_n^{(1)} \geq C_{n\alpha}^{(1)} \mid H_0\} = \alpha = \quad \text{significance level.} \qquad (8.3.47)$$

At this stage, we appeal to the weak convergence result in (8.3.21), so that for large n, we may approximate $C_{n\alpha}^{(1)}$ by $C_\alpha^{(1)}$, where

$$P\{ \sup_{0 \leq t \leq 1} |W(t)| \geq C_\alpha^{(1)} \} = \alpha. \qquad (8.3.48)$$

There are some well-known results on the fluctuations of Wiener processes which may be used in this context. For example, we know that for every $\lambda > 0$

$$P\{ \sup_{0 \leq t \leq T} |W(t)| < \lambda \}$$
$$= \sum_{k=-\infty}^{+\infty} (-1)^k [\Phi\{(2k + 1)\lambda/\sqrt{T}\} - \Phi\{(2k - 1)\lambda/\sqrt{T}\}],$$

$$(8.3.49)$$

where $\Phi(\cdot)$ is the standard normal d.f.; for the one-sided case, we have a much simpler expression. For all $\lambda \geq 0$,

$$P\{ \sup_{0 \leq t \leq T} |W(t)| < \lambda \} = 1 - 2[1 - \Phi(\lambda/\sqrt{T})]. \qquad (8.3.50)$$

Thus, for the one-sided case, $C_\alpha^{(1)} = \tau_{\alpha/2}$, whereas for the two-sided case, $C_\alpha^{(1)} \leq \tau_{\alpha/4}$, and the approximate equality sign holds for α not too large. There are some related results for drifted Wiener processes [i.e., for $W(t) + \mu(t)$, with linear $\mu(t)$] [viz., Anderson (1960)] which may even be used to study local asymptotic power of the test based on $Z_n^{(1)}$.

Let us next consider the case of RST based on the sequence in (8.3.44).

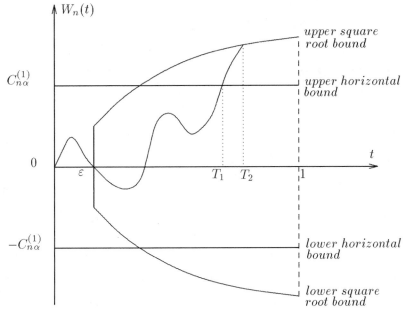

$T_i = $ stopping time for W_n according to the ith type bound, $i = 1, 2$
$\varepsilon \, (> 0) = $ lower truncation point for square root bound

Figure 8.3.1. *Stopping times for repeated significance tests*

Recall that by definition, for $m \leq n$,

$$T_m^* = W_n(m/n)\{(m/n)^{-1/2}\}. \qquad (8.3.51)$$

So, we need to use a square root boundary as has been suggested in Figure 8.3.1. There is a technical point which merits some consideration at this stage. Note that (8.3.21) does not ensure that

$$\{W_n(t)/q(t),\ 0 \leq t \leq 1\} \xrightarrow{\mathcal{D}} \{W(t)/q(t),\ 0 \leq t \leq 1\} \qquad (8.3.52)$$

for every $\{q(t),\ t \in [0,1]\}$. If $q(t)$ is bounded away from 0 (from below), then (8.3.52) holds. But, in (8.3.51), we have $q(t) = t^{1/2}$, so that $q(t) \to 0$ as $t \to \infty$. Also,

$$\int_0^1 q^{-2}(t)d(t) = \int_0^1 t^{-1}dt = \infty.$$

Hence, we are not in a position to use (8.3.21) to conclude on (8.3.52) for this specific $q(t)$. In fact, by the law of iterated logarithm for the Wiener process,

$$\lim_{t \downarrow 0} |W(t)|/\{-2t \log t\}^{1/2} = 1 \quad \text{a.s.} \tag{8.3.53}$$

so that $|W(t)/\sqrt{t}| \xrightarrow{\text{a.s.}} +\infty$ as $t \to 0$. To avoid this technical problem, often a small truncation (at 0) is made, so that one considers the statistic

$$Z_n^{(2)} = \max_{k_0 \leq k \leq n} |T_k^*|, \tag{8.3.54}$$

where

$$k_0/n \to \varepsilon > 0. \tag{8.3.55}$$

Then, we have, under H_0,

$$Z_n^{(2)} \xrightarrow{\mathcal{D}} \sup_{\varepsilon \leq t \leq 1} \{|W(t)|/\sqrt{t}\} \tag{8.3.56}$$

so that the cutoff point $C_{\alpha,\varepsilon}^{(2)}$ is defined by

$$P\{|W(t)/\sqrt{t}| \leq C_{\alpha,\varepsilon}^{(2)}, \ \varepsilon \leq t \leq 1\} = 1 - \alpha. \tag{8.3.57}$$

It is clear that the choice of ε has an important role in this context. The larger the value of ε, the smaller $C_{\alpha,\varepsilon}^{(2)}$ is. The stopping time T_2 is then defined as

$$T_2 = [\min\{k \geq k_0 : \ T_k^* > C_{\alpha,\varepsilon}^{(2)}\}] \wedge n. \tag{8.3.58}$$

For both T_1 and T_2 in (8.3.46) and (8.3.58), the corresponding stopping numbers are giving by nT_1 and nT_2, respectively. Although both the tests have significance level α ($0 < \alpha < 1$), their power properties (and the behavior of the stopping numbers) depend on the kind of alternatives and the choice of ε (> 0).

In some cases, only a (fixed) finite number of tests are made on the accumulating data. For example, for some prefixed k (≥ 2), at sample sizes $n_1, \ldots, n_k$, the statistics $T_{n_1}^*, \ldots, T_{n_k}^*$ are computed, so that the test statistic, similar to in (8.3.54) is given by

$$Z_n^{(3)} = \max_{j=1,\ldots,k} |T_j^*|. \tag{8.3.59}$$

This involves the joint distribution of the vector

$$[W_n(n_1/n)/\sqrt{n_1/n}, \ldots, W_n(n_k/n)/\sqrt{n_k/n}]. \tag{8.3.60}$$

If the n_j/n are prefixed, then the critical level $C_\alpha^{(3)}$ can be obtained by considering the multinormal distribution with null mean vector and dispersion matrix

$$(((n_i/n_{j'})^{-1/2})). \tag{8.3.61}$$

However, for $k \geq 3$, this task can become quite involved. If k is ≥ 8 and the n_i/n are scattered over $(\varepsilon, 1)$ evenly, $C_{\alpha,\varepsilon}^{(2)}$ may provide a reasonably close upper bound for $C_\alpha^{(3)}$. Again, the larger the value of ε, the better this approximation is. ∎

We conclude this section with some additional remarks on the weak convergence of $\{W_n\}$ (to W). In Chapter 3, we have briefly discussed the generalizations of the CLT to both triangular schemes of r.v.'s and some dependent sequences too. Extensions of the weak convergence result in (8.3.21) to such triangular schemes of r.v.'s or to dependent sequences have been worked under very general conditions. The conditions are essentially the same as in the case of the respective CLTs, but instead of requiring them to be true at the sample size n only, we may need that for each $t \in (0, 1]$ they hold for the respective partial sample size $n(t)$. For some details, we may refer to Sen (1981, Chap. 2, 1985) where various nonparametric statistics have been incorporated in the formulation of suitable weak invariance principles.

8.4 Weak convergence of empirical processes

We start with the (reduced) empirical (distribution) process

$$Z_n = \{Z_n(t), \quad 0 \le t \le 1\},$$

defined by (8.2.3). Recall that $G_n(t)$, $0 \le t \le 1$, is a step function having n jumps of magnitude $1/n$ at each of the order statistics $0 < Y_{n:1} < \cdots < Y_{n:n} < 1$. Therefore, Z_n has also n points of discontinuity, namely, the $Y_{n:i}$, $1 \le i \le n$, and at each of these points, the jump is of the magnitude $n^{-1/2}$ (which can be made small when $n \to \infty$). Thus, for every finite n, Z_n belongs to the $D[0, 1]$ space. Also, note that by (8.2.3), we have

$$EZ_n(t) = 0, \quad 0 \le t \le 1, \tag{8.4.1}$$

$$
\begin{aligned}
E[Z_n(s)Z_n(t)] &= E[I_{\{Y_1 \le s\}} - s][I_{\{Y_1 \le t\}} - t] \\
&= E[I_{\{(Y_1 \le s\}} I_{\{Y_1 \le t\}}\}] - st \\
&= \min(s, t) - st, \quad 0 \le s, t \le 1.
\end{aligned}
\tag{8.4.2}
$$

Moreover, for arbitrary m (≥ 1) and $0 \le t_1 < \cdots < t_m \le 1$, consider the vector $[Z_n(t_1), \ldots, Z_n(t_m)]^t = \mathbf{Z}_{nm}$ say. Then, for an arbitrary $\boldsymbol{\lambda} = (\lambda_1, \ldots, \lambda_m)^t$, we have

$$\boldsymbol{\lambda}^t \mathbf{Z}_{nm} = n^{-1/2} \sum_{i=1}^{n} \left(\sum_{j=1}^{m} \lambda_j [I_{\{Y_i \le t_j\}} - t_j] \right) = n^{-1/2} \sum_{i=1}^{n} U_i, \tag{8.4.3}$$

where the U_i are i.i.d. r.v.'s, with $EU_i = 0$ and

$$EU_i^2 = \boldsymbol{\lambda}^t \boldsymbol{\Gamma}_m \boldsymbol{\lambda}, \tag{8.4.4}$$

where $\boldsymbol{\Gamma}_m = ((t_i \wedge t_j - t_i t_j))$ is p.d.. Hence, by the multivariate CLT (viz., Theorem 3.3.9), we claim that as $n \to \infty$,

$$\mathbf{Z}_{nm} \xrightarrow{D} \mathbf{Z}_m = (Z_1, \ldots, Z_m)^t, \qquad (8.4.5)$$

where

$$\mathbf{Z}_m \sim N_m(\mathbf{0}, \boldsymbol{\Gamma}_m). \qquad (8.4.6)$$

In Section 8.3, we have introduced a special Gaussian function W, termed the Brownian motion or the Wiener process. Let us define a stochastic process $Z = \{Z(t),\ t \in [0,1]\}$ by letting

$$Z(t) = W(t) - tW(1), \quad 0 \le t \le 1, \qquad (8.4.7)$$

where $W = \{W(t),\ 0 \le t \le 1\}$ is a standard Wiener process on [0,1]. Then Z is a Gaussian process. Also,

$$EZ(t) = t - t = 0, \quad 0 \le t \le 1; \qquad (8.4.8)$$

$$\begin{aligned} E[Z(s)Z(t)] &= E[W(s)W(t)] + st E[W^2(1)] \\ &\quad - t E[W(s)W(1)] - s E[W(t)W(1)] \\ &= s \wedge t + st - st - st = s \wedge t - st. \end{aligned} \qquad (8.4.9)$$

As such, for arbitrary $m\ (\ge 1)$ and $0 \le t_1 < \cdots < t_m \le 1$, the vector $[Z(t_1), \ldots, Z(t_m)]^t$ follows a $N_m(\mathbf{0}, \boldsymbol{\Gamma}_m)$ distribution with $\boldsymbol{\Gamma}_m = ((t_j \wedge t_{j'} - t_j t_{j'}))$. Thus, by (8.4.5) and (8.4.7)–(8.4.9), the f.d.d. of Z_n converges to those of Z, as $n \to \infty$. Moreover, since W is tight and t belongs to a compact interval $[0, 1]$, Z is also tight. The process $Z = \{Z(t), 0 \le t \le 1\}$ is termed a **Brownian bridge** or a **tied-down Wiener process**. We may remark that $Z(0) = Z(1) = 0$, with probability 1, so that Z is tied down (to 0) at both ends of $[0, 1]$, whereas its perceptible fluctuations are in the open interval $0 < t < 1$, mostly in the central part. This might have been the reason why the terms bridge or tied-down part became associated with Z. We may also reverse the role of Z and W in (8.4.7) in a different representation. Consider the stochastic process $\xi = \{\xi(t),\ 0 \le t < \infty\}$, where

$$\xi(t) = (t+1)Z(t/(t+1)), \quad 0 \le t < \infty. \qquad (8.4.10)$$

Then using (8.4.8), (8.4.9) and (8.4.10), we obtain that

$$E\xi(t) = 0, \quad t \in [0, \infty), \qquad (8.4.11)$$

$$E[\xi(s)\xi(t)] = (s+1)(t+1)E\left[Z\left(\frac{s}{s+1}\right)Z\left(\frac{t}{t+1}\right)\right] = s \wedge t, \qquad (8.4.12)$$

for $s, t \in \mathbb{R}^+ = [0, \infty)$. Hence, ξ is a standard Wiener process on $\mathbb{R}^+$. We shall find this representation very useful in some applications.

Combining (8.4.5) with the above discussion, we notice that the f.d.d.'s of $\{Z_n\}$ converge to those of Z. So, to establish the weak convergence of Z_n to Z, all we need is to verify that $\{Z_n\}$ is tight. Toward this verification, we may note that $G_n(t) = n^{-1} \sum_{i=1}^n I_{\{Y_i \leq t\}}$, $0 \leq t \leq 1$, so that writing for $0 \leq s \leq t \leq 1$,

$$I_{\{Y_i \leq t\}} = I_{\{Y_i \leq s\}} + I_{\{s \leq Y_i \leq t\}}, \quad 1 \leq i \leq n,$$

and noting that

$$\mathrm{E}[I_{\{s \leq Y_i \leq t\}} \mid I_{\{Y_i \leq s\}}] = \begin{cases} 0, & I_{\{Y_i \leq s\}} = 1 \\ (t-s)/(1-s), & I_{\{Y_i \leq s\}} = 0, \end{cases} \qquad (8.4.13)$$

we obtain by some routine steps that for every n (≥ 1),

$$\{Z_n(t)/(1-t), \quad 0 \leq t < 1\} \quad \text{is a martingale.} \qquad (8.4.14)$$

As such, using the Kolmogorov Inequality for martingales, we obtain that for every $\lambda > 0$,

$$P\left\{ \sup_{0 \leq t \leq 1/2} |Z_n(t)| > \lambda \right\} \leq P\left\{ \sup_{0 \leq t \leq 1/2} \frac{|Z_n(t)|}{1-t} > \lambda \right\}$$

$$\leq 4\lambda^{-2} \mathrm{E}Z_n^2(1/2) = \lambda^{-2}. \qquad (8.4.15)$$

Thus, completing the other half by symmetry, we have

$$P\{ \sup_{0 \leq t \leq 1} |Z_n(t)| > \lambda \} \leq 2\lambda^{-2}, \quad \forall \lambda > 0. \qquad (8.4.16)$$

Then choosing λ adequately large, we obtain that (8.2.12) holds for Z_n. To verify (8.2.13), we proceed as in (8.3.20). Recall that for every $s \in (0,1)$ and $\delta > 0$, $s + \delta \leq 1$,

$$|Z_n(s+u) - Z_n(s)| = \left| (1-s-u)\frac{Z_n(s+u)}{(1-s-u)} - (1-s)\frac{Z_n(s)}{1-s} \right|$$

$$\leq \left| \frac{Z_n(s+u)}{1-s-u} - \frac{Z_n(s)}{1-s} \right| + \delta \left| \frac{Z_n(s)}{1-s} \right|, \quad \forall u \leq \delta. \qquad (8.4.17)$$

Therefore,

$$P\{ \sup_{0 \leq u \leq \delta} |Z_n(s+u) - Z_n(s)| > \varepsilon \}$$

$$\leq P\left\{ \sup_{0 \leq u \leq \delta} \left| \frac{Z_n(s+u)}{1-s-u} - \frac{Z_n(s)}{1-s} \right| > \varepsilon/2 \right\}$$

$$+ P\{|Z_n(s)| > (1-s)\varepsilon/2\delta\}. \qquad (8.4.18)$$

By (8.4.16), the second term on the right-hand side of (8.4.18) converges to 0 as $\delta \downarrow 0$. For the first term, we use (8.4.14) along with Theorem 8.3.1, and using the same technique as in (8.3.25)–(8.3.28), we obtain that it converges to 0 as $\delta \downarrow 0$ $(n \to \infty)$. Hence, the tightness of $\{Z_n\}$ holds. Therefore, we conclude that as $n \to \infty$

$$Z_n \xrightarrow{\mathcal{D}} Z. \qquad (8.4.19)$$

Let us exploit this basic weak convergence result in various statistical applications. Before that, we state some of the probability laws for same common functionals of Z which we may use for similar functionals of $\{Z_n\}$. First, we consider the case of

$$D^+ = \sup_{0 \le t \le 1} \{Z(t)\}. \qquad (8.4.20)$$

Note that by virtue of (8.4.10), we have for every $\lambda \ge 0$,

$$
\begin{aligned}
P\{D^+ \ge \lambda\} &= P\{Z(t) \ge \lambda \quad for \ some \quad t \in [0,1]\} \\
&= P\left\{ Z\left(\frac{u}{u+1}\right) \ge \lambda \ for \ some \ u \in [0, \infty) \right\} \\
&= P\left\{ (u+1)Z\left(\frac{u}{u+1}\right) \ge \lambda(u+1) \ for \ some \ u \in \mathbb{R}^+ \right\} \\
&= P\{\xi(t) \ge \lambda + \lambda t \ for \ some \ t \in \mathbb{R}^+\} \\
&= e^{-2\lambda^2}, \qquad (8.4.21)
\end{aligned}
$$

where the final step follows from a well-known result on the Brownian motion ξ (on $\mathbb{R}^+$) [viz. Anderson (1960)]. In a similar manner, it follows on letting

$$D = \sup_{0 \le t \le 1} |Z(t)| \qquad (8.4.22)$$

that for every $\lambda > 0$,

$$P\{D > \lambda\} = 2(e^{-2\lambda^2} - e^{-8\lambda^2} + e^{-18\lambda^2} - \cdots). \qquad (8.4.23)$$

If we consider a simple linear functional

$$L = \int_0^1 g(t)Z(t)dt, \qquad (8.4.24)$$

we immediately obtain that L is normal with 0 mean and

$$\mathrm{Var}(L) = \int_0^1 G^2(t)dt - \left\{ \int_0^1 G(t)dt \right\}^2, \qquad (8.4.25)$$

where

$$G(t) = \int_0^t g(u)du, \quad 0 \le t \le 1, \qquad (8.4.26)$$

and the square integrability of $G(t)$ is a part of the regularity assumptions.

Further, by (8.4.7), we have

$$L = \int_0^1 g(t)W(t)dt - W(1) \int_0^1 tg(t)dt, \qquad (8.4.27)$$

so that, by some simple steps, we have

$$\int_0^1 g(t)W(t)dt \sim N(0, \sigma_g^2), \qquad (8.4.28)$$

with

$$\sigma_g^2 = \int_0^1 G^2(t)dt + G^2(1) - 2G(1) \int_0^1 G(t)dt. \qquad (8.4.29)$$

Let us also consider the statistic

$$D_g^* = \int_0^1 g(t)Z^2(t)dt, \qquad (8.4.30)$$

where the weight function $g(\cdot)$, satisfies the condition that

$$\int_0^1 t(1-t)g(t)dt < \infty. \qquad (8.4.31)$$

In particular, for $g(\cdot) \equiv 1$, D^* has the same distribution as

$$\sum_{j \geq 1} (j^2\pi^2)^{-1} X_j^2, \qquad (8.4.32)$$

where the X_j are i.i.d. r.v's such that

$$X_j \sim N(0, 1). \qquad (8.4.33)$$

In a general setup, we may write

$$D_g^* \overset{D}{=} \sum_{j \geq 1} \lambda_j X_j^2, \qquad (8.4.34)$$

where the real numbers $\{\lambda_j, \ j \geq 1\}$ depend on $g(\cdot)$, the X_j are defined as in (8.4.33) and, under (8.4.31),

$$\sum_{j \geq 1} \lambda_j < \infty. \qquad (8.4.35)$$

Example 8.4.1 (One-sample goodness-of-fit tests): Suppose that $X_1, \ldots, X_n$ are n i.i.d. r.v.'s from a continuous d.f. F, defined on $\mathbb{R}$. We want to test for a null hypothesis $H_0 : F = F_0$, where F_0 has a specified form. Then, by making use of the probability integral transformation $Y = F_0(X)$, we have $Y_1, \ldots, Y_n$, n i.i.d. r.v.'s with a d.f. $G \ (= F \circ F_0^{-1})$, and the null hypothesis reduces to $H_0 : G = \text{Unif}(0,1)$ d.f. Let G_n be the empirical d.f. of the Y_i. Then, the classical Kolmogorov-Smirnov statistics for testing the goodness of fit of the d.f. F_0 are given by

$$D_n^+ = \sup_{0 \le t \le 1} \sqrt{n}[G_n(t) - t] = \sup_{x \in \mathbb{R}} \sqrt{n}[F_n(x) - F_0(x)], \qquad (8.4.36)$$

$$D_n^- = \sup_{0 \le t \le 1} \sqrt{n}[t - G_n(t)] = \sup_{x \in \mathbb{R}} \sqrt{n}[F_0(x) - F_n(x)] \qquad (8.4.37)$$

and

$$D_n = \sup_{0 \le t \le 1} |\sqrt{n}(G_n(t) - t)| = \sup_{x \in \mathbb{R}} |\sqrt{n}\{F_n(x) - F_0(x)\}|. \qquad (8.4.38)$$

Under the null hypothesis, the distribution of D_n^+, D_n^- or D_n is generated by the uniform order statistics, and, hence, it does not depend on F_0. Thus, these statistics are all exactly distribution-free (EDF). For small values of n, one can obtain this distribution by direct enumeration, but it becomes too laborious as $n \to \infty$. On the other hand, noting that

$$D_n^+ = \sup_{0 \le t \le 1} Z_n(t) \quad \text{and} \quad D_n = \sup_{0 \le t \le 1} |Z_n(t)|, \qquad (8.4.39)$$

we may as well use (8.4.19) along with (8.4.20) and (8.4.21) to conclude that D_n^+ (or D_n^-) has asymptotically the same distribution as D^+, and $D_n \xrightarrow{\mathcal{D}} D$. Thus, for large n, one may use (8.4.21) and (8.4.23) to provide a good approximation to the critical values of D_n^+ (D_n^-) and D_n which may be used for making one- and two-sided tests.

An alternative test for this goodness-of-fit problem may rest on the Cramér-von Mises statistic

$$\text{CvM}_n = \int_{-\infty}^{\infty} n[F_n(x) - F_0(x)]^2 dF_0(x) = \int_0^1 Z_n^2(t)dt, \qquad (8.4.40)$$

so that by virtue of (8.4.19), we may again refer to D_g^* in (8.4.30) with $g \equiv 1$, and conclude that the asymptotic distribution of CvM_n is given by the distribution of the statistic

$$V = \sum_{j \ge 1} (\frac{1}{j^2 \pi^2}) U_j, \qquad (8.4.41)$$

where the U_j are i.i.d. r.v.'s, each U_j having the central chi-squared d.f. with 1 degree of freedom. Some tabulations of the percentile points of U are available in the literature. ∎

Example 8.4.2 (Two-sample problem): Consider now two independent samples $X_1, \ldots, X_{n_1}$ and $Y_1, \ldots, Y_{n_2}$ drawn randomly from two populations with d.f.'s F and G, respectively, both defined on $\mathbb{R}$. We assume that both F and G are continuous a.e., and frame the null hypothesis (H_0 : $F \equiv G$) as the equality of F and G, against alternatives that F and G are not identical. This is the classical two-sample nonparametric model; more restrictive models relate to $G(x) = F\{(x - \theta)/\lambda\}$ where θ and λ are the location and scale factors and F is generally treated as unknown. Then the null hypothesis $F \equiv G$ reduces to $\theta = 0$, $\lambda = 1$, and in this setup, treating F as nuisance, one may frame alternatives in terms of θ and/or λ. There are various types of nonparametric tests for such semiparametric models. Although these tests may perform quite well for such shift/scale variation models, they may not be consistent against all alternatives $F \not\equiv G$. For this reason, often one prescribes Kolmogorov-Smirnov type of tests which remain consistent globally. To set the things in the proper perspectives, let

$$F_{n_1}(x) = n_1^{-1} \sum_{i=1}^{n_1} I_{\{X_i \leq x\}}, \quad x \in \mathbb{R}, \tag{8.4.42}$$

$$G_{n_2}(x) = n_2^{-1} \sum_{i=1}^{n_2} I_{\{Y_i \leq x\}}, \quad x \in \mathbb{R}. \tag{8.4.43}$$

Then the two-sample (one- and two-sided) Kolmogorov-Smirnov test statistics are

$$D_{n_1 n_2}^+ = \left(\frac{n_1 n_2}{n_1 + n_2} \right)^{1/2} \sup_x [F_{n_1}(x) - G_{n_2}(x)], \tag{8.4.44}$$

$$D_{n_1 n_2} = \left(\frac{n_1 n_2}{n_1 + n_2} \right)^{1/2} \sup_x |F_{n_1}(x) - G_{n_2}(x)|. \tag{8.4.45}$$

[The normalizing factor $(n_1 n_2/(n_1 + n_2))^{1/2}$ is included to produce proper nondegenerate distributions (under H_0) for large sample sizes.] Note that if the null hypothesis $H_0 : F = G$ holds, then all the $n = n_1 + n_2$ observations $X_1, \ldots, X_{n_1}, Y_1, \ldots, Y_{n_2}$ are from a common d.f. (F), and, hence, their joint distribution remains invariant under any permutation of the coordinates. This leads to a totality of $\binom{n}{n_1}$ $[= \binom{n}{n_2}]$ possible partitioning of the n observations into two subsets of n_1 and n_2 elements, each of which has the same (conditional) probability $\binom{n}{n_1}^{-1}$. On the other hand, $D_{n_1 n_2}^+$ or $D_{n_1 n_2}$ remains invariant under any (strictly) monotone transformation on the X's and Y's. Thus, if we use the (common) probability integral transformation $[Z = F(X)$ or $F(Y)]$, we may reduce these n observations to i.i.d. r.v.'s coming from the Unif(0,1) distribution. Hence, the permutation law stated above can be readily adapted to the n-fold uniform(0,1) d.f., so that these test statistics are EDF (under H_0). This feature may not be true if F and G are not continuous a.e., and also in the multivariate case, they are only permutationally (or conditionally) distribution-free (as the joint distribution of the coordinate-wise probability integral transformed variables depends generally on the unknown joint distribution of the vector).

It is clear that as n_1, n_2 both increase, the totality of such partitions [i.e., $\binom{n}{n_1}$] increases at an alarming rate, and, hence, the exact enumeration of the permutation distribution of $[D^+_{n_1 n_2}]$ or $[D_{n_1 n_2}]$ becomes prohibitively laborious. Fortunately, (8.4.19) can be very effectively used to simplify the large sample properties.

We assume that there exists a number ρ, $0 < \rho < 1$, such that

$$n_1/n \to \rho \quad \text{as} \quad n \to \infty. \tag{8.4.46}$$

Also, without any loss of generality, we assume that both F and G are uniform d.f. over $(0,1)$, i.e., $F(x) \equiv G(x) \equiv x$, $0 \le x \le 1$. As in (8.2.3), we define the two-sample reduced empirical processes by

$$Z^{(1)}_{n_1} = \{Z^{(1)}_{n_1}(t) = n_1^{1/2}[F_{n_1}(t) - t], \quad 0 \le t \le 1\}$$

and

$$Z^{(2)}_{n_2} = \{Z^{(2)}_{n_2}(t) = n_2^{1/2}[G_{n_2}(t) - t], \quad 0 \le t \le 1\},$$

respectively. Also, let

$$Z^*_n = \{Z^*_n(t), \quad 0 \le t \le 1\} \tag{8.4.47}$$

where

$$Z^*_n(t) = \left(\frac{n_2}{n}\right)^{1/2} Z^{(1)}_{n_1}(t) - \left(\frac{n_1}{n}\right)^{1/2} Z^{(2)}_{n_2}(t), \quad 0 \le t \le 1.$$

Then, by (8.4.44), (8.4.45) and (8.4.47), we obtain that

$$D^+_{n_1 n_2} = \sup_{0 \le t \le 1} Z^*_n(t); \quad D_{n_1 n_2} = \sup_{0 \le t \le 1} |Z^*_n(t)|. \tag{8.4.48}$$

Thus, if we are able to show that as $n \to \infty$,

$$Z^*_n \xrightarrow{D} Z \quad \text{(Brownian Bridge)}, \tag{8.4.49}$$

then (8.4.21) and (8.4.23) become adaptable here. For this, note that each of the $Z^{(i)}_{n_i}$ converges weakly to a Brownian bridge, say, Z_i, $i = 1, 2$, where $Z^{(1)}_{n_1}$ and $Z^{(2)}_{n_2}$ are independent and, hence, Z_1 and Z_2 are independent too. Moreover, for each $Z^{(i)}_{n_i}$, the tightness property follows as in the discussion before (8.4.19). Therefore, Z^*_n, being a linear function of $Z^{(1)}_{n_1}$ and $Z^{(2)}_{n_2}$, has also the tightness property. Thus, it suffices to show that the f.d.d.'s of $\{Z^*_n\}$ converge to those of a Brownian motion Z. Note that for an arbitrary $m (\ge 1)$ and $0 \le t_1 < \cdots < t_m \le 1$,

$$[Z^*_n(t_1), \ldots, Z^*_n(t_m)] = \sqrt{\frac{n_2}{n}} [Z^{(1)}_{n_1}(t_1), \ldots, Z^{(1)}_{n_1}(t_m)]$$

$$- \sqrt{\frac{n_1}{n}} [Z^{(2)}_{n_2}(t_1), \ldots, Z^{(2)}_{n_2}(t_m)], \tag{8.4.50}$$

so that using (8.4.5) for each of the two vectors on the right-hand side (along with their independence) we obtain that (8.4.5) holds for Z_*^n as well. Thus, (8.4.51) holds, and our task is accomplished. In this connection, note that in (8.4.46), we assumed that as $n \to \infty$, $n_1/n \to \rho$, $0 < \rho < 1$. If n_1/n converges to 0 or 1 as $n \to \infty$, then in (8.4.47) one of the terms drops out and the weak convergence result in (8.4.49) still holds if $\min(n_1, n_2) \to \infty$. However, $n_1 n_2/(n_1 + n_2)$ becomes $o(n_1 + n_2) = o(n)$, when $\rho \to 0$ or 1, and, hence, the convergence rate becomes slower. For this reason, it is desirable to limit ρ to the central part of $(0,1)$, i.e., to make n_1 and n_2 of comparable magnitudes. We may also consider the two-sample Cramér-von Mises' statistic

$$\int_{-\infty}^{\infty} \frac{n_1 n_2}{n_1 + n_2}[F_{n_1}(x) - G_{n_2}(x)]^2 dH_n(x) \qquad (8.4.51)$$

where $H_n(x) = (n_1/n)F_{n_1}(x) + (n_2/n)G_{n_2}(x)$, $x \in \mathbb{R}$. Again noting that under $H_0 : F = G$, H_n converges a.s. to F and proceeding as before, we may conclude that the Cramér-von Mises statistic in (8.4.51) converges in law to

$$\int_0^1 Z^2(t)dt, \qquad (8.4.52)$$

where $Z(t), t \in [0,1]$, is a Brownian bridge, so that as in (8.4.49) and (8.4.51), we may use the same distribution as that of V in (8.4.41).

Some other nonparametric tests for $H_0 : F = G$ are based on some specific functionals for which simpler asymptotic theory is available. For example, consider the two-sample Wilcoxon-Mann-Whitney statistic

$$W_{n_1 n_2} = n_1^{-1} \sum_{i=1}^{n_1} R_i - \frac{n+1}{2} \qquad (8.4.53)$$

where

$$R_i = \text{rank of } X_i \quad \text{among the } n \text{ observations}, \quad 1 \le i \le n_1. \qquad (8.4.54)$$

Note that by definition of the ranks,

$$n^{-1}R_i = H_n(X_i), \quad i = 1, \ldots, n_1, \qquad (8.4.55)$$

so that we have

$$(n+1)^{-1}W_{n_1 n_2} = \frac{n}{n+1}\frac{1}{n_1}\sum_{i=1}^{n_1} H_n(X_i) - \frac{1}{2}$$

$$= \frac{n}{n+1}\int H_n(x)dF_{n_1}(x) - \frac{1}{2}$$

$$= \frac{n}{n+1}\left\{\frac{n_1}{n}\int F_{n_1}(x)dF_{n_1}(x) + \frac{n_2}{n}\int G_{n_2}(x)dF_{n_1}(x)\right\} - \frac{1}{2}$$

$$
\begin{aligned}
&= \frac{n}{n+1}\left\{\frac{n_1}{n}\frac{1}{n_1}\frac{n_1(n_1+1)}{2n_1} + \frac{n_2}{n}\int G_{n_2}(x)dF_{n_1}(x) - F(x)\right\} - \frac{1}{2} \\
&= \frac{n_1+1}{2(n+1)} + \frac{n_2}{n+1}\int G_{n_2}(x)dF_{n_1}(x) - \frac{1}{2} \\
&= \frac{n_2}{n+1}\int G_{n_2}(x)dF_{n_1}(x) - \frac{n_2}{2(n+1)} \\
&= \frac{n_2}{n+1}\left\{\int G_{n_2}(x)dF_{n_1}(x) - \int F(x)dF(x)\right\}.
\end{aligned}
\tag{8.4.56}
$$

Thus, writing $G_{n_2} = F + (G_{n_2} - F)$ and $F_{n_1} = F + (F_{n_1} - F)$ we have from (8.4.56)

$$
\begin{aligned}
\frac{n^{1/2}}{n+1}W_{n_1 n_2} &= \frac{n_2}{n+1}\left\{\int F(x)d\{\sqrt{n}[F_{n_1}(x) - F(x)]\}\right\} \\
&\quad + \int \sqrt{n}\left\{G_{n_2}(x) - F(x)\right\}dF(x) \\
&\quad + \int \sqrt{n}\left[G_{n_2}(x) - F(x)\right]d\left\{F_{n_1}(x) - F(x)\right\}
\end{aligned}
\tag{8.4.57}
$$

where $\sqrt{n}\|G_{n_2} - F\| = O_p(1)$ and $\|F_{n_1} - F\| = o_p(1)$. Hence, the last term on the rhs of (8.4.57) is $o_p(1)$, whereas by partial integration of the first term, we have

$$
\begin{aligned}
\frac{n^{1/2}}{n+1}W_{n_1 n_2} &= \frac{n_2}{(n+1)}\left\{\int \sqrt{n}\{G_{n_2}(x) - F(x)\}dF(x)\right. \\
&\quad \left. - \int \sqrt{n}\left\{F_{n_1}(x) - F(x)\right\}dF(x)\right\} + o_p(1)
\end{aligned}
\tag{8.4.58}
$$

and, as such, we may appeal to the definition of $Z_{n_1}^{(1)}$ and $Z_{n_2}^{(2)}$ presented before (8.4.47), so that $n^{-1/2}W_{n_1 n_2}$ is distributed as a linear functional of a difference between two independent Brownian bridges, and then

$$
n^{-1/2}W_{n_1 n_2} \xrightarrow{\mathcal{D}} N(0, [12\rho(1-\rho)]^{-1})
\tag{8.4.59}
$$

where ρ is defined by (8.4.46). This method of attack based on the weak convergence of $Z_{n_1}^{(1)}$ and $Z_{n_2}^{(2)}$ works also neatly when $F \not\equiv G$ and also for general linear rank statistics involving some score function $a_n(k)$, $1 \le k \le n$, where

$$
a_n(k) \sim \psi(k/(n+1)), \quad 1 \le k \le n,
\tag{8.4.60}
$$

and ψ is a smooth function having finitely many points of discontinuity. In (8.4.56), we may need to replace $H_n(x)$ by $\psi\{\frac{n}{n+1}H_n(x)\}$, and a Taylor's expansion takes care of the situation. In this way, the weak invariance principles play a basic role in the asymptotic theory of nonparametric statistics. We shall discuss more about this in Section 8.6. ∎

Example 8.4.3 (Life-testing model): Consider the same two-sample model as in Example 8.4.2, but suppose now that the X_i and Y_j are the failure times (non-negative r.v.'s), so that F and G are both defined on $\mathbb{R}^+$. Consider a very simple case where the X_i's are the lifetimes of electric lamps of a particular brand (say, A) and the Y_i's, for a second brand (say, B). If we put all these n $(= n_1 + n_2)$ lamps to life testing at a common point of time (say, 0), then the successive failures occur sequentially over time, so that at any time point t (> 0), the observable random elements relate to

  i) actual **failure times** of the lamps if the failure times are $\leq t$,

 ii) operating status for other lamps which have not failed before t (i.e., the **censoring** event).

Thus, to collect the entire set of failure times, one may need to wait until all the failures have taken place. Often, based on time and cost considerations, it may not be possible to continue the experimentation all the way to the end, and one may need to curtail the study at an intermediate point (say, t^*). This relates to a **truncated** scheme. Note that even if t^* is fixed in advance, r_{t^*}, the number of failures occurring before t^* is a random variable. Alternatively, one may curtail the study after a certain number (say r_n) of failures have occurred. In this way, r_n may be fixed in advance, but T^*, the time period, becomes stochastic; such a scheme is termed a **censoring** one. In either case, the Kolmogorov-Smirnov and the Cramér-von Mises tests discussed in (8.4.48), (8.4.51) and (8.4.52) can be adapted by truncating the range $\mathbb{R}^+$ to $(0, t^*]$ or $(0, T^*]$. In the censoring scheme, r_n is nonstochastic, and, hence, such censored tests are all EDF under H_0. On the other hand, in the truncation scheme, these truncated tests are only conditionally EDF (given r_{t^*}) under H_0. However, in the large sample case, under H_0, $n^{-1} r_{t^*} \xrightarrow{\text{a.s.}} F(t^*)$, so that letting $r_n = nF(t^*)$, both the schemes can be studied in a common manner. Basically, they rest on the weak convergence of

$$Z_{n_1}^{(1)0} = \{Z_{n_1}^{(1)}(t), \quad 0 \leq t \leq F(t^*) = p^*\}$$

and

$$Z_{n_2}^{(2)0} = \{Z_{n_2}^{(2)}(t), \quad 0 \leq t \leq p^*\}$$

to $Z^0 = \{Z(t), 0 \leq t \leq p^*\}$, which is a direct corollary to the general result in the earlier example. ∎

In view of the accumulating nature of the data set, it is not uncommon to adapt a **time-sequential** scheme in such a life-testing model. In this scheme, one may want to monitor the study from the beginning with the objective that if at any early point of time there is a significant difference between the two-sample responses, the study may be curtailed at that time along with the rejection of the null hypothesis. This formulation is quite important in clinical trials/medical studies, where patients may be switched on to the better treatment if a significant treatment difference is detected. One of the nice properties of the Kolmogorov-Smirnov and the Cramér-von Mises tests is that once the critical values are determined by reference to a single truncation/censoring point, the tests are automatically adaptable to a time-sequential setup. For example, $|F_{n_1}(x) - G_{n_2}(x)|$ may be sequentially (in $x \geq 0$) monitored and if at any x, for the first time, this difference exceeds the critical level of $D^o_{n_1 n_2}$ ($= \sup_{x \leq t^*} |F_{n_1}(x) - G_{n_2}(x)|$), we stop the study along with the rejection of H_0. The distribution of the stopping time, in an asymptotic setup, may thus be related to the distribution of exit times for a Brownian bridge over an appropriate interval $J \subset [0,1]$. This explains the importance of the weak invariance principles in asymptotic theory of statistical inference. With rank statistics, the picture is somewhat more complicated as the form of the functional may itself depend on the truncation point. Nevertheless, the key solution is provided by the weak convergence results studied in Section 8.3 and 8.4. We provide a brief discussion of this important topic at the end of next section.

8.5 Weak convergence and statistical functionals

In Chapters 4 and 5, we observed that in a nonparametric setup, often, a parameter θ is expressed as a functional $\theta(F)$ of the underlying d.f. F. For example, **estimable parameters** or **regular functionals** in the Hoeffding (1948) sense are of the form

$$\int \cdots \int \psi(x_1, \ldots, x_m) dF(x_1) \cdots dF(x_m), \quad F \in \mathcal{F}, \qquad (8.5.1)$$

where m (≥ 1) is a finite positive integer and $\psi(\cdot)$ is a **kernel** (of specified form). In such a case, it may be quite natural to replace the unknown d.f. F by the sample d.f. F_n and estimate θ by $\theta(F_n)$, the **von Mises'** **functional**; unbiased estimators, i.e., **U-statistics** were also considered in Chapter 5, and they have some optimal properties too. There are some other functionals $\theta(F)$ which may not be expressible in the form of (8.5.1) with a fixed m (≥ 0). A very simple example is the quantile functional

$$\theta(F) = F^{-1}(p), \quad p \in (0,1), \quad F \in \mathcal{F}, \qquad (8.5.2)$$

where $\mathcal{F}$ is the class of all d.f.'s which admit a unique quantile. In such a case, we may as well consider an estimator $\theta(F_n)$ ($= T_n$, say), although to define it uniquely we may need to go by some convention (viz., Section 4.2). If we use the estimator $T_n = \theta(F_n)$ corresponding to (8.5.1), i.e.,

$$T_n = \int \cdots \int \psi(x_1, \ldots, x_m) dF_n(x_1) \cdots dF_n(x_m), \qquad (8.5.3)$$

we may write

$$dF_n(x_j) = dF(x_j) + d[F_n(x_j) - F(x_j)], \qquad 1 \le j \le m,$$

so that T_n can be decomposed into 2^m terms which can be recollected as

$$
\begin{aligned}
T_n \;=\; & \theta(F) + \binom{m}{1} \int \psi_1(x_1) d[F_n(x_1) - F(x_1)] \\
& + \binom{m}{2} \int \int \psi_2(x_1, x_2) d[F_n(x_1) - F(x_1)] d[F_n(x_2) - F(x_2)] \\
& + \cdots + \binom{m}{m} \int \cdots \int \psi(x_1, \ldots, x_m) \prod_{j=1}^{m} d[F_n(x_j) - F(x_j)],
\end{aligned}
$$

$$(8.5.4)$$

where

$$\psi_j(x_1, \ldots, x_j) = \mathrm{E}[\psi(X_1, \ldots, X_j) \mid X_i = x_i, i \le j], \qquad 1 \le j \le m. \quad (8.5.5)$$

In the literature, this is known as the **Hoeffding decomposition** of a symmetric function. For a general **statistical function**, $T_n = T(F_n)$, a similar expansion can be worked out whenever $\mathrm{E}_F T_n^2 < \infty$; however, it need not have $(m+1)$ orthonormal terms for some fixed m (≥ 1), and, second, the components, especially, the higher order ones, may not be that simple to be adaptable for further analysis. For this reason, often, we may look at a representation of T_n involving only one or two terms which would suffice for the desired asymptotic studies. For example, if we need only to study the (weak or strong) consistency of T_n [as an estimator of $T(F)$], we may express

$$T_n = T(F_n) = T(F + (F_n - F)), \qquad (8.5.6)$$

where (by Theorem 4.5.1) $\|F_n - F\| \xrightarrow{\text{a.s.}} 0$, so that the desired result follows if the functional $T(\cdot)$ is continuous in an appropriate norm. Let us assume that F is continuous, so that $Y = F(X)$ has the uniform (0,1) d.f., and the reduced empirical d.f. G_n, defined by (4.5.1), belongs to the $D[0, 1]$ space, for every $n \ge 1$. We may write then

$$T(H) = T(F^{-1}(F \circ H)) = \tau(F \circ H), \qquad (8.5.7)$$

where $\tau(\cdot)$ is the reduced functional. Then, writing $T_n = \tau(G_n)$, denoting by U the uniform(0,1) d.f., and noting that $\theta = T(F) = \tau(U)$, we have

$$T_n - \theta = \tau(G_n) - \tau(U). \qquad (8.5.8)$$

Therefore, if $\tau(\cdot)$, defined on the $D[0,1]$ space, is continuous relative to the Skorokhod metric in (8.2.16), then $\|G_n - U\| \xrightarrow{\text{a.s.}} 0$ ensures that $T_n - \theta(F) \xrightarrow{\text{a.s.}} 0$. Similarly, if we have in mind the study of the asymptotic normality of $n^{1/2}(T_n - \theta(F))$, we may write

$$Z_n = \{Z_n(t) = \sqrt{n}(G_n(t) - t), \quad 0 \le t \le 1\}$$

[as in (8.2.3)], and note that

$$n^{1/2}[T_n - \theta(F)] = n^{1/2}[\tau(U + n^{1/2}Z_n) - \tau(U)]. \qquad (8.5.9)$$

Thus, if the functional $\tau(\cdot)$ is differentiable in a suitable manner, so that the right-hand side of (8.5.9) can be expanded by a first order Taylor's expansion (albeit in a functional space), then we may write

$$n^{1/2}[T_n - \theta(F)] = \int \tau_1(U; t) dZ_n(t) + R_n, \qquad (8.5.10)$$

where

$$|R_n| = o(\|Z_n\|), \qquad (8.5.11)$$

and $\tau_1(U; \cdot)$ is the derivative of $\tau(\cdot)$ at U. Note that $\|Z_n\| = O_p(1)$ [by (8.4.16)], whereas $\tau_1(U; t)$ is a linear functional, so that

$$\sigma^2_{\tau_1} = \int_0^1 \tau_1^2(U; t) dt < \infty$$

ensures that

$$\int \tau_1(U; t) dZ_n(t) \sim N(0, \sigma^2_{\tau_1}). \qquad (8.5.12)$$

As such, by (8.5.10)–(8.5.12), one obtains that as $n \to \infty$,

$$n^{1/2}[T_n - \theta(F)]/\sigma_{\tau_1} \sim N(0, 1). \qquad (8.5.13)$$

$\tau_1(U; t)$ $[\equiv \tau_1^*(F; x)]$ is termed the **influence function** of $T(F)$ at x. The mode of differentiation in (8.5.9)–(8.5.11) needs some further clarification. This can, however, be made more precisely if we are permitted to use some basic results in functional analysis. However, consistent with our level of presentation, we state the following results in a bit less generality [viz., Fernholz (1983)].

Suppose that V and W are topological vector spaces, and let $L(V, W)$ be the set of continuous linear transformations form V to W. Let $\mathcal{S}$ be a class of compact subsets of V such that every subset consisting of a single point is in $\mathcal{S}$, and let A be an open subset of V. A function $T : A \to W$ is termed **Hadamard differentiable** at $F \in A$ if there exists a $T_F' \in L(U, V)$, such that for any $K \in \mathcal{S}$, uniformly for $H \subset K$,

$$\lim_{t \downarrow 0} t^{-1}\{T(F + tH) - T(F) - T_F'(tH)\}$$

$$= \lim_{t \downarrow 0} t^{-1} \{ R(T, F; tH) \} = 0. \qquad (8.5.14)$$

The **Hadamard** (or **compact**) derivative $T'_F(\cdot)$ reduces to the linear functional on the right-hand side of (8.5.10), whereas identifying t as $n^{-1/2}$, R_n in (8.5.10)–(8.5.11) corresponds to the remainder term $R(\cdot)$ in (8.5.14). There are other modes of differentiability (such as the Gateaux and Fréchet ones) which will not be discussed here.

However attractive such a differentiable approach may appear to be, there are certain limitations:

i) The functional $\tau = T \circ F^{-1}$ may depend on the unknown F in a rather involved manner, and verification of its continuity or Hadamand differentiability condition may require rather complicated analysis.

ii) As is the case with MLE, M- and R- estimators, the functional $T(\cdot)$ [or $\tau(\cdot)$] is defined implicitly as the root of some other (estimating) functionals. Refined treatment of such implicit functionals is even more delicate.

iii) In many cases $\tau(\cdot)$ turns out to be a bounded functional in order that Hadamard differentiability holds, and that may exclude many practically important cases.

iv) This approach is not quite in line with those in earlier chapters.

In view of these points, we shall not pursue this approach in further details.

A more convenient approach relates to a **first order representation** for a general statistical functional T_n, in the sense that

$$T_n - \theta(F) = n^{-1} \sum_{i=1}^{n} \Psi_F(X_i) + R_n, \qquad (8.5.15)$$

where $\Psi_F(x)$ generally depends on θ (through F),

$$E_F \Psi_F(X) = 0, \quad E_F \Psi_F^2(X) = \sigma_\Psi^2 < \infty, \qquad (8.5.16)$$

$$R_n = o_p(n^{-1/2}). \qquad (8.5.17)$$

We observed in Chapter 3 that the projection method (viz. Theorem 3.4.1) actually yields such a representation in many cases (including U-statistics, von Mises' functionals and many nonparametric statistics). In Chapter 5, we observed that for the MLE, (8.5.15) holds under the usual regularity conditions, where $\Psi_F(x) = -(\partial/\partial\theta) \log f(x; \theta)$. For R-estimators, (8.5.15) holds under more general regularity conditions than those pertaining to (8.5.14). Also, in general, (8.5.15) requires less complicated analysis than required for the Hoeffding decomposition, which goes further to decompose R_n into multiple orthogonal components of stochastically decreasing order of magnitudes. Also the first order case can be strengthened to the second order case by appealing to some relatively more stringent regularity conditions. We shall, therefore, be more inclined to the adaptation of (8.5.15) in various situations at hand. In this respect, the treatment for functionals of the form (8.5.2) needs a somewhat different approach, which we shall examine briefly.

Consider the reduced empirical process $Z_n = \{Z_n(t),\ t \in [0,1]\}$, defined by (8.2.3) and studied thoroughly in (8.4.1) – (8.4.19). Note that by virtue of (8.4.18) and (8.4.19), for every (fixed) p, $0 < p < 1$, and every $\varepsilon > 0$, $\eta > 0$, there exists a $\delta_0\ (> 0)$, such that for every $\delta \le \delta_0$

$$P\{\sup_{|t-p|<\delta} |Z_n(t) - Z_n(p)| > \varepsilon\} < \eta,\ \forall n \ge n_0. \tag{8.5.18}$$

On the other hand, for the uniform(0,1) d.f., if $\tilde{Y}_{n,p}$ stands for the sample p-quantile, then for every $\delta > 0$, we have, on noting that $G_n(\tilde{Y}_{n,p}) = p + o(n^{-1/2})$,

$$\lim_{n\to\infty} P\{|\tilde{Y}_{n,p} - p| > \delta\} = 0. \tag{8.5.19}$$

Thus, combining (8.5.18) and (8.5.19), we have for $n \to \infty$

$$n^{1/2}\{G_n(\tilde{Y}_{n,p}) - G(\tilde{Y}_{n,p}) - G_n(p) + p\} \xrightarrow{P} 0, \tag{8.5.20}$$

so that as n increases,

$$n^{1/2}[\tilde{Y}_{n,p} - p] \overset{P}{\approx} - n^{1/2}[G_n(p) - p], \tag{8.5.21}$$

as $G(y) = y$, $0 \le y \le 1$. On the other hand, if the d.f. F has a continuous and positive p.d.f. f [at $\xi_p : F(\xi_p) = p$], then for the sample p-quantile $\tilde{X}_{n,p}$, noting that $\tilde{Y}_{n,p} = F(\tilde{X}_{n,p})$, we have from (8.5.21),

$$\begin{aligned} n^{1/2}[\tilde{X}_{n,p} - \xi_p] &= n^{1/2}[\tilde{X}_{n,p} - \xi_p](\tilde{Y}_{n,p} - p)/(\tilde{Y}_{n,p} - p) \\ &= n^{1/2}(\tilde{Y}_{n,p} - p)\{[F(\tilde{X}_{n,p}) - F(\xi_p)]/(\tilde{X}_{n,p} - \xi_p)\}^{-1} \\ &= n^{1/2}(\tilde{Y}_{n,p} - p)\{f(\xi_p) + o_p(1)\}^{-1} \\ &= -f^{-1}(\xi_p)n^{1/2}[G_n(p) - p] + o_p(1), \end{aligned} \tag{8.5.22}$$

so that

$$\tilde{X}_{n,p} - \xi_p = n^{-1}\sum_{i=1}^{n} \Psi_F(X_i) + o_p\left(\frac{1}{\sqrt{n}}\right), \tag{8.5.23}$$

where

$$\Psi_F(X_i) = -\frac{1}{f(\xi_p)}[I_{\{X_i\le\xi_p\}} - p],\ 1 \le i \le n. \tag{8.5.24}$$

Thus, the first order representation in (8.5.15) holds for sample quantiles under the same regularity conditions, as in Theorem 4.3.3 pertaining to its asymptotic normality. We may, of course, get a stronger result under additional regularity conditions. Note that by virtue of (8.4.14), for every $n \ (\geq 1)$,

$$\{\exp[(1-t)^{-1}Z_n(t)], \quad t \in [0,1]\} \text{ is a submartingale.} \tag{8.5.25}$$

Hence, using the Kolmogorov-Hájek-Rényi-Chow Inequality and proceeding as in (8.4.17)–(8.4.18), we obtain that as $n \to \infty$,

$$\sup\{\sqrt{n}|G_n(t) - t - G_n(p) + p| : |t - p| \leq \frac{1}{\sqrt{n}} \log n\} = O(n^{-1/4} \log n) \quad \text{a.s.} \tag{8.5.26}$$

Note that for the uniform d.f.

$$\sup_{0 < p < 1} |\tilde{Y}_{n,p} - p| = \|G_n(t) - t\| = n^{-1/2}\|(1-t)W_n(t)\| \leq n^{-1/2}\|W_n(t)\| \tag{8.5.27}$$

where $\{W_n(t) = Z_n(t)/(1-t), 0 \leq t \leq 1\}$ is a martingale. Hence, using the Kolmogorov Inequality we can verify that

$$n^{-1/2}\|W_n(t)\|/\log n \xrightarrow{\text{a.s.}} 0$$

which by virtue of (8.5.27) implies that

$$|\tilde{Y}_{n,p} - p| < n^{-1/2} \log n \quad \text{a.s.}$$

Therefore, from the above two formulas, as $n \to \infty$,

$$\tilde{Y}_{n,p} - p = -[G_n(p) - p] + O(n^{-3/4} \log n) \quad \text{a.s.} \tag{8.5.28}$$

Thus, if we assume that the d.f. F admits an absolutely continuous p.d.f. f, such that $f'(x)$ is finite in a neighborhood of ξ_p, then from (8.5.28), we have for $n \to \infty$,

$$\tilde{X}_{n,p} - \xi_p = n^{-1} \sum_{i=1}^{n} \psi_F(X_i) + R_n, \tag{8.5.29}$$

where $\psi_F(x)$ is defined as in (8.5.24) and as $n \to \infty$,

$$R_n = O(n^{-3/4} \log n) \quad \text{a.s.} \tag{8.5.30}$$

In the literature, this is known as the **Bahadur** (1966) **representation of sample quantiles**. Equation (8.5.24) is a weaker version of (8.5.29), without requiring the finiteness (and existence) of $f'(x)$ at ξ_p. Also note that in (8.5.29), $R_n = O(n^{-3/4} \log n)$ and not $O_p(n^{-1})$ or $O(n^{-1} \log n)$ a.s., which could have been the case with some other smooth estimators. For discontinuous score functions, typically R_n is $O_p(n^{-3/4})$ or $O(n^{-3/4} \log n)$ a.s. [not $O_p(n^{-1})$]. We may refer to Sen (1981, Chap. 7) for a detailed account of this topic.

8.6 Weak convergence and nonparametrics

In nonparametric methods, rank-based procedures are generally adapted so as to induce invariance under appropriate groups of transformations and, thereby, to achieve more robustness. By definition, the ranks are closely related to the order statistics and the empirical d.f., and, therefore, weak convergence and related invariance principles for order statistics and empirical processes play a basic role in the asymptotic theory of nonparametric methods.

We may start with a remark that whether be it a single sample or a multisample model, the ranks are not independent r.v.'s, so that the laws of large numbers, probability inequalities and central limit theorems developed for independent summands (in Chapters 2 and 3) may fail to be directly applicable in nonparametrics. There have been several alternative tracks to bridge this gap, some of which are the following:

i) Express a nonparametric statistic in the form of a (generalized) U-statistic, and then use the methodology discussed in Chapter 5. This may not workout generally (viz., linear rank statistics/signed rank statistics when the score function is not a polynomial function).

ii) Consider an integral representation as in (8.4.56)–(8.4.57), and then use invariance principles for empirical processes to draw the desired conclusions. This method works out well for a larger class of nonparametric statistics, although it may require appropriate regularity conditions and elaborate analysis.

iii) Under appropriate hypotheses of invariance (with respect to suitable groups of transformations which map the sample space onto itself), there are generally some (sub)martingale or reversed martingale properties shared by various nonparametric statistics. As such, exploiting weak (or strong) invariance principles for such dependent sequences, asymptotic theory for a general class of nonparametric statistics can be obtained (under the hypothesis of invariance) under essentially minimal conditions. Using then the concept of **contiguity of probability measures**, the asymptotic theory can then be extended to cover alternative models.

Although in (i) or (ii), one need not be confined only to such local alternatives, in terms of applicability, they often dominate the asymptotics, so that the third approach may generally be the most simple one and may entail essentially the minimal regularity assumptions.

In this section, we intend to provide a very brief introduction to this third approach along with some applications. [For details, we may refer to Sen (1981).] Consider the two-sample problem treated in (8.4.53) through (8.4.59). Put it in the following general framework: $X_1, \ldots, X_n$ are independent r.v. with d.f.'s $F_1, \ldots, F_n$ and the null hypothesis H_0 states that $F_1 = \cdots = F_n = F$ (unknown). A general linear rank statistic is given by

$$T_n = \sum_{i=1}^{n} (c_i - \bar{c}_n) a_n(R_{ni}), \qquad (8.6.1)$$

where $\{c_i, \; i \geq 1\}$ is a sequence of (regression) constants, $\bar{c}_n = n^{-1} \sum_{i=1}^{n} c_i$, R_{ni} is the rank of X_i among $X_i, \ldots, X_n$, $i = 1, \ldots, n$, and

$$a_n(k) = E\psi(Y_{n:k}), \quad k = 1, \ldots, n, \qquad (8.6.2)$$

where $Y_{n:1} < \cdots < Y_{n:n}$ are the ordered r.v.'s of a sample of size n from the uniform (0,1) d.f., and ψ is a square integrable score function; $\psi(u) \equiv u - 1/2$, leads to the Wilcoxon statistic in (8.4.53). Recall that (even under H_0) T_n does not have independent summands. But (Exercise 8.6.1) note that by (8.6.2), for every $n \geq 1$,

$$\frac{k}{n+1} a_{n+1}(k+1) + \frac{n+1-k}{n+1} a_{n+1}(k) = a_n(k), \quad 1 \leq k \leq n. \qquad (8.6.3)$$

As such (Exercise 8.6.2), it is easy to show that

$$E\{T_{n+1} \mid \mathbf{R}_n, H_0\} = T_n \quad \text{a.e.}, \quad \forall n \geq 1, \qquad (8.6.4)$$

where $\mathbf{R}_n = (R_{n1}, \ldots, R_{nn})^t$. Given this martingale property of the T_n, $n \geq 1$, we can readily use the appropriate results for martingales (viz., inequalities, laws of large numbers, central limit theorems, etc.) to draw conclusions on the asymptotic behavior of T_n (under H_0) (Exercise 8.6.3). Thus, by invoking the asymptotic normality (under H_0) of T_n, i.e.,

$$T_n/(C_n A_n) \sim N(0,1) \qquad (8.6.5)$$

where

$$C_n^2 = \sum_{i=1}^{n} (c_i - \bar{c}_n)^2 \quad \text{and} \quad A_n^2 = \frac{1}{n-1} \sum_{i=1}^{n} [a_n(i) - \bar{a}_n]^2, \qquad (8.6.6)$$

we are able to obtain an asymptotic coverage probability by

$$P\{-C_n A_n \tau_{\alpha/2} \leq T_n \leq C_n A_n \tau_{\alpha/2} \mid H_0\} \cong 1 - \alpha. \qquad (8.6.7)$$

Consider now a simple regression model

$$F_i(x) = F(x - \theta - \beta c_i), \; 1 \leq i \leq n, \qquad (8.6.8)$$

so that under $\beta = 0$, the F_i are all the same. If each X_i is replaced by $X_i - bc_i$, the resulting ranks are denoted by $R_{ni}(b)$ and replacing the R_{ni} in (8.6.1) by the corresponding $R_{ni}(b)$, we denote the (aligned) rank statistic by $T_n(b)$, $b \in \mathbb{R}$. If $\psi(n)$ is nondecreasing in u, $0 \leq u \leq 1$, so that the

$a_n(i)$ are also increasing in i $(1 \leq i \leq n)$, we have

$$T_n(b) \quad \text{is nonincreasing in} \quad b \in \mathbb{R}. \tag{8.6.9}$$

Therefore, equating $T_n(b)$ to $-C_n A_n \tau_{\alpha/2}$ and $C_n A_n \tau_{\alpha/2}$, we get an upper and lower bound $\hat{\beta}_{U,n}$ and $\hat{\beta}_{L,n}$ respectively, where, by (8.6.7),

$$P\{\hat{\beta}_{L,n} \leq \beta \leq \hat{\beta}_{U,n}\} \cong 1 - \alpha. \tag{8.6.10}$$

Thus, we have a distribution-free confidence interval for β, and asymptotically it takes on a simple form. In fact, equating $T_n(b)$ to 0, we get the point estimate $\hat{\beta}_n$, termed the R-estimator of β. $\hat{\beta}_n$, is a translation-equivariant, robust, consistent estimator of β with the property that as $n \to \infty$,

$$C_n(\hat{\beta}_n - \beta)/A_n \xrightarrow{\mathcal{D}} N(0, \gamma^{-2}); \qquad \gamma = \int_{-\infty}^{\infty} \psi(F(x)) f^2(x) dx \tag{8.6.11}$$

The study of the asymptotic properties of (8.6.10) and (8.6.11) is facilitated by a **uniform asymptotic linearity result** (in b) of $T_n(b)$ (near β) – but this is outside the scope of the current treatise. We may refer to Chapters 4 and 5 of Sen (1981) for some of these details. In passing, we may, however, comment that by virtue of (8.6.9), proceeding as in the proof of the Glivenko-Cantelli Theorem (4.5.1), we may show that for every finite K,

$$\sup\{C_n^{-1} A_n^{-1} |T_n(\beta + C_n^{-1} b) - T_n(\beta) + b\gamma C_n| : |b| \leq K\}$$
$$\leq \max_{j \leq a}\{C_n^{-1} A_n^{-1} |T_n(\beta + C_n^{-1} b_j) - T_n(\beta) + b_j \gamma C_n|\} + \varepsilon/2,$$

$$\tag{8.6.12}$$

where a is a finite positive number, depending on ε and K. Then, for the point-wise convergence one may consider standard weak convergence results in nonparametrics.

For M-estimators of location and regression a very similar linearity theorem holds, and that provides the easy access to the study of asymptotic properties of the estimators [viz., Sen (1981, Chap. 8)]. There is another important aspect of nonparametrics that we may like to discuss briefly here. Typically, in an estimation problem (parametric or nonparametric), we have an estimator T_n of a parameter θ, such that

$$n^{1/2}(T_n - \theta)/\nu_\theta \xrightarrow{\mathcal{D}} N(0, 1) \tag{8.6.13}$$

where ν_θ^2, the asymptotic variance of T_n, generally depends on the unknown parameter θ or on the unspecified d.f. F. In a parametric setup, F is of known form, and, hence ν_θ can be estimated by various methods (e.g., MLE, etc.). In a nonparametric setup, $\nu_\theta = \nu(F)$ is itself a functional of the unknown d.f. F. If the form of this functional $\nu(\cdot)$ is known, a very simple and convenient method of estimating $\nu(F)$ would be to use the sample d.f. F_n for F and take the estimator as $v_n = \nu(F_n)$. This is generally related to the so-called **Delta-method** where expanding $\nu(F_n)$ around $\nu(F)$ (under appropriate smoothness conditions), consistency and other properties can be studied. The estimator is, in general, not unbiased, and, often, the bias can be of serious considerations. A more complicated situation may arise where the form of $\nu(\cdot)$ may lead to very high level of bias by the Delta-method or $\nu(\cdot)$ may not be at all of a simple form. In such a problem, **resampling plans** are often used to estimate $\nu(F)$ in a nonparametric fashion. Among these resampling methods, the two most popular ones are the **jackknife** and **bootstrap** methods. Let us briefly discuss them.

Suppose that T_n, as an estimator of θ, has a bias of the form

$$\mathrm{E}(T_n - \theta) = n^{-1}a(\theta) + n^{-2}b(\theta) + \cdots, \qquad (8.6.14)$$

where $a(\theta), b(\theta)$, etc., are unknown functionals. Then,

$$\mathrm{E}(T_{n-1} - \theta) = (n-1)^{-1}a(\theta) + (n-1)^{-2}b(\theta) + \cdots, \qquad (8.6.15)$$

so that

$$\mathrm{E}\{nT_n - (n-1)T_{n-1}\} = \theta - \frac{1}{n(n-1)}b(\theta) + O(n^{-3}). \qquad (8.6.16)$$

Thus, the bias is reduced to the order n^{-2}, instead of n^{-1}. Keeping this in mind, define for each i $(1 \leq i \leq n)$,

$$T_{n-1}^{(i)} = T(X_1, \ldots, X_{i-1}, X_{i+1}, \ldots, X_n), \qquad (8.6.17)$$

$$T_{ni} = nT_n - (n-U)T_{n-1}^{(i)}, \quad i = 1, \ldots, n. \qquad (8.6.18)$$

The T_{ni} are termed the **pseudovalues**. Let then

$$T_{nJ} = n^{-1}\sum_{i=1}^{n} T_{ni}. \qquad (8.6.19)$$

T_{nJ} is termed the **jackknifed version** of T_n. From (8.6.16) and (8.6.19), we have

$$\mathrm{E}(T_{nJ} - \theta) = -\frac{1}{n(n-1)}b(\theta) + O(n^{-2}) = O(n^{-2}), \qquad (8.6.20)$$

so that jackknifing can reduce the **bias** of T_n; this was the primary objective of jackknifing when proposed nearly 40 years ago. Let then

$$v_{nJ}^2 = \frac{1}{n-1}\sum_{i=1}^{n}[T_{ni} - T_{nJ}]^2. \qquad (8.6.21)$$

This is termed the **jackknifed variance estimator**. Under appropriate regularity conditions,

$$v_{nJ}^2 \xrightarrow{\text{P}} v^2(F). \tag{8.6.22}$$

To appreciate (8.6.22), consider the simplest case:

$$T_n = \bar{X}_n = \frac{1}{n}\sum_{i=1}^n X_i.$$

Then $T_{ni} = X_i$, $1 \le i \le n$, $T_{nJ} = T_n$ and

$$v_{nJ}^2 = S_n^2 = \frac{1}{n-1}\sum_{i=1}^n (X_i - \bar{X}_n)^2$$

which converges a.s. to $\sigma^2 = \mathrm{Var}(X) = v^2(F)$, as $n \to \infty$, whenever $\sigma^2 < \infty$. A very similar motivation emerges whenever T_n can be well approximated by a linear statistic. In this context, let us refer to the first order approximation in (8.5.15), and define R_{ni} and R_{nJ} as in (8.6.18)–(8.6.19) (with T_n replaced by R_n). Then, for each i, $1 \le i \le n$,

$$T_{ni} = \theta(F) + \psi_F(X_i) + R_{ni}, \tag{8.6.23}$$

so that

$$T_{nJ} = \theta(F) + \bar{\psi}_n + R_{nJ}; \quad \bar{\psi}_n = \frac{1}{n}\sum_{n-1} \psi_F(X_i). \tag{8.6.24}$$

Therefore, by (8.6.21) and (8.6.24), we have

$$\begin{aligned} v_{nJ}^2 &= \frac{1}{n-1}\sum_{i=1}^n \{\psi_F(X_i) - \bar{\psi}_n\}^2 + \frac{1}{n-1}\sum_{i=1}^n [R_{ni} - R_{nJ}]^2 \\ &\quad + \left[\frac{2}{n-1}\sum_{i=1}^n [\psi_F(X_i) - \bar{\psi}_n][R_{ni} - R_{nJ}]\right]. \end{aligned} \tag{8.6.25}$$

The first term on the right-hand side of (8.6.25) converges a.s. to $\nu^2(F) = \sigma_\psi^2 = \mathrm{E}\psi_F^2(X_i)$ as $n \to \infty$. Thus, a sufficient condition for (8.6.22) to hold is that as $n \to \infty$

$$\frac{1}{n-1}\sum_{i=1}^n [R_{ni} - R_{nJ}]^2 \xrightarrow{\text{a.s.}} 0. \tag{8.6.26}$$

Though (8.6.26) is stronger than (8.5.17), it can be verified under quite general condition. We shall not, however, enter into these details. But we prescribe (8.6.26) as a key solution to (8.6.22).

Let us recall that $T_n = T(X_1, \ldots, X_n)$ and F_n is the empirical d.f. of the X_i. Let $X_1^*, \ldots, X_n^*$ be n independent observations drawn with replacement in an equal probability sampling (SRSWOR) scheme form $X_1, \ldots, X_n$; we say that $X_1^*, \ldots, X_n^*$ are (conditionally) i.i.d. r.v.'s drawn form F_n. Let then

$$T_n^* = T(X_1^*, \ldots, X_n^*). \tag{8.6.27}$$

Also let us draw M such independent sets of n observations, denoted by $\mathbf{X}_i^* = (X_{i1}^*, \ldots, X_{in}^*)^t$, and let the corresponding T_n^* be denoted by T_{ni}^*, $i = 1, \ldots, M$. Let then

$$G_{nM}^*(y) = M^{-1} \sum_{i=1}^{M} I_{\{n^{1/2}(T_{ni}^* - T_n) \leq y\}}, \quad y \in \mathbb{R}, \tag{8.6.28}$$

and

$$G_n(y) = P\{n^{1/2}(T_n - \theta) \leq y\}, \quad y \in \mathbb{R}. \tag{8.6.29}$$

Also, let

$$v_{nB}^2 = \frac{1}{M} \sum_{i=1}^{M} n(T_{ni}^* - T_n)^2. \tag{8.6.30}$$

Then, under appropriate regularity conditions, for large M (and n),

$$\|G_{nM}^* - G_n\| \xrightarrow{\mathrm{P}} 0, \tag{8.6.31}$$

and the **bootstrap** variance estimator $v_{MB}^{*2} \xrightarrow{\mathrm{P}} v^2(F)$. Perhaps, it will be convenient for us to motivate this result through the functional approach is Section 8.5. We denote the empirical d.f. for $X_1^*, \ldots, X_n^*$ by F_n^*, so that corresponding to the M copies of the bootstrap samples, we have the empirical d.f.'s $F_{n1}^*, \ldots, F_{nM}^*$. As in (8.5.6), we take $T_n = T(F_n)$, so that

$$T_{ni}^* = T(F_{ni}^*), \quad i = 1, \ldots, M, \tag{8.6.32}$$

which are (conditionally on $X_1, \ldots, X_n$ being given) i.i.d. r.v.'s. Thus, by (8.6.28) and (8.6.32), we have

$$\begin{aligned}
v_{nB}^{*2} &= \frac{1}{M} \sum_{i=1}^{M} n[T(F_{ni}^*) - T(F_n)]^2 \\
&= \int_{-\infty}^{\infty} x^2 dG_{nM}^*(y) \\
&= \int_{-\infty}^{\infty} x^2 dG_n(y) + \int_{-\infty}^{\infty} x^2 d[G_{nM}^*(y) - G_n(y)] \\
&= \mathrm{E}\{n(T_n - \theta)^2\} + \frac{1}{\sqrt{M}} \int_{-\infty}^{\infty} x^2 d\{\sqrt{M}[G_{nM}^*(y) - G_n(y)]\},
\end{aligned} \tag{8.6.33}$$

where the first term converges to $\nu^2(F)$, and under quite general regularity conditions, conditionally on F_n, when n and M are large,

$$M^{1/2}(G_{nM}^* - G_n) \xrightarrow{\mathcal{D}} \quad \text{Gaussian } W \text{ on } \mathbb{R}, \tag{8.6.34}$$

where W is tied down at both extremities. Thus, by (8.6.33) and (8.6.34), we have for large M (and n),

$$v_{nB}^{*2} = \mathrm{E}\{n(T_n - \theta)^2\} + O_p\left(\frac{1}{\sqrt{M}}\right) \xrightarrow{\mathrm{P}} \nu^2(F). \qquad (8.6.35)$$

Actually, (8.6.31) is a direct consequence of (8.6.34), so that for both (8.6.31) and (8.6.35), the weak convergence result in (8.6.34) provides the basic key. On the other hand, noting that $n^{1/2}(F_n - F)$ converges weakly to a Gaussian process, so that $n^{1/2}\|F_n - F\| = O_p(1)$, by imposing appropriate smoothness conditions of F, (8.6.34) can be studied as in Section 8.4 under the conditional setup (given F_n). We omit these details. There are some situations (viz., sample quantiles) where jackknifing may not work out well, but bootstrapping does better. But, in general, they perform equally well in a majority of regular cases.

8.7 Strong invariance principles

The invariance principle studied in Section 8.2 [see (8.2.15)] is termed the weak invariance principle. There are some important applications in Statistics (especially, in sequential analysis) where a stronger mode is more convenient. Following Skorokhod (1956) we may consider this as follows.

Let $\{X_i,\ i \geq 1\}$ be a sequence of i.i.d. r.v.'s with a d.f. F, defined on $\mathbb{R}$, and for simplicity of presentation, we take $\mathrm{E}X = 0$ and $\mathrm{Var}(X) = 1$. Consider then the partial sum sequence

$$S_n = X_1 + \cdots + X_n, n \geq 1; \quad S_0 = 0. \qquad (8.7.1)$$

As in Section 8.3, let us consider a Wiener process $W = \{W(t),\ t \in [0, \infty)\}$. Then the **Skorokhod embedding of Wiener process** asserts that there exists a sequence $\{T_i,\ i \geq 1\}$ of non-negative and independent r.v.'s, such that for every $n \geq 1$,

$$\begin{pmatrix} Y_1 \\ Y_2 \\ \vdots \\ Y_n \end{pmatrix} \overset{D}{=} \begin{pmatrix} W(T_1) \\ W(T_1 + T_2) - W(T_1) \\ \vdots \\ W(\sum_{i=1}^{n} T_i) - W(\sum_{i=1}^{n-1} T_i) \end{pmatrix} \qquad (8.7.2)$$

and

$$\mathrm{E}(T_i) = \mathrm{Var}(T_i) = 1, \quad i \geq 1. \qquad (8.7.3)$$

This embedding has been extended not only to nonidentically distributed r.v.'s, but also to martingales and some other dependent sequences. Komlós, Major and Tusnády (1975) have shown that whenever the X_i are independent with mean 0 and variances equal to 1 and have finite moment generating functions (in a neighborhood of 0), as $n \to \infty$,

$$S_n = W(n) + O(\log n) \quad \text{a.s.}, \tag{8.7.4}$$

while Strassen (1967), under a fourth moment condition on the X_i, showed that as $n \to \infty$,

$$S_n = W(n) + O[(n \log \log n)^{1/4} (\log n)^{1/2}] \quad \text{a.s.} \tag{8.7.5}$$

Strassen's results also pertains to martingales under a very mild regularity condition on their variance function.

Almost sure invariance principles have also been established for empirical processes. Consider a two-parameter Gaussian process $Y = \{Y(s,t) : 0 \leq s < \infty, \ 0 \leq t < \infty\}$, such that $EY = 0$ and

$$EY(s,t)Y(s',t') = (s \wedge s')(t \wedge t'), \quad \forall (s,t), (s',t'). \tag{8.7.6}$$

Y is termed **Brownian sheet**. Let then

$$Y^* = \{Y^*(s,t), \quad 0 \leq s \leq 1, \ 0 \leq t < \infty\}$$

be defined by

$$Y^*(s,t) = Y(s,t) - sY(1,t), 0 \leq s \leq 1, 0 < t < \infty. \tag{8.7.7}$$

Y^* is termed a **Kiefer process**. Consider now the reduced empirical process $Z_n = \{Z^*(s,t); \ 0 \leq s \leq 1, \ t \in [0,\infty)\}$ defined by

$$Z^*(s,t) = Z_{[t]}(s), \quad s \in [0,1], \quad t \in \mathbb{R}^+, \tag{8.7.8}$$

where $Z_n \equiv 0$ for $n \leq 0$. Then the following holds:

$$\sup_{0 \leq s \leq 1} |Z^*(s,t) - Y^*(s,t)| = O((\log t)^2), \quad \text{a.s. as} \quad t \to \infty. \tag{8.7.9}$$

Note that in (8.7.4)–(8.7.5), $\{S_n\}$ and W are not necessarily defined on a common probability space, and similarly in (8.7.9), Z^* and Y^* may not be defined on a common probability space. However, these a.s. invariance principles allow us to replace the S_n (or Z^*) by W (or Y^*) for diverse probabilistic analyses, and, hence, they have great scope in various asymptotic analyses. For same of these applications, we may refer to Sen (1981).

8.8 Concluding notes

The topics covered in this chapter may deserve a more sophisticated rigorous treatment, but that would have been a violation of the uniformity of the level of presentation of the current treatise. We hope that the reader could understand our motivation in providing, mostly, a basic survey of these specialized (and advanced) topics at an intermediate level. The reader may skip a greater part of this chapter and yet may have a comprehensive account

of the major undercurrents. The examples cited throughout this chapter
are, of course, geared more toward potential applications. Therefore, we
suggest that one should look into them, even if one may not be interested
in the theoretical aspects. Nevertheless these topics, treated in more generality in Parthasarathy (1967) and Billingsley (1968), help in acquiring a
better understanding of the modern methodology in asymptotic methods
in statistics, and we sincerely hope that the survey outlined in this chapter
makes it convenient to look into the directions for further reading with a
view to incorporating more of these basic tools into the main stream of
large sample methods in statistics. However, we like to make it a point of
distinction: mathematical and probabilistic tools are indispensable in the
study of large sample methods, although the emphasis should be primarily
on the applicable methodology, not merely on abstract concepts which may
not be appealing to statisticians at large.

8.9 Exercises

Exercise 8.3.1: Let W_n be defined as in (8.3.2)–(8.3.4). Then show that
$\{W_n(t), t \geq 0\}$ is a martingale. Hence, for every (fixed) s (≥ 0), $\{W_n(s + t) - W_n(s), t \geq 0\}$ is a martingale, so that $\{[W_n(s + t) - W_n(s)]^2, t \geq 0\}$
is a submartingale.

Exercise 8.3.2 (Continued): Show that (8.3.24) holds, i.e.,

$$\sup_{0 \leq u \leq \delta} |W_n(s + u) - W_n(s)| \overset{D}{=} \max_{0 \leq j \leq m} n^{-1/2}|S_j|,$$

where $m = [n\delta]$. Hence, use Theorem 2.4.1 to verify that (8.3.20) holds
when the X_j [in (8.3.1)] have finite moments up to the order r, for some
$r > 2$.

Exercise 8.3.3: Verify (8.3.28) for the right-hand side of (8.3.25).

Exercise 8.3.4: Use Exercises 8.3.1 and 8.3.2 to verify the tightness part
of Theorem 8.3.2.

Exercise 8.4.1: Verify (8.4.14).

Exercise 8.4.2: Verify (8.4.29). (You may use partial integration.)

Exercise 8.4.3: For Z_n^* defined by (8.4.47), verify that the tightness of
$Z_{n_1}^{(1)}$ and $Z_{n_2}^{(2)}$ ensure the same for Z_n^*.

Exercise 8.4.4: Verify (8.4.59) from (8.4.58).

Exercise 8.5.1: Verify (8.5.26).

Exercise 8.6.1: Verify (8.6.3).

Exercise 8.6.2: Verify (8.6.4).

Exercise 8.6.3: Verify (8.6.5).

References

Agresti, A. (1990). *Categorical Data Analysis*. New York: John Wiley.

Anderson, T.W. (1960). A modification of the sequential probability ratio test to reduce the sample size. *Annals of Mathematical Statistics, 31*, 165-197.

Anscombe, F.J. (1948). The transformation of Poisson, Binomial and negative binomial data. *Biometrika, 35*, 246-254.

Anscombe, F.J. (1952). Large sample theory of sequential estimation. *Proceedings of the Cambridge Philosophical Society, 48*, 600-607.

Armitage, P. (1975). *Sequential Medical Trials*. New York: John Wiley.

Bahadur, R.R. (1966). A note on quantiles in large samples. *Annals of Mathematical Statistics, 37*, 577-580.

Barlow, R., Bartholomew, D., Bremmer, J.M. and Brunk, H.D. (1972). *Statistical Inference under Order Restrictions*. New York: John Wiley.

Berry, A.C. (1941). The accuracy of the Gaussian approximation to the sum of independent variates. *Transactions of the American Mathematical Society, 49*, 122-136.

Billingsley, P. (1968). *Convergence of Probability Measures*. New York: John Wiley.

Bhapkar, V.P. (1966). A note on the equivalence of two test criteria for hypotheses in categorical data. *Journal of the American Statistical Association, 61*, 228-235.

Blom, G. (1958). *Statistical Estimates and Transformed Beta Variables*. New York: John Wiley.

Brown, B.M. (1971a). Martingale central limit theorems. *Annals of Mathematical Statistics, 42*, 59-66.

Brown, B.M. (1971b). A note on convergence of moments. *Annals of Mathematical Statistics, 42*, 777-779.

Chernoff, H. (1954). On the distribution of the likelihood ratio. *Annals of Mathematical Statistics, 25*, 573-578.

Chow, Y.S. (1960). A martingale inequality and the law of large numbers. *Proceedings of the American Mathematical Society, 11*, 107-111.

Chow, Y.S. and Robbins, H. (1965). On the asymptotic theory of fixed-width sequential confidence intervals for the mean. *Annals of Mathematical Statistics, 36*, 457-462.

Chow, Y.S. and Teicher, H. (1978). *Probability Theory: Independence, Interchangeability, Martingales*. New York: Springer-Verlag.

Cox, D.R. (1972). Regression models and life tables (with discussion). *Journal of the Royal Statistical Society,* **B 74**, 187-220.

Cox, D.R. and Oakes, D. (1984). *Analysis of Survival Data.* London: Chapman and Hall.

Cramér, H. (1946). *Mathematical Methods of Statistics.* Princeton, NJ: Princeton University Press.

Daniels, H.A. (1945). The statistical theory of the strength of bundles of threads. In *Proceedings of the Royal Society,* **A 183**, 405-435

Dantzig, G.B. (1940). On the non-existence of tests of Student's hypothesis having power functions independent of σ. *Annals of Mathematical Statistics,* **11**, 186-192.

David, H.A. (1970). *Order Statistics.* New York: John Wiley.

Dvoretzky, A. (1971). Asymptotic normality of sums of dependent random variables. In *Proceedings of the Sixth Berkeley Symposium on Mathematical Statistics and Probability, vol.* **2**, 513-536. Berkeley: University of California Press.

Eicker, F. (1967). Limit theorems for regression with unequal and dependent errors. In *Proceedings of the Fifth Berkeley Symposium on Mathematical Statistics and Probability, vol.* **1**, 59-82. Berkeley: University of California Press.

Elandt-Johnson, R.C. (1971). *Probability Models and Statistical Methods in Genetics.* New York: John Wiley.

Fahrmeir, L. and Kaufmann, H. (1985). Consistency and asymptotic normality of the maximum likelihood estimator in generalized linear models. *Annals of Statistics,* **13**, 342-368.

Feller, W. (1971). *An Introduction to Probability Theory and its Applications, vol* **2**, *2nd edition.* New York: John Wiley.

Fernholz, L.T. (1983). *Von Mises Calculus for Statistical Functionals.* Lecture Notes in Statistics, **19**. New York: Springer-Verlag.

Finney, D.J. (1978). *Statistical Method in Biological Assay, 3rd edition.* London: Charles Griffin.

Fisher, R.A. (1922). On the mathematical foundations of theoretical Statistics. *Philosophical Transactions of the Royal Society,* **222**, 309-368. Reprinted in *Contributions to Mathematical Statistics* (by R.A. Fisher, 1950). New York: John Wiley.

Gangopadhyay, A.K. and Sen, P.K. (1990). Bootstrap confidence intervals for conditional quantile functions. *Sankhya,* **A 52**, 346-363.

Gangopadhyay, A.K. and Sen, P.K. (1992). Contiguity in nonparametric estimation of a conditional functional. In *Nonparametric Statistics and Related Topics (ed. A.E. Saleh),* 141-162. Amsterdam: North-Holland.

Gayen, A.K. (1951). The frequency distribution of the product moment correlation coefficient in random samples of any size drawn from non-normal universes. *Biometrika,* **38**, 219-247.

Ghosh, J.K. (1971). A new proof of the Bahadur representation and an application. *Annals of Mathematical Statistics,* **42**, 1957-1961.

Ghosh, J.K. and Sen, P.K. (1985). On the asymptotic performance of the log-likelihood ratio statistics for the mixture model and related results. In *Proceedings of the Berkeley Symposium in honour of Jerzy Neyman and Jack Kiefer (eds. L. LeCam and R.A. Olshen)*, 789-806. Belmont, CA: Wadsworth.

Gnedenko, B.V. (1943). Sur la distribution limite du terme maximum d'une série aléatoire. *Annals of Mathematics*, **44**, 423-453.

Gnedenko, B.V. (1969). *The Theory of Probability*. Moscow: Mir Publishers.

Godambe, V.P. and Heyde, G.C. (1987). Quasi-likelihood and optimal estimation. *International Statistical Review*, **55**, 231-244.

Grizzle, J.E., Starmer, C.F and Koch, G.G. (1969). The analysis of categorical data by linear models. *Biometrics*, **25**, 489-504.

Gumbel, E.J. (1958). *Statitics of Extremes*. New York: Columbia University Press.

Hájek, J. (1968). Asymptotic normality of simple linear rank statistics under alternatives. *Annals of Mathematical Statistics*, **39**, 325-346.

Hájek, J. (1972). Local asymptotic minimax and admissibility in estimation. In *Proceedings of the Sixth Berkeley Symposium on Mathematical Statistics and Probability, vol.* **1**, 175-194. Berkeley: University of California Press.

Hájek, J. and Rényi, A. (1950). Generalisation of an inequality of Kolmogorov. *Acta Mathematica Academiae Scientiarum Hungaricae*, **6**, 281-283.

Hájek, J. and Šidák, Z. (1967). *Theory of Rank Tests*. New York: Academic Press.

Hewitt, E. and Savage, L.J. (1955). Symmetric measures on Cartesian products. *Transactions of the American Mathematical Society*, **80**, 470-501.

Hoeffding, W. (1948). A class of statistics with asymptotically normal distribution. *Annals of Mathematical Statistics*, **19**, 293-325.

Hoeffding, W. (1963). Probability inequalities for sums of bounded random variables. *Journal of the American Statistical Association*, **58**, 13-30.

Hogg, R.V. and Craig, A.T. (1970). *Introduction to Mathematical Statistics, 3rd edition.* New York: Macmillan.

Huber, P.J. (1964). Robust estimation of a location parameter. *Annals of Mathematical Statistics*, **35**, 73-101.

Huber, P.J. (1981). *Robust Statistics.* New York: John Wiley.

Inagaki, N. (1973). Asymptotic relations between the likelihood estimating function and the maximum likelihood estimator. *Annals of the Institute of Statistical Mathematics*, **25**, 1-26.

Jurečková, J. (1977). Asymptotic relations of M-estimates and R-estimates in linear regression model. *Annals of Statistics*, **5**, 464-472.

Keating, J.P., Mason, R.L. and Sen, P.K. (1993). *Pitman's Measure of Closeness: A Comparison of Statistical Estimators.* Philadelphia: SIAM.

Koch, G.G., Imrey, P.B., Singer, J.M., Atkinson, S.S. and Stokes, M.E. (1985). *Analysis of Categorical Data.* Montréal: Les Presses de l'Uni-versité de Montréal.

Komlós, J., Major, P. and Tusnády, G. (1975). An approximation of partial sums of independent r.v.'s and the sample d.f. I. *Zeitschrift für Wahrscheinlichkeitstheorie und Verwandte Gebiete*, **32**, 111-131.

LeCam, L. (1956). On the asymptotic theory of estimation and testing hypotheses. In *Proceedings of the Third Berkeley Symposium on Mathematical Statistics and Probability, vol.* **1**, 129-156. Berkeley: University of California Press.

LeCam, L. (1986). *Asymptotic Methods in Statistical Decision Theory.* New York: Springer-Verlag.

LeCam, L. and Yang, G.C. (1990). *Asymptotics in Statistics: Some Basic Concepts*. New York: Springer-Verlag.

Lehmann, E.L. (1953). The power of rank tests. *Annals of Mathematical Statistics*, **24**, 23-43.

Liang, K.-Y. and Zeger, S.L. (1986). Longitudinal data analysis using generalized linear models. *Biometrika*, **73**, 13-22.

Loynes, R.M. (1970). An invariance principle for reversed martingales. *Proceedings of the American Mathematical Society*, **25**, 56-64.

McCullagh, P. and Nelder, J.A. (1989). *Generalized Linear Models, 2nd edition*. London: Chapman and Hall.

McLeish, D.L. (1974). Dependent central limit theorems and invariance principles. *Annals of Probability*, **2**, 620-628.

Mood, A.M. (1941). On the joint distribution of the median in samples from a multivariate population. *Annals of Mathematical Statistics*, **12**, 268-278.

Mosteller, F. (1946). On some useful "inefficient" statistics. *Annals of Mathematical Statistics*, **17**, 377-408.

Nelder, J.A. and Wedderburn, R.W.M. (1972). Generalized linear models. *Journal of the Royal Statistical Society*, **A 135**, 370-384.

Neyman, J. (1949). Contributions to the theory of the χ^2 test. In *Proceedings of the First Berkeley Symposium in Mathematical Statistics and Probability*, 239-273. Berkeley: University of California Press.

Neyman, J. and Scott, E.L. (1948). Consistent estimates based on partially consistent observations. *Econometrica*, **16**, 1-32.

Parthasarathy, K.R. (1967). *Probability Measures on Metric Spaces*. New York: Academic Press.

Pfanzagl, J. (1982). *Contributions to a General Asymptotic Statistical Theory*. New York: Springer-Verlag.

Pitman, E.J.G. (1937). The closest estimate of statistical parameters. *Proceedings of the Cambridge Philosophical Society*, **33**, 212-222.

Pratt, J.W. (1981). Concavity of the log likelihood. *Journal of the American Statistical Association*, **76**, 103-106.

Puri, M.L. and Sen, P.K. (1971). *Nonparametric Methods in Multivariate Analysis*. New York: John Wiley.

Puri, M.L. and Sen, P.K. (1985). *Nonparametric Methods in General Linear Models*. New York: John Wiley.

Pyke, R. and Root, D. (1968). On convergence in r-mean normalized partial sums. *Annals of Mathematical Statistics*, **39**, 379-381.

Rao, C.R. (1973). *Linear Statistical Inference and its Applications, 2nd edition*. New York: John Wiley.

Redner, R. (1981). Note on the consistency of the maximum likelihood estimate for nonidentifiable distributions. *Annals of Statistics*, **9**, 225-228.

Sarhan, A.E. and Greenberg, B.G., eds. (1962). *Contributions to Order Statistics*. New York: John Wiley.

Searle, S.R. (1971). *Linear Models*. New York: John Wiley.

Sen, P.K. (1959). On the moments of the sample quantiles. *Calcutta Statistical Association Bulletin*, **9**, 1-20.

Sen, P.K. (1960). On some convergence properties of U-statistics. *Calcutta Statistical Association Bulletin,* **10**, 1-18.

Sen, P.K. (1963). On the estimation of relative potency in distribution (-direct) assays by distribution-free methods. *Biometrics,* **19**, 532-552.

Sen, P.K. (1964). On some properties of the rank weighted means. *Journal of the Indian Society of Agricultural Statistics,* **16**, 51-61.

Sen, P.K. (1968a). Estimation of regression coefficients based on Kendall's tau. *Journal of the American Statistical Association,* **63**, 1379-1389.

Sen, P.K. (1968b). Robustness of some nonparametric procedures in linear models. *Annals of Mathematical Statistics,* **39**, 1913-1922.

Sen, P.K. (1976). Weak convergence of progressively censored likelihood ratio statistics and its role in asymptotic theory of life testing. *Annals of Statistics,* **4**, 1247-1257.

Sen, P.K. (1981). *Sequential Nonparametrics: Invariance Principles and Statistical Inference.* New York: John Wiley.

Sen, P.K. (1985). *Theory and Applications of Sequential Nonparametrics.* Philadelphia: SIAM.

Sen, P.K., Bhattacharyya, B.B. and Suh, M.W. (1973). Limiting behavior of the extrema of certain sample functions. *Annals of Statistics,* **1**, 297-311.

Serfling, R.J. (1980). *Approximation Theorems of Mathematical Statistics.* New York: John Wiley.

Silvapulle, M.J. (1981). On the existence of maximum likelihood estimators for the binomial response models. *Journal of the Royal Statistical Society,* **B 43**, 310-313.

Silvapulle, M.J. and Burridge, J. (1986). Existence of maximum likelihood estimates in regression models for grouped and ungrouped data. *Journal of the Royal Statistical Society,* **B 48**, 100-106.

Silvey, S.D. (1959). The Lagrangian multiplier test. *Annals of Mathematical Statistics,* **30**, 389-407.

Singer, J.M. and Sen, P.K. (1985). M-methods in multivariate linear models. *Journal of Multivariate Analysis,* **17**, 168-184.

Singer, J.M. and Sen, P.K. (1986). A note on the asymptotic behavior of maximum likelihood estimators. In *Proceedings of the 6th Brazilian Symposium of Probability and Statistics (SINAPE),* 205-211. Rio de Janeiro: Associação Brasileira de Estatística.

Skorokhod, A.V. (1956). Limit theorems for stochastic processes. *Theory of Probability and Applications,* **1**, 261-290.

Stein, C. (1945). A two sample test for a linear hypothesis whose power is independent of the variance. *Annals of Mathematical Statistics,* **16**, 243-258.

Strassen, V. (1967). Almost sure behavior of sums of independent random variables and martingales. In *Proceedings of the Fifth Berkeley Symposium in Mathematical Statistics and Probability, vol.* **2**, 315-343. Berkeley: University of California Press.

Thompson, J.R. and Tapia, R.A. (1990). *Nonparametric Function Estimation, Modelling and Simulation.* Philadelphia: SIAM.

Tucker, H.G. (1967). *A Graduate Course in Probability*. New York: Academic Press.

van Beeck, P. (1972). An application of Fourier methods to the problem of sharpening the Berry-Esséen inequality. *Zeitschrift für Wahrscheinlichkeitstheorie und Verwandte Gebiete*, **23**, 187-197.

van Zwet ,W.R. (1965). *Convex Transformations of Random Variables*. Amsterdam Mathematical Centre Transactions, 7.

von Bahr, B. (1965). On the convergence of moments in the central limit theorem. *Annals of Mathematical Statistics*, **36**, 808-818.

von Mises, R. (1947). On the asymptotic distribution of differentiable statistical functionals. *Annals of Mathematical Statistics*, **18**, 309-348.

Wald, A. (1943). Tests of statistical hypotheses concerning several parameters when the number of observations is large. *Transactions of the American Mathematical Society*, **54**, 426-482.

Wald, A. (1949). Note on the consistency of the maximum likelihood estimate. *Annals of Mathematical Statistics*, **20**, 595-601.

Wedderburn, R.W.M. (1976). On the existence an uniqueness of the maximum likelihood estimates for certain generalized linear models. *Biometrika*, **63**, 27-32.

Zeger, S.L. and Liang, K.-Y. (1986). Longitudinal data analysis for discrete and continuous outcomes. *Biometrics*, **42**, 121-130.

Zolotarev, M. (1967). A sharpening of the inequality of Berry-Esséen. *Zeitschrift für Wahrscheinlichkeitstheorie und Verwandte Gebiete*, **8**, 332-342.

Index